WASTE TREATMENT IN THE METAL MANUFACTURING, FORMING, COATING, AND FINISHING INDUSTRIES

ADVANCES IN INDUSTRIAL AND HAZARDOUS WASTES TREATMENT SERIES

Advances in Hazardous Industrial Waste Treatment (2009)
edited by Lawrence K. Wang, Nazih K. Shammas, and Yung-Tse Hung

Waste Treatment in the Metal Manufacturing, Forming, Coating, and Finishing Industries (2009)
edited by Lawrence K. Wang, Nazih K. Shammas, and Yung-Tse Hung

Heavy Metals in the Environment (2009)
edited by Lawrence K. Wang, J. Paul Chen, Nazih K. Shammas, and Yung-Tse Hung

RELATED TITLES

Handbook of Industrial and Hazardous Wastes Treatment (2004)
edited by Lawrence K. Wang, Yung-Tse Hung, Howard H. Lo, and Constantine Yapijakis

Waste Treatment in the Food Processing Industry (2006)
edited by Lawrence K. Wang, Yung-Tse Hung, Howard H. Lo, and Constantine Yapijakis

Waste Treatment in the Process Industries (2006)
edited by Lawrence K. Wang, Yung-Tse Hung, Howard H. Lo, and Constantine Yapijakis

Hazardous Industrial Waste Treatment (2007)
edited by Lawrence K. Wang, Yung-Tse Hung, Howard H. Lo, and Constantine Yapijakis

WASTE TREATMENT IN THE METAL MANUFACTURING, FORMING, COATING, AND FINISHING INDUSTRIES

EDITED BY
LAWRENCE K. WANG
NAZIH K. SHAMMAS
YUNG-TSE HUNG

CRC Press
Taylor & Francis Group
Boca Raton London New York

CRC Press is an imprint of the
Taylor & Francis Group, an **informa** business

CRC Press
Taylor & Francis Group
6000 Broken Sound Parkway NW, Suite 300
Boca Raton, FL 33487-2742

© 2009 by Taylor & Francis Group, LLC
CRC Press is an imprint of Taylor & Francis Group, an Informa business

No claim to original U.S. Government works
Printed in the United States of America on acid-free paper
10 9 8 7 6 5 4 3 2 1

International Standard Book Number-13: 978-1-4200-7223-5 (Hardcover)

This book contains information obtained from authentic and highly regarded sources. Reasonable efforts have been made to publish reliable data and information, but the author and publisher cannot assume responsibility for the validity of all materials or the consequences of their use. The authors and publishers have attempted to trace the copyright holders of all material reproduced in this publication and apologize to copyright holders if permission to publish in this form has not been obtained. If any copyright material has not been acknowledged please write and let us know so we may rectify in any future reprint.

Except as permitted under U.S. Copyright Law, no part of this book may be reprinted, reproduced, transmitted, or utilized in any form by any electronic, mechanical, or other means, now known or hereafter invented, including photocopying, microfilming, and recording, or in any information storage or retrieval system, without written permission from the publishers.

For permission to photocopy or use material electronically from this work, please access www.copyright.com (http://www.copyright.com/) or contact the Copyright Clearance Center, Inc. (CCC), 222 Rosewood Drive, Danvers, MA 01923, 978-750-8400. CCC is a not-for-profit organization that provides licenses and registration for a variety of users. For organizations that have been granted a photocopy license by the CCC, a separate system of payment has been arranged.

Trademark Notice: Product or corporate names may be trademarks or registered trademarks, and are used only for identification and explanation without intent to infringe.

Library of Congress Cataloging-in-Publication Data

Waste treatment in the metal manufacturing, forming, coating, and finishing industries / edited by Lawrence K. Wang, Nazih K. Shammas, Yung-Tse Hung.
    p. cm. -- (Advances in industrial and hazardous wastes treatment series ; 2)
Includes bibliographical references and index.
ISBN-13: 978-1-4200-7223-5
ISBN-10: 1-4200-7223-4
1. Metalworking industries--Waste disposal. 2. Metals--Finishing--Waste disposal. 3. Metals--Environmental aspects. I. Wang, Lawrence K. II. Shammas, Nazih K. III. Hung, Yung-Tse. IV. Title. V. Series.

TD899.M45.W36 2009
671.028'6--dc22
    2008024407

Visit the Taylor & Francis Web site at
http://www.taylorandfrancis.com

and the CRC Press Web site at
http://www.crcpress.com

Contents

Preface vii
Editors ix
Contributors xi

Chapter 1 Waste Minimization and Cleaner Production 1
Nazih K. Shammas and Lawrence K. Wang

Chapter 2 Waste Treatment in the Iron and Steel Manufacturing Industry 37
Gupta Sudhir Kumar, Debolina Basu, Yung-Tse Hung, and Lawrence K. Wang

Chapter 3 Treatment of Nonferrous Metal Manufacturing Wastes 71
Nazih K. Shammas and Lawrence K. Wang

Chapter 4 Management, Minimization, and Recycling of Metal Casting Wastes 151
An Deng, Yung-Tse Hung, and Lawrence K. Wang

Chapter 5 Waste Treatment in the Aluminum Forming Industry 199
Lawrence K. Wang and Nazih K. Shammas

Chapter 6 Treatment of Nickel-Chromium Plating Wastes 233
Lawrence K. Wang, Nazih K. Shammas, Donald B. Aulenbach, and William A. Selke

Chapter 7 Waste Treatment and Management in the Coil Coating Industry 261
Lawrence K. Wang and Nazih K. Shammas

Chapter 8 Waste Treatment in the Porcelain Enameling Industry 307
Lawrence K. Wang and Nazih K. Shammas

Chapter 9 Treatment of Metal Finishing Industry Wastes 345
Nazih K. Shammas and Lawrence K. Wang

Chapter 10 Removal of Heavy Metals from Soil 381
Nazih K. Shammas

Chapter 11 Cleanup of Metal Finishing Brownfield Sites 433
Nazih K. Shammas

Index **479**

Preface

Environmental managers, engineers, and scientists who have had experience with industrial and hazardous waste management problems have noted the need for a handbook series that is comprehensive in its scope, directly applicable to daily waste management problems of specific industries, and widely acceptable by practicing environmental professionals and educators.

Many standard industrial waste treatment and hazardous waste management texts adequately cover a few major industries, for conventional in-plant pollution control strategies, but no one book, or series of books, focuses on new developments in innovative and alternative cleaner production technologies, waste minimization methodologies, environmental processes, design criteria, effluent standards, performance standards, pretreatment standards, managerial decision methodology, and regional and global environmental conservation.

The entire *Advances in Industrial and Hazardous Wastes Treatment* book series emphasizes (a) in-depth presentation of environmental pollution sources, waste characteristics, control technologies, management strategies, facility innovations, process alternatives, costs, case histories, effluent standards, and future trends for each industrial or commercial operation, and (b) in-depth presentation of methodologies, technologies, alternatives, regional effects, and global effects of each important industrial pollution control practice that may be applied to all industries, such as industrial ecology, pollution prevention, in-plant hazardous waste management, site remediation, groundwater decontamination, and stormwater management.

In a deliberate effort to complement other industrial waste treatment and hazardous waste management texts published by Taylor & Francis and CRC, this book, *Waste Treatment in the Metal Manufacturing, Forming, Coating, and Finishing Industries*, covers many new industries and new waste management topics, such as waste minimization, cleaner production, iron and steel manufacturing waste treatment, nonferrous metals manufacturing waste treatment, foundry waste treatment, aluminum forming waste treatment, nickel-chromium plating waste treatment, coil coating waste treatment, porcelain enameling waste treatment, metal finishing waste treatment, heavy metals removal from soil, metal finishing brownfield cleanup, and more are presented in detail. Special efforts were made to invite experts to contribute chapters in their own areas of expertise. Since the field of industrial hazardous waste treatment is very broad, no one can claim to be an expert in all industries; collective contributions are better than a single author's presentation for a handbook of this nature.

This book is to be used as a college textbook as well as a reference book for the environmental professional. It features the major metal manufacturing, forming, coating and finishing industries and hazardous pollutants that have significant effects on the environment. Professors, students, and researchers in environmental, civil, chemical, sanitary, mechanical, and public health engineering and science will find valuable educational materials here. The extensive bibliographies for each metal-related industrial waste treatment or practice should be invaluable to environmental managers or researchers who need to trace, follow, duplicate, or improve on a specific industrial hazardous waste treatment practice.

A successful modern industrial hazardous waste treatment program for a particular industry will include not only traditional water pollution control but also air pollution control, noise control, soil conservation, site remediation, radiation protection, groundwater protection, hazardous waste management, solid waste disposal, and combined industrial–municipal waste treatment and management. In fact, it should be a total environmental control program. Another intention of this book is to provide technical and economical information on the development of the most feasible total environmental control program that can benefit both industry and local municipalities. Frequently, the most economically feasible methodology is a combined industrial–municipal waste treatment.

Lawrence K. Wang, New York
Nazih K. Shammas, Massachusetts
Yung-Tse Hung, Ohio

Editors

Lawrence K. Wang has over 25 years of experience in facility design, plant construction, operation, and management. He has expertise in water supply, air pollution control, solid waste disposal, water resources, waste treatment, hazardous waste management and site remediation. He is a retired dean/director of both the Lenox Institute of Water Technology and Krofta Engineering Corporation, Lenox, Massachusetts, and a retired vice president of Zorex Corporation, Newtonville, New York. Dr. Wang is the author of over 700 technical papers and 19 books, and is credited with 24 U.S. patents and 5 foreign patents. He received his BSCE degree from National Cheng-Kung University, Taiwan, his MS degrees from both the Missouri University of Science and Technology and the University of Rhode Island, and his PhD degree from Rutgers University, New Jersey.

Nazih K. Shammas has been an environmental expert, professor and consultant for over forty years. He is an ex-dean and director of the Lenox Institute of Water Technology, and advisor to Krofta Engineering Corporation, Lenox, Massachusetts. Dr. Shammas is the author of over 250 publications and eight books in the field of environmental engineering. He has experience in environmental planning, curriculum development, teaching and scholarly research, and expertise in water quality control, wastewater reclamation and reuse, physicochemical and biological treatment processes and water and wastewater systems. He received his BE degree from the American University of Beirut, Lebanon, his MS from the University of North Carolina at Chapel Hill, and his PhD from the University of Michigan at Ann Arbor.

Yung-Tse Hung has been a professor of civil engineering at Cleveland State University since 1981. He is a fellow of the American Society of Civil Engineers. He has taught at 16 universities in eight countries. His primary research interests and publications have been involved with biological wastewater treatment, industrial water pollution control, industrial waste treatment, and municipal wastewater treatment. He is now credited with over 450 publications and presentations on water and wastewater treatment. Dr. Hung received his BSCE and MSCE degrees from National Cheng-Kung University, Taiwan, and his PhD degree from the University of Texas at Austin. He is the editor of *International Journal of Environment and Waste Management*, *International Journal of Environmental Engineering*, and *International Journal of Environmental Engineering Science*.

Contributors

Donald B. Aulenbach
Rensselaer Polytechnic Institute
Troy, New York

Debolina Basu
Centre for Environmental Science
  and Engineering
Indian Institute of Technology Bombay
Mumbai, India

An Deng
College of Civil Engineering
Hohai University
Nanjing, China

Yung-Tse Hung
Department of Civil and Environmental
  Engineering
Cleveland State University
Cleveland, Ohio

Gupta Sudhir Kumar
Centre for Environmental Science
  and Engineering
Indian Institute of Technology Bombay
Mumbai, India

William A. Selke
Lenox Institute of Water Technology
Lenox, Massachusetts

Nazih K. Shammas
Lenox Institute of Water Technology
Lenox, Massachusetts

Lawrence K. Wang
Lenox Institute of Water Technology
Lenox, Massachusetts

1 Waste Minimization and Cleaner Production

Nazih K. Shammas and Lawrence K. Wang

CONTENTS

1.1	Introduction and Background	2
1.2	Good Housekeeping	4
	1.2.1 Function of a Good Housekeeping Program	5
	1.2.2 Creation of a Good Housekeeping Program	5
	1.2.3 Good Housekeeping: What to Do	6
1.3	Strategy for Waste Reduction	7
	1.3.1 Phase I	7
	1.3.2 Phase II	7
	1.3.3 Phase III	8
1.4	Planning for Waste Reduction	9
1.5	Audit Review	12
	1.5.1 Raw Materials and Utilities	12
	1.5.2 Processes and Integrated Source Control	12
	1.5.3 End-of-Pipe Emission Control Systems	13
	1.5.4 Final Emissions and Discharges	13
	1.5.5 Storage and Handling	14
1.6	Cleaner Production	14
	1.6.1 Barriers to Cleaner Production	14
	1.6.2 Program as a Response to Barriers	16
	1.6.3 Goals for Cleaner Production Programs	16
1.7	Metal Finishing	17
	1.7.1 Industry Profile	17
	1.7.2 Effective BMPs	17
	1.7.3 Waste Minimization in Electroplating	18
1.8	Primary Metals	19
	1.8.1 Industry Profile	19
	1.8.2 Effective BMPs	20
1.9	Case Studies	21
	1.9.1 Recycling Zinc in Viscose Rayon Plants by Two-Stage Precipitation	21
	1.9.1.1 The Significance	21
	1.9.1.2 The New Process	21
	1.9.1.3 The Economics	22
	1.9.1.4 Areas of Application	22
	1.9.2 Pollution Abatement in a Copper Wire Mill	22
	1.9.2.1 The Significance	22
	1.9.2.2 The New Process	23

                1.9.2.3   The Economics 25
                1.9.2.4   Areas of Application 25
        1.9.3 Gas-Phase Heat Treatment of Metals 25
                1.9.3.1   Cleaner Production 25
                1.9.3.2   Advantages 26
                1.9.3.3   Economic Benefits 26
        1.9.4 New Technology: Galvanizing of Steel 26
                1.9.4.1   Cleaner Production 27
                1.9.4.2   Advantages 27
                1.9.4.3   Economic Benefits 28
        1.9.5 Waste Reduction in Electroplating 28
                1.9.5.1   Cleaner Production 28
                1.9.5.2   Advantages 28
        1.9.6 Waste Reduction in Steelwork Painting 28
                1.9.6.1   Cleaner Production 29
                1.9.6.2   Advantages 30
                1.9.6.3   Economic Benefits 30
        1.9.7 Recovery of Copper from Printed Circuit Board Etchant 30
                1.9.7.1   Cleaner Production 31
                1.9.7.2   Advantages 32
                1.9.7.3   Economic Benefits 32
        1.9.8 Chrome Recovery and Recycling in the Leather Industry 32
                1.9.8.1   Cleaner Production 32
                1.9.8.2   Advantages 33
                1.9.8.3   Economic Benefits 33
        1.9.9 Minimization of Organic Solvents in Degreasing and Painting of Metals 33
                1.9.9.1   Cleaner Production 34
                1.9.9.2   Advantages 35
                1.9.9.3   Economic Benefits 35
References 35

1.1 INTRODUCTION AND BACKGROUND

For many years a large part of industrial pollution control has been carried out essentially on an end-of-pipe basis, and a wide range of unit processes (physical, chemical, and biological) have been developed to service the needs of the industry. Such end-of-pipe systems range from low intensity to high intensity arrangements, from low technology to high technology, and from low cost to high cost. Most end-of-pipe systems are destructive processes in that they provide no return to the operating company in terms of increased product yield or lower operating cost, except in those circumstances where reduced charges would then apply for discharge to a municipal sewer.

It should be noted that in all cases the size (and hence cost) of end-of-pipe treatment has a direct relationship to both the volume of effluent to be treated and the concentration of pollutants contained in the discharge. For example, the size of most physicochemical reactors (balancing, neutralizing, flocculation, sedimentation, flotation, oxidation, reduction, etc.) is determined by hydraulic factors such as surface loading rate and retention time.

The size of most biological reactors is determined by pollution load, for example, kg BOD (biochemical oxygen demand) or COD (chemical oxygen demand) per kg MLVSS (mixed liquor volatile suspended solids) per day in the case of suspended growth type systems, and kg BOD or COD per m^3 of media or reactor volume in the case of fixed-film type systems.

Waste Minimization

It is evident therefore that the reduction of emissions by action at source can have a significant impact on the size and hence the cost of an end-of-pipe treatment system. On this basis, it should be established practice in industry that no capital expenditure for end-of-pipe treatment should be made until all waste reduction opportunities have been exhausted. This has not often been the case, and many treatment plants have been built that are both larger and more complicated than is necessary.

Increased environmental pressure and awareness now require industry to meet tighter environmental standards on a global basis. In many countries, such requirements generally cannot be met by using conventional end-of-pipe solutions without seriously impacting on the economic viability of the individual industries. Accordingly, much more emphasis has to be placed on waste reduction as a necessary first step to reduce to a minimum the extent of the end-of-pipe treatment to be provided. A full understanding of the nature of all wastestreams (aqueous, gaseous, or solid) and the exact circumstances by which they are generated must be developed in order to achieve cleaner production and to eliminate or minimize pollution before it arises. This is a necessity for industry today.

Waste minimization is a policy mandated by the U.S. Congress in the 1984 Hazardous and Solid Waste Amendments[1] (HSWA) to the Resource Conservation and Recovery Act (RCRA).[2,3] The U.S. Environmental Protection Agency (U.S. EPA) has established an Office of Pollution Prevention to promote waste reduction. On February 26, 1991, U.S. EPA published a pollution prevention strategy aimed at providing guidance and direction for incorporating pollution prevention into U.S. EPA programs.[4]

Pollution prevention practices have become part of the U.S. National Pollutant Discharge Elimination System (NPDES) program, working in conjunction with best management practices (BMPs) to reduce potential pollutant releases. Pollution prevention methods have been shown to reduce costs as well as pollution risks through source reduction and recycling/reuse techniques.[5]

Best management practices are inherently pollution prevention practices. Traditionally, BMPs have focused on good housekeeping measures and good management techniques intending to avoid contact between pollutants and water media as a result of leaks, spills, and improper waste disposal. However, based on the authority granted under the regulations, BMPs may encompass the entire universe of pollution prevention, including production modifications, operational changes, materials substitution, materials and water conservation, and other such measures.[5]

U.S. EPA endorses pollution prevention as one of the best means of pollution control. In 1990, the Pollution Prevention Act[6] was enacted and set forth a national policy that "pollution should be prevented or reduced at the source whenever feasible; pollution that cannot be prevented should be recycled in an environmentally safe manner, whenever feasible; pollution that cannot be prevented or recycled should be treated in an environmentally safe manner whenever feasible; and disposal or other release into the environment should be employed only as a last resort and should be conducted in an environmentally safe manner."

Significant opportunities exist for industry to reduce or prevent pollution through cost-effective changes in production, operation, and raw materials use. In addition, such changes may offer industry substantial savings in reduced raw materials, pollution control, and liability costs, as well as protect the environment and reduce health and safety risks to workers. Where pollution prevention practices can be both environmentally beneficial and economically feasible, one would consider their implementation to be prudent.

Improvement in environmental performance and production efficiency in both the short and in the long term are expected to be achieved by means of the following steps[7,8]:

1. *Effective management and training.* This is the introduction of a sustained approach to pollution control and environmental management. It will be achieved as a result of senior management's commitment to
    a) Specific objectives for overall environmental performance, including specific performance targets on a process by process basis including utilities

b) Cradle to grave philosophy in product design
   c) A management structure that positively links production, pollution control, and the environment with clearly defined responsibilities and lines of communication to managing director level, supported by
      i. An initial audit of present production methods, housekeeping practices, procedures and factory support services to identify opportunities for waste reduction and optimized end-of-pipe treatment
      ii. Regular environmental audits to ensure standards are being maintained
      iii. Monitoring programs and procedures designed to continuously assess process efficiency and environmental performance
      iv. A database with relevant information and documentation on performance and on efficient use of resources and reduction of waste production
      v. Training procedures for technical and operational personnel
      vi. General environmental awareness programs at all levels within the company hierarchy.
2. *In-house process control.* This comprises the achievement of optimum efficiency in relation to production and processing methods including the introduction, where feasible, of cleaner processes (alternative technology) or processing methods (substitute materials and/or reformulations, process modifications, and equipment redesign).
3. *Good housekeeping.* This involves the rethinking of localized habitual practice and the identification and implementation of new practices and procedures.
4. *Water conservation/reuse/recycle.* In this, the aim is to achieve optimum efficiency in relation to water use, looking at the possible elimination of use, the regulation of use to only specific requirements, sequential use, or reuse and in-process recycling.
5. *Waste recovery and/or reuse.* This comprises the identification and implementation of opportunities to recover process chemicals and materials for direct reuse or for reuse elsewhere through renovation or conversion technology.

As a result of the foregoing the industrial facility will do the following:

1. Decrease costs for raw materials, energy, and waste treatment/disposal
2. Improve the working environment, thus decreasing costs associated with workers' health
3. Acquire the favorable image of a company that protects the environment
4. Create a potential for production expansion by being one step ahead of environmental regulations

The country as a whole will benefit from:

1. Decreased pollution loadings
2. Decreased consumption of raw materials and energy
3. Decreased costs associated with workers' safety and health

1.2 GOOD HOUSEKEEPING

Good housekeeping is essentially the maintenance of a clean, orderly work environment. Maintaining an orderly facility means that materials and equipment are neat and well kept to prevent releases to the environment. Maintaining a clean facility involves the expeditious remediation of releases to the environment. Together, these terms—clean and orderly—define a good housekeeping program.[5]

Maintaining good housekeeping is the heart of a facility's overall pollution control effort. Good housekeeping cultivates a positive employee attitude and contributes to the appearance of sound management principles at a facility. Some of the benefits that may result from a good housekeeping program include ease in locating materials and equipment; improved employee morale; improved manufacturing and production efficiency; lessened raw, intermediate, and final product losses due to spills, waste, or releases; fewer health and safety problems arising from poor materials and equipment management; environmental benefits resulting from reduced releases of pollution; and overall cost savings.

1.2.1 Function of a Good Housekeeping Program

Good housekeeping measures can be easily and simply implemented. Some examples of commonly implemented good housekeeping measures include the orderly storage of bags, drums, and piles of chemicals; prompt cleanup of spilled liquids to prevent significant runoff to receiving waters; expeditious sweeping, vacuuming, or other cleanup of accumulations of dry chemicals to prevent them from reaching receiving waters; and proper disposal of toxic and hazardous wastes to prevent contact with and contamination of storm water runoff.

The primary impediment to a good housekeeping program is a lack of thorough organization. To overcome this obstacle, a three-step process can be used[5]:

1. Determine and designate an appropriate storage area for every material and every piece of equipment
2. Establish procedures requiring that materials and equipment be placed in or returned to their designated areas
3. Establish a schedule to check areas to detect releases and ensure that any releases are being mitigated

The first two steps act to prevent releases that would be caused by poor housekeeping. The third step acts to detect releases that have occurred as a result of poor housekeeping.

1.2.2 Creation of a Good Housekeeping Program

As with any new or modified program, the initial stages will be the greatest hurdle; ultimately, however, good housekeeping should result in savings that far outweigh the efforts associated with initiation and implementation. Generally, a good housekeeping plan should be developed in a manner that creates employee enthusiasm and thus ensures its continuing implementation. The first step in creating a good housekeeping plan is to evaluate the organization of the facility site. In most cases, a thorough release identification and assessment has already generated the needed inventory of materials and equipment and has determined their current storage, handling, and use locations. This information, together with that from further assessments, can then be used to determine if the existing location of materials and equipment is adequate in terms of space and arrangement.

Cramped spaces and those with poorly placed materials increase the potential for accidental releases due to constricted and awkward movement in these areas. A determination should be made as to whether materials can be stored in a more organized and safer manner (e.g., stacked, stored in bulk as opposed to individual containers, and so on). The proximity of materials to their place of use should also be evaluated. Equipment and materials used in a particular area should be stored nearby for convenience, but should not hinder the movement of workers or equipment. This is especially important for waste products. Where waste conveyance is not automatic, waste receptacles should be located as close as possible to the waste generation areas, thereby preventing inappropriate disposal leading to environmental releases.

Appropriately designated areas (e.g., equipment corridors, worker passageways, dry chemical storage areas) should be established throughout the facility. The effective use of labeling is an integral part of this step. Signs and adhesive labels are the primary methods used to assign areas.

Many facilities have developed innovative labeling approaches, such as color coding the equipment and materials used in each particular process. Other facilities have stenciled outlines to assist in the proper positioning of equipment and materials.

Once a facility site has been organized in this manner, the next step is to ensure that employees maintain this organization. This can be accomplished through explaining organizational procedures to employees during training sessions, distributing written instructions, and, most importantly, demonstrating by example.

Support of the program must be demonstrated, particularly by responsible facility personnel. Shift supervisors and others in positions of authority should act quickly to initiate activities to rectify poor housekeeping. Generally, employees will note this dedication to the good housekeeping program and will typically begin to initiate good housekeeping activities without prompting. Although initial implementation of good housekeeping procedures may be challenging, these instructions will soon be followed by employees as standard operating procedures.

Despite good housekeeping measures, the potential for environmental releases remains. Thus, the final step in developing a good housekeeping program involves the prompt identification and mitigation of actual or potential releases. Where potential releases are noted, measures designed to prevent release can be implemented. Where actual releases are occurring, mitigation measures such as those described below may be required.

Mitigative practices are simple in theory: the immediate cleanup of an environmental release lessens chances of spreading contamination and lessens impacts due to contamination. When considering choices for mitigation methods, a facility must consider the physical state of the material released and the media to which the release occurs. Generally, the ease of implementing mitigative actions should also be considered. For example, diet, crushed stone, asphalt, concrete, or other covering may top a particular area. Consideration as to which substance would be easier to clean in the event of a release should be evaluated.

Conducting periodic inspections is an excellent method to verify the implementation of good housekeeping measures. Inspections may be especially important in the areas identified in the release identification and assessment step where releases have previously occurred.

It may not always be possible to immediately correct poor housekeeping. However, deviations should occur only in emergencies. The routines and procedures established as a part of the program should allow for adequate time to conduct good housekeeping activities.[5]

1.2.3 Good Housekeeping: What to Do

1. Integrate a recycling/reuse and conservation program in conjunction with good housekeeping. Include recycle/reuse opportunities for common industry wastes such as paper, plastic, glass, aluminum, and motor oil, as well as facility-specific substances such as chemicals, used oil, dilapidated equipment, and so on into the good housekeeping program. Provide reminders of the need for conservation measures including turning off lights and equipment when not in use, moderating heating/cooling and conserving water.
2. When reorganizing, keep pathways and walkways clear with no protruding containers.
3. Create environmental awareness by developing a regular (e.g., monthly) good housekeeping day.
4. Develop slogans and posters for publicity. Involve employees and their families by inviting suggestions for slogans and allowing children to develop the facility's good housekeeping posters.
5. Provide suggestion boxes for good housekeeping measures.
6. Develop a competitive program that may include company-wide competition or facility-wide competition. Implement an incentive program to spark employee interest (i.e., one half day off for the shift that best follows the good housekeeping program).

Waste Minimization

7. Conduct inspections to determine the implementation of good housekeeping. These may need to be conducted more frequently in areas of most concern.
8. Pursue an ongoing information exchange throughout the facility, the company, and other companies to identify beneficial good housekeeping measures.
9. Maintain necessary cleanup supplies (i.e., gloves, mops, brooms, and so on).
10. Set job performance standards that include aspects of good housekeeping.

1.3 STRATEGY FOR WASTE REDUCTION

Pollution prevention initiatives tend to progress in three separate stages, beginning with a waste audit and associated training and awareness raising, which brings forward the most easily implemented and cost-effective waste reduction measures, as described below. The strategy should be for each company to move through the first stage and get started on a long-term and sustained pollution prevention effort involving all the three stages.

A way to classify wastestreams is to consider them "intrinsic," "extrinsic," or somewhere in-between. Intrinsic wastes are inherent in the fundamental process configuration, whereas extrinsic ones are associated with the auxiliary aspects of the operation.

Intrinsic wastes are built into the original product and process design. These represent impurities present in the reactants, byproducts, coproducts, residues inherent in the process configuration, and spent materials employed as part of the process. Becoming free of intrinsic wastes requires modifying the process system itself, often significantly. Such changes tend to require a large amount of research and development, major equipment modifications, improved reaction (e.g., catalytic) or separation technology—and time.

Extrinsic wastes are more functional in nature and are not necessarily inherent to a specific process configuration. These may occur as a result of unit upsets, selection of auxiliary equipment, fugitive leaks, process shutdown, sample collection and handling, solvent selection, or waste handling practices. Extrinsic wastes can be, and often are, reduced readily through administrative controls, additional maintenance or improved maintenance procedures, simple recycling, minor materials substitution or equipment changes, operator training, managerial support, and changes in auxiliary aspects of the process.

A recent study of programs for existing facilities of several companies reveals that a pollution-prevention initiative will tend to progress in stages.[9] After a training period and an audit of the wastes in the process, the first reduction efforts emphasize the simple, obvious, and most cost-effective alternatives and are generally directed at extrinsic wastes.

1.3.1 Phase I

Phase I efforts include good housekeeping and standard operating practices, waste segregation (separating hazardous wastes from trash), simple direct recycling of materials without treatment, and the other practices noted above. Emphasis is on the operation rather than the underlying system. Activities carried out during this period usually generate a good and immediate economic return on any pollution-prevention investment (return on investment, ROI).

1.3.2 Phase II

If the program continues and additional reductions are desired, more expensive and more complex projects begin to emerge (Phase II). These are often associated with equipment modifications, process modifications and process control and may include the addition or adaptation of auxiliary equipment for simple source treatment, possibly for recycle. This phase usually has little immediate ROI, and more inclusive approaches to assessing the economics of the operation (estimating costs for waste handling, long-term liability, risk) are needed to justify the continued pollution-prevention operation.

1.3.3 Phase III

The program becomes mature (Phase III) when it starts to address the intrinsic wastes through more complex recycling and reuse activities, more fundamental changes to the process, changes in the raw material or catalysts, or reformulation of the product. Emphasis has now shifted to the process itself.

Because of the long payback required for some of these Phase III changes, they are best introduced as a new unit or process is being developed. Justifying fundamental changes to the process as part of the pollution-prevention program *per se* is particularly difficult—the first construction-cost estimate of process plants involving new technology is usually less than half of the final cost, with many projects experiencing even worse performance.

The project will progress in stages, beginning with a waste audit carried out by an audit team. The audit team consists of a waste audit expert, a sector specialist, a financial expert, an economist/marketing expert, and an expert in product life-cycle assessments. The audit team also supports the project in its different stages.

The following seven outputs will be produced by the audit team[9]:

1. Availability of material balances for selected unit process operations (Table 1.1)
2. Obvious waste reduction measures identified and implementation initiated (improved housekeeping) (Table 1.2)
3. Long-term waste reduction options identified (emphasis minimization of hazardous waste) (Table 1.3)
4. Financial and environmental evaluation of waste reduction options (Table 1.4)
5. Development and implementation started on a plan to reduce wastes and increase production efficiency (Table 1.5)
6. Recommendations for equipment modifications and/or process changes to reduce wastes (Table 1.6)
7. Opportunities identified for product reformulation (Table 1.7)

TABLE 1.1
Availability of Material Balances for Selected Unit Process Operations

Activities

Undertake audit preparatory work:
  1. Introduce the audit to top management
  2. Select and train waste audit team
  3. Identify laboratory and other equipment resources
  4. Select scope of audit
  5. Collect existing site plans and process diagrams
  6. Preliminary survey

Determination of raw material inputs to unit operations
Record water usage
Evaluation of waste recycling
Quantify process outputs
Quantify wastewater streams
Quantify gaseous and particulate emissions
Quantify offsite waste disposal
Assemble input and output data for unit operations
Prepare material balance

Source: From UNIDO, Project Document, United Nations Industrial Development Organization, Industrial Sectors and Environment Division, Vienna, Austria, April 1995.

TABLE 1.2
Obvious Waste Reduction Measures Identified and Implementation Initiated (Improved Housekeeping)

Activities

Identify opportunities for improvements in specifications and ordering procedures for raw materials
Identify opportunities for improved materials receiving operations
Identify opportunities for improvements in materials storage
Identify opportunities for improvements in material and water transfer and handling
Identify opportunities for improved process control
Identify opportunities for improved cleaning procedures
Compile a prioritized implementation plan of the most obvious waste reduction measures identified in Table 1.3

Source: From UNIDO, Project Document, United Nations Industrial Development Organization, Industrial Sectors and Environment Division, Vienna, Austria, April 1995.

TABLE 1.3
Long-Term Waste Reduction Options Identified (Emphasis Minimization of Hazardous Waste)

Activities

Based on the material balance obtained in Table 1.1 for each unit operation, locate sources of hazardous waste
Identify opportunities for increased recycling through waste segregation
Identify potential for changes in process conditions
Identify opportunities for reduced raw material use
Identify opportunities for raw material substitution

Source: From UNIDO, Project Document, United Nations Industrial Development Organization, Industrial Sectors and Environment Division, Vienna, Austria, April 1995.

Measurement equipment such as flow measurement gauges, sampling equipment and effluent analysis equipment is necessary for carrying out the audits. A budget provision is made to cover one set of equipment. The equipment will remain in the custody of the industrial facility.

1.4 PLANNING FOR WASTE REDUCTION

Waste reduction should be geared towards increasing production efficiency in existing industrial plants; that is, one must know what is going on inside the factory walls. In-depth knowledge about the production is essential for the implementation of a preventive approach to environmental protection that involves waste segregation, simple recycling, process control, equipment modifications, source treatment, complex recycling, process changes, raw material changes, and even product reformulation.

Countries need to build the technical and scientific institutional capacity to develop, absorb, and diffuse pollution prevention techniques and cleaner production processes essential for a successful program. This could be done by the following[9]:

1. Demonstrating the financial and economic advantages and environmental benefits of such a program
2. Providing technical support for the design, establishment, operation, evaluation, and monitoring of pollution prevention techniques and cleaner production processes and technologies

TABLE 1.4
Financial and Environmental Evaluation of Waste Reduction Options

Activities

Determine financial implications of audit options
Calculate annual operating cost for existing processes including waste treatment and/or disposal costs
Determine potential savings for each waste reduction option:
   1. Reduced raw materials costs
   2. Reduced waste treatment costs
   3. Reduced waste disposal costs
   4. Reduced utility costs
   5. Reduced maintenance costs

Determine investment required for each waste reduction option
Determine financial attractiveness of each option and rank options
Evaluate the environmental impacts of each option:
   1. Effect on volume and pollutant concentration in wastes
   2. Potential cross-media effects
   3. Changes in toxicity, degradability, and treatability of wastes
   4. Reduced use of nonrenewable resources, including energy
   5. Likelihood of unsafe incidents

Prioritize options according to financial and environmental impacts

Source: From UNIDO, Project Document, United Nations Industrial Development Organization, Industrial Sectors and Environment Division, Vienna, Austria, April 1995.

TABLE 1.5
Development and Implementation Started on a Plan to Reduce Wastes and Increase Production Efficiency

Activities

Organize seminar to present the results of the waste audit and its evaluation and tangible waste reductions achieved so far to plant management and to draft waste reduction plan
Establish a monitoring program to run alongside the waste reduction plan to facilitate measurement of actual improvements
Establish an internal waste charging system (cost centres at each waste-generating location)
Establish training program for:
   1. Managerial and supervisory staff
   2. Technical staff
   3. Plan operations

Establish a database on waste discharges, resource use, and reduction of waste production and resource consumption

Source: From UNIDO, Project Document, United Nations Industrial Development Organization, Industrial Sectors and Environment Division, Vienna, Austria, April 1995.

Recognizing the need to prevent pollution and minimize waste, governments, through their environmental protection agencies, should continue their catalytic role to promote, (with industry, research organizations and other relevant institutions) the establishment of a network that will allow the transfer of environmental protection technology.

The United Nations Industrial Development Organization (UNIDO) defines "cleaner production" as "the conceptual and procedural approach to production that demands that all phases of the life-cycle

TABLE 1.6
Recommendations for Equipment Modifications and/or Process Changes to Reduce Wastes

Activities

Evaluate the following equipment and/or process changes, from the standpoint of environment and safety, which could reduce wastes:
1. Reduction in transfer distances between raw material storage and process and between individual unit operations
2. Improvements in materials handling equipment (conveyors, pumps, transfer points)
3. Improved process control (monitoring and instrumentation); more automation
4. Replacement of batch operations with continuous flow or optimized sequencing of batch operations
5. Waste segregation
6. Introduction of water reuse technology or sequential water reuse.

Determine financial implications of equipment/process modification:
1. Determine investment costs
2. Revise operating costs
3. Determine financial implications of options evaluated above

Source: From UNIDO, Project Document, United Nations Industrial Development Organization, Industrial Sectors and Environment Division, Vienna, Austria, April 1995.

TABLE 1.7
Opportunities Identified for Product Reformulation

Activities

1. Undertake a detailed evaluation of the industry subsector, of which the audited facility is part, within the country or region, to determine the size and type of market for its product and supplier/customer relationships
2. Evaluate the environmental impacts of the product after it leaves the plant gate (whether as an input to subsequent industrial processing or following disposal by the consumer)
3. Identify alternative products or modified products that minimize environmental impact through enhanced recycling potential or biodegradability
4. Determine the financial implications to the plant of product reformulation or modification

Source: From UNIDO, Project Document, United Nations Industrial Development Organization, Industrial Sectors and Environment Division, Vienna, Austria, April 1995.

of products must be addressed with the objective of the prevention or minimization of short- and long-term risks to humans and the environment. A total societal commitment is required for effecting this comprehensive approach to achieving the goal of sustainable societies."[9]

The UNIDO program links existing sources of information on low and nonwaste technologies and promotes cleaner production worldwide through four primary activities: the International Cleaner Production Information Clearing house (ICPIC), expert working groups, a newsletter, and training activities.

UNIDO has developed a manual on waste reduction auditing[10] suitable for industrial firms. The manual contains the basic methodology that will be used when assisting any industrial facility in identifying and implementing waste minimization opportunities. However, the importance of integrating an environmental strategy into the corporate strategy must be emphasized. Waste auditing is merely a tool to discover new opportunities for improvement. Without a comprehensive environmental policy embedded in the corporate policy, there will not be a sustained effort towards cleaner production.

The UNIDO project "Demonstration of Cleaner Production Techniques"[11] demonstrates that the concept of preventing wastes at their source as opposed to end-of-pipe treatment is as applicable and profitable in developing countries as in developed countries. The experience gained as well as the demonstrations produced will be of great value in the promotion and implementation of a Cleaner Production Program.

1.5 AUDIT REVIEW

The audit review should cover five main areas: raw materials and utilities, processes and integrated source control, end-of-pipe emission control systems, final emissions and discharges, and storage and handling.[9,12,13] Risk category is usually identified as high, medium or low (H/M/L).

1.5.1 RAW MATERIALS AND UTILITIES

1. Are all raw materials used onsite documented in an inventory? Provide a schedule and identify the sources of raw materials.
2. Has one individual been nominated to be responsible for the maintenance of the inventory?
3. Are records kept on the quantities of raw materials used and unit costs?
4. Has an environmental assessment been carried out on all the raw materials used?
5. Has environmental assessment documentation been provided?
6. Has a risk category for each raw material used been identified?
7. Has the potential for using alternative, less damaging materials been considered?
8. Has the potential for the optimum use of raw materials through conservation of resources to minimization of losses been considered?
9. Has the potential for reuse/recycle/recovery been considered for all materials in use or likely to be introduced?
10. Are disposal requirements and implications considered before introducing any materials?

1.5.2 PROCESSES AND INTEGRATED SOURCE CONTROL

1. Are all processes used onsite documented in an inventory? Provide a schedule of processes and identify the risk category.
2. Has an individual been nominated to be responsible for the maintenance of this inventory? Identify a nominated individual and identify the risk category.
3. Has an environmental impact assessment been carried out for all unit processes? Provide details of the assessments and identify the risk category for each process.
4. Have all hazards associated with the use of process materials been identified, for example, identifying a schedule of risks? Identify a risk category on a hazard-by-hazard basis.
5. Has the potential for using alternative, less damaging processes been considered? Identify changes already introduced and identify the potential for further change.
6. Has consideration been given to the conservation of water through application of integrated source control on a process-by-process basis, for example, conservation of water, reuse of water, recycling of water?
7. Has consideration been given to the avoidance or minimization of waste through the application of integrated source control on a process-by-process basis, that is, minimization of process solution losses through redesign of working procedures or minimization of process solution losses through the application of direct recovery procedures?
8. Has consideration been given to the recovery of materials through the application of integrated source control on a process-by-process basis, for example, direct or indirect recovery of materials by sidestream treatment, process solution enhancement through sidestream removal of contaminants, conversion of waste to byproduct of value?
9. Are records kept of specific raw material usage on a process-by-process basis?

Waste Minimization

1.5.3 End-of-Pipe Emission Control Systems

1. Are design details and specifications for end-of-pipe emission control systems fully documented in an inventory? Provide details of all end-of-pipe control systems (for aqueous emissions, gaseous emissions, and waste). Identify the risk category.
2. Has an individual been nominated to be responsible for the maintenance of this inventory? Identify a nominated individual and identify the risk category.
3. Are end-of-pipe emission control systems monitored on a regular basis to ensure compliance with design requirements (inputs and outputs)? Provide monitoring information over the last 12 months. Identify the risk category on a system-by-system basis.
4. Have all end-of-pipe systems been regularly checked for integrity and correctness of operation? Provide reports for the last 12 months. Identify a risk category in relation to integrity on a system-by-system basis.
5. Are alternative processes available that would further reduce environmental impact on a technical and economic basis? Identify potential opportunities. Identify the risk category.

1.5.4 Final Emissions and Discharges

1. Are all emissions and discharges documented in an inventory, for example, process effluent domestic wastewater, cooling water, stack emissions, hazardous wastes, nonhazardous wastes? Provide a schedule of emissions. Identify the risk category.
2. Has one individual been nominated responsible for the maintenance of this inventory? Identify a nominated individual. Identify the risk category.
3. Are emissions and discharges to the sewer, surface water or groundwater controlled by regulations? Provide details of the relevant regulations. Provide details of the specific emission standards required. Identify the risk category.
4. Are final emissions and discharges to the sewer, surface water, or groundwater fully quantified and characterized on an ongoing basis? Provide monitoring data on the relevant emissions and discharges for the last 12 months. Identify the risk category.
5. Do emissions and discharges to the sewer, surface water, or groundwater fully comply with relevant regulations? Provide data on the extent of compliance. Identify the risk category on an emission-by-emission basis.
6. Are emissions and discharges to atmosphere controlled by regulations? Provide details of the relevant regulations. Provide details of the specific emission standards required. Identify the risk category.
7. Are the final emissions and discharges to atmosphere fully quantified and characterized on an ongoing basis? Provide monitoring data on relevant emissions and discharges for the last 12 months.
8. Do emissions and discharges to atmosphere fully comply with relevant regulations? Provide data on the extent of compliance. Identify the risk category on an emission-by-emission basis.
9. Are emissions and discharges of waste to offsite disposal controlled by regulations? Provide details of the relevant regulations. Provide details of the specific controls and requirements. Identify the risk category.
10. Are emissions and discharges to offsite disposal fully quantified and characterized on an ongoing basis? Provide monitoring data on all disposal arrangements for the last 12 months. Identify the risk category.
11. Do emissions and discharges of waste to offsite disposal fully comply with relevant regulations? Provide data on the extent of compliance. Identify the risk category on a waste type basis.
12. Are the contractors who are responsible for disposal competent? Provide evidence. Identify the risk category.

13. Do all waste-handling procedures comply with existing legislation? Provide confirmation of compliance. Identify the risk category.
14. Are records kept of the fate of wastes produced onsite? Provide documentation for the last 12 months. Identify the risk category.
15. Are records kept on the amount of waste generated per unit of production? Provide specific waste generation data for the last 12 months. Identify the risk category.
16. Are contingency/emergency plans in place in the event of accidental emission/discharge? Provide documentary evidence. Identify the risk category.

1.5.5 Storage and Handling

1. Does an inventory exist for all materials (raw materials, products, byproducts, and waste materials) stored onsite? Provide a schedule of materials stored onsite. Identify the risk category.
2. Have all legal requirements associated with storage and handling of materials been identified? Provide schedules of applicable legal requirements. Provide details on how the regulations are enforced. Identify the risk category.
3. Are raw process and waste materials stored in a safe and appropriate manner; for example, are bulk acids in tanks bunded with secondary containment, are flammable materials in a fire-protected, ventilated store, are powders and pellets in areas fitted with dust extraction segregation of noncompatible materials? Provide details of existing storage arrangements, inducing plans and specifications. Identify risk areas. Identify the risk category.
4. Has consideration been given to the requirements for segregation of incompatible materials? Provide details on the type of wastes stored in specific areas. Identify risk areas. Identify the risk category.
5. Are all stored materials labeled clearly and correctly? Identify a schedule of omissions. Identify the risk category.
6. Has consideration been given to the measures required to contain or monitor for spills or leaks, for example, provision of adequate bund capacity, use of sealants, provision of blind gully pots, atmospheric vapor/gas monitoring, groundwater monitoring, surface water monitoring? Provide details on existing arrangements for all storage areas, including drawings and specifications where available. Identify risk areas. Identify the risk category.
7. Has the integrity of raw material, process, and waste storage areas been checked on a regular basis, for example, ground quality monitoring, inspection of tanks, containers, bunds, and so on? Provide details and records. Identify the risk category.

1.6 CLEANER PRODUCTION

Since the late 1980s, several developed countries have made major public sector commitments to build awareness of cleaner production, also referred to as pollution prevention and waste minimization. These commitments, most notably in Denmark, the Netherlands, the U.K. and the U.S., have led the private sector to investigate and implement pollution prevention measures for existing processes and products. As a result, cleaner production is now seen in these countries as a potentially cost-effective complement to pollution abatement in meeting environmental standards.

There have been several efforts to transfer the experience of developed countries in this field to developing countries. All of these efforts are examples of technology transfer (i.e., the transfer of knowledge, skills, equipment and so on) to achieve a particular objective: the reduction of pollution intensity in the industrial sector of developing countries.

1.6.1 Barriers to Cleaner Production

National pollution control programs implemented by UNIDO aim to influence national policies on the reduction of industrial pollution in developing countries as well as to change the approach of individual entrepreneurs to this problem.[14] National environmental policies for the reduction of industrial pollution consist of discharge standards and implementation schedules based on the pollution abatement potential of end-of-pipe technologies. They do not recognize the considerable potential of source reduction for meeting discharge standards and for minimizing the costs of installing and operating pollution abatement technologies. In turn, enterprises, particularly small and medium enterprises (SMEs), are not concerned about environmental matters (or even waste minimization), and when they are confronted with government regulations respond in one of two ways.[15] Either they make no investment in end-of-pipe technology, claiming that it is impossible given their financial situation, or they install the technology to signify compliance with environmental regulations and then fail to carry out the necessary operational and maintenance activities that would actually reduce pollutants.

These national policies and entrepreneurial approaches reflect the dominant strategy for industrial environmental management in developed countries. Primary reliance on end-of-pipe pollution abatement has been the basis for industrial environmental management in most developed countries since the late 1970s. Although it has been effective in reducing pollution from major sources and in many situations was the only way to meet regulatory deadlines, end-of-pipe treatment has been an expensive approach and has not managed to reduce pollution from all sources. More recently, some developed countries, and industries in those countries, have been calling for cleaner production as the first choice for reducing pollution, including that from the industrial sector. Although a few companies recognized the importance of the preventive approach in the 1970s, only in the late 1980s did governments in a few developed countries begin to encourage its general application.

The problem for environmental management institutions and industrial establishments in developing countries (and in developed countries as well, but to a lesser extent) is that they are not aware of the potential of preventive measures, such as the reduction of excess process inputs and the utilization of nonproduct outputs to meet environmental norms. In some cases, these countries do not have information about cleaner production techniques and technologies and in other cases they do not have the professional staff that can convey the information or adapt it to a given industrial situation. In still other cases, they do not think that cleaner production techniques and technologies are appropriate for their situations, because they are heavily invested in pollution control technology.

The limited utilization of cleaner production techniques and technologies in developing countries, in spite of their significant potential for waste minimization because of old and inefficient equipment, has a number of causes[16,17]:

1. A legislative and regulatory regime that does not assign priority to cleaner production
2. Confusion over the difference between cleaner technology and end-of-pipe control
3. A lack of knowledge (or awareness) of the financial and environmental benefits of no-cost and low-cost changes, primarily good housekeeping but also small modifications to existing equipment
4. The unsuitability of some techniques and technologies for developing countries or for certain types or sizes of industry
5. The lack of information about process-specific technology options
6. A broken supply chain for many simple source-reduction technologies
7. The perception by enterprises that local environmental consulting engineers and research institutions provide inappropriate advice and information
8. A lack of technical personnel at the plant level to install and maintain techniques and technologies
9. Costs of the technology (usually not a significant constraint)
10. Cultural factors
11. The slow rate of new investment among SMEs, which lowers the rate of diffusion of new technologies

1.6.2 Program as a Response to Barriers

The Environment and Energy Branch of UNIDO and the Industry and Environment Program Activity Centre of UNEP supported pollution control programs in approximately 20 countries over a five-year period. UNIDO/UNEP played a coordinating and catalytic role in cleaner production by being a source of information on cleaner production, supporting demonstrations of cleaner production techniques and technologies, training industry and government officials, and providing policy advice on environmental management. They worked primarily with SMEs[15] in the private sector, which became the core of a network of institutions and trained local experts involved in pollution prevention activities.

The programs did several things to facilitate the transfer of technical information and technology from developed to developing countries[17]:

1. They disseminated information on cleaner production by serving as an information clearing house, publishing newsletters and holding marketing seminars in order to increase awareness.
2. They conducted sectoral and cross-sectoral in-plant demonstrations of cleaner production to show the potential of waste minimization in the country.
3. They trained in-plant personnel and consulting engineers on how to conduct waste reduction audits in order to increase the in-country capacity for such activities.
4. They prepared and distributed country-specific technical reports (a waste audit manual in the appropriate language, sector-specific guidelines and fact sheets) to allow factories interested in cleaner production to pursue relevant activities on their own.
5. They held conferences and meetings to increase awareness on the part of key policy-makers from ministries of environment and industry, environmental management agencies, and financial institutions, in the hope that they will support the adoption of appropriate institutional policies.

1.6.3 Goals for Cleaner Production Programs

1. *Demonstration.* One goal is the implementation of in-plant demonstrations that exploit the readily available source reduction measures for existing processes and products and that can inspire a small number of "innovative" enterprises to implement similar measures. Cleaner production programs in countries like the U.S. and the Netherlands[18] can be judged to have succeeded in achieving this goal, and it would probably be reasonable to assume that similar cleaner production programs initiated in developing countries, such as the NCPC program, will also succeed.
2. *Dissemination.* A second goal is the dissemination of the results of demonstration projects to a large number of plants in the industrial sector, in order to obtain a multiplier effect.
3. *Integration.* A third goal, and the one that is clearly the most significant indicator of the penetration of cleaner production techniques and technologies, is the integration of waste minimization considerations into all aspects of standard industrial practice. Only in this way will cleaner production, and for that matter any environmentally superior technology,[19] become a sustained, continuous effort to reduce the resource and pollution intensity of existing and new processes and products.

These three goals should not be seen as mutually exclusive. Each may be appropriate in a given context and form part of a continuum for measuring the success of cleaner production programs. In the short term, successful demonstration projects are necessary where there has never been a cleaner production program. In the intermediate term, dissemination of the results is necessary to stimulate enterprises to investigate and implement cleaner production measures. In the long term, integration of cleaner production into all aspects of entrepreneurial decision-making is necessary for a sustained effort.

1.7 METAL FINISHING

1.7.1 Industry Profile

The category of metal finishing includes manufacturers that take raw metal stock and subject it to various treatments to produce a product at, or closer to, its finished stage. Manufacturers classified as metal finishers perform similar operations that fall under a variety of standard industrial classification codes, including fabricated metal products; machinery, except electrical and electronic machinery, equipment, and supplies; transportation equipment; measuring, analyzing, and controlling instruments; photographic, medical, and optical goods; watches and clocks; and miscellaneous manufacturing industries.[5] The processes used to treat raw metal stock and, correspondingly, the wastes produced are the common link among the metal finishing category members. Some of these processes are especially amenable to BMPs; that is, implementation of BMPs is relatively easy and results in a significant reduction in the discharge of pollutants. Listed below are processes common among metal finishers and the targeted pollutants that enter wastewater streams.[5]

1. *Electroplating.* Typical wastes produced include spent process solutions containing copper, nickel, chromium, brass, bronze, zinc, tin, lead, cadmium, iron, aluminum, and compounds formed from these metals.
2. *Electroless plating.* The most common wastes produced are spent process solutions containing copper and nickel.
3. *Coating.* Depending on the coating material that is being applied, wastes of concern include spent process solutions containing hexavalent chromium, and active organic and inorganic solutions.
4. *Etching and chemical milling.* Typical solutions used in etching and milling that ultimately enter the wastestream and are of concern include chromic acid and cupric chloride.
5. *Cleaning.* Various organic and inorganic compounds enter the wastewater stream from cleaning operations.

The sources of the targeted pollutants are process solutions and raw materials that enter the wastewater stream primarily through rinsing or cleaning processes. A work piece that is removed from a process or cleaning solution is typically subjected to rinsing directly afterwards, carrying excess process contaminants, referred to as dragout, into the rinse tank. The dragout concentrates pollutants in the rinse tank, which is typically discharged into the sewer system.

Another pathway by which targeted pollutants enter the wastewater stream is through the disposal of spent batch process solutions into the sewer system. Spent solutions consist of aqueous wastes and may contain accumulated solids as well. Spent solutions are typically bled at a controlled rate into the wastewater stream. Other sources of pollutants in wastewater streams include cleanup of spills and washdown of fugitive aerosols from spray operations.

1.7.2 Effective BMPs

Numerous practices have been developed to eliminate or minimize discharges of pollutants from the metal finishing industry. Successful source reduction measures have been implemented to eliminate cyanide plating baths, as well as substitute more toxic solvents with less toxic cleaners.

In many cases, cleaning with solvents has been eliminated altogether through the use of water-based cleaning supplemented with detergents, heating, and/or agitation. Other source reduction measures have been implemented to minimize the discharges of toxic materials. For example, drain boards and splash plates have been commonly installed to prevent drips and spills. Additionally, the design of immersion racks or baskets and the positioning of parts on these racks or baskets have also been optimized to prevent trapping of solvents, acids/caustics, or plating baths.

The utilization of recycle and reuse measures has also been commonly used. Many facilities have been able to minimize water use and conserve rinsewaters and plating baths by measures including the following[20,21]:

1. Utilizing a dead rinse, resulting in the concentration of plating bath pollutants. This solution may be reused directly or further purified for reuse.
2. Conserving waters through countercurrent rinsing techniques.
3. Utilizing electrolytic recovery, customized resins, selective membranes, and adsorbents to separate metal impurities from plating baths, acid/caustic dips, and solvent cleaning operations.

These operations and measures not only extend the useful life of solutions, but also prevent or reduce the discharge of pollutants from these operations. Two industries have implemented best management practices that resulted in substantial cost savings and pollutant reductions. Emerson Electric implemented a program that resulted in savings of more than USD 910,000/yr (in terms of 2007 USD)[22] and reductions in solvents, oxygen-demanding pollutants, and metals. Best management practices implemented by a furniture manufacturer in the Netherlands resulted in a reduction in metals discharged and a decrease in water use. A detailed discussion of these programs is provided in the following paragraphs.

Emerson Electric, a manufacturer of power tools, implemented a Waste and Energy Management Program to identify opportunities for pollution prevention. An audit resulted in the following actions[5]:

1. Development of an automated electroplating system that reduced process chemical usage by 25%, process batch dumps by 20%, and wastewater treatment cost by 25%.
2. Installation of a water-based electrostatic immersion painting system to replace a solvent-based painting system. The water-based system resulted in a waste solvent reduction of more than 95%.
3. Installation of an ultrafiltration system that recovers 30 kg/d (65 lb/d) of waste oil and purifies 1135 kg/d (2500 lb/d) of alkaline cleaning solution for reuse, which resulted in a reduction of 5-day biochemical oxygen demand (BOD_5) loadings to the treatment system of 200 kg/month (370 lb/month). This avoided the need for installation of additional treatment.
4. Installation of an alkaline and detergent and steam degreasing system, which resulted in a reduction in waste solvents by 80%.

In addition to the reduction of pollutants, Emerson realized annual costs savings of USD 835,000 (in terms of 2007 USD)[22] in reduced raw material use, USD 2900 in reduced water use, and USD 68,500 in reduced waste disposal.

A furniture manufacturer in the Netherlands reduced metals in its effluent by switching to cyanide-free baths, allowing for longer drip times, using spray rinsing, reusing water, and implementing a closed cooling system. These best management practices, complemented by the installation of treatment technology, reduced metals in the effluent from 945 to 37 kg/yr. Water use also decreased from 330,000 to 20,000 m^3/yr.

1.7.3 Waste Minimization in Electroplating

The Michigan Department of Environmental Quality recommended the following procedures for waste minimization in the electroplating industry[23]:

1. Slow line to an 8 sec count for removal from baths, which drastically reduces the dragout and in turn reduces waste in rinsewater. Rinsewater flow can now be reduced and ultimately the amount of sludge generated can be reduced.

2. Hold the rack for a 10 sec count over the bath, during which time the majority of drips will fall. By doing this you will reduce waste in rinsewater.
3. Put drip catchers between the baths to catch and return any solution to the bath. This will also eliminate most of the buildup between the baths and ultimately reduce the cleanup time and waste generated.
4. Use the rinse bath water again in a different area. For example, if there is a line with a chromic acid etch bath followed by counter-flow rinse baths and a neutralizer bath followed by counter-flow rinse baths, use the dirtiest rinse after the neutralizer bath and pipe it to the rinse baths after the chromic acid tank. This saves water and reduces sludge.
5. Spraying or aerating the rinses uses less water and does a better job. Also, counter-flow rinses will save water.
6. Assess wastewater treatment chemicals, and replace the chemicals that create large volumes of sludge with chemicals that do not.
7. If there is a three-bath rinse after a metal bath, leave the first rinse as a dead bath and use as make-up for the metal bath.
8. Cost out a dryer for the sludge to reduce the volume of sludge.
9. Look at metal recovery online and either reuse or sell it as scrap.
10. Look at sending your waste to a smelter who recovers metals from dried sludge. Separate wastewater treatments may be needed for metal separation.

1.8 PRIMARY METALS

1.8.1 INDUSTRY PROFILE

Primary metal industries include facilities involved in smelting and refining of metals from ore, pig, or scrap; rolling, drawing, extruding, and alloying metals; manufacturing castings, nails, spikes, insulated wire, and cable; and production of coke. Major subcategories include blast furnaces, steel works, rolling and finishing mills; iron and steel foundries; primary and secondary smelters and refiners of nonferrous metals such as copper, lead, zinc, aluminum, tin, and nickel; establishments engaged in rolling, drawing, and extruding nonferrous metals; and facilities involved in nonferrous castings and related fabricating operations. The main processes common to metal forming operations and the wastes that are typically generated are discussed in the following[5]:

1. *Sintering.* This process agglomerates iron-bearing materials (generally fines) with iron ore, limestone, and finely divided fuel such as coke breeze. The fine particles consist of mill scale from hot rolling operations and dust generated from basic oxygen furnaces, open hearth furnaces, electric arc furnaces, and blast furnaces. These raw materials are placed on a traveling grate of a sinter machine. The surface of the raw materials is ignited by a gas and burned. As the bed burns, carbon dioxide, cyanides, sulfur compounds, chlorides, fluorides, and oil and grease are released as gas. Sinter may be cooled by air or a water spray at the discharge end of the machine, where it is then crushed, screened, and collected for feeding into blast furnaces. Wastewater results from sinter cooling operations and air scrubbing devices that utilize water.
2. *Iron making.* Molten iron is produced for steel making in blast furnaces using coke, iron ore, and limestone. Blast furnace operations use water for noncontact cooling of the furnace, stoves, and ancillary facilities and to clean and cool the furnace top gases. Other water, such as floor drains and drip legs, contribute a lesser portion of the process wastewaters.
3. *Steel making.* Steel is an iron alloy containing less than 1% carbon. Raw materials needed to produce steel include hot metal, pig iron, steel scrap, limestone, burned lime, dolomite fluorspar, and iron ores. In steel-making operations, the furnace charge is melted and

refined by oxidizing certain constituents, particularly carbon, in the molten bath, to specified levels. Processes include the open hearth furnace, the electric hearth furnace, the electric arc furnace, and the basic oxygen furnace, all of which generate fumes, smoke, and waste gases. Wastewaters are generated when semiwet or wet gas collection systems are used to cleanse the furnace off gases. Particulates and toxic metals in the gases constitute the main source of pollutants in process wastewaters.
4. *Casting operations.* This subcategory includes both ingot casting and continuous casting processes. Casting refers to the procedure of turning molten metal into a specified shape. Molten metal is distributed into an oscillating, water-cooled mold, where solidification takes place. As the metal solidifies into the mold, the cast product is typically cooled using water, which is subsequently discharged.
5. *Forming operations.* Forming is achieved by passing metal through cylindrical rollers, which apply pressure and reduce the thickness of the metal. Rolling reduces ingots to slabs or blooms. Secondary operations reduce slabs or blooms to billets, plates, shapes, strips, and other forms. Cooling and lubricating compounds are used to protect the rolls, prevent adhesion, and aid in maintaining the desired temperature. Hot rolling generates wastewaters laden with toxic organic compounds, suspended solids, metals, and oil and grease. Cold rolling operations, occurring at temperatures below the recrystallization point of the metal, require more lubrication. The lubricants used in cold rolling include more concentrated oil–water mixtures, mineral oil, kerosene-based lubricants (neat oils), or graphite-based lubricants, which are typically recycled to reduce oil use and pollutant discharges. Subsequent operations may include drawing or extrusion to manufacture tube, wire, or die casting operations. In these operations, similar pollutants are discharged. Contaminated wet scrubber wastewaters may also be generated from extrusion processes but to a lesser degree than in iron- and steel-making and sintering operations.
6. *Acid pickling.* Steel products are immersed in heated acid solutions to remove surface scale during pickling operations. This generates wastewater from three sources:
   a) Rinsewater used to clean the product after immersion in pickling solution
   b) Spent pickling solution or liquor
   c) Wastewater from wet fume scrubbers

   The first source accounts for the largest volume of wastewater but the second source is very acidic and contains high concentrations of iron and heavy metals.
7. *Alkaline cleaning.* This process is used when vegetable, mineral, and animal fats and oils must be removed from the metal surface prior to further processing. Large-scale production or situations where a cleaner product is required may use electrolytic cleaning. The alkaline cleaning bath typically contains a solution of water, carbonates, alkaline silicates, phosphates, and sometimes wetting agents to aid cleaning. Alkaline cleaning results in the discharge of wastewaters from the cleaning solution tank, and subsequent rinsing steps. Potential contaminants include dissolved metals, solids, and oils.

1.8.2 Effective BMPs

Primary metals manufacturing operations have experienced source reduction and recycle/reuse benefits similar to those available to metal finishing operations, including conserving waters through countercurrent rinsing techniques, and utilizing electrolytic recovery, customized resins, selective membranes, and adsorbents to separate metal impurities from acid/caustic dips and rinsewaters to thereby allow for recycle and reuse.

Some very unique opportunities are also exclusively available to the primary metals industry. For example, the use of dry air control devices and dry cast quench operations have been adopted at some facilities to avoid the generation of contaminated wastewater. Additionally, many facilities are finding markets for byproducts (e.g., sulfides resulting from nonferrous smelting operations can be

Waste Minimization

converted to sulfuric acid and subsequently sold) which avoids the need to discharge these contaminants.[24,25]

California Steel Industries, Inc., located in Fontana, CA, reclaimed wastes to increase profits and address water use issues. The facility, a steel mill, is situated in an area that does not have a ready supply of process water. Also, the offsite recycling facility used to dispose of spent process pickle liquor was soon to become unavailable. As a result of these concerns, the company constructed an onsite recycling facility designed to recover ferrous chloride for resale and to reuse water and hydrogen chloride for use in steel processing operations. Environmental benefits include the recovery and resale of 20 to 25 t/d of ferrous chloride, 13,440 L/d of hydrogen chloride, and 49,200 L/d of water. In addition, corporate liability was minimized because spent liquor was no longer sent to a disposal facility.

1.9 CASE STUDIES

1.9.1 Recycling Zinc in Viscose Rayon Plants by Two-Stage Precipitation

1.9.1.1 The Significance

Over 22.7 million kg (50 million lb) of zinc sulfate are used annually in the U.S. for the manufacture of approximately 454 million kg (one billion lb) of viscose rayon. Zinc is used as a regeneration retardant in the acid spinning bath. Because it is not consumed in any of the viscose reactions, these 22.7 million kg (50 million lb) of zinc represent process losses, through dragout by the filaments to the subsequent wash streams, filter backwashing, splashes, leaks, and the washing of equipment.[14]

The effects of zinc as a pollutant are well documented. Concentrations as low as 1.0 mg/L have been shown to be harmful to fish. In addition, there is some evidence indicating that zinc has a synergistic property when associated with copper.

Although it has been known that zinc can be precipitated from acid wastestreams by the use of lime, the resultant sludge has been of low zinc assay, contaminated with other compounds, and with very poor settling characteristics. In commercial operations, the sludge presented a disposal problem and recovery of zinc suitable for recycle was impossible.

In a U.S. EPA demonstration grant with the American Enka Company, a process for precipitating a dense sludge with high zinc assay was proven. The zinc in the sludge was recovered and recycled to the rayon manufacturing plant. This recycling of zinc was shown to have no ill effects on the rayon yarn.

There are ten viscose rayon manufacturing plants in the U.S., all of which are believed to use zinc sulfate in their spinning bath. This process greatly enhances the economics of removing this source of zinc pollution, allowing neutralization of the acid stream and recovery of the zinc while generating a good profit for industrial yarns and at a moderate cost for textile yarns.

1.9.1.2 The New Process

The key to this zinc recovery process is a two-stage precipitation,[26] with the second precipitation taking place under careful pH control, using sodium hydroxide in contact with circulating slurry of zinc hydroxide crystals. All of the zinc precipitates in the second step, most of the impurities in the first.

The elements of the process are as follows. Acid and alkaline wastestreams are collected in a neutralization tank. Here sufficient lime is added to raise the pH to 6.0. At this point, no zinc hydroxide will precipitate but a portion of the iron, calcium sulfate, and other impurities will form a light precipitate. With a coagulant aid, the mixture is sent to a clarifier where a clear overflow containing the dissolved zinc is obtained.

This clear overflow is contacted in a reactor with a circulating stream of previously precipitated sludge containing zinc hydroxide. The pH is raised subsequently to 9.5 to 10.0 with sodium hydroxide. The bulk of the zinc precipitates onto the existing crystals in the circulating slurry. At steady-state conditions, the withdrawal rate of the circulating slurry stream is made equivalent to that of the zinc being added. This dense sludge is then settled. The settled sludge of 4 to 7% zinc assay is converted back to zinc sulfate with sulfuric acid and sent back to the spinning bath. If desired, the sludge can be filtered or centrifuged to 18% solids before dissolving with acid.

The zinc content of the overflow water from the densator-reactor is set by the pH–solubility relationship of zinc in water and results in a zinc content of 0.5 to 1 mg/L at pH = 10. Once the precipitated zinc is removed from the wastewater, the pH can be readjusted to a lower value.

1.9.1.3 The Economics

The conventional technique for removing zinc from the spinning acid wastestream has been direct lime precipitation to ~pH 10, with no zinc recovery. The economics of this approach are compared to the American Enka zinc recycle process.

The economics of recovery are a very strong function of the amount of zinc used in the preparation of the yarn and the ratio of acid to zinc in the spinning bath. In manufacturing industrial yarns and tire cords, it is common to use 4.5 to 7.5 kg of zinc per 100 kg of yarn. This high concentration of zinc makes recovery extremely attractive. Textile yarns use less zinc, and although recovery is still the most economic solution, it offers less of a return. These two cases are presented as extremes, with many plants falling between the two values.

The use of two-stage precipitation combined with zinc recycle offers a saving of 2007 USD 498,000 over neutralization for a plant producing industrial yarns and a saving of USD 88,400 for textile yarns. Many plants produce a mix of the two and the results would therefore fall between these values. The costs associated with the more extensive sludge handling and storage in neutralization and precipitation only are not included. The cost of installing the complete neutralization and zinc recycle system would have negligible economic impact on the rayon industry, running from USD 1.14/100 kg profit to a USD 0.37/100 kg cost compared to selling prices of USD 86 to 100/100 kg of staple, USD 200 to 230/100 kg of tire yarn, and USD 290 to 430/100 kg of filament. Zinc oxide manufacturers face the loss of the bulk of a 22.7 million kg (50 million lb/yr) market as this product is reused rather than wasted.[14]

1.9.1.4 Areas of Application

This technology, with only small modifications to conform to local plant conditions, could have immediate application in any viscose rayon plant with soluble zinc in the plant wastestream. The techniques of initially precipitating the impurities, which would prohibit zinc recycle as well as the use of a sludge recirculation process to obtain a dense sludge, are excellent examples of good process engineering being applied to a waste problem.

In a broader sense this technology could have application to any wastestream containing soluble zinc in a form that can be precipitated by lime or caustic addition. The possibility of recycling the precipitated zinc would depend upon the nature of the process considered and may require further work. Examples of other areas that produce zinc-containing wastes are groundwood pulp, metal plating, zinc refining, and recirculating water systems.

1.9.2 POLLUTION ABATEMENT IN A COPPER WIRE MILL

1.9.2.1 The Significance

All wire drawing operations require cleaning of the metal surfaces before drawing to prevent surface impurities from being pulled into the drawn wire. This cleaning or "pickling" is usually accomplished by the use of sulfuric or hydrochloric acid. To maintain good pickling activity the solution must be replaced when it reaches a minimum concentration. This depleted pickling solution is then a waste disposal problem.

Waste Minimization

The metal must also be washed free of pickling solution. The resulting rinsewaters contain metal salts. Because of the low concentration of these contaminants the rinses are difficult to treat economically.

In the case of the production of copper wire, additional complications are present because of the chemical reduction of cupric oxide to a cuprous oxide coating, which cannot be removed by sulfuric acid. This coating has normally been treated by a "secondary pickle" of chromic acid–sulfuric acid, chromic acid–ammonium bifluoride mixtures, or by nitric acid. All of these techniques produce additional pollutants. Each of the three to four drawing steps required to produce fine copper wire from copper rod requires these pickling and rinse steps.

The waste from such an operation, if treated by conventional precipitation techniques without an examination of the manufacturing process itself, would impose a severe cost on the manufacturing operation and produce large amounts of sludge for disposal.

In a U.S. EPA demonstration grant, the Volco Brass and Copper Company, of Kenilworth, NJ, with Lancy Laboratories as consultants, demonstrated that water consumption could be reduced by 90% from 757,000 L/d to 75,700 L/d (200,000 gal/d to 20,000 gal/d) by chemical rinsing and water reuse. The sulfuric acid pickle was regenerated and high purity metallic copper recovered by continuous electrolysis,[27,28] thereby eliminating the dumping of spent pickle liquor. Hydrogen peroxide was proven to be an improved secondary pickle and the chromates and fluorides previously used were eliminated. Total solids leaving the plant in the rinsewaters were reduced from 1136 kg/d (2500 lb/d) to less than 45 kg/d (100 lb/d). Metal losses in the effluent were reduced to less than 0.45 kg/d (1 lb/d) compared to the previous 273 to 318 kg/d (600 to 700 lb/d). A comparison of the effluent quality before and after the process modification is shown in Table 1.8.[14]

1.9.2.2 The New Process

The pollution control system that is integrated into the manufacturing process consists of three basic steps:

1. The regeneration and copper recovery system for the primary pickle bath
2. The chemical rinse system
3. The use of hydrogen peroxide plus proprietary additives for the secondary pickle.

Figure 1.1 illustrates the final process. Block A shows the work flow through the new system. After the hot sulfuric acid pickle and the secondary pickle of 2.5% hydrogen peroxide in sulfuric acid, the work passes through a chemical rinse step that neutralizes the acid dragout. It also precipitates

TABLE 1.8
Comparison of Effluent Quality before and after Process Modification

Parameter	Old Quality	New Quality
Water usage (L/min)	570	38
pH	3.8	7.5–8.5
Total Cr (mg/L)	90	0
Zn (mg/L)	200	1
Cu (mg/L)	100	1
Suspended solids (mg/L)	30	20
Dissolved solids (mg/L)	1500	800

Source: From UNIDO, *Case Studies of Cleaner Production and Site Remediation*, Training Manual DTT-5-4-95, United Nations Industrial Development Organization, Industrial Sectors and Environment Division, Vienna, Austria, April 1995.

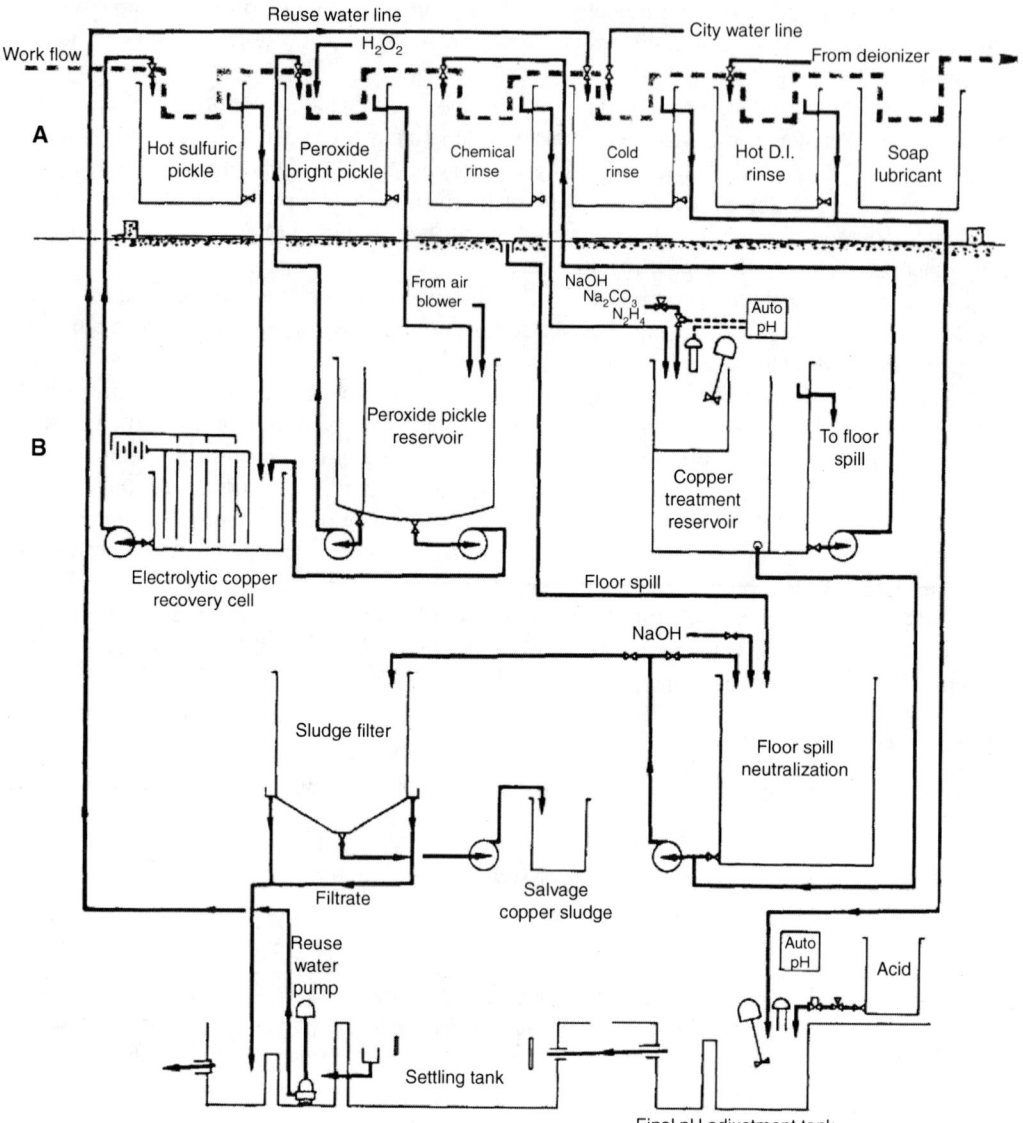

FIGURE 1.1 Illustration of the final copper recovery process. DI, deionized water.

any copper salts by reduction of cupric (Cu^{2+}) ions to cuprous (Cu^{+}) ions, which are insoluble at the pH of the chemical rinse. The work then goes to a cold rinse using city water, a hot rinse using deionized water, and finally a lubricant bath prior to the drawing operation.

Block B shows the electrolytic copper recovery cell, which recovers metallic copper and regenerates sulfuric acid from the metal salts in the hot sulfuric acid pickle solution. It was originally felt that trace metals (zinc, tin, lead) would interfere with the recovery of pure copper. By controlling current density at 50 to 100 A/m^2, however, pure copper can be recovered while maintaining the copper concentration in the pickle bath at 15 g/L.

The secondary pickle reservoir is also shown in Block B. Copper sulfate accumulates in this bath and eventually crystallizes out. These crystals can be recovered and sold as a copper-rich sludge or added to the electrolytic copper recovery loop.

Waste Minimization

The chemical rinse reservoir is maintained at the proper pH and composition by the addition of caustic, sodium carbonate, and a reducing agent, in this case hydrazine. The sludge draw off along with the flow from the floor spill neutralization first goes to a sludge filter to recover salvage copper sludge and then to a final sump for discharge.

The rinse flows go to a pH adjustment tank, a settling tank, and finally to the rinsewater sump, where the bulk of the flow is recirculated to the first water rinse tank.

1.9.2.3 The Economics

The economics for this project are presented in comparison to the previous operating situation with essentially no waste treatment, and to estimated costs if a conventional precipitation and neutralization waste treatment system had been installed without modifying the manufacturing process itself. The approach taken for this project gave a major reduction in pollutants, including sludge, at a slight profit, whereas the isolated installation of a waste treatment system would have resulted in a major cost to the company.

Several changes were made in the plant operation simultaneously with the installation of the pollution abatement system resulting in a total of 2007 USD 130,000 annual savings cost in the drawing operation. The credit of 2007 USD 39,000 annually for increased die life taken for this project is an estimate by the Volco staff.[14]

1.9.2.4 Areas of Application

This process is currently being used at five other installations manufacturing copper and copper-alloy products. The chemical rinse technique is applicable to electroplating operations and has gained wide acceptance there. Any facility utilizing a fluoride–chromate bright pickle should consider the use of a hydrogen peroxide–sulfuric acid mixture as an alternative to treatment.

1.9.3 GAS-PHASE HEAT TREATMENT OF METALS

Chartered Metal Industries Toolroom in Singapore produces a wide range of standard and customized products to support manufacturers in the metal industries. Their production includes high-volume, batch-run precision parts, prototype components, subassemblies, tooling, fixtures, and gauges.

The hardening, carburizing, and nitrocarburizing of steel are heat treatment processes usually carried out in baths of molten salts, such as nitrites, nitrates, carbonates, cyanides, chlorides, or caustics. The combination of chemicals and high temperature means that there are risks of explosion, burns, and poisoning. Environmental problems arise from the resulting vapors and the removal, transport, and disposal of the toxic salts. Disposal of cyanide salt costs 2007 USD 4300/t. Neutralization[27] of quench water, oil, cleaning water and washing water should be carried out before discharge to the sewer, but is not always carried out. Off-gases can be cleaned by passing the exhaust gases through a chemical scrubber, although this also is not always done.

1.9.3.1 Cleaner Production

The new process avoids these problems by gas-phase treatment using a fluidized bed of alumina particles (Figure 1.2). A mixture of air, ammonia, nitrogen, natural gas, pg (liquefied petroleum gas), and other gases are used as the fluidizing gas to carry out the heat treatment. The bed is heated by electricity or gas and quenching is also carried out in a fluidized bed.[14]

Fluidized beds have been used for some years in a variety of roles: heat exchange, gas absorption, chemical reaction, and combustion. In this case the mixture of gases produces the fluidizing atmosphere for heat treatment of the material immersed in the fluidized bed. Hydrocarbon gases are used for carburizing, ammonia for nitriding, and nitrogen for neutral hardening. The hot exhaust gases are used for heat exchange.

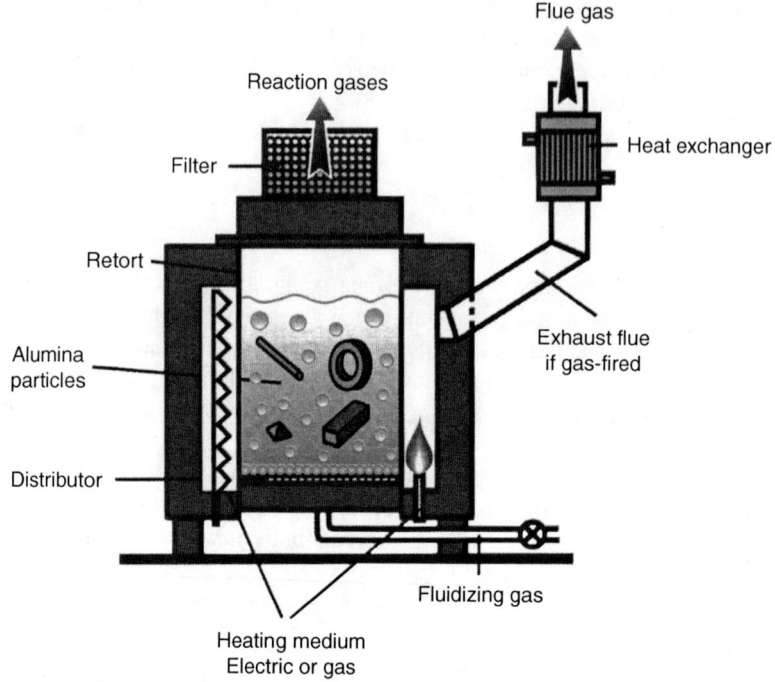

FIGURE 1.2 Fluidized bed. (From Wang, L.K. et al. *Case Studies of Cleaner Production and Site Remediation*, Training Manual DTT-5-4-95, United Nations Industrial Development Organization, Industrial Sectors and Environment Division, Vienna, Austria, April 1995.)

1.9.3.2 Advantages

The most obvious advantages are the reduction in effluents and the improved working atmosphere. The safety aspects have also been improved to a very large extent and the quality of the product in many cases is superior to that produced by the older methods. All forms of heat treatment are amenable to fluidized bed techniques, but austempering is the most cost effective, in spite of the nitrate bath method being less troublesome than other traditional methods.

1.9.3.3 Economic Benefits

The installation consists of four fluid beds (Figure 1.2) used to replace their existing salt bath lines.[14]

	Cost Saving in 2007 USD
Energy	47,000 USD/yr
Salt & maintenance	66,000 USD/yr
Total	113,000 USD/yr
Capital investment	234,000 USD
Payback	2 yr

1.9.4 New Technology: Galvanizing of Steel

Galvanizing is an antirust treatment for steel. The traditional technique consists of chemically pretreating the steel surface, then immersing it in 10 to 16 m long baths of molten zinc at 450°C (Figure 1.3). The process involves large quantities of expensive materials, which increases the cost of the finished steel. In addition there are significant quantities of waste arising from the chemical and zinc baths. There is also the problem of fumes from these operations.

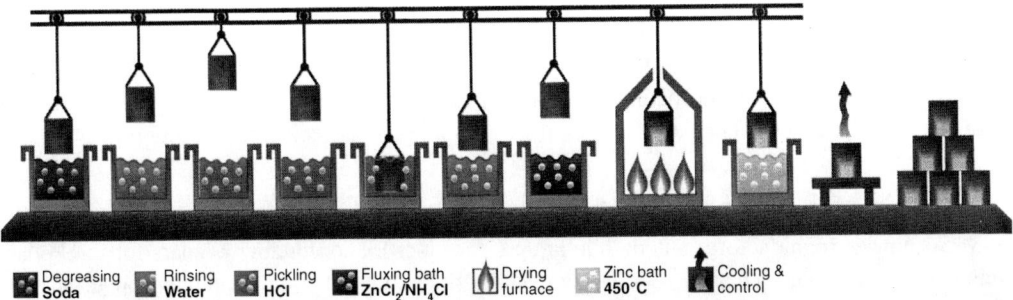

FIGURE 1.3 Sketch of a classical hot dip. (From Wang, L.K. et al. *Case Studies of Cleaner Production and Site Remediation*, Training Manual DTT-5-4-95, United Nations Industrial Development Organization, Industrial Sectors and Environment Division, Vienna, Austria, April 1995.)

1.9.4.1 Cleaner Production

The company's objective was to galvanize steel products of constant cross-section, such as reinforcing and structural steel, tubes, wire, and so on, on a more compact production line, using up to two to three times less zinc, with reduced energy consumption and the suppression of all forms of pollution.

The raw steel is fed in automatically. The process can be operated continuously or in batches, depending on the material to be coated. The surface preparation is performed by controlled shot blasting (Figure 1.4). The steel is heated by induction and enters the coating chamber through a window profiled to match the cross-section of the steel. The zinc is melted in an inert atmosphere by an electric furnace and flows into the galvanizing unit. The liquid zinc is held in suspension by an electromagnetic field. The speed of the production line is controlled by computer. Measuring the thickness of the coating using electromagnetic methods allows precise control of the process.[14]

The first stage of the project was to develop the technology for coilable material, that is, wire and thin rod. The company later developed the technology to handle rigid steel.

The technology that enabled the cleaner production included induction heating to melt the zinc, the use of an electromagnetic field to control the distribution of the molten zinc, and computer control of the process.

1.9.4.2 Advantages

These include the following:

1. Total suppression of conventional plating waste
2. A smaller inventory of zinc
3. Better control of the quality and thickness of the zinc coating
4. Reduced labor requirements
5. Reduced maintenance
6. Safer working conditions

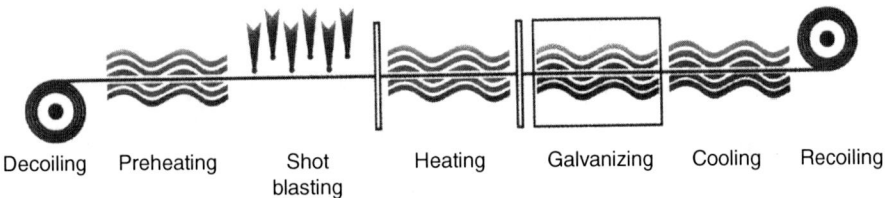

FIGURE 1.4 Sketch of the prototype line. (From Wang, L.K. et al. *Case Studies of Cleaner Production and Site Remediation*, Training Manual DTT-5-4-95, United Nations Industrial Development Organization, Industrial Sectors and Environment Division, Vienna, Austria, April 1995.)

1.9.4.3 Economic Benefits

Capital cost was reduced by two-thirds compared to a traditional dip-coating process. The lower operating costs resulted in the coating process comprising 18% of the steel cost, compared to 60% with traditional methods. The payback period was three years for replacing existing plant.[14]

1.9.5 WASTE REDUCTION IN ELECTROPLATING

FSM Sosnowiec manufactures automobile lamps, door locks, and window winders for the Polish-manufactured Fiat cars. The lamp bodies are made of zinc–aluminum alloy and then copper–nickel–chromium plated. The door locks and window winders are made of steel and then zinc plated. The wastestreams contain cyanide and the heavy metals chromium (VI), copper, zinc, and nickel. The company carries out the traditional treatments of detoxification, neutralization, and dewatering.[29]

1.9.5.1 Cleaner Production

A pollution prevention audit was carried out to reduce environmental pollution, improve working conditions, and improve efficiency. One of the results was that low concentration plating and pacifying is now being introduced. All of the rinsing systems have been modified so that some of the circulating (overflow) rinses have been changed to static rinses. A similar system has been installed for nickel and cyanide. The final rinse tank in each rinsing sequence has been equipped with ion exchange columns, which permit water recycling and raw materials recovery.

1.9.5.2 Advantages

These include the following:

1. A decrease in both water and raw materials consumption
2. A reduction in both wastestream quantities as follows[14]:
    a) Chromic acid, 80%
    b) Copper, 95%
    c) Cyanide, 80%
    d) Nickel, 98%
    e) Zinc, 96%
    f) Wastewater, 93%
3. Purification of the wastewater to the following levels:
    a) Chromium, 0.1 mg/L
    b) Copper, 0.1 mg/L
    c) Nickel, 1.0 mg/L
    d) Cyanide, 2.0 mg/L
    e) Zinc, 0.9 mg/L

1.9.5.3 Economic Benefits

	Cost Saving in 2007 USD/yr
Total savings	251,000 USD/yr
Capital investment	47,000 USD
Payback period	2 months

1.9.6 WASTE REDUCTION IN STEELWORK PAINTING

The Ostrowiec Steelworks of Poland consists of eleven departments; the main production departments are steelworks processing, steel construction, machinery building, and the foundry. The manufacture of steel products is carried out using production-line methods. The final operations required for almost all products are surface treatment and painting. In the machinery-building department these

Waste Minimization

operations are carried out with shot-blasting machines and manually operated spray booths. The original painting method was air-atomized spraying, which has the lowest transfer efficiency of the coating methods and yields large quantities of waste.

1.9.6.1 Cleaner Production

A pollution prevention audit was carried out to improve the environment and efficiency and working conditions in the painting areas.[30] The objective of this program was to reduce the quantity of wastes and costs of painting by a combination of improvements to the technology and good housekeeping. The overall aim was to improve the quality of coating, to reduce the amount of paint raw material and to reduce the quantities of wastes.

The existing painting method was compared with two more advanced painting technologies. The transfer efficiencies for the different methods are as follows:

1. Air-atomized spray (conventional), 30 to 50%
2. Airless spray, 65 to 70%
3. Pressure-atomized electrostatic spray, 85 to 90%

In conventional spraying, compressed air is used both to atomize the paint and to carry it to the surface to be painted (Figure 1.5). With airless spraying the paint is pumped under high pressure to a small jet where the high velocity is sufficient to induce atomization. The lack of any expanding compressed air stream eliminates unwanted spray mist, reduces the loss of paint by overspray, and most of the paint adheres to the work surface (Figure 1.6). With pressure-atomized electrostatic

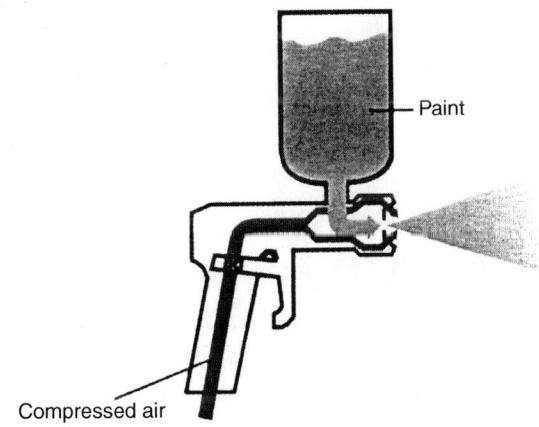

FIGURE 1.5 Air atomized paint spraying.

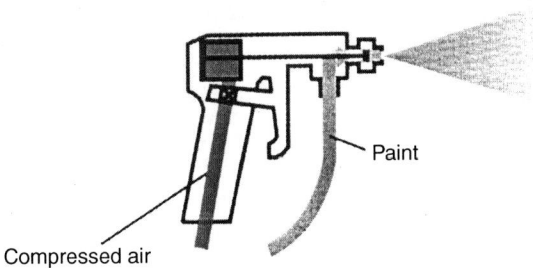

FIGURE 1.6 Airless or high pressure paint spraying. (From Wang, L.K. et al. *Case Studies of Cleaner Production and Site Remediation*, Training Manual DTT-5-4-95, United Nations Industrial Development Organization, Industrial Sectors and Environment Division, Vienna, Austria, April 1995.)

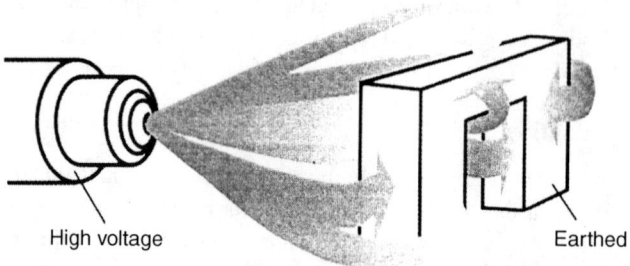

FIGURE 1.7 "Wrap around" effect of electrostatic paint spraying. (From Wang, L.K. et al. *Case Studies of Cleaner Production and Site Remediation*, Training Manual DTT-5-4-95, United Nations Industrial Development Organization, Industrial Sectors and Environment Division, Vienna, Austria, April 1995.)

spray, paint is delivered at high pressure as before, but it is fed to an insulated nozzle. An electrostatic charge of about 100 kV is applied to this nozzle. The charging of the paint particles assists the atomization and causes them to repel each other. Additionally, the charged paint moves along the field lines to the earthed work piece. As the electrostatic field lines envelop the object the paint particles cannot fly straight past, but "wrap" themselves uniformly around the surface. It is this effect that gives the high paint efficiency and reduces waste (Figure 1.7). Note that electrostatic hand spray guns require a small mains transformer and a much reduced current to avoid accidental electrical shock. Comparison of raw material consumption and waste quantities of the different methods are as follows[14]:

	Air-Atomized Spray	Airless Spray	Pressure-Atomized Electrostatic Spray
Paints (m^3)	8.0	6.8	5.6
Solvents (m^3)	6.5	1.6	1.6
Wastes (kg)	2400	1400	500

1.9.6.2 Advantages

These include the following:

1. Reduction of high disposal costs
2. Reduced running costs
3. Decreased financial liability by generating a smaller quantity of hazardous wastes
4. Improved public perception and acceptance in the business community
5. Reductions in the effluent concentrations of about 45% for sludge and 75% for organic solvents

1.9.6.3 Economic Benefits

The cost savings in 2007 USD for airless spray and pressure-atomized electrostatic spray are as follows[14]:

	Airless Spray	Pressure-Atomized Electrostatic Spray
Total savings (USD/yr)	50,000	51,200
Capital investment (USD)	6,200	17,000
Payback (months)	1.5	4

1.9.7 Recovery of Copper from Printed Circuit Board Etchant

Praegitzer Industries Inc., founded in 1981, is a leading designer and manufacturer of advanced circuit boards. The company employs 500 people in three locations.

Waste Minimization

In the manufacture of printed circuit boards, the unwanted copper is etched away by acid solutions of cupric chloride (Equation 1.1). As the copper dissolves, the effectiveness of the solution tails and it must be regenerated. The traditional way of doing this is to oxidize the cuprous ion produced with acidified hydrogen peroxide. During the process the volume of solution increases steadily and the copper in the surplus liquor is precipitated as copper oxide and usually landfilled.

In the etching process:

$$Cu + CuCl_2 \rightarrow 2CuCl \tag{1.1}$$

1.9.7.1 Cleaner Production

The original proposal for recovering the copper in high-quality form came from the U.K. Electricity Research Council. Using an electrolytic technique involving a divided cell (Figure 1.8), simultaneous regeneration of the etching solution and recovery of the unwanted copper is possible. A special membrane allows hydrogen and chloride ions through, but not the copper. The copper is transferred via a bleed valve and recovered at the cathode as pure flakes (Equation 1.2).

In the electrolytic process:

$$2CuCl \rightarrow CuCl_2 + Cu \tag{1.2}$$

This process was enabled by the development of a suitable cell-dividing material; a process development where the excess etchant is pumped to the recovery circuit and the copper is obtained in a recoverable form.

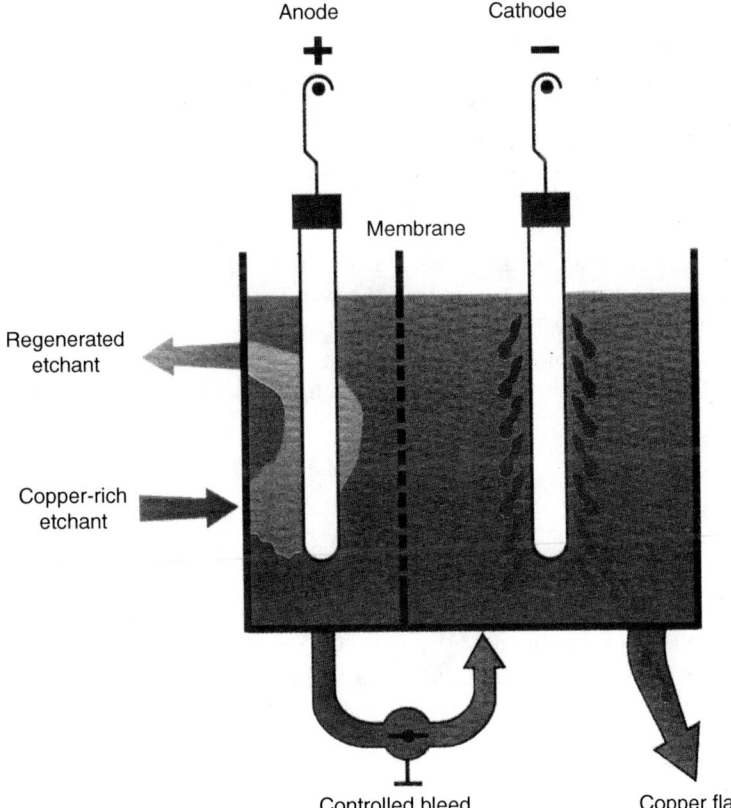

FIGURE 1.8 Electrolytic process. (From Wang, L.K. et al. *Case Studies of Cleaner Production and Site Remediation*, Training Manual DTT-5-4-95, United Nations Industrial Development Organization, Industrial Sectors and Environment Division, Vienna, Austria, April 1995.)

1.9.7.2 Advantages

These include the following:

1. Improvement in the quality of the circuit boards
2. Elimination of virtually all the disposal costs
3. Maintenance of the etching solution at its optimum composition
4. Recovery of the copper in a high-value form
5. No hazardous chemicals need to be handled

1.9.7.3 Economic Benefits

Based on 50 t of copper recovered per year, the cost saving in 2007 USD is as follows[14]:

Copper (USD/yr)	65,000
Materials (USD/yr)	104,000
Disposal (USD/yr)	32,500
Total (USD/yr)	500
Capital investment (USD)	286,000
Payback	17 months

1.9.8 CHROME RECOVERY AND RECYCLING IN THE LEATHER INDUSTRY

The Greek and Dutch governments have a framework of bilateral collaboration in the field of environmental protection. One result of this has been that clean technology has been applied in a full-scale cooperative R&D project between the two countries at the Germanakos tannery. The project was carried out with the support of the Commission of the European Communities.

The Germanakos SA tannery near Athens in Greece was founded in 1978. Today it produces good quality upper leather from cattle hides, processing 2200 t/yr and with an annual turnover of 2007 USD 11 million and a staff[14] of 65.

Tanning is a chemical process that converts putrescible hides and skins into a stable material. Vegetable, mineral, and other tanning agents may be used—either separately or in combination—to produce leather with different qualities and properties. Trivalent chromium (Cr^{3+}) is the major tanning agent, because it produces modern, thin, light leather suitable for shoe uppers, clothing, and upholstery. However, recent limits for discharge to the environment have limited chromium discharge to levels as low as 2 mg/L in wastewaters.

1.9.8.1 Cleaner Production

The technology developed involves the recovery of chromium from the spent tannery liquors and its reuse.

Tanning of hides is carried out with basic chromium sulfate, $Cr(OH)SO_4$, at a pH of 3.5 to 4.0. After tanning the solution is discharged by gravity to a collection pit. The liquor is sieved during this transfer to remove particles and fibers that have come from the hides. The liquor is then pumped to the treatment tank and a calculated quantity of magnesium oxide is added with stirring until the pH reaches at least 8. The stirrer is switched off and the chromium precipitates as a compact sludge of $Cr(OH)_3$. After settling the clear liquid is decanted off. The remaining sludge is dissolved by adding a calculated quantity of concentrated sulfuric acid (H_2SO_4) until a pH of 2.5 is reached. The liquor now contains $Cr(OH)SO_4$ and is pumped back to a storage tank for reuse (Figure 1.9).

In conventional chrome tanning processes 20 to 40% of the chrome used is discharged into wastewaters. In the new process 95 to 98% of the waste Cr^{3+} can be recycled.

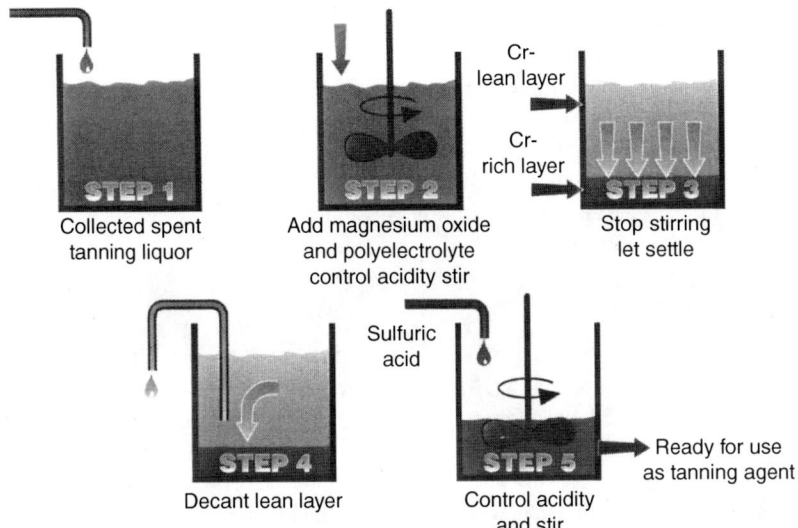

FIGURE 1.9 Five-step batch process for chromium recycling. (From Wang, L.K. et al. *Case Studies of Cleaner Production and Site Remediation*, Training Manual DTT-5-4-95, United Nations Industrial Development Organization, Industrial Sectors and Environment Division, Vienna, Austria, April 1995.)

1.9.8.2 Advantages

These include the following:

1. Very little change to production process
2. More consistent product quality
3. Easier monitoring of the amounts of water and process chemicals used
4. Much reduced chromium content in the effluent waters.

1.9.8.3 Economic Benefits

For the Germanakos tannery, which has a chrome recycling capacity of 12 m³/d, the approximate costs were as follows[14]:

	Cost Saving in 2007 USD
Savings (USD/yr)	95,880
Operating cost (USD/yr)	39,260
Total net savings (USD/yr)	56,620
Capital investment (USD)	52,000
Payback	11 months

Savings can be made with any plant processing more than 1.7 m³/d.

1.9.9 MINIMIZATION OF ORGANIC SOLVENTS IN DEGREASING AND PAINTING OF METALS

Thorn Jarnkonst of Sweden produces lighting fixtures from aluminum or steel sheets for indoor and outdoor use. The production amounts to 750,000 units. They employ about 650 people. In 1988 the company merged with Thorn EMI, the main branch being located in England.

Metal working, degreasing, and painting are the main phases in this production process. The degreasing of the metal sections has been carried out in the past by using the volatile organic compound trichloroethylene, which is a pollutant and an environmental hazard.

The painting plant consisted of an automatic liquid lacquer line, with different colors using different organic solvents. The air pollution and the accumulated remaining products were a considerable problem, both within the plant and externally.

When the company planned to expand production the local authorities ordered the company to reduce its current air emissions. As a result the company intended to install equipment to capture the trichloroethylene and incinerate the solvents from the painting plant.[31] However, an independent research organization, by carrying out a pollution prevention audit, suggested an alternative approach having environmental benefits.

1.9.9.1 Cleaner Production

The pollution prevention audit started with an analysis of the material flow in the degreasing process. It was shown that by better housekeeping, the need for trichloroethylene degreasing could be reduced by 50%, but this has now been cut to zero. The cutting of aluminum sheets required cutting fluids, which were difficult to remove without the use of chlorinated solvents. A change to biodegradable cutting oils allowed an alkaline degreasing procedure in place of the previous trichloroethylene method.

The degreasing is carried out in a new piece of equipment in the form of a totally enclosed tunnel, 30 m long. The metal products are suspended from an overhead conveyor and then pass through five zones where they are sprayed with various liquids (Figure 1.10). The stages carried out are degreasing, water rinse, iron phosphating to aid the adherence of the paint, water rinse, a deionized water rinse, and drying. The liquid runs off the metal items into tanks below, where it is recirculated back to the spray nozzles.

Electrostatic powder painting uses polymer-based paints that do not have any solvent in their formulation. A long-term problem was changing to a different color of paint. This is now accomplished by changing the whole module with containers of different colors. The company has now installed a new electrostatic powder painting line having 12 automatic powder guns. The paint is

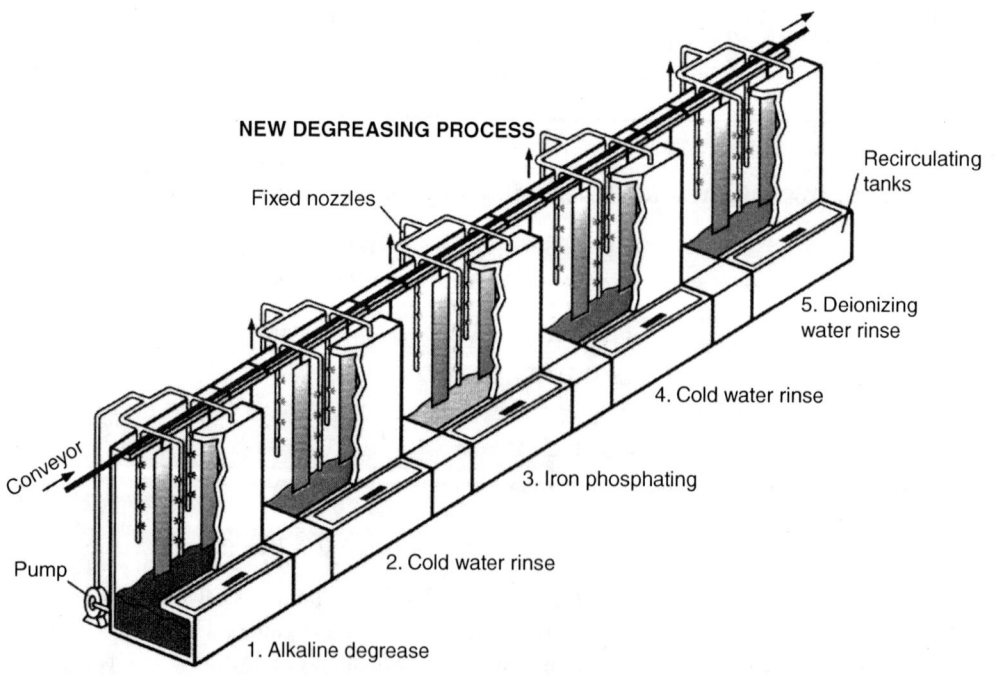

FIGURE 1.10 Schematic of the new degreasing process. (From Wang, L.K. et al. *Case Studies of Cleaner Production and Site Remediation*, Training Manual DTT-5-4-95, United Nations Industrial Development Organization, Industrial Sectors and Environment Division, Vienna, Austria, April 1995.)

positively charged relative to the metal items. Now only 5% of the colors have organic solvents and are used only for the painting of short production runs in special colors or for retouching of the automatically sprayed items where necessary. Manual spraying is carried out in a ventilated booth fitted with two electrostatic guns.

1.9.9.2 Advantages

These include the following:

1. Changed degreasing techniques:
   a) The environmental advantages that have been achieved are external and also within the workplace.
   b) The company more than adequately meets the demands from the authorities.
   c) The water purification plant, which is also used for other process baths, can be used for alkaline degreasing too and results in little additional water pollution from the degrease stage.
2. Changed painting techniques:
   a) There is a large reduction in the discharge of organic solvents.
   b) Hazardous waste is reduced.
   c) There is an improved work environment.
   d) Production has been enabled to expand without conflicting with environmental demands.

1.9.9.3 Economic Benefits

The alkaline degrease turned out to be USD 32,800 cheaper a year than the trichloroethylene degrease and did not require the installation of recovery equipment.

The powder painting techniques have led to considerably lower working costs. The following costs for solvent painting have disappeared with the use of powder painting.[14]

	Cost Savings in 2007 USD
Paint (USD/yr)	268,000
Cleaning (USD/yr)	81,000
Disposal (USD/yr)	61,000
Pumping (USD/yr)	43,000
Labor (USD/yr)	146,000
Total (USD/yr)	599,000
Capital investment (USD)	559,000
Payback	11 months

REFERENCES

1. HSWA, Federal Hazardous and Solid Wastes Amendments (HSWA), U.S. EPA, Washington, DC, November 1984, available at http://www.epa.gov/osw/laws-reg.htm, 2007.
2. Federal Register, Resource Conservation and Recovery Act (RCRA), 42 US Code s/s 6901 et seq. 1976, U.S. Government, Public Laws, available at www.access.gpo.gov/uscode/title42/chapter82_.html, January 2004.
3. South Carolina DHEC, Reduction of Hazardous Waste Legislative Summary: 2004 Report, South Carolina Department of Health and Environmental Control, Columbia, SC, February 2005.
4. WEF, Waste Minimization and Waste Reduction, Water Environment Federation, Alexandria, VA, available at http://www.wef.org/GovernmentAffairs/PolicyPositionStatements/Waste_Minimization+_Reduction. htm, 2007.
5. U.S. EPA, Guidance Manual for Developing Best Management Practices (BMPs), EPA 833-B-93-004, U.S. EPA, Washington, DC, October 1993.

6. U.S. EPA, Pollution Prevention Act of 1990, United States Code Title 42, U.S. EPA, Washington, DC, available at http://www.epa.gov/p2/pubs/p2policy/act1990.htm, October 2006.
7. U.S. EPA, Waste Minimization Opportunity Assessment Manual, EPA/625/7-88/003, U.S. EPA, Cincinnati, OH, July 1988.
8. University of Maryland, Waste Minimization/Pollution Prevention Plan, University of Maryland, Baltimore, MD, available at http://www.ehs.umaryland.edu/Waste/wastemin.htm, 2007.
9. UNIDO, Project Document, United Nations Industrial Development Organization, Industrial Sectors and Environment Division, Vienna, Austria, April 1995.
10. UNIDO, Waste Reduction Auditing Manual, US/GLO/91/103, United Nations Industrial Development Organization, Industrial Sectors and Environment Division, Vienna, Austria, 1991.
11. UNIDO, Demonstration of Cleaner Production Techniques, US/INT/91/X79, United Nations Industrial Development Organization, Industrial Sectors and Environment Division, Vienna, Austria, 1991.
12. Cheremisinoff, P.N., *Waste Minimization and Cost Reduction for the Process Industries*, William Andrew, Norwich, NY, 1995, pp. 331, available at http://www.williamandrew.com/titles/1388.htm, 2007.
13. Friedman, F.B., *Practical Guide to Environmental Management*, 9th ed., Environmental Law Institute, Washington, 2003.
14. Wang, L.K., Krouzek, J.V., and Kounitson, U., *Case Studies of Cleaner Production and Site Remediation*, Training Manual DTT-5-4-95, United Nations Industrial Development Organization, Industrial Sectors and Environment Division, Vienna, Austria, April 1995.
15. Dautzenberg, A. and Groene, T., *Waste Prevention in SMEs; From Desire to Company Practice*, Milieu, 1993, pp. 87–91.
16. Foecke, T., McDonald, K. and Nelson, K., *Sri Lanka–Minnesota Scoping Mission on Institution-Building for Waste Minimization*, final report prepared for the World Environment Centre, Washington, DC, March 1994.
17. Heaton, G.R., *Financing the Process of Technological Change: Innovative Mechanisms for the Transfer of Environmentally Sound Technologies*. UNIDO, Vienna, Austria, 1995.
18. Chodak, M. and Dobes, V., *Pollution Prevention Activities in the Netherlands*, report prepared for the Province of North Holland, June 1993.
19. Heaton, G.R., Repetto, R. and Sobin, R., *Backs to the Future: U.S. Government Policy Towards Environmentally Critical Technology*, World Resources Institute, Washington, DC, 1992.
20. North Carolina DEHNR, *General Waste Minimization Options for Metal Cleaning*, NC Department of Environment, Health and Natural Resources, available at http://www.p2pays.org/ref%5C01/00021.htm, 2007.
21. Mississippi DEQ, Waste Minimization Plan. The Mississippi Department of Environmental Quality Jackson, MS, available at www.deq.state.ms.us, 2007.
22. U.S. ACE, Yearly Average Cost Index for Utilities, in *Civil Works Construction Cost Index System Manual*, 110-2-1304, U.S. Army Corps of Engineers, Washington, DC, 2007, p. 44, available at http://www.nww.usace.army.mil/cost, 2007.
23. Michigan DEQ, Waste Management Guidance—Waste Minimization, Michigan Department of Environmental Quality, Lansing, MI, March 1998, available at http://www.michigan.gov/deq, 2007.
24. U.S. EPA, Waste Minimization in Metal Parts Cleaning, EPA/530SW89049, U.S. EPA, Washington, DC, August 1989.
25. U.S. EPA, Guides to Pollution Prevention—Research and Educational Institutions, EPA/625/7-90/010, U.S. EPA, Cincinnati, OH, June 1990.
26. Wang, L.K., Vaccari, D.A., Li, Y., and Shammas, N.K., Chemical precipitation, in *Physicochemical Treatment Processes*, Wang, L.K., Hung, Y.T., and Shammas, N.K., Eds., Humana Press, Totowa, NJ, 2005, pp. 141–198.
27. Wang, L.K., Hung, Y.T., and Shammas, N.K., Eds., *Physicochemical Treatment Processes*, Humana Press, Totowa, NJ, 2005.
28. Wang, L.K., Hung, Y.T., and Shammas, N.K., Eds., *Advanced Physicochemical Treatment Processes*, Humana Press, Totowa, NJ, 2006.
29. Wang, L.K., Shammas, N.K., and Hung, Y.T., Eds., *Biosolids Treatment Processes*, Humana Press, Totowa, NJ 2007.
30. Virginia WMP, Fact Sheet, Waste Minimization Program on Source Reduction Techniques for Local Governments, Virginia Waste Minimization Program, Richmond, VA, available at http://www.p2pays.org/ref/11/10314.htm, 2007.
31. Conway R.A., Warner, D.J., Wiles, C.C., and Duckett, E.J. *Hazardous and Industrial Solid Waste Minimization Practices*, ASTM, West Conshohocken, PA, 1989, pp. 209.

2 Waste Treatment in the Iron and Steel Manufacturing Industry

*Gupta Sudhir Kumar, Debolina Basu,
Yung-Tse Hung, and Lawrence K. Wang*

CONTENTS

2.1	Introduction	38
2.2	Industrial Processes in the Iron and Steel Industry	38
2.3	Coke Making	39
	2.3.1 Process Description	39
	2.3.2 Sources of Process Wastes	42
	2.3.2.1 Emissions	42
	2.3.2.2 Effluents	43
	2.3.2.3 Hazardous Wastes	43
	2.3.3 Treatment Techniques	44
2.4	Ironmaking	44
	2.4.1 Process Description	44
	2.4.2 Sources of Process Waste	46
	2.4.2.1 Emissions	46
	2.4.2.2 Effluents	48
	2.4.2.3 Byproducts	48
	2.4.3 Treatment Techniques	49
2.5	Steel Making	50
	2.5.1 Basic Oxygen Furnace Process	50
	2.5.1.1 Process Description	50
	2.5.1.2 Sources of Process Waste	50
	2.5.1.3 Treatment Techniques	52
	2.5.2 Electric Arc Furnace	53
	2.5.2.1 Process Description	53
	2.5.2.2 Sources of Process Waste	55
	2.5.2.3 Treatment Techniques	56
2.6	Refining and Casting	57
	2.6.1 Process Description	57
	2.6.2 Sources of Process Waste	58
	2.6.2.1 Emissions	58
	2.6.2.2 Effluent	59
	2.6.2.3 Byproducts	59
	2.6.3 Treatment Techniques	59
2.7	Forming and Finishing	60
	2.7.1 Process Overview	60

		2.7.2	Sources of Process Waste	61
			2.7.2.1 Emissions	61
			2.7.2.2 Effluents	61
			2.7.2.3 Byproducts	64
			2.7.2.4 Hazardous Wastes	64
		2.7.3	Treatment Techniques	64
	2.8	Pollution Prevention Measures		65
		2.8.1	High-Rate Recycle	66
		2.8.2	Countercurrent Cascade Rinsing	66
		2.8.3	Acid Reuse, Recycle, and Recovery Systems	67
		2.8.4	Extension of Process Solution Life	67
	2.9	Process Modifications		68
		2.9.1	Effluent-Free Pickling Process with Fluid Bed Hydrochloric Acid Regeneration	68
		2.9.2	Nitric-Acid-Free Pickling	68
		2.9.3	Effluent-Free Exhaust Cleaning	68
		2.9.4	Elimination of Coke with Cokeless Technologies	68
		2.9.5	Reducing Coke Oven Emissions	69
References				70

2.1 INTRODUCTION

The iron and steel industry is currently on an upsurge because of strong global and local demands. It plays a critical role in the infrastructural and overall economic development of a country. The versatility of steel can be understood from its wide range of applications in the construction, transportation, and process industries. There has been a remarkable growth in world crude steel production, from 189 million metric tons in 1950 to 1244 million metric tons in 2006 (International Iron and Steel Institute, IISI). However, the steel production process is an energy-, raw-material, and labor-intensive process, accounting for major environmental releases.[1–7]

Environmental regulations have always had a profound effect on all stages of manufacturing and forming processes of the iron and steel industry. Taking into consideration the high cost of new equipment and the relatively long lead time required to bring them into the industry, any change in production method or product takes place at a slow pace. The installation of major pieces of new steel-making equipment may cost millions of dollars and require additional retrofitting of other equipment.[8–16] However, in spite of the competition from substitute materials, which forces steelmakers to invest in cost-saving and quality-enhancing technologies, it has always remained a challenge for the industry to develop and maintain cleaner yet efficient steel-making processes. It may therefore be expected that in the long run it would be likely for the steel industry to take up simplified and continuous manufacturing technologies that reduce the capital costs for new plant construction and allow smaller plants to operate efficiently.

2.2 INDUSTRIAL PROCESSES IN THE IRON AND STEEL INDUSTRY

Steel is an alloy of iron, usually containing less than 1% carbon. The process of steel production occurs in several sequential steps. The basic oxygen furnace (BOF) and the electric arc furnace (EAF) are two types of technology in use today for steel making. Although these technologies use different input materials, the output from both furnace types is in the form of molten steel that is further processed into steel mill products. The BOF input materials include molten iron, scrap, and oxygen. In the EAF, the input materials are electricity and scrap. BOFs are typically used for high tonnage production of carbon steels, and EAFs are used to produce carbon steels, low-tonnage alloy, and specialty steels. During the manufacturing of steel using a BOF, coke making and iron making precede steel making; however these steps are not needed for steel making with an EAF. Coke, which acts as a fuel and carbon source, is produced by heating coal in the absence of oxygen at high temperatures in coke ovens. Pig iron is produced by heating the coke, iron ore, and limestone in a blast furnace. In the BOF, molten iron from the blast furnace is combined with flux and scrap steel, followed by the injection of high-purity oxygen. In an EAF process, the input material consist primarily of scrap steel, which is melted and refined by passing an electric current from electrodes through the scrap. The molten steel from either process is formed into ingots or slabs and rolled into finished products. Rolling operations may require reheating, rolling, cleaning, and coating the steel. The process of coke making, iron making, steel making, and subsequent forming and finishing operations is collectively referred to as fully integrated production. Figure 2.1 presents a detailed flowsheet of the iron and steel manufacturing process. This chapter describes the various processes involved in steel making, the sources and types of wastes generated from these manufacturing processes, and the technical advancement necessary for pollution prevention and economic operation of the iron and steel plant.[17–20]

2.3 COKE MAKING

2.3.1 Process Description

Coke is a residue obtained after heating coal to very high temperatures (1650 to 2200°F) in the absence of oxygen, and removing all its volatile components.[14] It is further used as a reductant for blast furnace iron making, having good permeability that allows the free flow of gases within the furnace shaft. Nearly 1.3 to 1.35 t of bituminous (coking) coal is required for the production of 1 t of coke.[1] The U.S. integrated iron and steel industry uses the byproduct process for the manufacture of almost all coke. Byproduct coke ovens allow the collection of volatile material emitted during the coking process. Coking is carried out in brick ovens called batteries, which consist of coking chambers, heating flues, and regenerative chambers. The coking chambers are located alternatively with heating chambers, with the regenerative chambers located underneath. Pulverized coal is charged into the oven through the openings provided at the top. The necessary heat for distillation of the volatile components is supplied by the external combustion of recovered coke oven gas, blast furnace gas, and natural gas through flues located between the ovens.

Finally, the coke produced is removed through doors on either end of the oven and pushed out into a quenching car to be transported to a quenching tower, where it is sprayed onto the coke mass to cool it. The coke is sized and sent to the blast furnace or for storage. The "foul" gas obtained during the coking operation is processed to recover byproducts such as tar, light oils, ammonia, and naphthalene. Foul gas cleaning involves the spraying of weak ammonia, which condenses some tar and ammonia from the gas. The remaining gas is cooled by passing through a condenser and is then compressed by an exhauster. Any remaining coal tar is removed by a tar extractor, either by impingement against a metal surface or collection by an electrostatic precipitator.

Ammonia is removed by passing the gas through a saturator, where ammonia reacts with sulfuric acid to form ammonium sulfate, which is crystallized and removed. In the Phosam process ammonia is scrubbed directly from coke oven gas with phosphoric acid and then stripped.[19] The gas is further cooled to condense naphthalene. The light oils are removed in an absorption tower and subsequently refined. The last cleaning step is the removal of hydrogen sulfide in a scrubbing tower. The purified gas may be used as fuel for the coke ovens or in other plant combustion processes. However, the nonrecovery process for byproduct gas may also be used, in which the unpurified gas is burned within the process rather than being recovered. The energy recovered in the form of heat from the waste gases is passed through a waste heat boiler to generate steam for electricity production or process use. Figure 2.2 gives a flow diagram of the coke-making process and Table 2.1 provides the inputs and outputs of the coke-making process. Table 2.2 gives an overview of the key environmental and energy facts of coke making.

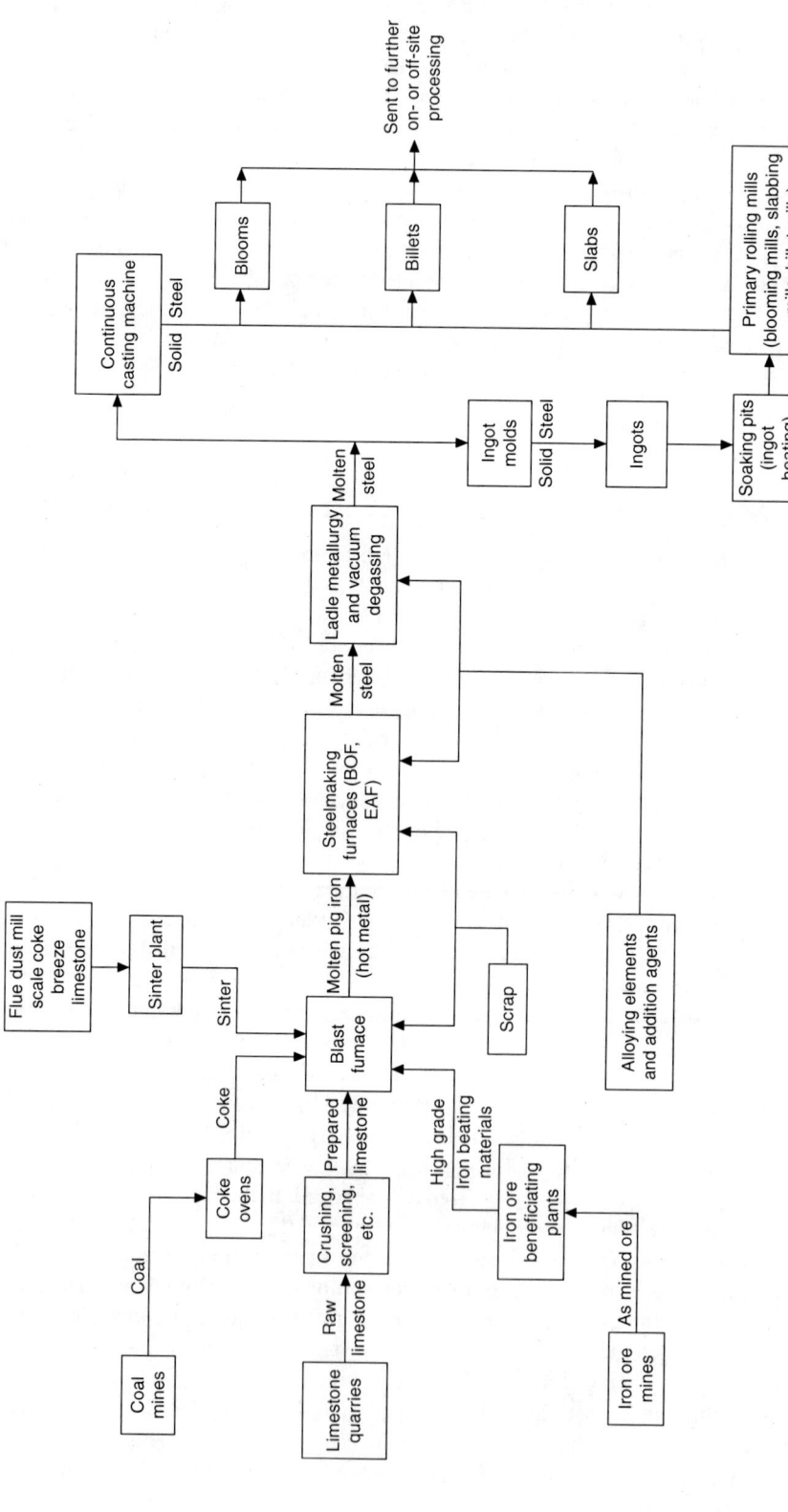

FIGURE 2.1 Iron and steel manufacturing process overview. (From U.S. EPA, Development Document for the Proposed Effluent Limitations Guidelines and Standards for the Iron and Steel Manufacturing Point Source Category, EPA-821-B-00-011, U.S. EPA, Washington, DC, 2000.)

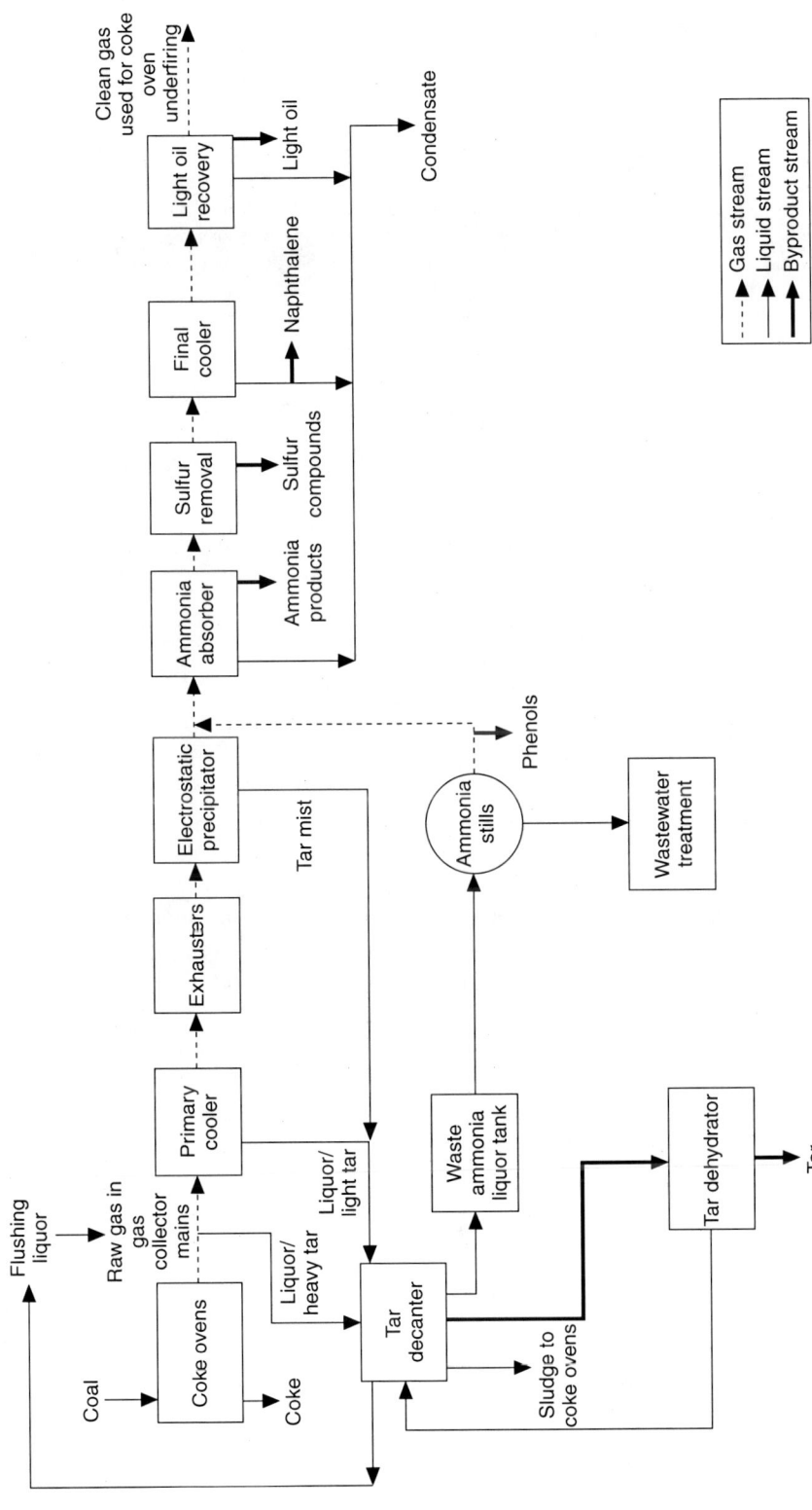

FIGURE 2.2 Coke-making flow diagram. (From U.S. EPA, Development Document for Final Effluent Limitations Guidelines and Standards for the Iron and Steel Manufacturing Point Source Category, EPA-821-R-02-004, U.S. EPA, Washington, DC, 2002.)

TABLE 2.1
Inputs and Outputs of the Coke-Making Process

Inputs	Outputs
Coal	Coke
Heat (from coke oven gas, blast furnace gas, natural gas)	Coke oven gas and byproducts including coal tar, light oil, and ammonia liquor
Electricity	Ammonia, phenol, cyanide, and hydrogen sulfide
Water	Charging, pushing, and quenching emissions
	Products of combustion (SO_2, NO_x, CO, particulate)
	Oil (K143 and K144)
	Ammonia still lime sludge (K060)
	Decanter tank tar sludge (K087)
	Tar residues (K141, K142)
	Benzene releases in byproduct recovery operations
	Naphthalene residues generated in the final cooling tower
	Sulfur and sulfur compounds recovered from coke oven gas
	Wastewater from cleaning and cooling (contains zinc, ammonia still lime, decanter tank tar, or tar distillation residues)
	Air pollution control (APC) dust

Source: From U.S. EPA, Profile of the Iron and Steel Industry, EPA 310-R-95-005, U.S. EPA, Washington, DC, 1995.

2.3.2 SOURCES OF PROCESS WASTES

2.3.2.1 Emissions

The process of coke making emits particulate matters, volatile organic compounds (VOCs), carbon monoxide, and other pollutants. The various sources of emissions include the following:

1. Fugitive particulate emissions during material handling (coal preparation)
2. Significant emissions of particulate matter and VOCs due to oven charging
3. VOC emissions from distillation within the oven during the coking cycle from leaks in the doors, charge lids, and offtake caps

TABLE 2.2
Overview of Key Environmental and Energy Facts for Coke-Making

Energy	Emissions	Effluents	Byproducts/Hazardous Wastes
3.4×10^6 Btu per ton of coke produced	Major pollutants: particulate, VOCs, CO	Largest sources: waste ammonia liquor, ammonia distillation, crude light oil recovery	Major byproducts: tar, light oils, ammonia, naphthalene
	Largest sources: coke handling, charging, pushing, quenching	Typical wastewater volume: 100 gal/t of coke	Hazardous wastes: 7 RCRA-listed wastes (K060, K087, K141 to K145)
			Largest source: coke oven gas cleaning

Source: From Energetics, Inc., Energy and Environmental Profile of the U.S. Iron and Steel Industry, DOE/EE-0229, U.S. Department of Energy, Washington, DC, 2000.

Waste Treatment in the Iron and Steel Manufacturing Industry

4. Particulate emissions due to pushing of coke from the oven into the quench car
5. Particulate from the coke mass during coke quenching
6. Trace organic compounds and dissolved solids from the quench water entrained in the steam plume rising from the tower
7. Emissions from the underfire or combustion stack due to combustion of gas in the coke oven flues
8. VOC emissions from the processing steps for separating ammonia, coke oven gas, tar, phenol, light oils, and pyridine from the foul gas

2.3.2.2 Effluents

The major consumption of water in coke plants is for cooling purposes in a variety of cooling and condensing operations. For the coke quenching operation alone, about 120 to 900 gal of water are required per ton of coke.[1] The various sources of process wastewater include the following:

1. Excess ammonia liquor from the primary cooler tar decanter
2. Barometric condenser wastewater from the crystallizer, the final coolers, light oil recovery operations, ammonia still operation, coke oven gas condensates, desulfurization processes, and air pollution control operations

About 100 gal of process wastewater is typically generated from 1 t of coke produced.[15] These wastewaters from byproduct coke making contain high levels of oil and grease, ammonia nitrogen, sulfides, cyanides, thiocyanates, phenols, benzenes, toluene, xylene, other aromatic volatile components, and polynuclear aromatic compounds. They may also contain toxic metals such as antimony, arsenic, selenium, and zinc. Water-to-air transfer of pollutants may take place due to the escape of volatile pollutants from open equalization and storage tanks and other wastewater treatment systems in the plant.

2.3.2.3 Hazardous Wastes

There are seven Resource Conservation and Recovery Act (RCRA) listed hazardous wastes associated with coke making, as listed below:

1. K060: ammonia still lime sludge
2. K087: decanter tank tar sludge
3. K141: process residues from coal tar recovery operations
4. K142: tar storage tank residues
5. K143: process residues from the recovery of light oil
6. K144: wastewater sump residues from light oil refining
7. K145: residues from naphthalene collection and recovery operations

Process residues from coal tar recovery (K141) are generated when the uncondensed gas from the coke oven collecting main enter the primary cooler. The condensates from the primary cooler flow into the tar collecting sump and discharged to the flushing liquor decanter. Tar storage tank residues (K142) are the residuals of the crude coal tar. These residues are recycled to the oven or landfilled. Residues from light oil processing units (K143) are built up in the oil scrubber and oil stripping still over time. Resin is also accumulated due to cleaning of the wash oil used in the light oil recovery process is resin. The residue from either a wash oil purifier or a wash oil decanter called wash oil muck is removed and recycled to the coke oven, reclaimed offsite, or used as blast furnace or boiler fuel. Wastewater sump residues (K144) accumulated in the bottom of a sump allowing oil and water to separate during light oil recovery are either recycled to the oven or landfilled offsite. Residues from naphthalene collection and recovery (K145) at the bottom of a skimmer sump where naphthalene is mechanically skimmed off the surface or in the hot and cold sumps used for collecting, or surge vessels, and on the surfaces of the cooling tower are recycled to the decanter or sometimes to the oven.

2.3.3 Treatment Techniques

Various pollution control equipment are needed to trap the fine particles of coke generated during charging, pushing, loading, and transporting operations. This solid waste, comprising fine particles, should be properly landfilled. For effective control of charging emissions, goosenecks and collecting main passages should be cleaned frequently to prevent obstructions. Emissions due to the combustion of gas in the coke oven are controlled by conventional gas cleaning equipment such as electrostatic precipitators and fabric filters. In fulfillment of requirements under the National Emission Standards for Hazardous Air Pollutants (NESHAPs) rule, existing coke ovens need to provide coke byproduct plants with inert gas blanketing systems for eliminating 95 to 98% of benzene emissions and preventing emissions of other VOCs as well.[6]

Conventional wastewater treatment techniques consist of physical/chemical treatments, including oil separation, dissolved gas flotation, and ammonia distillation (for removal of free cyanides, free sulfides, and ammonia) followed by biological treatment (for organics removal) and residual ammonia nitrification. Almost all residuals from coke-making operations are either recovered as crude byproducts (e.g., as crude coal tar, crude light oil, ammonium sulfate, or other sulfur compounds) and sold or recycled to the coke ovens for recovery of carbon values (e.g., coal tar decanter sludge, coke plant wastewater treatment sludge).[15] The constituents of concern and their average measured concentrations for K141 through K145 are presented in Table 2.3.

2.4 IRONMAKING

2.4.1 Process Description

In the blast furnace iron-making process, iron ore is reduced by removing oxygen, followed by melting of the resulting iron. Agglomeration processes such as pelletization and sintering help in

TABLE 2.3
Constituents of Concern and Average Measured Concentrations for Wastes K141 through K145[a]

Constituent	K141 (Process Residues from Coal Tar Recovery)	K142 (Tar Storage Tank Residues)	K143 (Residues from Light Oil Processing)	K144 (Wastewater Treatment Sludges from Light Oil Refining)	K145 (Residues from Naphthalene Collection and Recovery)
Benzene	3850	260	1600	3000	1000
Benz (a) anthracene	7850	6600	69	68	22
Benzo (a) pyrene	8450	6500	34	65	7
Benzo (b) fluoranthene	5450	7500	59	75	26
Chrysene	7950	6000	59	6	22
Dibenz (a,h) anthracene	1750	1000	38	15	1
Indeno (1,2,3-cd) pyrene	6140	2900	40	37	4
Naphthalene	95,000	55,000	52000	27,000	140,000

[a] Concentrations measured in mg/kg.

Source: From U.S. EPA, Federal Register, Part III, U.S. EPA, Washington, DC, August 18, 1992.

producing coarse particles of suitable sized iron ore for easy charging into the blast furnace. In pelletization, an unbaked "green" pellet is formed from iron ore concentrate combined with a binder. These green pellets are hardened by heat treatment in an oxidizing furnace. Pelletizing is usually done at the mine site. Sintering is a crucial process in the steel mill, using natural fine iron-bearing materials as well as those recovered from ore handling and other iron and steel operations, for example, ore fines from screening operations, water treatment plant sludges, and air pollution control dusts, and fusing them into porous sinters suitable for charging to blast furnaces. In sintering, both iron ore fines and other iron-bearing materials (iron-bearing scale, dusts, and slag) are thoroughly mixed with fluxes (lime or dolomite) and approximately 5% of a finely divided fuel, such as coke breeze or anthracite.[20]

The mix is loaded onto a traveling grate called sinter strand, which is in the form of a shallow trough with small holes in the bottom. The bed of materials on the grate is ignited by passing under an ignition burner fired with natural gas and air. As the grate moves slowly towards the discharge end, windboxes on the underside of the strand pull down the combustion gases through the material bed into a duct to gas cleaning equipment. As the coke fines burn in the bed, the generated heat sinters the fine particles. The temperature of the bed reaches around 1300 to 1480°C. Average production rates of 22 to 43 metric tons/m^2/d of grate area are expected, depending upon the characteristics of the ore materials and the sintering conditions.[13]

The fused sinter mass is cooled, crushed, screened, and sent to be charged, along with the ore, to the blast furnace. Approximately 2.5 t of raw materials, including water and fuel, are required to produce 1 t of sinter product.[16] Figures 2.3 and 2.4 illustrate the sintering and blast furnace iron-making processes, respectively. Blast furnaces are used to produce pig iron, which represents about three-quarters of the charge to basic oxygen steel-making furnaces. The chemical composition of pig iron typically comprises the following[1]:

1. Carbon (4.0 to 4.5%)
2. Silicon (0.3 to 1.5%)
3. Manganese (0.25 to 2.2%)
4. Phosphorus (0.04 to 0.20%)
5. Sulfur before desulfurization (0.03 to 0.8%)
6. Iron (>90%)

Iron ore, coke, flux (limestone and dolomite), and sinter are fed into the top of the blast furnace; heated air augmented with gaseous, liquid, or powdered fuel is injected into its base. As the charge

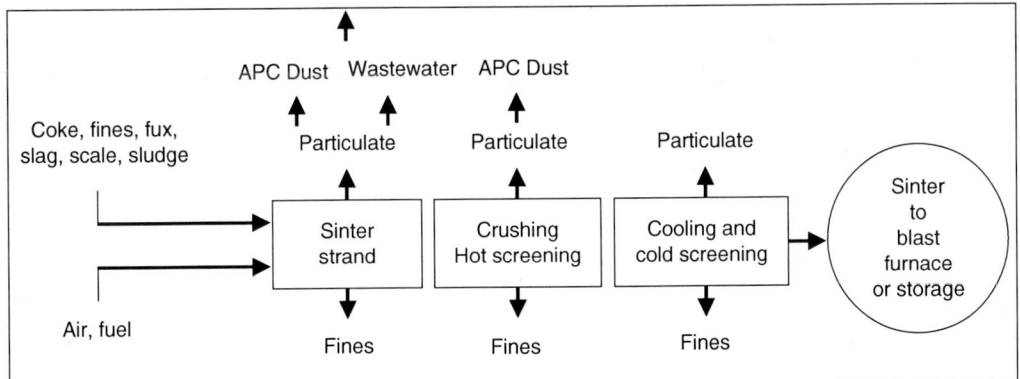

FIGURE 2.3 Sintering flow diagram. (From Energetics, Inc., Energy and Environmental Profile of the U.S. Iron and Steel Industry, DOE/EE-0229, U.S. Department of Energy, Washington, DC, 2000.)

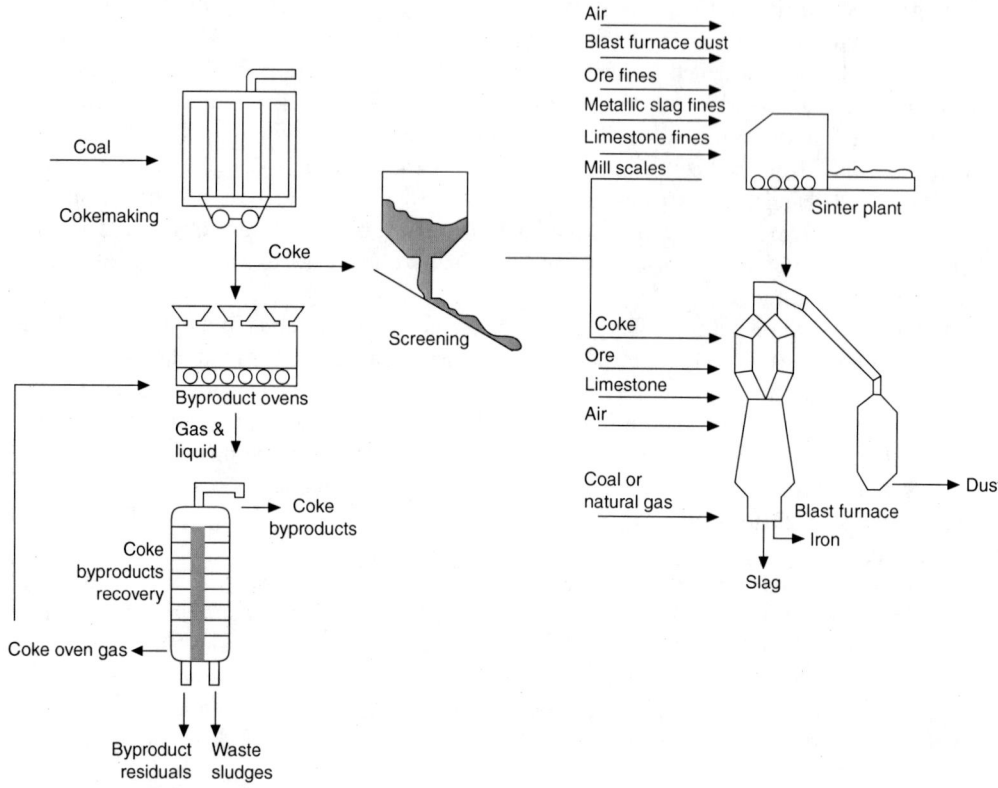

FIGURE 2.4 Flow diagram of the iron-making operation. (From U.S. EPA, Profile of the Iron and Steel Industry, EPA 310-R-95-005, U.S. EPA, Washington, DC, 1995.)

materials descend, the reducing gas (containing carbon monoxide) generated by the burning coke flows upward, converting the iron oxide (FeO) in the ore to iron (Fe). The coke also provides the structural support for the unmelted burden materials. The combustion of the coke generates sufficient heat to melt the iron, which accumulates in the bottom of the furnace (hearth). The major function of the flux is to combine with unwanted impurities, that is, ash in the coke and gangue in the ores, to make a drainable fluid slag. Unreacted reducing gas (blast furnace gas) is collected at the top, cleaned, and used as a fuel. The molten iron called the "hot metal" is tapped into refractory-lined cars for transport to the basic oxygen furnace. The iron may be processed at desulfurization stations to minimize sulfur compounds before charging in the basic oxygen furnace. Molten slag, which floats on top of the molten iron, is also tapped and processed for sale as a byproduct. The production of one net ton of iron requires approximately 1.5 to 1.7 t of ore or other iron-bearing material, 0.35 to 0.55 t of coke, 0.25 t of limestone or dolomite, and 1.6 to 2.0 t of air.[1,16] The inputs and outputs of the iron-making operation are shown in Table 2.4, with a brief overview of key environmental and energy facts of sintering and iron making in Table 2.5 and Table 2.6, respectively.

2.4.2 Sources of Process Waste

2.4.2.1 Emissions

Emissions from sinter plants are generated from raw material handling, windbox exhaust, sinter discharge (associated sinter crushers and hot screens), and from the cooler and cold screen. The primary source of particulate emissions, mainly irons oxides, magnesium oxide, sulfur oxides, carbonaceous compounds, aliphatic hydrocarbons, and chlorides, are due to the windbox exhaust.

TABLE 2.4
Inputs and Outputs of the Iron-Making Operation

Inputs	Outputs
Iron ore (primarily as pellets)	Molten iron
Coke (coal)	Slag
Sinter	Blast furnace gas
Limestone	Residual sulfur dioxide or hydrogen sulfide
Heated air (from coke oven gas, blast furnace gas, natural gas, fuel oil)	Air pollution control (APC) dust and/or waste treatment plant sludge
Electricity	Process wastewater
Natural gas	Kish
Coal	
Oxygen	
Water	

Source: From U.S. EPA, Profile of the Iron and Steel Industry, EPA 310-R-95-005, U.S. EPA, Washington, DC, 1995.

Contaminants such as fluorides, ammonia, and arsenic may also be present. At the discharge end, emissions are mainly iron and calcium oxides. A wide variety of organic and heavy metal hazardous air pollutants (HAPs) may be released during sinter operations from the coal/coke on the sinter grate and iron, respectively. The heavy metal HAPs include cadmium, chromium, lead, manganese, and nickel. Total HAPs releases from individual sinter manufacturing operations may exceed 10 t/yr.[6] The typical components of the dust generated during sintering practice include iron, carbon, sulfur, Fe_2O_3, SiO_2, Al_2O_3, CaO, and MgO.

Large quantities of carbon monoxide and sulfur dioxide are also emitted during iron making. The primary source of blast furnace particulate emissions is due to the contact of molten iron and slag with the air above their surface during casting (removal of the molten iron and slag from the furnace). Emissions are also generated by drilling and plugging the taphole, which is an opening at the base of the furnace to allow iron and slag to flow out into runners that lead to transport ladles. Heavy emissions result from the use of an oxygen lance to open a clogged taphole. Another potential source of emissions is the blast furnace top. No serious emission problem is created if charging is through a sealed system. However, minor emissions may occur during charging from imperfect bell seals in the double bell system. Occasionally, a cavity may form in the blast furnace charge, causing

TABLE 2.5
Overview of Key Environmental and Energy Facts: Sintering

Energy	Emissions	Effluents	Byproducts/Hazardous Wastes
1.55×10^6 Btu per ton of sinter	Largest source: windbox	Largest source: wet air pollution control devices	Dust/sludge
	Particulate: iron and sulfur oxides, carbonaceous compounds, aliphatic hydrocarbons, chlorides	Typical wastewater volume: 120 gal/t of sinter	No RCRA-listed hazardous wastes

Source: From Energetics, Inc., Energy and Environmental Profile of the U.S. Iron and Steel Industry, DOE/EE-0229, U.S. Department of Energy, Washington, DC, 2000.

TABLE 2.6
Overview of Key Environmental and Energy Facts: Blast Furnace Iron Making

Energy	Emissions	Effluents	Byproducts/Hazardous Wastes
16.1×10^6 Btu per ton of iron (gross)	Largest source: removal of iron and slag from furnace (casting)	Largest source: gas cooling water and scrubber water for gas cleaning	Total generation (t/yr) Slag: 14×10^6 Dust/sludge: $\sim 1.1 \times 10^6$
12.1×10^6 Btu per ton of iron (net)	Particulate: iron oxides, MgO, carbonaceous compounds	Typical water flows: 6000 gal/t of iron	Reuse: near 100% (slag); ~40% (dust/sludge)

Source: From Energetics, Inc., Energy and Environmental Profile of the U.S. Iron and Steel Industry, DOE/EE-0229, U.S. Department of Energy, Washington, DC, 2000.

a collapse of part of the charge above it. The resulting pressure surge in the furnace opens a relief valve to the atmosphere to prevent damage to the furnace by the high pressure created and is referred to as a "slip."[14]

During hot metal desulfurization, used to remove or alter sulfur compounds in the hot metal, the exhaust gases are found to bear particulate matter. Emissions may also result from slag handling. Sulfur dioxide is formed when the sulfur in the slag is exposed to air. The presence of moisture can result in the formation of hydrogen sulfide. Most sulfur emissions associated with slag handling result from quenching operations. It has been reported that highly toxic polychlorinated dibenzofurans (PCDFs) are dominant in the stack flue gases of sinter plants.[20]

2.4.2.2 Effluents

In an integrated mill, the blast furnace is one of the largest water users. Its main use is for noncontact cooling of various parts of the furnace and auxiliaries. Additional water is used for furnace moisture control, dust control, and slag granulation. Contact water use is primarily associated with blast furnace gas-cleaning operations for recovering the fuel value of the off gas. Nearly all of the wastewater generated from blast furnace operations is direct contact water from the gas coolers and high-energy scrubbers used to clean the blast furnace gas. Typical water requirements are 6000 gal/t of iron.[15]

Water in a sintering plant is mainly used for controlling the moisture content of the presinter mix, for dust control, and sinter product cooling. Wastewaters are generated from wet air pollution control devices, that is, electrostatic precipitator or wet venturi-type scrubber technology on the windbox and discharge ends of the sinter machines. Applied flows for wet air pollution control devices are typically 1000 gal/t, with discharge rates of 50 to 100 gal/t for the better controlled plants.[1] The principal pollutants of wastewater generated from blast furnace operations include total suspended solids, ammonia nitrogen, cyanides, phenolic compounds, copper, lead, nickel, zinc, selenium, arsenic, chromium, and cadmium.

2.4.2.3 Byproducts

The primary byproducts generated during the production of molten iron include blast furnace gas, slag, air pollution control dust (flue dust varying in size from about 6 mm to only a few microns), and waste treatment plant sludge (blast furnace filter cake). The blast furnace gas or top gas is a heated, dust-laden, combustible gas that can be used as a fuel throughout the plant. Water treatment plant sludge is generated as a result of wet-scrubbing systems containing relatively high levels of zinc and lead. This sludge needs to be treated before recycling as feedstock to the sinter plant or blast furnace to maintain an acceptable level of zinc and lead in the furnace. Between 2.0 and 3.0 net tons of this gas are generated for each ton of pig iron produced.[1] Blast furnace gas contains about 40% carbon monoxide and carbon dioxide combined. The dust and sludge are composed of oxides of iron, calcium, silicon, magnesium, manganese, and aluminum. Blast furnace slag comprises about 20 to 40% of molten iron production by weight. Lower grade ores yield higher slag fractions, sometimes as high as 500 to 1000 lb of slag for each ton of pig iron produced.[1]

2.4.3 Treatment Techniques

During the use of oil-bearing mill scale as a revert material, the VOC emissions that are generated are evaporated off the sinter strand into the windbox prior to incineration. Cyclone cleaners followed by a dry or wet electrostatic precipitator, high pressure drop wet scrubber, or baghouse are used to control the sinter strand windbox emissions. Crusher and hot screen emissions, which are the next largest emission source, are usually controlled by hooding and a baghouse or scrubber. Baghouses are used to capture particulates generated during conveyor transport and loading or unloading of sinter plant feedstocks and product. The air pollution control dust that is collected by the baghouses is either recycled as feedstock to the sinter plant or landfilled. The iron making process is the highest-emitting process among those considered, responsible for approximately 40% of total emissions for both criteria pollutants and CO_2.[1] Nitrogen oxide (NO_x) emitted from sinter plants is controlled by selective catalytic reduction (SCR) using NH_3. V_2O_5/TiO_2-based catalysts are considered to be state of the art for this application.[20] Casting emissions are controlled by evacuation through retrofitted capture hoods to a gas cleaner, or by flame suppression techniques. Emissions controlled by hoods and an evacuation system are usually vented to a baghouse. In hot metal desulfurization, exhaust gases are discharged through a series of baghouses to control airborne particulate matter. Emissions from the blast furnace are controlled by a wet venture scrubber or another control device.[1,14]

Wastewater treatment comprises sedimentation for removal of heavy solids, recycle of clarifiers or thickener overflows, and metals precipitation treatment for blowdowns. Standard treatment includes sedimentation in thickeners or clarifiers, cooling with mechanical draft cooling towers, and high-rate recycle. Low-volume blowdowns (<70 gal/t) are either consumed in slag cooling at furnaces with adjacent slag pits, or treated in conventional metals precipitation systems. A few mills practice alkaline chlorination to treat ammonia nitrogen, cyanides, and phenolic compounds.[15]

About 60% of the particulate is removed from the blast furnace gas stream by dry cyclonic vortex separation (i.e., dust catcher) of the heavy particles (flue dust). Fine particulates are subsequently removed in a two-stage cleaning operation consisting of a wet scrubber (primary cleaner), which removes about 90% of the remaining particulate, and a high-energy venturi impact scrubber or electrostatic precipitator (secondary cleaner), which removes up to 90% of the particulate eluding the primary cleaner. During the two-stage blast furnace gas cleaning process the fine particles removed by the gas washer become entrained in a liquid–solid stream that continues on to the treatment plant for settling and solids separation. The concentrated sludge can be dewatered further by mechanical filtration.[1]

At some plants the blast furnace dust is recycled as feedstock to the sinter plant. At plants without sintering operations, blast furnace dust is sometimes mixed with other byproduct residues, briquetted, and recycled back to the blast furnace. In other plants, the dust is landfilled or stockpiled.[1] Several techniques are available for removing the zinc and lead. The majority of blast furnace sludge is land disposed as solid waste or stockpiled. Because of the similarity between wastewater sludges generated by sinter plants and blast furnaces, these streams are commingled and cotreated.[1] The blast furnace slag is cooled and processed to be reused for various applications such as onsite in-land reclamation and landfill construction.

2.5 STEEL MAKING

2.5.1 Basic Oxygen Furnace Process

2.5.1.1 Process Description

The basic oxygen furnace steel-making process refines a charge of molten pig iron and ambient scrap into steel using very high purity oxygen. This results in a reduction of the carbon content of less than 1% for steel from "hot metal" containing about 4% carbon. Other elements in the hot metal such as silicon, phosphorus, sulfur, and manganese, are transferred to a slag phase. The basic raw materials required to make steel in the oxygen steel-making process include:

1. Hot metal (pig iron) from the blast furnace
2. Steel scrap (20 to 35%)
3. Other metallic iron sources (e.g., DRI, ore, oxides)
4. Fluxes (e.g., lime)

After the hot metal and scrap are charged, oxygen is injected into the BOF. The fluxes are then added to control sulfur and phosphorus and prevent erosion of the furnace refractory lining. The principle active ingredients from the fluxes are CaO (from burnt lime) and MgO (from dolomitic lime). Burnt lime consumption ranges from 40 to 100 lb/net ton of steel produced, while dolomitic lime requirements range from 30 to 80 lb/ton.[1] The energy required to raise the fluxes, scrap, and hot metal to steel-making temperatures is provided by oxidation of various elements in the charge materials, particularly iron, silicon, carbon, and manganese. No external heat source is needed, as the temperature increase caused by the oxidation reactions is countered by the addition of scrap and other coolants. During processing, the carbon in the iron is oxidized and released as CO (about 90%) and CO_2 (about 10%). These gaseous oxides exit the furnace carrying small amounts of iron oxide and lime dust. BOFs are classified according to the location of oxygen injection:

1. *Top blown.* Oxygen is injected above the hot metal bath by means of a retractable, water-cooled lance.
2. *Bottom blown.* Oxygen is injected under the molten metal bath, usually carrying pulverized additives.
3. *Combination blown.* Oxygen is injected both above and below the bath.

Bottom stirring is accomplished by the introduction of inert gas under the bath. Silicon, manganese, iron, and phosphorus form oxides that combine with the fluxes to create a liquid slag floating on top of the steel bath, removing sulfur and phosphorus from the metal. At the end of the cycle, raw steel is tapped into a ladle where it is deoxidized and alloying elements are added to adjust the composition to final levels, or to concentrations suitable for further ladle treatment processes. Figure 2.5 illustrates BOF steel-making and Table 2.7 below gives the inputs and outputs of the steel-making operation. An overview of key environmental and energy facts of steel making is presented in Table 2.8.

2.5.1.2 Sources of Process Waste

Emissions
The most significant emissions from BOF steel-making occur during the oxygen blow period. The predominant compounds emitted are iron oxides (including FeO and Fe_2O_3), although heavy metals are also present. Tapping emissions include iron oxides, sulfur oxides, and other metallic oxides, depending on the grade of scrap used. Hot metal transfer emissions are mostly iron oxides.[6,14] The particulate-laden combustion gases and fume (a very fine iron oxide containing high and variable amounts of zinc) released during oxygen blow periods are removed from the furnace by evacuation

Waste Treatment in the Iron and Steel Manufacturing Industry

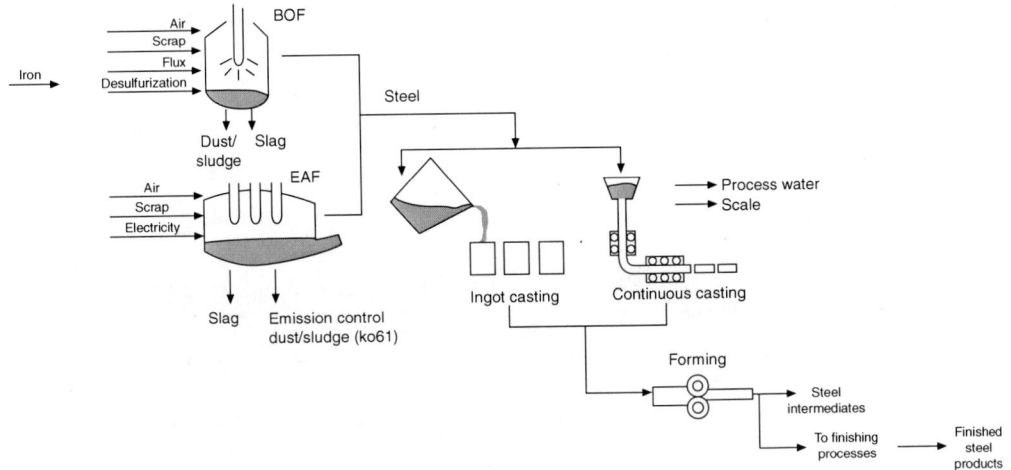

FIGURE 2.5 Flow diagram of steel-making process. (From U.S. EPA, Profile of the Iron and Steel Industry, EPA 310-R-95-005, U.S. EPA, Washington, DC, 1995.)

through a large collection main. The BOF gas, consisting mainly of CO, leaves the BOF at a temperature of between 1600 to 1800°C. Most of the hazardous air pollutants (HAPs) generated in the BOF are heavy metals, including cadmium, chromium, lead, manganese, and nickel.[1] Standard treatment of BOF-generated effluents consists of sedimentation in clarifiers or thickeners and recycle of at least 90% of the applied water. Blowdown treatment consists of metals precipitation.

Effluents
Water is mainly used for cooling in the vessel hood, ductwork, trunnion, and oxygen lance. Both closed-loop and evaporative systems are used for component cooling. The gases and submicron fumes that are released during BOF steelmaking are quenched with water to reduce their temperature and

TABLE 2.7
Inputs and Outputs of the Steel-Making Operation

Inputs	Outputs
Molten iron	Molten steel
Metal scrap	Air pollution control (APC) dust and sludge
Other metallic iron sources	Metal dusts (consisting of iron particulate, zinc, and other metals associated with the scrap, and flux)
Ore	Slag
Iron oxide materials and waste oxides	Kish
Oxygen	Carbon monoxide and carbon dioxide
Alloy materials (e.g., aluminum, manganese, chromium, nickel)	Nitrogen oxides and ozone
Fluxes (e.g., lime)	
Electricity and natural gas for auxiliary processes	
Nitrogen	
Argon	
Water	

Source: From U.S. EPA, Profile of the Iron and Steel Industry, EPA 310-R-95-005, U.S. EPA, Washington, DC, 1995.

TABLE 2.8
Overview of Key Environmental and Energy Facts: Steel Making

Energy	Emissions	Effluents	Byproducts/Hazardous Wastes
0.9×10^6 Btu per net ton of raw steel	Largest source: oxygen blow	Largest source: BOF off-gas control systems	Total generation (t/yr) Slag: 6.0×10^6 Dust: 0.3×10^6 Sludge: 1.3×10^6
	Particulate: iron oxides, heavy metals, fluorides	Typical water flows: 1000 gal/t	Reuse: less than 50%

Source: From Energetics, Inc., Energy and Environmental Profile of the U.S. Iron and Steel Industry, DOE/EE-0229, U.S. Department of Energy, Washington, DC, 2000.

volume prior to being treated in air pollution control systems. The three major off-gas control systems result in the generation of wastewater streams containing total suspended solids and metals (primarily lead and zinc, but also arsenic, cadmium, copper, chromium, and selenium). In the open combustion and suppressed combustion systems, about 1100 and 1000 gal of water per ton of steel are used, respectively.[15]

Byproducts

BOF steel-making byproducts include BOF slag, air pollution control (APC) dust, and water treatment plant (WTP) sludge. BOF slag is composed of calcium silicates and ferrites combined with fused oxides of iron, aluminum, manganese, calcium, and magnesium. After removing the molten BOF slag from the furnace, it is cooled and processed to recover the high metallic portions (iron and manganese) for use in the sinter plant or blast furnaces. The remaining nonferrous fraction is crushed and sized for reuse either within the steel works or externally. Owing to the difference in composition of BOF slag compared with blast furnace slag, the oxides present in BOF slag can result in volume expansion of up to 10% when hydrated. Hence its use is more limited than blast furnace slag. BOF slag outputs are approximately 20% by weight of the steel output.[1] Entrained steel in the slag is typically recovered and returned to the furnace. A typical BOF slag composition is as follows:

1. CaO, 48%
2. FeO, 26%
3. SiO_2, 12%
4. MgO, 6 to 7%
5. MnO, 5%
6. Al_2O_3, 1 to 2%
7. P_2O_5, 1%

The marketable slag makes up about 10 to 15% of the steel output, or 210 to 300 lb/t of steel.[1] BOF dust and sludge generated during the cleaning of gases emitted from the BOF represent two of the three largest-volume wastes typically land disposed by the iron and steel industry.

2.5.1.3 Treatment Techniques

The hot BOF gases are typically treated by one of three air pollution control methods:

1. *Semiwet.* Water is added for conditioning furnace off-gas temperature and humidity prior to processing the gas in electrostatic precipitators or baghouses.
2. *Wet—open combustion.* Excess air is admitted to the off-gas collection system, allowing combustion of carbon monoxide prior to high-energy wet scrubbing for air pollution control.

Waste Treatment in the Iron and Steel Manufacturing Industry

3. *Wet—suppressed combustion.* Excess air is not admitted to the off-gas collection system prior to high-energy wet scrubbing for air pollution control, thus suppressing combustion of carbon monoxide.[15]

Charging and tapping emissions are controlled by a variety of evacuation systems and operating practices. Charging hoods, tapside enclosures, and full furnace enclosures are used in the industry to capture these emissions and send them to either the primary hood gas cleaner or a second gas cleaner.[15,16] Pollution prevention opportunities for the reduction of heavy metals at the BOF are limited as heavy metals are an inherent part of the iron ore material stream, so the higher the iron production, the greater will be the use of the ore.

The cleaning of BOF gas is done by quenching the mixture of gas and particulate with water in the collection main to reduce the temperature. This quenching process removes the larger particles from the gas stream and entrains them in the water system. After settling in the classifier, these coarse solids can be easily dewatered via a long sloping screw conveyor or reciprocating rake and deposited in bins or hoppers. These solids are referred to as classifier sludge. The fine particulate matter remaining in the gas stream is forced through venturi scrubbers, where it is entrained in a wastewater stream and sent to thickener/flocculation tanks for settling and solids removal. This underflow slurry can be dewatered using mechanical filtration. In dry cleaning systems, the particulate matter collected in the electrostatic precipitator or baghouse is managed as a dust.[1]

The rising cost of scrap and waste disposal, the scarcity of onsite landfill space, and potential environmental liabilities make it an economic necessity to recover iron units from dust and sludge. However, recycling to the blast furnace raises the hot metal phosphorus content to undesirable levels. Also, the increasing use of galvanized scrap could increase dust and sludge zinc content. Zinc is known to form a circuit in the furnace, resulting in extra coke consumption and also increasing the risk of scaffolding.[7] The quantity of zinc that can be charged to the blast furnace lies between 0.2 and 0.9 lb/t of hot metal.[1] BOF dust and sludge that is not recycled is usually landfilled.

2.5.2 Electric Arc Furnace

2.5.2.1 Process Description

Electric arc steel-making furnaces produce carbon and alloy steels from scrap metal along with variable quantities of direct reduced iron (DRI), hot briquetted iron, and cold pig iron. Hot metal may also be added if available. The charge is melted in cylindrical, refractory-lined electric arc furnaces (EAFs) equipped with carbon electrodes (one for DC furnaces, three for AC furnaces). During charging, the roof is removed to place scrap metal and other iron-bearing materials into the furnace. Alloying agents and fluxes are added through doors on the side of the furnace. The electrodes are lowered into the furnace to about an inch above the metal and current is applied, generating heat to melt the scrap. Modern electric arc furnaces use an increasing amount of chemical energy to supplement the melting process. The chemical energy contribution is derived by burning elements or compounds in an exothermic manner. Sources that provide chemical energy include[3]:

1. Oxy-fuel burners and oxygen lancing
2. Charge carbon
3. Foaming carbon
4. Exothermic constituents in scrap
5. Exothermic constituents in alternate iron sources

Oxy-fuel burners are used to introduce combinations of natural gas, oil, or even coal into the furnace to displace electricity use. The reaction of carbon with oxygen within the bath to produce CO results in a significant energy input into the process. The injection of a carbon source also promotes

TABLE 2.9
Inputs and Outputs of the Steel-Making Operation (EAF)

Inputs	Outputs
Scrap metal	Molten steel
Direct reduced iron	Slag
Hot briquetted iron	Carbon monoxide
Cold pig iron	Nitrogen oxides and ozone
Hot metal	EAF emission control dust and sludge (K061)
Alloy materials (e.g., aluminum, manganese, chromium, nickel)	
Fluxes (e.g., lime)	
Electricity	
Oxygen	
Nitrogen	
Natural gas	
Oil	
Coal or other carbon source	
Water	

Source: From U.S. EPA, Profile of the Iron and Steel Industry, EPA 310-R-95-005, U.S. EPA, Washington, DC, 1995.

the formation of a foamy slag, which retains energy that is transferred to the bath. The generation of the CO within the bath is necessary to flush out dissolved gases (nitrogen and hydrogen) in the steel, as well as flush oxide inclusions from the steel into the slag.[1] Some EAFs use ferromanganese as a catalyst in the melt to add energy and help stabilize the melt. The efficiency of manganese combustion can be between 90 and 100%.[3] Residence time in the furnace for a 100% scrap charge ranges from about 45 min to several hours.[1] When the charge is fully molten it is refined to remove unwanted materials (e.g., phosphorus, sulfur, aluminum, silicon, manganese, and carbon), tapped from the tilted furnace, and sent for secondary treatment prior to casting. Because scrap metal rather than molten iron is the primary material charged, EAF steel producers avoid the coke-making and iron-making process steps. Figure 2.5 illustrates electric arc furnace steel making and the Table 2.9 gives its major inputs and outputs. An overview of key environmental and energy facts of electric arc furnace process of steel making is presented in Table 2.10.

TABLE 2.10
Overview of Key Environmental and Energy Facts: Electric Arc Furnace Steel Making

Energy	Emissions	Effluents	Byproducts/Hazardous Wastes
5.2 to 5.6 × 10^6 Btu per net ton of raw steel	Largest sources: melting and refining	Largest source: wet/semiwet air cleaning systems	EAF slag: 50–75% reused
	Particulate: iron oxide (melting); calcium oxide (refining)		K061: EAF dust/sludge Major components: Fe, Zn, Cr, and their oxides Total generation: ~900,000 t/yr

Source: From Energetics, Inc., Energy and Environmental Profile of the U.S. Iron and Steel Industry, DOE/EE-0229, U.S. Department of Energy, Washington, DC, 2000.

2.5.2.2 Sources of Process Waste

Emissions

All phases of the EAF operation result in primary or secondary emissions. Primary emissions include those produced during EAF melting and refining operations, whereas secondary emissions are from charging, tapping, and escape of fumes. The major constituents in EAF emissions are particulate matter and gases (carbon monoxide, SO_x, and NO_x). Carbon monoxide is produced in large quantities in the EAF from oxygen lancing and slag foaming activities as well as from the use of pig iron or DRI in the charge. Large amounts of CO and hydrogen are generated at the start of meltdown as oil, grease, and other combustible materials evolve from the surface of the scrap. In the presence of sufficient oxygen these compounds will burn to emit more of CO_2. NO_x is formed in furnace operations when nitrogen passes through the arc between electrodes and also during burner use in EAFs. Levels of about 36 to 90 g of NO_x per ton of steel have been reported.[1]

The organic compounds present in scrap mixes are burned off as VOCs in the furnace or destroyed by preheating followed by afterburning. However, in the absence of sufficient oxygen these hydrocarbon compounds enter the off-gas system. Iron and zinc oxide are the predominant particulate constituents emitted during melting, and nitrogen oxides and ozone are generated in minor amounts. During the refining process small amounts of calcium oxide may also be emitted from the slag. Melting emissions account for about 90% of total EAF emissions. The remaining 10% of emissions are generated during charging and tapping. Iron oxides and oxides from the fluxes are the primary constituents of the slag handling emissions. During tapping, iron oxide is the major particulate compound emitted.[14]

Effluents

Although the significant water requirement in EAFs is for noncontact cooling purpose, few furnaces also discharge significant process wastewater. Most electric arc furnaces are operated with dry air cleaning systems with no process wastewater discharges. Other noncontact water applications include water-cooled ductwork, roof, sidewalls, doors, lances, panels, cables, and arms. These systems usually incorporate evaporative cooling towers or closed cooling loops.[1] A small number of wet and semiwet air cleaning systems are also in use.

The pollutants of concern are the same as in wet basic oxygen furnaces, but the concentration of metals (primarily lead and zinc, but also arsenic, cadmium, copper, chromium, and selenium) in wastewater is higher because of the higher percentage of scrap charged. Wastewater treatment operations are similar to those for the wet basic oxygen furnaces, including sedimentation in clarifiers or thickeners and recycle of the water.[14]

Byproducts

The two major byproducts generated during EAF steel making are slag and dust. As a result of the oxidation of phosphorus, silicon, manganese, carbon, and other materials during melting, a slag containing some of these oxidation products is formed on top of the molten metal. Electric arc furnaces produce between 110 and 420 lb of slag for every ton of molten steel made, with an average value of about 230 lb/t.[1] EAF dust is made up of the particulate matter and gases produced during the EAF process. The particulate matter removed from emissions in a dry system is referred to as EAF dust and the particulate matter removed by a wet system is the EAF sludge.

Hazardous wastes

The dust (or sludge) removed from EAF emissions is a listed hazardous waste, K061. The primary component is iron or iron oxides; a typical EAF dust contains 24% iron by weight. In cases where lower grades of scrap are used (generally for carbon steel production), EAF dust can contain large amounts of zinc and lead (as high as 44% ZnO and 4% PbO). Similarly, stainless steel production yields dust with high percentages of chromium and nickel oxides (as high as 12% Cr_2O_3 and 3% NiO). EAF dust also contains cadmium in concentrations on the order of about 0.1% by weight. Other possible EAF dust components include other metals and flux.[1]

TABLE 2.11
Concentration of 14 Regulated Elements in Electric Arc Furnace Dust

Element	Total Concentration (mg/kg)
Antimony	5.0–294.0
Arsenic	10.2–400.0
Barium	24–400
Beryllium	<0.5–8.1
Cadmium	1.4–4,988
Chromium	<0.05–106,000
Lead	1.3–139,000
Mercury	<0.001–41
Nickel	<10–22,000
Selenium	0.07–600
Silver	2.5–71.0
Thallium	0.8–50.0
Vanadium	24–475
Zinc	3,900–32,000

Source: From U.S. EPA, Final BDAT Background Document for K061, U.S. EPA, Washington, DC, August 1988.

The primary leachable hazardous constituents of EAF emission control dust/sludge are lead, cadmium, and hexavalent chromium. Generally, 20 to 40 lb of EAF dust per ton of steel are generated, depending on the mill's specific operating practices, with an average of about 35 lb/t of steel melted.[1] Table 2.11 shows the typical ranges of concentration of each of these elements in EAF dust.

2.5.2.3 Treatment Techniques

Primary emissions are controlled using a direct evacuation system, whereas the secondary emissions are controlled by canopy hoods or auxiliary tapping hoods. A direct evacuation system (DES) consisting of ductwork attached to a separate hole in the furnace roof draws emissions to a gas cleaner, thereby helping to control CO, NO_x, VOC, and particulate emissions. The canopy hood is mainly useful for capturing emissions during charging and tapping. Particulate collection is achieved with a baghouse, or scrubbers and electrostatic precipitators as required. Particulate matter removed from EAF emissions using these cleaning methods is a hazardous waste called "EAF dust."

Cooled and solidified slag is crushed and screened to recover metallics for recycle or reuse and the lower metallic aggregate is used in construction applications. The slag produced in EAFs is either reused or landfilled. EAF dust is conveyed into a gas cleaning system.[1]

The treatment options available for the 14 elements antimony, arsenic, barium, beryllium, cadmium, chromium, lead, mercury, nickel, selenium, silver, thallium, vanadium, and zinc are as follows[15]:

1. Transporting the dust to an offsite processor for thermal treatment and removal of zinc, chemical fixation, glassification, or fertilizer manufacture
2. Onsite processing by agglomerating or briquetting and directly recycling back through the EAF (to concentrate the zinc content)
3. Onsite processing in a separate processing facility to glassify or vitrify the heavy metal content
4. Onsite processing using hydrometallurgical or pyrometallurgical processes to upgrade the zinc values to zinc oxide or metallic zinc

In this process EAF dust, other zinc-bearing wastes, recycled materials, coke or coal, lime, and silica are mixed and fed to a rotary furnace. The zinc and other volatile nonferrous metals in the feed are entrained in the furnace off-gas and are carried from the furnace to an external dust collection system. The resulting oxide (zinc calcine) is a crude zinc-bearing product that is further refined at zinc smelters. A byproduct of the process is a nonhazardous, iron-rich slag that can be used in road construction. Solidification technologies change the physical form of the waste to produce a solid structure in which the contaminant is mechanically trapped.

Technologies for onsite recycling of the dust back into an EAF (e.g., briquetting, pelletizing, and pneumatic injection) are still being developed. They have the potential to recover some of the iron oxide values in the dust while concentrating the zinc values. Concentrating the zinc values reduces final recycling costs because smaller quantities of dust will be shipped offsite, and the resulting dust has a higher zinc concentration (improving the cost efficiency of subsequent zinc recovery treatment).

2.6 REFINING AND CASTING

2.6.1 Process Description

Ladle metallurgical furnace (LMF) processes are used to refine molten steel from the BOF or EAF prior to ingot or continuous casting. These processes include the following steps:

1. *Reheating.* Arc reheating or oxygen injection is used to adjust the temperature of steel to levels needed for uninterrupted sequential casting.
2. *Refining.* The first step in refining is deoxidation of the steel with ferromanganese, ferrosilicon, silicomanganese, and aluminum. The second step is desulfurization, which is required for steel grade, which requires low sulfur content. The third step involves addition of ferroalloys and fluxes to the molten steel. The fourth step involves inclusion modification in which the steel in the ladle is stirred by argon gas bubbling to obtain a homogeneous bath temperature, composition and the removal of nonmetallic inclusions. Inclusions responsible for clogging nozzles during the continuous casting process may be removed by calcium treatment.[1]
3. *Degassing.* In vacuum degassing, molten steel is subjected to a vacuum for composition and temperature control, deoxidation, degassing (hydrogen removal), decarburization, and the removal of impurities. While the molten steel is under vacuum, elements that have a relatively higher vapor pressure (such as manganese and zinc) volatilize and exit with the gases.[14] The argon–oxygen decarburization (AOD) is the predominant method used for the manufacture of stainless steel.[1]

In continuous casting, the molten steel is solidified into a semifinished shape (i.e., billet, bloom, or slab) for subsequent rolling in the finishing mill. Continuous casting eliminates the need for conventional processes like the ingot teeming process and directly casting steel into semifinished shapes.[14] The continuous process has higher yields, quality, and productivity compared with the ingot process, as well as higher energy efficiency.[1] The molten steel is delivered in ladles and poured into a reservoir or tundish, from which it is released into the molds of the casting machine. As it descends through the molds, the metal is cooled and emerges with a hardened outer shell. As the semifinished shapes proceed on the runout table, the center also solidifies, providing the semifinished shape.[14]

In ingot casting, the molten steel is teemed into ingot molds followed by cooling and stripping the ingots out of the molds. The ingots are then heated to a uniform temperature in soaking pits to prepare them for rolling. The heated ingots are removed from the pits and rolled into slabs, blooms, or billets. Figure 2.6 illustrates the refining and casting processes, Table 2.12 gives the major inputs and outputs of the process, and Table 2.13 presents an overview of the key environmental and energy facts of refining and casting.

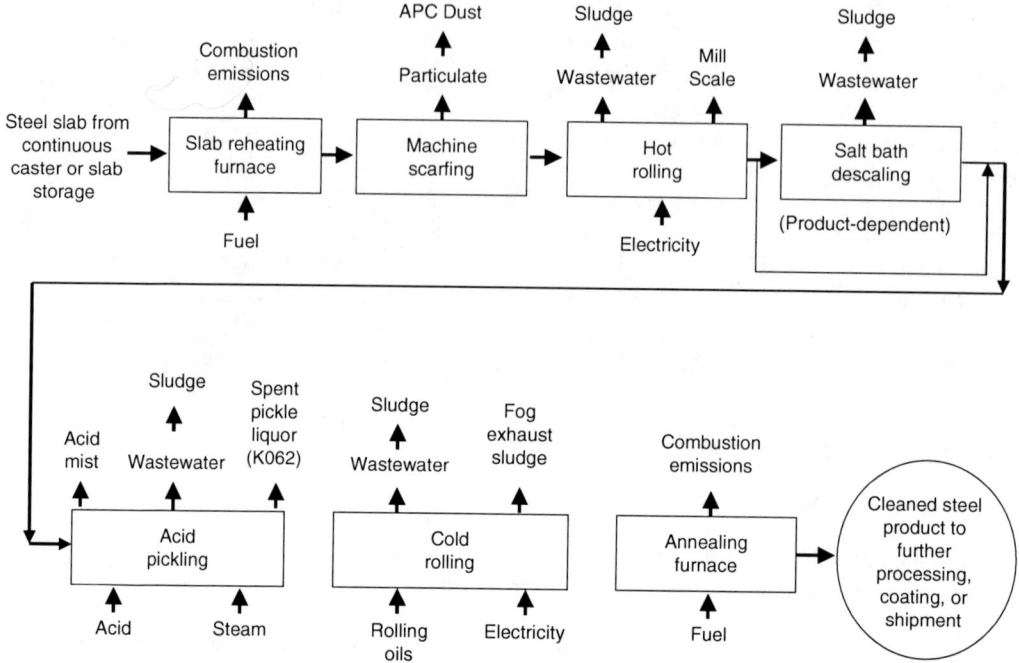

FIGURE 2.6 Refining and casting flow diagram. (From Energetics, Inc., Energy and Environmental Profile of the U.S. Iron and Steel Industry, DOE/EE-0229, U.S. Department of Energy, Washington, DC, 2000.)

2.6.2 Sources of Process Waste

2.6.2.1 Emissions

During ingot casting, particulate emissions are produced during the teeming of molten steel into ingot molds. The major emissions include iron and other oxides (FeO, Fe_2O_3, SiO_2, CaO, and MgO). These emissions are controlled by side draft hoods that are vented to a baghouse.[1] Operational changes in ingot casting such as bottom pouring instead of top pouring can reduce total emissions.

TABLE 2.12
Inputs and Outputs of the Refining and Casting Process

Inputs	Outputs
Molten steel	Semifinished steel shapes
Alloying elements	Process wastewater
Deoxidants	Scale
Fluxes	Sludge
Fuel (natural gas, coke oven gas, blast furnace gas)	Waste oil and grease
Electricity	Air pollution control (APC) dust
Oxygen	
Argon	
Water	

Source: From U.S. EPA, Profile of the Iron and Steel Industry, EPA 310-R-95-005, U.S. EPA, Washington, DC, 1995.

TABLE 2.13
Overview of Key Environmental and Energy Facts: Refining and Casting

Energy	Emissions	Effluents	Byproducts/Hazardous Wastes
Energy use per ton of cast steel: ingot casting[a], 2.78×10^6 Btu	Largest source: teeming into molds	Sources: vacuum degassing, continuous casting cooling water	Mill scale, sludge
Continuous casting, 0.29×10^6 Btu	Particulate: iron and other oxides	Typical wastewater volume per ton of steel Degassing: 25 gal Continuous casting: <25 gal	

[a] Includes soaking pits.
Source: From Energetics, Inc., Energy and Environmental Profile of the U.S. Iron and Steel Industry, DOE/EE-0229, U.S. Department of Energy, Washington, DC, 2000.

Bottom pouring exposes much less of the molten steel to the atmosphere than top pouring, thereby reducing the formation of particulate.[6]

2.6.2.2 Effluent

Among all the refining processes, only vacuum degassing uses process water and generates effluent streams. Vacuum degassing involves direct contact between gases removed from the steel and condenser water. The principal pollutants contained in the effluent are low levels of total suspended solids (TSS) and metals (particularly lead and zinc, but also chromium, copper, and selenium) that volatilize from the steel. Applied water rates for vacuum degassing are typically around 1250 gal/t of steel; with discharge rates of 25 gal/t achieved through high-rate recycle.[15]

Water use in the continuous casting process may be categorized as primary, secondary, and auxiliary. The primary cooling process is a closed-loop, nonevaporative, noncontact cooling of the molten steel shell in the mold (or molds on a multistrand machine) employed to obtain high surface and strand quality. Secondary or spray cooling occurs as the strand moves out of the mold with contact water sprays covering its surface. Auxiliary cooling is noncontact cooling of the casting equipment. Direct contact water cooling system is also used for flume flushing to transport mill scale from the caster runout table.[1]

Applied water rates for the contact systems are typically about 3600 gal/t of cast product; discharge rates for the better controlled casters are less than 25 gal/t. The principal pollutants are total suspended solids, oil, and grease, and low levels of particulate metals.[15]

2.6.2.3 Byproducts

In comparison to the iron-making and steel-making process, wastes generation from refining processes are very small, including solid wastes from the ladle metallurgy facility, argon bubbling APC dust and nozzle blockages. The major byproducts of continuous casting are scale and sludge.

2.6.3 TREATMENT TECHNIQUES

Certain refining processes, including ladle metallurgy, generate particulate emissions. These emissions are typically collected in baghouses as air pollution control dust. Standard treatment for vacuum degassing wastewater includes processing the total recirculating flow or a portion of the flow in clarifiers for TSS removal, cooling with mechanical draft cooling towers, and high-rate recycle. Blowdowns are usually cotreated with steel-making or continuous casting wastewaters for metals removal.[15] Cooling wastewater treatment includes settling basins (scale pits) for scale recovery, oil skimmers, mixed- or single-media filtration, and high-rate recycle.[15]

The air pollution control dusts from refining are nonhazardous and may be processed, recycled, or landfilled. Scale settling basins are provided for periodic collection of scale generated during casting and subsequently wash off from the steel. Fine-grained solids that do not settle out in the scale settling basins are removed by settling, flocculation/clarification processes, or by filtration, depending on the level of water treatment required and the degree of water recycle practiced. The scale is usually recycled and reused within the mill for sintering. Scale may also be landfilled or even charged to an electric arc furnace. Sludge generated during continuous casting is processed and recycled onsite or landfilled.[1]

2.7 FORMING AND FINISHING

2.7.1 PROCESS OVERVIEW

In forming operations the ingots, slabs, billets, and blooms obtained after casting are further processed to produce strip, sheets, plate, bar, rod, or other structural shapes through various hot forming and sometimes cold forming operations, depending on the final product. In hot forming operations, preheated (typically in the range of 1800°F), solidified steel is reduced in cross-section through a series of forming steps by applying mechanical pressure through work rolls to produce semifinished shapes for further hot or cold rolling, or finished shapes. The hot forming mills can be grouped into one of the following four types[18]:

1. Primary mills
2. Section mills
3. Flat mills (plate, hot strip, and sheet)
4. Pipe and tube mills (seamless and butt-weld)

Preheating of steel helps the slabs to undergo a surface preparation step called "scarfing," which removes defects prior to entering the rolling mill by removing a thin layer of the steel surface through localized melting and oxidation. Surface scale is removed from the heated slab by a scale breaker and water sprays prior to its entry into this mill.

Finishing processes clean the surface of the semifinished, hot-rolled steel products before forming or cold rolling or coating operations. Mill scale, rust, oxides, oil, grease, and soil are chemically removed from the surface of the steel using solvent cleaners, pressurized water, air blasting, abrasives, alkaline agents, salt baths, or acid pickling. Salt bath descaling is a finishing process that uses the physical and chemical properties of molten salt baths to remove heavy scale from the surface of selected stainless and high-alloy steels in subsequent water quenching steps. The two salt bath descaling operations are[1]:

1. *Oxidizing (or Kolene®)*. This removes scale using molten salt baths other than those containing sodium hydride.
2. *Reducing (or Hydride®)*. This removes scale using molten salt baths containing sodium hydride.

These two salt bath *descaling processes* may be either batch or continuous and are conducted prior to combination acid pickling (hydrofluoric and nitric acids). Descaling may also be performed using an electrolytic solution of sodium sulfate. The other mechanical descaling operation known as blast cleaning uses abrasives such as sand, steel, or iron grit to clean the steel surface. A compressed air blast cleaning apparatus or rotary-type blasting cleaning machine is used to bring the abrasives in contact with the steel.[18]

The acid pickling process chemically removes oxides and scale from the surface of the steel by the action of water solutions of inorganic acids. It is widely used because of its comparatively low operating costs and ease of operation. Carbon steel is usually pickled with hydrochloric acid; stainless steels are pickled with sulfuric, hydrochloric, nitric, and hydrofluoric acids.[18] The pickling process uses various organic chemicals to inhibit the acid from attacking the base metal. Wetting agents may be used for effective contact of the acid solution with the metal surface. The pickling bath ends with the steel being passed through one or more rinse operations. Alkaline cleaners may also be used to remove mineral oils, grease, and animal fats and oil (used in some rolling solutions) from the steel surface prior to cold rolling. Common alkaline cleaning agents include caustic soda, soda ash, alkaline silicates, and phosphates.[18]

Steel that has been hot-rolled and pickled may be cold-rolled immediately at ambient temperatures before further oxidation can occur. This is done to impart the desired mechanical and surface properties in the steel, and for cold working of the pipe and tube. The two main types of cold mill products are cold-rolled sheets/coils for sale or for further processing in galvanizing and coatings lines, and cold-rolled coils for subsequent tinning.[1]

Cold rolling hardens the steel, which must then be heated in an annealing furnace to make it more ductile. The annealing process involves heating the strip to about 1300°F in an inert atmosphere to prevent oxidation, and then allowing it to cool such that the crystal structure of the steel changes.[1] In batch annealing, gas burners are used to indirectly heat stacked coils, whereas in continuous annealing the coils are unwound and passed through an extended furnace. After the steel has undergone the annealing process, it is run through a temper mill to produce the desired flatness, metallurgical properties, and surface finish.

Steel-coating operations, such as hot coating and electroplating, improve resistance to corrosion and improve appearance. Hot coating operations involve immersing precleaned steel into molten baths of zinc, zinc/aluminum alloy, aluminum, chromium, lead, antimony, tin/lead alloy, and zinc/nickel alloy. Electroplated steel production uses electrodes to deposit a metal coating (zinc, chromium, tin, nickel, brass, cobalt, copper, nickel/tin alloy, zinc/nickel alloy, and zinc/iron/aluminum alloy) onto the surface of the steel. Figure 2.7 illustrates the forming and finishing processes and Table 2.14 gives the major inputs and outputs for steel sheet, a typical steel product. The key environmental and energy facts of forming and finishing are presented in Table 2.15.

2.7.2 Sources of Process Waste

2.7.2.1 Emissions

Significant emissions from forming and finishing are limited to operations like reheat furnaces, scarfing, and pickling. Emissions from reheat furnaces include products of combustion, most of which are well controlled. Hand- or machine-scarfing of semifinished steel results in the release of a fine iron oxide fume. The major pollutants emitted during scarfing are iron and other oxides (FeO, Fe_2O_3, SiO_2, CaO, and MgO). Acid mists arise from the hot acid baths used in acid pickling operations. Emissions from teeming and handscarfing are localized and usually uncontrolled.[16] Hydrogen chloride (from hydrochloric acid pickling) is the primary hazardous air pollutant associated with pickling, with emissions from surface pickling typically over 10 t/yr/facility.[6]

2.7.2.2 Effluents

In hot rolling operations, wastewater is generated from direct cooling of mill stand work rolls, descaling of steel prior to rolling, maintenance of steel surface cleanliness and transportation of scale to scale pits. In finishing operations (i.e., pickling, cold reduction, annealing, temper, cleaner, and coating lines), water and sources of wastewater are rinsewater, pickle liquor, and wet air pollution control (WAPC) devices.[19] Water use and discharge rates from hot forming operations vary greatly depending upon the type of hot forming mill and the shapes produced. Applied process

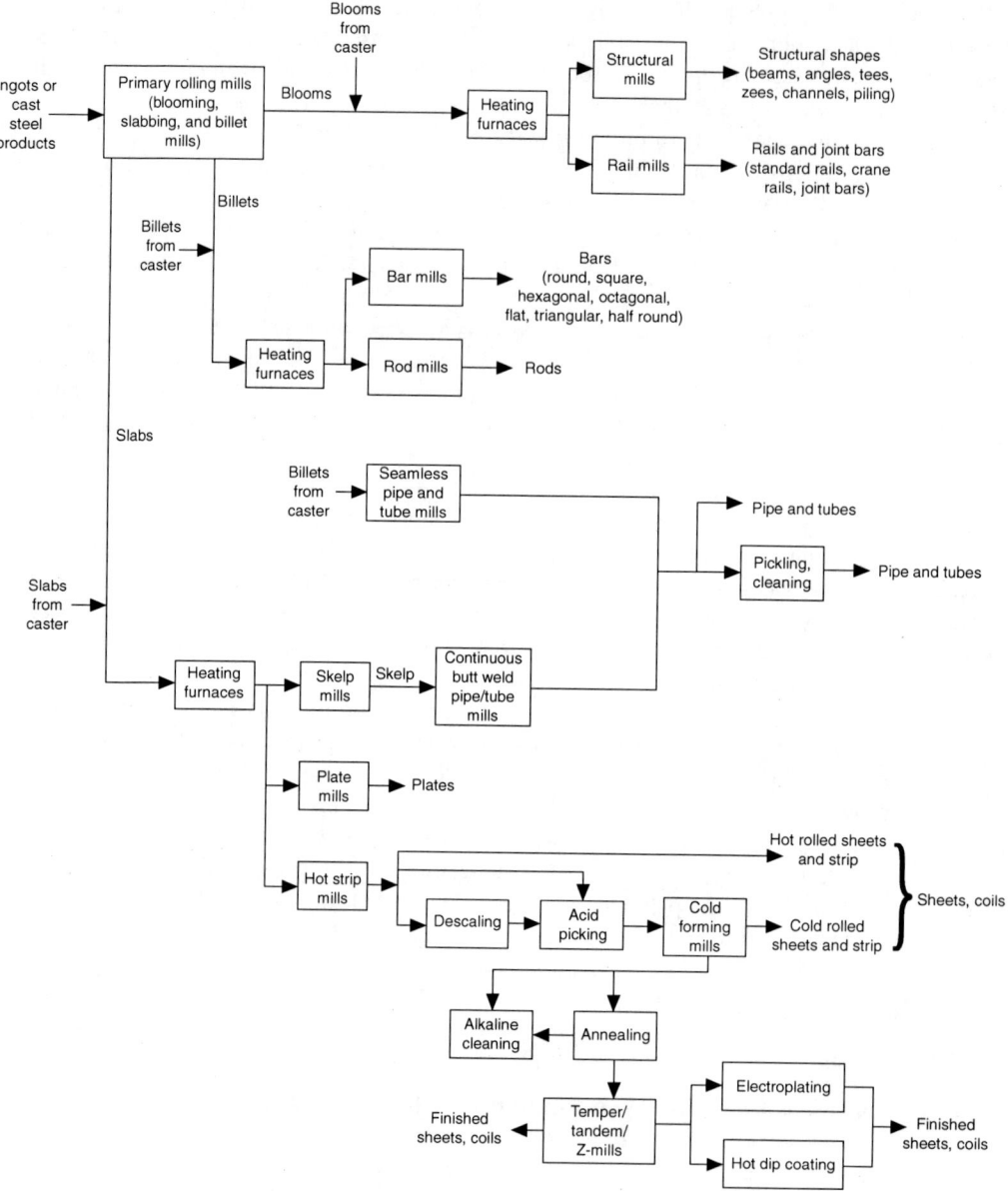

FIGURE 2.7 Forming and finishing flow diagram (sheet production). (From U.S. EPA, Development Document for Final Effluent Limitations Guidelines and Standards for the Iron and Steel Manufacturing Point Source Category, EPA-821-R-02-004, U.S. EPA, Washington, DC, 2002.)

water rates typically range from 1500 gal/t for specialty plate mills to more than 6000 gal/t for hot strip mills. Discharge rates can approach zero for mills equipped with high-rate recycle systems.[15]

Typical process wastewaters from finishing operations include rinses and spent concentrates from alkaline cleaners, pickling solutions, plating solutions, and electrochemical treating solutions. Salt bath descaling wastewaters originate from quenching and rinsing operations conducted after processing in the molten salt baths. The principal pollutants in these effluents are TSS, cyanides, dissolved iron, hexavalent and trivalent chromium, and nickel. Wastewater flows normally range

TABLE 2.14
Inputs and Outputs for Steel Sheet

Inputs	Outputs
Semifinished steel shapes (slabs, billets, blooms)	Cleaned steel products (e.g., sheets, plates, bars, pipe)
Process cooling, rinsing, and cleaning water	Process wastewater containing mill scale, oils, other pollutants, and low levels of metals
Pickling acids	Wastewater sludge
Molten salts	Air pollution control (APC) dust
Alkaline cleaners	Spent pickle liquor (K062)
Fuel	Spent pickle liquor rinse water sludge
Electricity	
Oxygen	
Nitrogen	
Hydrogen	

Source: From U.S. EPA, Profile of the Iron and Steel Industry, EPA 310-R-95-005, U.S. EPA, Washington, DC, 1995.

from 300 to 1800 gal/t, and are cotreated with wastewaters from other finishing operations.[15] Acid pickling discharge water flows for the different pickling processes are as follows:

1. Between 280 and 1020 gal/t for hydrochloric acid processes
2. Between 90 and 500 gal/t for sulfuric acid processes, and up to 1500 gal/t for combination acid processes.[1]

TABLE 2.15
Overview of Key Environmental and Energy Facts: Forming and Finishing

Energy	Emissions	Effluents	Byproducts/Hazardous Wastes
Energy use per net ton of product	Largest sources: machine scarfing, hydrochloric acid pickling (acid mists), reheat furnace (NO_x)	Largest sources: direct cooling and descaling	Mill scale Total generation: 3.7×10^6 t/yr
Reheat furnace: 1.6×10^6 Btu average; 1.4×10^6 Btu for modern furnaces	Particulate: iron and other oxides	Typical wastewater volumes (gal/t)	Reuse: most recycled
Hot rolling: 0.8×10^6 Btu	Typical acid mist generation: >10 tons/year per facility	Hot forming: 0 to >6000	Rolling sludge Total generation: about 1.0×10^6 t/yr Reuse: minimal
Acid pickling: 1.2×10^6 Btu		Descaling: 300 to 1800	K062, Spent pickle liquor Hazardous components: lead, nickel, chromium
Cold rolling: 0.7×10^6 Btu		Pickling: 70 to 1000	Total generation: about 6 million t/yr Reuse: some recycled
Cleaning/annealing: 1.0×10^6 Btu			

Source: From Energetics, Inc., Energy and Environmental Profile of the U.S. Iron and Steel Industry, DOE/EE-0229, U.S. Department of Energy, Washington, DC, 2000.

Pickling done before coating may use a mildly acidic bath; such spent liquor is not considered hazardous. Waste pickle liquor flows typically range between 10 and 20 gal/t of pickled product. Rinsewater flows may range from less than 70 gal/t for bar products to more than 1000 gal/t for certain flat-rolled products. The principal pollutants in rinsewater include TSS, dissolved iron, and metals. For carbon steel operations, the principal metals are lead and zinc; for specialty and stainless steels the metals include chromium and nickel.[15]

Spent pickle liquor in the acid pickling wastewaters is listed as hazardous waste K062, regulated under RCRA, as it contains considerable residual acidity and high concentrations of dissolved iron salts.[2] Exhausted pickling baths are mainly composed of nitrate (150 to 180 g/L), fluoride (60 to 80 g/L), iron (III) (30 to 45 g/L), chromium (III) (5 to 10 g/L), and nickel (II) (3 to 5 g/L).

Process wastewater from cold forming operations results from using synthetic or animal-fat based rolling solutions. The rolling solutions may be treated and recycled at the mill, used on a once-through basis, or a combination of the two. The principal pollutants are TSS, oil and grease (emulsified), and the metals lead and zinc for carbon steels and chromium and nickel for specialty and stainless steels. Chromium may also be a contaminant from cold rolling of carbon steels resulting from wear on chromium-plated work rolls. Toxic organic pollutants including naphthalene, other polynuclear aromatic compounds, and chlorinated solvents have been found in cold rolling wastewaters. Process wastewater discharge rates may range from less than 10 gal/t for mills with recirculated rolling solutions to more than 400 gal/t for mills with direct application of rolling solutions.[15]

2.7.2.3 Byproducts

The main byproducts associated with forming and finishing are scales (typically iron oxides), an oily sludge that results from lubricating the rolls (water treatment plant sludges), and air pollution control dusts associated with treating effluents and cleaning exhaust gases. Coarse scale is separated from the sludge and collected in scale pits. The sludge, which is produced from the treatment of mill scale pit overflows, consists of oils, greases, and fine-grained solids that are collected in settling basins or other separation equipment. This sludge cannot typically be added to the sinter plant because of opacity problems. A number of sludge treating processes are used to de-oil fine sludges to as low as 0.1% at a reasonable cost. However, most oily sludge (containing more than 3% oil) is landfilled rather than recycled. The quantities of scale generated vary, but tend to range between 10 and 80 lb/t for nonoily scale and 4 to 60 lb/t for oily scale. Scale is usually sold or recycled and reused within the plant (usually for sintering), although recycling the oily scale without first deoiling it may cause problems with the opacity of the gaseous stream emitted from the plant or other operational problems. Another byproduct associated with cold rolling is fog exhaust sludge generated from the mist or fog produced during cold rolling. Fog exhaust systems are utilized to allow continuous observation of the strip during processing. Airborne particulates combine with steam and oil mist generated during cold rolling and are discharged to a settling chamber. The settled material is a sludge that is generally landfilled.[1]

2.7.2.4 Hazardous Wastes

Spent pickle liquor is considered a hazardous waste (K062) because it contains considerable residual acidity and high concentrations of dissolved iron salts. For example, spent pickle liquor and waste acid from the production of stainless steel is considered hazardous. The hazardous constituents in K062 are lead, nickel, and hexavalent chromium. Waste pickle liquor sludge generated by lime stabilization of spent pickle liquor is not considered hazardous unless it exhibits one or more of the characteristics of hazardous waste. An estimated 6 million tons of spent pickle liquor are generated annually in the U.S.[1]

2.7.3 Treatment Techniques

Control techniques for removal of acid mists from the exhausted air include packed towers and wet scrubbers. Wet scrubbing has been identified as the control technology achieving Maximum

Achievable Control Technology (MACT) standards. Machine scarfing operations generally use an electrostatic precipitator or water spray chamber for control; most hand scarfing operations are uncontrolled.[16] High removal efficiencies (>95%) can be achieved for hydrochloric and sulfuric acids, whereas the efficiency is lower in the case of hydrofluoric acid systems used in stainless steel pickling. About half of spent pickle liquor is managed for recovery of iron, chromium, and nickel.[14] Metals recovery treatment options include a high temperature metals recovery (HTMR) facility or processing onsite using chemical precipitation or other techniques. Spent pickle liquor may be sold as treatment aids for municipal and centralized wastewater treatment systems or as a replacement for ferric chloride solution used in the manufacture of fine ferric oxide powder. This waste can also be discharged or landfilled in a nonhazardous waste landfill once it is neutralized with lime and "delisted," or it can be injected into deep wells. U.S. EPA estimates that 40% of mills using sulfuric acid treat and then dispose of the wastes to receiving bodies of water. Another 45% have the spent liquor hauled offsite by private contractors, who treat the waste with lime stabilization or other methods, and then dispose of it in landfills or lagoons. The remaining 15% of mills use deep water injection, discharge the waste to a publicly owned treatment works (POTW), or engage in acid recovery. It has been estimated that approximately 80% of spent pickle liquor industry-wide is either recycled through acid regeneration plants or used in municipal wastewater treatment.[1]

The principal pollutants are total suspended solids, oil, and grease. Low levels of metals, that is, chromium, copper, lead, nickel, and zinc, are found in particulate form. Cooling and descaling water is normally discharged from the mill into scale pits where the heavier solid particles settle out. The semicleaned water is typically sent on to a treatment plant containing straining devices, solids removal, and/or deep bed filtration to remove fine particulate. Wastewater treatment may also include the collection of fine mill scale, grease, hydraulic fluids, and rolling oils. The process water is then either recycled back to the mill and/or discharged. Descaling wastewaters are usually cotreated with wastewaters from other finishing operations (e.g., combination acid pickling or cold rolling).[15]

In-process controls for acid pickling include countercurrent rinsing, use of indirect heating versus direct steam sparging for acid solutions, and recycle and reuse of fume scrubber blowdown. Some steel mills are equipped with acid recovery or regeneration systems for spent sulfuric and hydrochloric acids, respectively. After elementary neutralization, which raises the pH above 2.0, rinsewaters are usually cotreated with wastewaters from cold rolling, alkaline cleaning, hot coating, and electroplating operations.[15] Conventional treatment of cold rolling wastewaters includes chemical emulsion breaking, dissolved gas flotation for gross oil removal, and cotreatment with other finishing wastewaters for removal of toxic metal dissipated by a system of flood lubrication.[15]

Lubricants applied to the product being rolled must serve the dual purpose of lubricating and cooling. The water treatment plant sludge for cold rolling therefore contains more oil and grease, which are recovered for subsequent reuse (e.g., as a fuel) or recycle rather than for disposal. Solid waste generation in finishing facilities typically consists of central treatment plant (CTP) sludge. This metallic sludge, which contains fine-grained iron oxide, can be further dewatered by mechanical filtration or by use of sludge drying beds. The dewatered sludge is typically landfilled. The treatment of finishing facility effluents also generates both insoluble and soluble oils, which can be processed and sold for reuse.[1]

2.8 POLLUTION PREVENTION MEASURES

The iron and steel industry needs to opt for technologies that help to either prevent or reduce the generation and discharge of process wastes. The various preventive measures to be adopted for reducing the environmental impacts are as follows:

1. Reduction of dust emissions at furnaces by covering iron runners and using nitrogen blankets during tapping of the blast furnace

2. Use of pneumatic transport, enclosed conveyor belts or self-closing conveyor belts, wind barriers and other dust suppression measures to reduce the formation of fugitive dust
3. Use of low-NO_x burners to reduce NO_x emissions from burning fuel in ancillary operations; use of dry SO_x and dust removal systems in flue gases
4. Recycling of iron-rich materials such as iron ore fines, pollution control dust, and scale in a sinter plant
5. Recovery of thermal energy in the gas from the blast furnace before using it as a fuel; increasing fuel efficiency and reducing emissions by improving blast furnace charge distribution; recovery of energy from sinter coolers and exhaust gases
6. Use of a continuous process for casting steel to reduce energy consumption
7. Wastewater minimization

The processes used in manufacturing steel products use a significant amount of water, and wastewater minimization is necessary both in terms of water use and pollutant discharge loadings. These technologies achieve these reductions by retarding pollutant buildup and improving water quality to allow greater reuse; reducing the volume of wastewater treated and discharged; prolonging process bath life, enabling sites to spend less on process bath makeup and reducing bath treatment and disposal costs; and improving treated effluent quality by enhanced wastewater treatment.[19] The various types of water minimization techniques are given in the following sections.

2.8.1 High-Rate Recycle

High-rate recycle systems consist of a water recirculation loop that recycles 95% or more of the water from a process for reuse. They are used for product cooling, cleaning, and air pollution control, in operations like blast furnace iron making, sintering, basic oxygen furnace steel making, vacuum degassing, continuous casting and hot forming operations. However, during the recycling operation a portion of the water is discharged to prevent concentration buildup of contaminants within the system. These blowdown streams are either treated at an end-of-pipe treatment system or discharged to surface water or publicly owned treatment works (POTWs). The high-rate recycle consists of solids removal devices, cooling devices, and water softening technologies to improve water quality prior to reuse. Improvement in the water quality helps to increase recycle rates significantly. This in turn decreases the pollutant loading, blowdown discharges rates, and the amount of fresh water added as makeup.

2.8.2 Countercurrent Cascade Rinsing

Countercurrent cascade rinsing involves a series of consecutive rinse tanks in which water flows from one tank to another in the direction opposite to that of the product flow. Fresh water flows into the rinse tank located farthest from the process tank and overflows to the rinse tanks closer to the process tank. Over a certain period of time, the first rinse becomes contaminated with dragout solution and reaches a stable concentration much lower than the process solution. The second rinse stabilizes at a lower concentration, which enables less rinsewater to be used compared to a one-rinse tank. The greater the number of countercurrent cascade rinse tanks, the less will be the amount of rinsewater needed to adequately remove the process solution. This differs from a single, once-through rinse tank where the rinsewater is discharged without any recycle or reuse. Countercurrent cascade rinsing is used in steel finishing operations, including acid pickling, alkaline cleaning, electroplating, and hot dip coating, as the steel needs to be relatively contaminant-free for processing. However, such systems have a higher capital cost compared to once-through rinsing systems and require more space. Also, the relatively low flow rate through the rinse tanks require the use of air or mechanical agitation for dragout removal.

2.8.3 ACID REUSE, RECYCLE, AND RECOVERY SYSTEMS

Acid reuse, recycle, and recovery systems are extensively used in the acid pickling industry. Typical industrial acid reuse and recovery systems include the following:

1. *Fume scrubber water recycle.* The steel finishing industry uses fume scrubbers to capture acid gases from pickling tanks. Scrubber water, which may contain a dilute caustic solution, is neutralized and recirculated continuously to adsorb the acid. Makeup water is added to replace water lost through evaporation and water that is blown down to end-of-pipe metals treatment.
2. *Hydrochloric acid regeneration.* This process is used to treat the spent pickle liquor containing free hydrochloric acid, ferrous chloride, and water that is obtained from steel finishing operations. The liquor is concentrated by heating to remove some of the water, followed by thermal decomposition in a "roaster" at temperatures (925 to 1050°C) sufficient for complete evaporation of water and decomposition of ferrous chloride into iron oxide (ferric oxide, Fe_2O_3) and hydrogen chloride (HCl) gas.[19] The iron oxide is separated for offsite recovery or disposal. The hydrogen chloride gas is reabsorbed in water (sometimes rinsewater or scrubber water) to produce hydrochloric acid solution, which is reused in the pickling operation.
3. *Sulfuric acid recovery.* Recovery of sulfuric acid takes place by pumping the spent pickle liquor high in iron content into a crystallizer, where the iron is precipitated (under refrigeration or vacuum) as ferrous sulfate heptahydrate crystals. The water is removed as the crystals are formed and the free acid content of the solution increases to a level usable in the pickling operation. The byproduct ferrous sulfate heptahydrate, referred to as "copperas," is commercially marketable as a coagulant used for water and wastewater treatment. The Blow–Knox–Ruthner process may also be used for sulfuric acid recovery. In this process, the waste liquor is concentrated by evaporation and discharged to reactors where anhydrous hydrogen chloride gas is bubbled to react with the ferrous sulfate, producing sulfuric acid and ferrous chloride. Ferrous chloride is then separated from the sulfuric acid (which is returned to the pickling line) and converted to iron oxide in direct fired roaster. This is followed by the liberation of HCl, which may be recovered by scrubbing, stripping, and recycling to the reactor.[8]
4. *Acid purification and recycle.* This technology is used to process various acid pickling solutions such as the sulfuric acid and nitric/hydrofluoric acids used in stainless steel finishing mills. Acid is purified by adsorption on a bed of alkaline anion exchange resin that separates the acid from the metal ions. Acid is desorbed from the resin using water. The metal-rich, mildly acidic solution passes through the resin and is collected at the top of the bed. Water is then pumped downward through the bed and desorbs the acid from the resin. The purified acid solution is collected at the bottom of the bed and recycled back to the process. Acid purification and recycle reduces nitrate discharges and the overall volume of acid pickling wastewater discharged.

2.8.4 EXTENSION OF PROCESS SOLUTION LIFE

Prolonging the life of solutions reduces the additional investment of fresh process solutions and time spent replacing spent process solutions. The technologies to extend process solution life are as follows:

1. *In-tank filtration.* Steel finishing electroplating and alkaline cleaning operations use in-tank filters to extend process bath life by removing contaminants in the form of suspended solids. Solids are usually disposed of offsite. Devices such as granular activated carbon filters remove dissolved contaminants, such as organic constituents.

2. *Magnetic separation of fines in cold-rolling solution.* Magnetic separators are used to extend the life of cold-rolling solutions. The most effective systems use vertical or horizontal configurations of magnetic rods to remove metal fines.
3. *Evaporation with condensate recovery.* With this technology, steel finishing mills can recover electroplating chemicals such as chrome, nickel, and copper that are lost to electroplating rinsewater. There are two basic types of evaporators: atmospheric and vacuum. In a vacuum evaporator, evaporated water can be recovered as a condensate and reused on site. Generally deionized water is preferred as rinsewater to prevent unwanted contaminants from returning and accumulating in the electroplating process bath in addition to the dragout.

2.9 PROCESS MODIFICATIONS

2.9.1 Effluent-Free Pickling Process with Fluid Bed Hydrochloric Acid Regeneration

This pickling process is operated such that no wastewater is discharged from a hydrochloric acid pickling line. Spent pickle liquor is fed via a settling tank and venturi loop into the fluidized bed reactor. The fluidized bed consists of granulated iron oxide. Residual acid and water are evaporated at 850 °C and the iron chloride is converted to hydrochloric acid gas. Growth and the new formation of iron oxide grains in the fluidized bed are controlled so that a dust-free granulated product is obtained. Because the fluidized bed process operates at approximately 850°C, rinse and scrubber water from the pickle line is used at the regeneration plant to cool fluidized bed off-gases, which contain hydrochloric acid vapor and a small amount of iron oxide dust. The off-gases are cooled to approximately 100°C in a venturi scrubber. The thermal energy of the off-gases helps to concentrate the pickling liquor by evaporation before it is fed to the reactor.[19]

From the venturi scrubber, the cooled gas stream goes to the absorber, where hydrogen chloride is absorbed with rinsewater from the pickling line and fresh water to produce hydrochloric acid. The acid is recycled directly to the pickling process or placed in a storage tank for later use. Having passed through the scrubbing stages and mist collector, the fluidized bed off-gases are virtually free of hydrochloric acid and are released to the atmosphere.

2.9.2 Nitric-Acid-Free Pickling

Nitric-acid-free pickling requires the same equipment as conventional acid pickling processes and is also compatible with acid regeneration. This technology uses a nitric-acid-free solution that contains an inorganic mineral acid base, hydrogen peroxide, stabilizing agents, wetting agents, brighteners, and inhibitors.[19]

2.9.3 Effluent-Free Exhaust Cleaning

Wet air pollution control (WAPC) devices are used to treat exhaust gases from stainless steel pickling operations, thereby generating wastewater, which are treated using the selective catalytic reduction (SCR) technology in which anhydrous ammonia is injected into the gas stream prior to a catalyst to reduce NO_x to nitrogen and water. The most common types of catalysts are a metal oxide, a noble metal, or zeolite.

2.9.4 Elimination of Coke with Cokeless Technologies

Some cokeless technologies in use or under development include the Japanese Direct Iron Ore Smelting (DIOS) process, in which molten iron is produced directly with coal and sinter feed ore, the HIsmelt process, where ore fines and coal are used to achieve a production rate of 8 t/hr using ore directly in the smelter, and the Corex process, which has an integral coal desulfurizing step, making it amenable to a variety of coal types.[14]

2.9.5 REDUCING COKE OVEN EMISSIONS

These technologies in use or under development reduce the quantity of coke needed by changing the method by which coke is added to the blast furnace or by substituting a portion of the coke with other fuels, thereby reducing coking emissions. Pulverized coal injection substitutes pulverized coal for about 25 to 40% of coke in the blast furnace. Nonrecovery coke battery allows the combustion of the gases from the coking process, thus consuming the byproducts that are usually recovered. The Davy Still Autoprocess is a precombustion cleaning process in which coke oven battery process water is utilized to strip ammonia and hydrogen sulfide from coke oven emissions. Another option involves the use of alternative fuels such as natural gas, oil, and tar/pitch instead of coke into the blast furnace.[14]

Various treatment technologies are used at the iron and steel plant for recycle system water treatment prior to recycle and reuse, or end-of-pipe wastewater treatment prior to discharge to surface water or a POTW. The physical/chemical treatment technologies extensively used include equalization, tar removal, free and fixed ammonia stripping, cooling technologies, cyanide treatment technologies, oily wastewater treatment technologies, carbon dioxide injection, metals treatment technologies, solids separation technologies, and polishing technologies.

Ammonia stripping also removes cyanide, phenols, and other VOCs typically found in coke-making wastewater. Phenols may also be removed by conversion into nonodorous compounds or into crude phenol or sodium phenolate by either biological means (phenol concentration <25 mg/L) or by physical processes.[21] However, the Koppers dephenolization process is considered to be quite effective as it lowers the phenol content by 80 to 90% in ammonia still wastes. In this process a stream stripping process followed by mixing in a solution of caustic soda results in renewal of pure phenol with the flue gas.[8]

Blast furnace, vacuum degassing, continuous casting, and hot forming operations use cooling methods in recirculation systems. Byproduct recovery coke-making plants commonly use cooling prior to biological treatment systems, because high temperatures are detrimental to the biomass. Cyanide treatment technologies include alkaline and breakpoint chlorination using sodium hypochlorite or chlorine gas in a carefully controlled pH environment to remove cyanide and ammonia. In cyanide precipitation, cyanide combines with iron to form an insoluble iron–cyanide complex that can be precipitated and removed by gravity settling. Ozone oxidation results in the conversion of cyanide to cyanate. Oily wastewaters from hot forming and cold rolling operation are treated by gravity flotation, oil/water separation, emulsion breaking, followed by dissolved air flotation and ultrafiltration.[21–23] Carbon dioxide injection is one method of removing scale-forming metal ions that accumulate in water recirculation systems from BOF recycle water.

Strong reducing agents such as sulfur dioxide, sodium bisulfite, sodium metabisulfite, and ferrous sulfate are used in the iron and steel finishing sites to reduce hexavalent chromium to the trivalent form, which allows the metal to be removed from solution by chemical precipitation.[21–23] Metal-containing wastewaters may also be treated by chemical precipitation or ion-exchange.

Solid wastes, including scale, biosolids, precipitate from cyanide and chemical precipitation systems, and solids from filtration backwash may be treated using scale pits, classifiers, clarifiers, and the microfiltration technique.[19,24,25] Polishing technologies include multimedia filters following clarification to remove small concentrations (<20 mg/L) of entrained suspended solids, or carbon adsorption to remove trace concentrations of organic pollutants remaining in coke-making wastewater following biological treatment. Biological denitrification (anaerobic) can be used to treat coke-making wastewater following biological nitrification. Steel mill sludge thickening and dewatering may be accomplished using gravity thickeners, rotary vacuum filters, centrifugation, sludge drying, belt and pressure filters. However, it has been identified that rolling mill sludges are not amenable to vacuum filters and centrifuges.[9,24,25]

REFERENCES

1. Energetics, Inc., Energy and Environmental Profile of the U.S. Iron and Steel Industry, DOE/EE-0229, U.S. Department of Energy, Washington, DC, 2000.
2. Gálvez, J.L., Dufour, J., Negro, C., and López-Mateos, F. Determination of iron and chromium fluorides solubility for the treatment of wastes from stainless steel mill, *Chem. Eng. J.*, doi:10.1016/j.cej 03.014, 2007.
3. Heard, R.A. and Roth, J.L. Optimizing energy in electric furnace steel-making, *Iron and Steel Engineer*, April 1998.
4. International Iron and Steel Institute (IISI), International Iron and Steel Institute, available at http://www.worldsteel.org, December 2008.
5. Grieshaber, K.W., Philipp, C.T., and Bennett, G.F., Process for recycling spent potliner and electric arc furnace dust into commercial products using oxygen enrichment, *Waste Management*, 14(3–4), 267–276, 1994.
6. Marsosudiro, P.J., Pollution Prevention in the Integrated Iron and Steel Industry and Its Potential Role in MACT Standards Development, 94-TA28.02, U.S. EPA, Washington, DC, 1994.
7. Makkonen, H.T., Heino, J., Laitila, L., Hiltunen, A., Poylio, E., and Harkki, J., Optimisation of steel plant recycling in Finland: dusts, scales and sludge, *Resources, Conservation and Recycling*, 35, 77–84, 2002.
8. Nemerow, N.L. and Dasgupta, A., *Industrial and Hazardous Waste Treatment*, Van Nostrand Reinhold, New York, 1991.
9. Patzelt, R.R. and Hassick, D.E., Dewatering of steel-mill sludges by belt-press filtration, Section 13—Steel and Foundry Wastes, in *Proceedings of the 39th Industrial Waste Conference*, Purdue University, 469–485, 1984.
10. U.S. Department of Energy, Office of Industrial Technologies, Hydrochloric Acid Recovery Systems for Galvanizers and Steel Manufacturers (NICE³ fact sheet), DOE/CH10093–233, 1993.
11. U.S. EPA, Final BDAT Background Document for K061, U.S. EPA, Washington, DC, August 1988.
12. U.S. EPA, Federal Register, Part III, U.S. EPA, Washington, DC, August 18, 1992.
13. U.S. EPA, Report to Congress on Metal Recovery, Environmental Regulation & Hazardous Waste, EPA530-R-93-018, U.S. EPA, Washington, DC, 1994.
14. U.S. EPA, Profile of the Iron and Steel Industry, EPA 310-R-95-005, U.S. EPA, Washington, DC, 1995.
15. U.S. EPA, Draft Iron and Steel Regulatory Review, 40 CFR Part 420, Effluent Limitations Guidelines and Standards for the Iron and Steel Manufacturing Point Source Category, EPA 821-R-95-037, U.S. EPA, Washington, DC, 1995.
16. U.S. EPA, Compilation of Air Pollutant Emission Factors, Vol. I: Stationary Point and Area Sources, AP-42, 5th ed., U.S. EPA, Washington, DC, 1995.
17. U.S. EPA, Iron and Steel Chapter of Sector Notebook Data Refresh—1997, U.S. EPA, Washington, DC, 1998.
18. U.S. EPA, Development Document for the Proposed Effluent Limitations Guidelines and Standards for the Iron and Steel Manufacturing Point Source Category, EPA-821-B-00-011, U.S. EPA, Washington, DC, 2000.
19. U.S. EPA, Development Document for Final Effluent Limitations Guidelines and Standards for the Iron and Steel Manufacturing Point Source Category, EPA-821-R-02-004, U.S. EPA, Washington, DC, 2002.
20. Wang, L., Lee, W., Tsai, P., Lee, W., and Chang-Chien, G., Emissions of polychlorinated dibenzo-p-dioxins and dibenzofurans from stack flue gases of sinter plants, *Chemosphere*, 50(9), 1123–1129, 2003.
21. Wang, L.K., Hung, Y.T., and Shammas, N.K., Eds., *Physicochemical Treatment Processes*, Humana Press, Totowa, NJ, 2005.
22. Wang, L.K., Hung, Y.T., and Shammas, N.K., Eds., *Advanced Physicochemical Treatment Processes*, Humana Press, Totowa, NJ, 2006.
23. Wang, L.K., Hung, Y.T., and Shammas, N.K., Eds., *Advanced Physicochemical Treatment Technologies*, Humana Press, Totowa, NJ, 2007.
24. Wang, L.K., Shammas, N.K., and Hung, Y.T., Eds., *Biosolids Treatment Processes*, Humana Press, Totowa, NJ, 2007.
25. Wang, L.K., Shammas, N.K., and Hung, Y.T., Eds., *Biosolids Engineering and Management*, Humana Press, Totowa, NJ, 2008.

3 Treatment of Nonferrous Metal Manufacturing Wastes

Nazih K. Shammas and Lawrence K. Wang

CONTENTS

3.1	Industry Description	72
3.2	Nonferrous Metal Processing Industry	74
	3.2.1 Aluminum	74
	3.2.1.1 Industry Size and Geographic Distribution	74
	3.2.1.2 Product Characterization	74
	3.2.1.3 Industrial Process Description	76
	3.2.1.4 Material Inputs and Pollution Outputs	79
	3.2.2 Copper	81
	3.2.2.1 Industry Size and Geographic Distribution	81
	3.2.2.2 Product Characterization	81
	3.2.2.3 Industrial Process Description	82
	3.2.2.4 Material Inputs and Pollution Outputs	85
	3.2.3 Lead	86
	3.2.3.1 Industry Size and Geographic Distribution	86
	3.2.3.2 Product Characterization	87
	3.2.3.3 Industrial Process Description	87
	3.2.3.4 Raw Material Inputs and Pollution Outputs	89
	3.2.4 Zinc	91
	3.2.4.1 Industry Size and Geographic Distribution	91
	3.2.4.2 Product Characterization	91
	3.2.4.3 Industrial Process Description	91
	3.2.4.4 Material Inputs and Pollution Outputs	94
	3.2.5 Columbium and Tantalum	96
	3.2.6 Silver	96
	3.2.7 Tungsten	96
	3.2.8 Beryllium	96
	3.2.9 Selenium	96
3.3	Wastewater Characterization	96
	3.3.1 Primary Aluminum	97
	3.3.2 Secondary Aluminum	99
	3.3.3 Primary Columbium and Tantalum	101
	3.3.4 Primary Copper	101
	3.3.5 Secondary Copper	104
	3.3.6 Primary Lead	109
	3.3.7 Secondary Lead	109

	3.3.8 Secondary Silver	110
	3.3.9 Primary Tungsten	110
	3.3.10 Primary Zinc and Cadmium	115
3.4	Pollutant Removability and Treatment	119
3.5	Management of Chemicals in the Wastestream	119
	3.5.1 Chemical Release and Transfer Profile	135
	3.5.2 Summary of the Toxicity of Top Chemicals	135
	3.5.2.1 Chlorine	144
	3.5.2.2 Copper	144
	3.5.2.3 Hydrochloric Acid	144
	3.5.2.4 Lead	144
	3.5.2.5 Zinc and Zinc Compounds	145
	3.5.3 Comparison of Toxic Release Inventory between Industries	145
	3.5.4 Pollution Prevention Opportunities	145
	3.5.4.1 Process Equipment Modification	147
	3.5.4.2 Raw Materials Substitution or Elimination	147
	3.5.4.3 Precious Metals Recovery	147
	3.5.5 Important Pollution Prevention Case Studies	147
	3.5.5.1 The Use of Electric Induction to Replace Fossil Fuel Combustion	148
	3.5.5.2 Processing Nonferrous Metal Hydroxide Sludge Wastes	148
References		149

3.1 INDUSTRY DESCRIPTION

The nonferrous metals industry encompasses establishments that engage in the following: primary and secondary smelting and refining of nonferrous metal from ore or scrap; rolling, drawing, and alloying; and the manufacturing and casting of basic metal products such as nails, spikes, wire, and cable. Primary smelting and refining produces metals directly from ores, and secondary refining and smelting produces metals from scrap and process waste. Scrap is bits and pieces of metal parts, bars, turnings, sheets, and wire that are off-specification or worn out but capable of being recycled.[1] The industry does not include the mining and beneficiation of metal ores; rolling, drawing, or extruding metals; or scrap metal collection and preliminary grading.[2]

Two metal recovery technologies are generally used to produce refined metals. Pyrometallurgical technologies are processes that use heat to separate desired metals from other less or undesirable materials. These processes capitalize on the differences between constituent oxidation potential, melting point, vapor pressure, density, or miscibility when melted. Examples of pyrometallurgical processes include drying, calcining, roasting, sintering, retorting, and smelting. Hydrometallurgical technologies differ from pyrometallurgical processes in that the desired metals are separated from undesirables using techniques that capitalize on differences between constituent solubilities or electrochemical properties while in aqueous solutions. Examples of hydrometallurgical processes include leaching, chemical precipitation, electrolytic recovery, membrane separation, ion exchange, and solvent extraction.

During pyrometallic processing, an ore, after being concentrated by beneficiation (crushing, washing, and drying) is sintered, or combined by heat, with other materials such as baghouse dust and flux. The concentrate is then smelted, or melted, in a blast furnace in order to fuse the desired metals into impure molten bullion. This bullion then undergoes a third pyrometallic process to refine the metal to the desired level of purity. Each time the ore or bullion is heated, waste materials are created. Air emissions such as dust may be captured in a baghouse and are either disposed of or returned to the process depending upon the residual metal content. Sulfur is also captured, and

TABLE 3.1
Global Size of the Nonferrous Metal Industry (1000 USD)

	2001	2002	2003	2004	2005
U.S.	1,168,979	1,205,751	1,525,103	1,959,119	2,732,399
Germany	1,026,226	987,855	1,188,257	1,586,034	1,782,619
U.K.	539,568	578,023	693,718	967,331	1,256,697
France	514,214	539,964	562,873	899,418	1,124,048
Mozambique	586	297	—	—	1,022,005
Canada	547,290	500,397	502,896	689,104	851,403
Netherlands	278,752	312,556	426,390	616,114	799,880
Belgium	275,577	268,371	295,804	554,500	535,881
Japan	124,748	150,086	215,931	314,197	533,532
Mexico	240,521	202,321	258,524	383,034	501,177
Spain	371,858	407,224	438,838	337,198	499,065
Australia	167,667	210,146	260,276	309,567	378,943
Italy	111,081	109,916	127,044	293,584	367,374
Switzerland	157,352	153,179	175,809	256,276	317,029
Chile	93,607	103,244	121,045	186,278	245,939
Thailand	43,885	49,214	76,483	134,662	224,893
Poland	98,152	105,213	142,727	177,777	223,185
Singapore	138,611	147,928	145,768	194,402	219,283
Sweden	98,071	107,138	140,437	176,010	217,977
Korea, Republic	28,725	30,580	61,183	176,905	200,589

when concentrations are above 4% it can be turned into sulfuric acid, a component of fertilizers. Depending upon the origin of the ore and its residual metals content, various metals such as gold and silver may also be produced as byproducts.

There are an estimated 800 plants in the U.S. involved in the primary or secondary recovery of nonferrous metals. These plants represent 61 subcategories. However, many of these subcategories are small, represented by only one or two plants, or do not discharge any wastewater. This chapter focuses on 296 facilities that produce the major nonferrous metals [aluminum, columbium (niobium), tantalum, copper, lead, silver, tungsten, and zinc]. The volume of wastewater discharged in this industry varies from 0 to 540 m³/T (0 to 160,000 gal/t) of metal produced.[1,3] The global size of the industry is reflected in Table 3.1 (reported in 1000 USD) for the top 20 export countries for nonferrous base metal waste and scrap.[4] Here T = metric ton = 1000 kg = 2204.6 lbs, t = 2000 lbs.

Nonferrous metal facilities are distributed throughout the U.S. Most sites are located near ore production facilities, near adequate transportation facilities, or near adequate power supplies.

Table 3.2 presents an industry summary for the nonferrous metals industry indicating the number of subcategories and the number and type of dischargers. Table 3.3 presents best practicable technology (BPT) limitations that have been promulgated and reported in the Federal Register.[3,5]

Production operations are subject to a number of regulations, including those imposed by the Resource Conservation and Recovery Act (RCRA),[6,7] the Clean Water Act (CWA),[8] and the Clean Air Act (CAA).[9] A number of RCRA-listed hazardous wastes are produced during primary refining operations that require the heating of ores to remove impurities. Specific pretreatment standards under the CWA apply to the processes associated with copper and aluminum. Lastly, large amounts of sulfur are released during copper, lead, and zinc smelting operations, which are regulated under the CAA.

TABLE 3.2
Nonferrous Metal Industry Summary

Item	Number
Total subcategories	61
Phase I	26
Phase II	35
Subcategories studied	12
Discharges in industry	
Direct	129
Indirect	79
Zero discharge	215

Source: From U.S. EPA, Treatability Manual, Volume II Industrial Descriptions, report EPA-600/2-82-001b, U.S. EPA, Washington, DC, September 1981.

3.2 NONFERROUS METAL PROCESSING INDUSTRY

The nonferrous metals industry is divided into 61 subcategories by the type and source of the metal to be smelted and/or refined and by similar wastewater sources. Twenty-six of these subcategories are grouped as Phase I. The remaining subcategories have been identified as Phase II of the nonferrous metals subcategories (Table 3.2).

3.2.1 ALUMINUM

The information on aluminum is collected from the U.S. EPA,[1,2] the Aluminum Association,[10] the U.S. Trade Commission[11] and the U.S. Department of Commerce.[12]

3.2.1.1 Industry Size and Geographic Distribution

The U.S. aluminum smelting industry consists of 23 smelting facilities operated by 13 firms, which employ approximately 20,000 people. The secondary smelting industry operates an estimated 68 plants, with 3600 employees. The majority of primary aluminum producers are located either in the Northwest (39.1% of U.S. capacity) or the Ohio River Valley (31.1% of U.S. capacity), while most secondary aluminum smelters are located in Southern California and the Great Lakes Region. The reason for the difference in plant locations is due to the energy-intensive nature of the primary aluminum smelting process and the cost of fuels. Primary smelters are located in the Northwest and Ohio River Valley to take advantage of the abundant supplies of hydroelectric and coal-based energy, and secondary smelters locate themselves near major industrial and consumer centers to take advantage of the large amounts of scrap generated. Secondary smelting uses 95% less energy to produce the same product as primary reduction. On average, a third of primary production costs are attributable to the cost of energy.

3.2.1.2 Product Characterization

The primary and secondary aluminum industry produces ingots of pure (greater than 99%) aluminum that serve as feedstock for other materials and processes. Within the U.S., the leading end-users of aluminum come from three industries: containers and packaging, transportation, and building and construction. Examples of materials produced with aluminum are sheet metal; aluminum plate and foil; rod, bar, and wire; beverage cans; automobiles; aircraft components; and window/door frames.

TABLE 3.3
BPT Limitations for the Nonferrous Metals Industry[a]

Parameter	Primary Aluminum Smelting (kg/T of Product)	Secondary Aluminum Smelting — Chlorine Demagging (kg/T of Magnesium Removed)	Secondary Aluminum Smelting — Wet Processing (kg/T of Product)	Primary Copper Smelting[b]	Primary Copper Refining (kg/T)	Secondary Copper[b,c] (mg/L)	Primary Lead[b,c] (mg/L)	Primary Zinc (kg/T of Product)
COD	—	6.5	1.0	—	0.05	25	25	0.21
TSS	1.5	175	1.5	—	—	10	—	—
Oil and grease	—	—	—	—	—	—	—	—
Ammonia (as nitrogen)	—	—	0.01	—	—	—	—	—
pH, pH units	6.0–9.0	7.5–9.0	7.5–9.0	—	6.0–9.0	6.0–9.0	6.0–9.0	6.0–9.0
Fluoride	1.0	—	0.4	—	—	—	—	—
Aluminum	—	—	1.0	—	—	—	—	—
Arsenic	—	—	—	—	0.00003	—	—	—
Cadmium	—	—	0.003	—	0.0008	0.25	—	0.0004
Copper	—	—	—	—	0.00026	—	0.5	—
Lead	—	—	—	—	—	—	0.5	0.04
Selenium	—	—	—	—	—	—	—	—
Zinc	—	—	—	—	0.0003	5	5	0.04

Dashes indicate parameters not regulated for BPT in this subcategory.

[a] Average daily value over 30 consecutive days; maximum daily value permitted is twice the average amount, except for the secondary aluminum subcategory for which maximum values are not available. T = metric ton = 1000 kg = 2204.6 lbs.

[b] There shall be no discharge or process wastewater pollutants to navigable waters except for a volume equivalent to the precipitation in excess of the 10-year 24-hr rainfall event.

[c] BPT limitations established for the volume of wastewater equal to the difference between the mean precipitation for that month that falls within the impoundment and the mean evaporation from the pond water surface.

Source: From 2-36, U.S. EPA, Effluent guidelines and standards for nonferrous metals, 40CFR421; 39FR12822, April 8, 1974; Amended by 40FR8514, February 27, 1975; 40FR48348, October 15, 1975; 41FR 54850, December 15, 1976, U.S. EPA, Washington, DC, 1974–1976.

At present, the automotive sector is the largest end-user. The next largest end-user is the beverage can stock. Automotive use of aluminum is expected to sky-rocket as the sector increases its use of aluminum to increase fuel efficiency.

3.2.1.3 Industrial Process Description

This section specifically contains a description of commonly used production processes, associated raw materials, the byproducts produced or released, and the materials either recycled or transferred offsite. This discussion, coupled with schematic drawings of the identified processes, provides a concise description of where wastes may be produced in the process.

Primary aluminum processing

Primary aluminum producers generally use a three step process to produce aluminum alloy ingots. First, alumina is extracted from bauxite ore using the Bayer process (Figure 3.1). In the Bayer process, finely crushed bauxite is mixed with an aqueous sodium hydroxide (caustic soda) solution to form slurry. The slurry is then reacted at a high temperature under steam pressure in a vessel known as a digester, and creates a mixture of dissolved aluminum oxides and bauxite residues. During the reaction a majority of the impurities such as silicon, iron, titanium, and calcium oxides drop to the bottom of the digester and form sludge. The remaining sodium aluminate slurry is then flash cooled by evaporation and sent for clarification. During clarification, agents such as starch are added to help any fine impurities that remain in the slurry, such as sand, to drop out, further purifying the sodium aluminate solution. The solution is then fed into a precipitation tank to be crystallized. In the precipitator the solution is allowed to cool with the addition of a small amount of aluminum hydroxide "seed." The seed stimulates the precipitation of solid crystals of aluminum hydroxide and sodium hydroxide.

The aluminum hydroxide crystals settle to the tank bottom, and are removed. The crystals are then washed to remove any caustic soda residues, vacuum dewatered, and sent on for calcination. In the calciners (a type of rotating kiln) the aluminum hydroxide is roasted for further dewatering.

In the second step, the aluminum oxide (alumina) produced during the Bayer process is reduced to make pure molten aluminum. Alumina is a fine white powder, and consists of about equal weights of aluminum and oxygen. The strong chemical bond that exists between the aluminum and oxygen makes separating them difficult—pyrometallurgical separation requires a temperature of about 1980°C (3600°F). However, alumina will dissolve when placed in the molten metal cryolite at around only 950°C (1742°F). Once dissolved, the aluminum oxide is readily separated into aluminum and oxygen by an electric current. The Hall–Heroult process, as this type of electrolytic reduction is known begins with the placement of the alumina into electrolytic cells, or "pots," filled with molten cryolite (Figure 3.2). Although the process requires large amounts of electricity (13 or 15 kW of electricity per kg of aluminum produced), only a low voltage is needed. This allows the pots to be laid out in a series along one long electrical circuit to form what is known as a "potline." Within each pot a positive electric current is passed through the cryolite by means of a carbon anode submerged in the liquid cryolite. The oxygen atoms, separated from aluminum oxide, carry a negative electrical charge and are attracted to the carbon anodes. The carbon and the oxygen combine immediately to form carbon dioxide and carbon monoxide. These gases bubble free of the melt. The aluminum (which is more than 99% pure) collects at the bottom of the pot, is siphoned off, placed into crucibles, and then transferred to melting/holding furnaces.

The third step consists of either mixing the molten aluminum with other metals to form alloys of specific characteristics, or casting the aluminum into ingots for transport to fabricating shops.[1,13] Casting involves pouring molten aluminum into molds and cooling it with water. At some plants, the molten aluminum may be batch treated in furnaces to remove oxide, gaseous impurities, and active metals such as sodium and magnesium before casting. Some plants add a flux of chloride and fluoride salts and then bubble chlorine gas, usually mixed with an inert gas, through the molten

Treatment of Nonferrous Metal Manufacturing Wastes

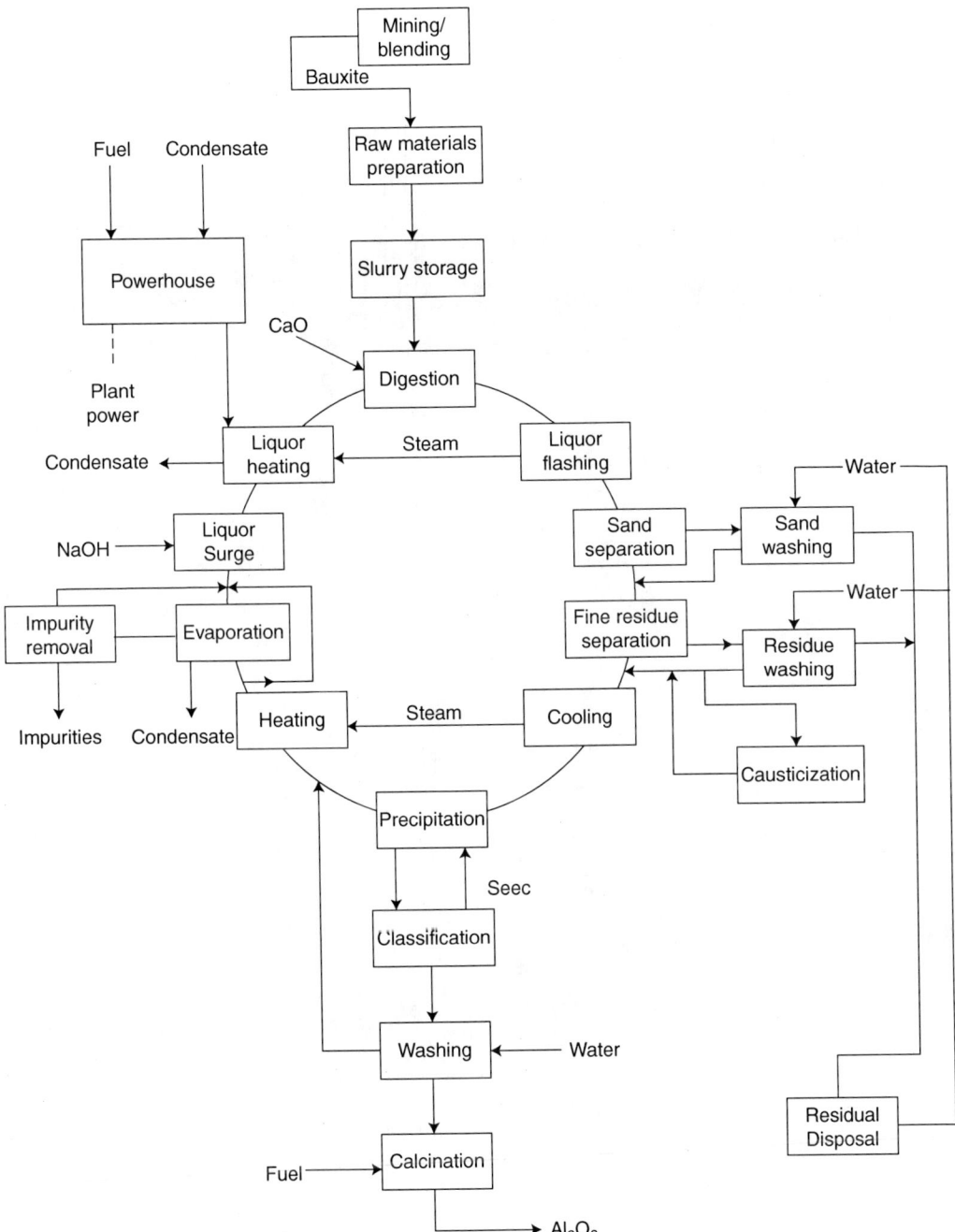

FIGURE 3.1 Bayer process for alumina refining. (From U.S. EPA, Profile of the Nonferrous Metals Industry, publication EPA/310-R-95-010, U.S. EPA, Washington, DC, September 1995.)

mixture. Chloride reacts with the impurities to form HCl, Al_2O_3, and metal chloride emissions. Dross forms to float on the molten aluminum and is removed before casting.

Two types of anodes may be used during the reduction process: either an anode paste or a prebaked anode. Because the carbon is consumed during the refining process (about 0.5 kg of carbon is consumed for every kg of aluminum produced), if anode paste (Soderberg anode) is used, it needs

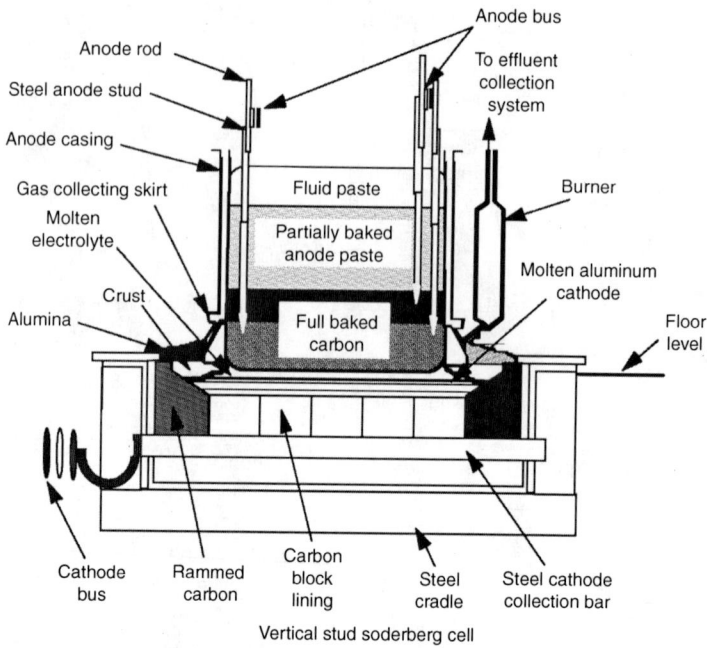

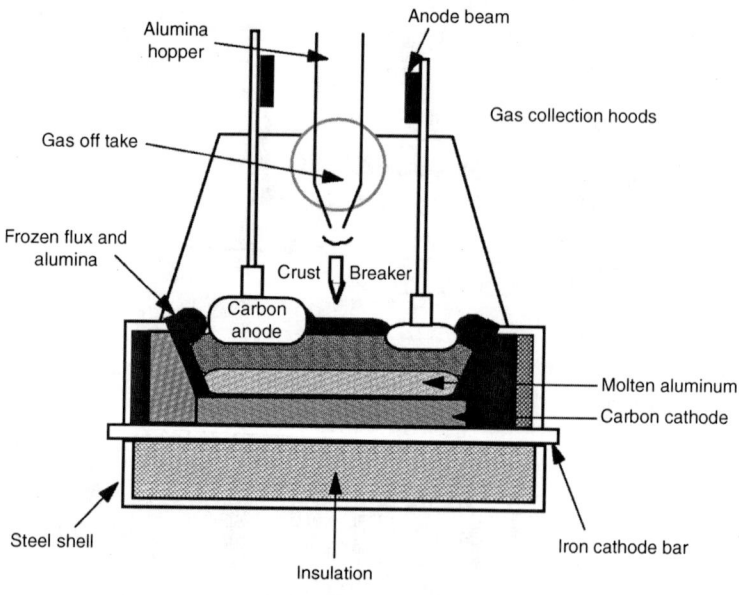

FIGURE 3.2 Aluminum anodes. (From U.S. EPA, Profile of the Nonferrous Metals Industry, publication EPA/310-R-95-010, U.S. EPA, Washington, DC, September 1995.)

to be continuously fed through an opening in the steel shell of the pot. The drawback to prebaked anodes is that they require that a prebaked anode fabricating plant be located nearby or onsite. Most aluminum reduction plants include their own facilities to manufacture anode paste or prebaked anode blocks. These prebaked blocks, each of which may weigh about 300 kg, must be replaced after 14 to 20 d of service.

The waste materials produced during the primary production of aluminum are fluoride compounds. Fluoride compounds are principally produced during the reduction process. One reason that prebaked anodes are favored is that the closure of the pots during smelting facilitates the capture of fluoride emissions, although many modern smelters use other methods to capture and recycle fluorides and other emissions.

The pots used to hold the aluminum during smelting range in size from 9 to 15 m (30 to 50 ft) long, 2.7 to 3.6 m (9 to 12 ft) wide, and 0.9 to 1.2 m (3 to 4 ft) high, and are lined with refractory brick and carbon. Eventually, the carbon linings crack and must be removed and replaced. However, during the aluminum reduction process iron cyanide complexes form in the carbon portion of the liners. When the linings are removed they are "spent," and are considered to be RCRA-listed hazardous waste.

Secondary aluminum processing
In the secondary production of aluminum, scrap is usually melted in gas- or oil-fired reverberatory furnaces of 14,000 to over 45,000 kg capacities. The furnaces have one or two charging wells separated from the main bath by a refractory wall that permits only molten metal into the main bath. The principal processing of aluminum-base scrap involves the removal of magnesium by treating the molten bath with chlorine or with various fluxes such as aluminum chloride, aluminum fluoride, or mixtures of sodium and potassium chlorides and fluorides. To facilitate handling, a significant proportion of the old aluminum scrap, and in some cases new scrap, is simply melted to form sweated pig that must be processed further to make specification-grade ingot.

Another method of secondary aluminum recovery uses aluminum drosses as the charge instead of scrap. Traditionally, the term dross was defined as a thick liquid or solid phase that forms at the surface of molten aluminum, and is a byproduct of melting operations. It is formed with or without fluxing and the free aluminum content of this byproduct can vary considerably. Most people in the industry have generally referred to dross as being lower in aluminum content, while the material with higher aluminum content is referred to as "skim," or "rich" or "white dross." If a salt flux is used in the melting process, the byproduct is usually called a "black dross" or "salt cake." Drosses with about 30% metallic content are usually crushed and screened to bring the content up to about 60 to 70%. They are then melted in a rotary furnace, where the molten aluminum metal collects on the bottom of the furnace and is tapped off. Salt slags containing less than 30% metallic may be leached with water to separate the metallic. In addition to this classic dross-recycling process, a new dross treatment process using a water-cooled plasma gas arc heater (plasma torch) installed in a specially designed rotary furnace has been patented recently. The new process eliminates the use of salt flux in the conventional dross treatment process, and reports recovery efficiencies of 85 to 95%.

3.2.1.4 Material Inputs and Pollution Outputs

The material inputs and pollution outputs resulting from primary and secondary aluminum processing are presented in Table 3.4.

Primary aluminum processing
Primary aluminum processing activities result in air emissions, process wastes, and other solid-phase wastes. Large amounts of particulates are generated during the calcining of hydrated aluminum oxide, but the economic value of this dust for reuse in the process is such that extensive controls are used to reduce emissions to relatively small quantities. Small amounts of particulates are emitted from the bauxite grinding and materials handling processes. Emissions from aluminum reduction processes are primarily gaseous hydrogen fluoride and particulate fluorides, alumina, carbon monoxide, volatile organics, and sulfur dioxide from the reduction cells, and fluorides, vaporized organics and sulfur dioxide from the anode baking furnaces. A variety of control devices such as wet scrubbers are used to abate emissions from reduction cells and anode baking furnaces.

TABLE 3.4
Process Materials Inputs/Pollution Outputs for Aluminum

Process	Material Input	Air Emissions	Process Wastes	Other Wastes
Bauxite refining	Bauxite, sodium hydroxide	Particulates		Residue containing silicon, iron, titanium, calcium oxides, and caustic
Alumina clarification and precipitation	Alumina slurry, starch, water		Wastewater containing starch, sand, and caustic	
Alumina calcination	Aluminum hydrate	Particulates and water vapor		
Primary electrolytic aluminum smelting	Alumina, carbon anodes, electrolytic cells, cryolite	Fluoride, both gaseous and particulates, carbon dioxide, sulfur dioxide, carbon monoxide, C_2F_6, CF_4, and perfluorinated carbons (PFC)		Spent potliners
Secondary scrap aluminum smelting	Aluminum scrap, oil or gas, chlorine or other fluxes (aluminum chloride, aluminum fluoride, sodium and potassium chlorides, and fluorides)	Particulates and HCl/Cl_2		Slag containing magnesium and chlorides
Secondary aluminum dross recycling	Aluminum dross, water	Particulates	Wastewater, salts	

Source: From U.S. EPA, Profile of the Nonferrous Metals Industry, publication EPA/310-R-95-010, U.S. EPA, Washington, DC, September 1995.

Wastewaters generated from primary aluminum processing are produced during clarification and precipitation, although much of this water is fed back into the process to be reused.

Solid-phase wastes are generated at two stages in the primary aluminum process; red mud is produced during bauxite refining, and spent potliners result from the reduction process. Red mud normally contains significant amounts of iron, aluminum, silicon, calcium, and sodium. The types and concentrations of minerals present in the mud depend on the composition of the ore and the operating conditions in the digesters. Red mud is managed onsite in surface impoundments, and has not been found to exhibit any of the characteristics of hazardous waste. The process does however, generate hazardous waste. The carbon potliners used to hold the alumina/cryolite solution during the electrolytic aluminum reduction process eventually crack and need to be removed and replaced. When the liners are removed they are "spent," and are considered to be RCRA-listed hazardous waste.

Secondary aluminum processing

Secondary aluminum processing also results in air emissions, wastewaters, and solid wastes. Atmospheric emissions from reverberatory (chlorine) smelting/refining represent a significant fraction of the total particulate and gaseous effluents generated in the secondary aluminum industry. Typical furnace effluent gases contain combustion products, chlorine, hydrogen chloride, and metal chlorides of zinc, magnesium, and aluminum, aluminum oxide and various metals and metal compounds, depending on the quality of the scrap charges. Emissions from reverberatory (fluorine) smelting/refining are similar to those from reverberatory (chlorine) smelting/refining. The use of AlF_3 rather than chlorine in the demagging step reduces demagging emissions. Fluorides are emitted as gaseous fluorides or as dusts. Baghouse scrubbers are usually used for fluoride emission control.

Solid-phase wastes are also generated during secondary scrap aluminum smelting. The slag generated during smelting contains chlorides resulting from the use of fluxes and magnesium. Wastewaters are also generated during secondary aluminum processing when water is added to the smelting slags to aid in the separation of metallics. The wastewaters are also likely to be contaminated with salt from the various fluxes used.

3.2.2 COPPER

The information on copper is collected from U.S EPA,[1,2] the U.S. Department of Commerce,[12] the U.S. Department of the Interior, the Bureau of Mines,[14] and the International Copper Association.[15]

3.2.2.1 Industry Size and Geographic Distribution

Copper ore is mined in both the Northern and Southern Hemispheres but is primarily processed and consumed by countries in the Northern Hemisphere. The U.S. is both a major producer (second only to Chile) and consumer of copper.[1]

The domestic primary unwrought, or unworked, integrated copper industry consists of mines, concentrators, smelters, refineries, and electrowinning plants (the nonferrous metals industry encompasses facilities engaging in primary smelting and refining, but not mining). Of the 65 mines actively producing copper in the U.S., 33 list copper as the primary product. The remaining 32 mines produce copper either as a byproduct or coproduct of gold, lead, zinc, or silver. Nineteen of the 33 active mines that primarily produce copper are located in Arizona, which accounts for 65% of domestically mined copper ore. The remaining mines are located throughout New Mexico and Utah, which together account for 28% of domestic production, and Michigan, Montana, and Missouri accounting for the remainder.[14] Five integrated producers produce over 90% of domestic primary copper.

According to the U.S. Bureau of Mines, 441,000 t of copper are recovered yearly by leaching/electrowinning methods.[14] Although solution operations are conducted throughout the Southwestern U.S., almost 75% of the facilities[14] are located in Arizona. There are two facilities in New Mexico, one in Utah, and one in Nevada.

Of recycled, or secondary copper, 56% is derived from new scrap, while 44% comes from old scrap. Domestically, the secondary copper smelting industry is led by four producers. Like the secondary aluminum industry, these producers buy the scrap they recycle on the open market, in addition to using scrap generated in their own downstream productions. The secondary copper industry is concentrated in Georgia, South Carolina, Illinois, and Missouri.

3.2.2.2 Product Characterization

Because of the superior electrical conductivity of copper, the leading domestic consumer of refined copper is wire mills, accounting for 75% of refined copper consumption. Brass mills producing copper and copper alloy semifabricated shapes are the other major domestic consumers at 23%. The dominant end-users of copper and copper alloy are the construction and electronic products industries, accounting for 65% of copper end-usage. Transportation equipment such as radiators also account for a fair amount of copper end-usage at 11.6%.[1] Copper and copper alloy powders are used for brake linings and bands, bushings, instruments, and filters in the automotive and aerospace industries, for electrical and electronic applications, for antifouling paints and coatings, and for various chemical and medical purposes. Copper chemicals, principally copper sulfate and the cupric and cuprous oxides, are widely used as algaecides, fungicides, wood preservatives, copper plating, pigments, electronic applications, and numerous special applications.

3.2.2.3 Industrial Process Description

This section specifically contains a description of commonly used production processes, associated raw materials, the byproducts produced or released, and the materials either recycled or transferred offsite. This discussion, coupled with schematic drawings of the identified processes, provide a concise description of where wastes may be produced in the process.[16,17]

Primary copper processing

Copper is mined in both open pits and underground mines, depending upon the ore grade and the nature of the ore deposit. Copper ore typically contains less than 1% copper and is in the form of sulfide minerals. Once the ore is delivered above the ground, it is crushed and ground to a powdery fineness, after which it is concentrated for further processing. In the concentration process, ground ore is slurried with water, chemical reagents are added, and air is blown through the slurry. The air bubbles attach themselves to the copper minerals and are then skimmed off the top of the flotation cells. The concentrate contains between 20 and 30% copper. The "tailings," or gangue minerals, from the ore fall to the bottom of the cells and are removed, dewatered by "thickeners," and transported as slurry to a tailings pond for disposal. All water used in this operation, from dewatering thickeners and the tailings pond, is recovered and recycled back into the process.

Copper can be produced either pyrometallurgically or hydrometallurgically depending upon the ore type used as a charge. The ore concentrates, which contain copper sulfide and iron sulfide minerals, are treated by pyrometallurgical processes to yield high-purity copper products. Oxide ores, which contain copper oxide minerals that may occur in other parts of the mine, together with other oxidized waste materials, are treated by hydrometallurgical processes to yield high-purity copper products. Both processes are illustrated in Figure 3.3.

Copper conversion is accomplished by a pyrometallurgical process known as "smelting." During smelting the concentrates are dried and fed into one of several different types of furnaces. There the sulfide minerals are partially oxidized and melted to yield a layer of "matte," a mixed copper–iron sulfide, and "slag," an upper layer of waste.

The matte is further processed by a process known as "converting." The slag is tapped from the furnace and stored or discarded in slag piles onsite. A small amount of slag is sold for railroad ballast and for sand blasting grit. A third product of the smelting process is sulfur dioxide, a gas that is collected, purified, and made into sulfuric acid for sale or for use in hydrometallurgical leaching operations.

Following smelting, the copper matte is fed into a converter. During this process the copper matte is poured into a horizontal cylindrical vessel (approximately 10 × 4 m) fitted with a row of pipes. The pipes, known as "tuyeres," project into the cylinder and are used to introduce air into the converter. Lime and silica are added to the copper matte to react with the iron oxide produced in the process to form slag. Scrap copper may also be added to the converter. The furnace is rotated so that the tuyeres are submerged, and air is blown into the molten matte, causing the remainder of the iron sulfide to react with oxygen to form iron oxide and sulfur dioxide.

Following the "blow," the converter is rotated to pour off the iron silicate slag. Once all of the iron is removed, the converter is rotated back and given a second blow, during which the remainder of the sulfur is oxidized and removed from the copper sulfide. The converter is then rotated to pour off the molten copper, which at this point is called "blister" copper (so named because if allowed to

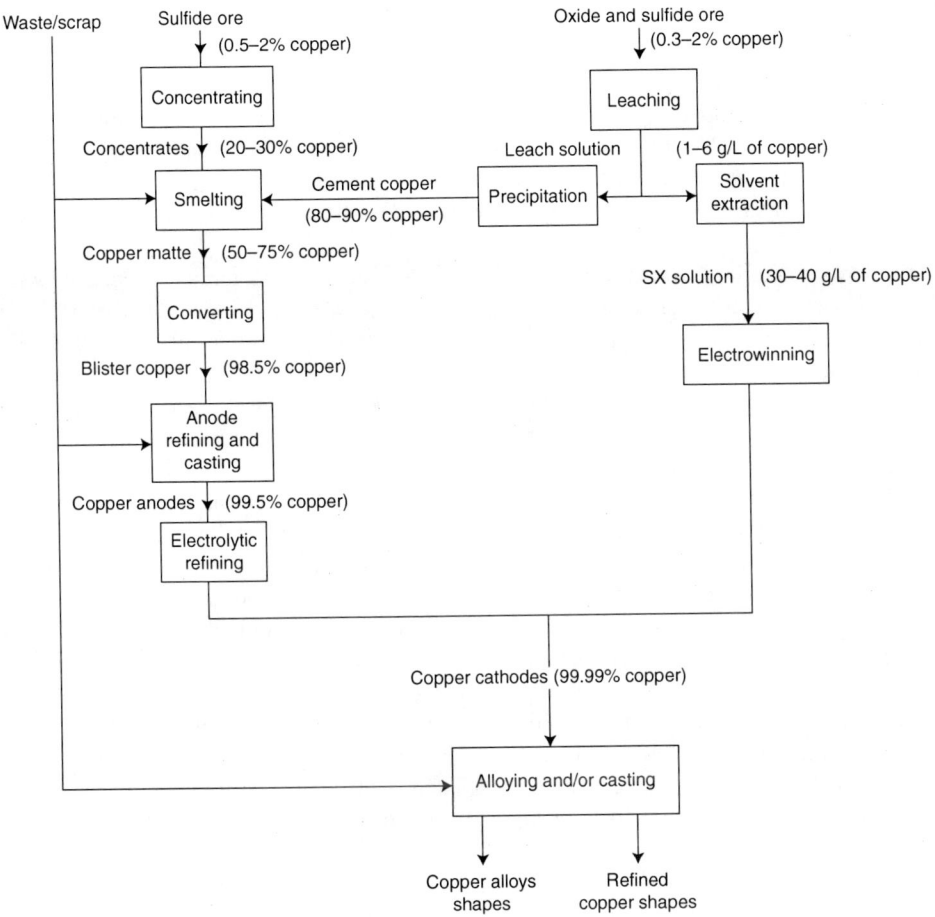

FIGURE 3.3 Copper production process. (From U.S. Congress, Copper Technology and Competitiveness, Congress of the United States, Office of Technology Assessment, Washington, DC, 1994.)

solidify at this point, it will have a bumpy surface due to the presence of gaseous oxygen and sulfur). Sulfur dioxide from the converters is collected and fed into the gas purification system together with that from the smelting furnace, and made into sulfuric acid. Owing to its residual copper content, slag is recycled back to the smelting furnace.

Blister copper, containing a minimum of 98.5% copper, is refined to high-purity copper in two steps. The first step is "fire refining," in which the molten blister copper is poured into a cylindrical furnace, similar in appearance to a converter, where first air and then natural gas or propane are blown through the melt to remove the last of the sulfur and any residual oxygen from the copper. The molten copper is then poured into a casting wheel to form anodes pure enough for "electrorefining."

In electrorefining, the copper anodes are loaded into electrolytic cells and interspaced with copper "starting sheets," or cathodes, in a bath of copper sulfate solution. When a DC current is passed through the cell the copper is dissolved from the anode, transported through the electrolyte, and redeposited on the cathode starting sheets. When the cathodes have built up to a sufficient thickness they are removed from the electrolytic cell and a new set of starting sheets is put in their place. Solid impurities in the anodesus metals such fall to the bottom of the cell as a sludge, where they are ultimately collected and processed for the recovery of precio as gold and silver. This material is known as "anode slime."

The cathodes removed from the electrolytic cell are the primary product of the copper producer and contain >99.99% copper. These may be sold to wire-rod mills as cathodes or processed further to a product called "rod." In manufacturing rod, cathodes are melted in a shaft furnace and the molten copper is poured onto a casting wheel to form a bar suitable for rolling into a 3/8-in.-diameter continuous rod. This rod product is shipped to wire mills, where it is extruded into various sizes of copper wire.

In the hydrometallurgical process, the oxidized ores and waste materials are leached with sulfuric acid from the smelting process. Leaching is performed *in situ* or in specially prepared piles by distributing acid across the top and allowing it to percolate down through the material, where it is collected. The ground under the leach pads is lined with an acid-proof, impermeable plastic material to prevent leach liquor from contaminating groundwater. Once the copper-rich solutions are collected they can be processed by either of two processes—the "cementation" process or the "solvent extraction/electrowinning" process (SXEW).

In the SXEW process, the pregnant leach solution (PLS) is concentrated by solvent extraction. In solvent extraction, an organic chemical that extracts copper but not impurity metals (iron and other impurities) is mixed with the PLS. The copper-laden organic solution is then separated from the leachate in a settling tank. Sulfuric acid is added to the pregnant organic mixture, which strips the copper into an electrolytic solution. The stripped leachate, containing the iron and other impurities, is returned to the leaching operation where its acid is used for further leaching. The copper-rich strip solution is passed into an electrolytic cell known as an "electrowinning" cell. An electrowinning cell differs from an electrorefining cell in that it uses a permanent, insoluble anode. The copper in solution is then plated onto a starting sheet cathode in much the same manner as it is on the cathode in an electrorefining cell. The copper-depleted electrolyte is returned to the solvent extraction process where it is used to strip more copper from the organic. The cathodes produced from the electrowinning process are then sold or made into rod in the same manner as those produced from the electrorefining process.

Electrowinning cells are used also for the preparation of starting sheets for both the electrorefining and electrowinning processes. Here copper is plated onto either stainless steel or titanium cathodes. When sufficient thickness has built up, the cathodes are removed and the copper plating on both sides of the stainless steel or titanium is stripped off. After straightening and flattening, these copper sheets are fabricated into starting sheet cathodes by mechanically attaching copper strips to be used as hangers when they are in the electrolytic cell. Both the starting sheet and the strips become part of the final product. The same care in achieving and maintaining purity must be maintained with these materials as is practiced for the electrodeposited copper.

An activity that is carried out concurrently with primary copper production is sulfur fixation. As mentioned above, in the pyrometallurgical process most of the sulfur in the ore is transformed into sulfur dioxide (although a portion is discarded in the slag). The copper smelting and converting processes typically generate over 0.5 t of sulfur dioxide per ton of copper concentrate. In order to meet CAA emission standards, sulfur dioxide releases must be controlled. This is accomplished by elaborate gas collection and filtration systems, after which the sulfur dioxide contained in the off-gases is made into sulfuric acid. In general, if the sulfur dioxide concentration exceeds 4% it will be converted into sulfuric acid, an ingredient in fertilizer. Fugitive gases containing less than 4% sulfuric acid are either released to the atmosphere or scrubbed to remove the sulfur dioxide. The sulfur recovery process requires the emissions to flow through a filtering material in the air emissions scrubber to capture the sulfur. Blowdown slurry is formed from the mixture of the filtering material and sulfur emissions. This slurry contains not only sulfur, but cadmium and lead, metals that are present in copper ore. The acid plant blowdown slurry/sludge that results from thickening of blowdown slurry at primary copper facilities is regulated by the RCRA as hazardous waste.

Secondary copper processing

The primary processes involved in secondary copper recovery are scrap metal pretreatment and smelting. Pretreatment includes cleaning and concentration to prepare the material for the smelting furnace. Pretreatment of the feed material can be accomplished using several different procedures, either separately or in combination. Feed scrap is concentrated by manual and mechanical methods such as sorting, stripping, shredding, and magnetic separation. Feed scrap is sometimes briquetted in a hydraulic press. Pyrometallurgical pretreatment may include sweating, burning of insulation (especially from scrap wire), and drying (burning off oil and volatiles) in rotary kilns. Hydrometallurgical methods include flotation and leaching with chemical recovery.

After pretreatment, the scrap is ready for smelting. Although the type and quality of the feed material determines the processes the smelter will use, the general fire-refining process is essentially the same as for the primary copper smelting industry.

3.2.2.4 Material Inputs and Pollution Outputs

The material inputs and pollution outputs resulting from primary and secondary copper processing are presented in Table 3.5.

TABLE 3.5
Process Materials Inputs/Pollution Outputs for Copper

Process	Material Input	Air Emissions	Process Wastes	Other Wastes
Copper concentration	Copper ore, water, chemical reagents, thickeners		Flotation wastewaters	Tailings containing waste minerals such as limestone, and quartz
Copper leaching	Copper concentrate, sulfuric acid		Uncontrolled leachate	Heap leach waste
Copper smelting	Copper concentrate, siliceous flux	Sulfur dioxide, particulate matter containing arsenic, antimony, cadmium, lead, mercury, and zinc		Acid plant blowdown slurry/sludge, slag containing iron sulfides, silica
Copper conversion	Copper matte, scrap copper, siliceous flux	Sulfur dioxide, particulate matter containing arsenic, antimony, cadmium, lead, mercury, and zinc		Acid plant blowdown slurry/sludge, slag containing iron sulfides, silica
Electrolytic copper refining	Blister copper		Process wastewater	Slimes containing impurities such as gold, silver, antimony, arsenic, bismuth, iron, lead, nickel, selenium, sulfur, and zinc
Secondary copper processing		Particulates	Slag granulation waste	Slag

Source: From U.S. EPA, Profile of the Nonferrous Metals Industry, publication EPA/310-R-95-010, U.S. EPA, Washington, DC, September 1995.

Primary copper processing

Primary copper processing results in air emissions, process wastes, and other solid-phase wastes. Particulate matter and sulfur dioxide are the principal air contaminants emitted by primary copper smelters. Copper and iron oxides are the primary constituents of the particulate matter, but other oxides, such as arsenic, antimony, cadmium, lead, mercury, and zinc, may also be present, with metallic sulfates and sulfuric acid mist. Single-stage electrostatic precipitators are widely used in the primary copper industry to control these particulate emissions. Sulfur oxides contained in the off-gases are collected, filtered, and made into sulfuric acid.

Large amounts of water are used in the copper concentration process, although disposal of liquid wastes is rarely a problem because the vast majority of the water is recycled back into the process. Once the wastewater exits the flotation process it is sent to a sediment control pond where it is held long enough for most of the sediment to settle.

The seepage and leaking of sulfuric acid solutions used in leaching can also produce liquid wastes; however, this potential is offset by the copper producer's interest in collecting as much of the copper-bearing leachate as possible. Older operations generally do not have protective liners under the piles and experience some loss of leachate. New leaching operations use impermeable membranes to confine leach solutions and channel them to collection ponds.

Electrolytic refining does produce wastewaters that must be treated and discharged, reused, or disposed in some manner. Many facilities use a wastewater treatment operation to treat these wastes.

Primary copper processing primarily generates two solid-phase wastes: slag and blowdown slurry/sludge. Slag is generated during the smelting, converting, fire-refining, and electrolytic refining stages. Slag from smelting furnaces is higher in copper content than the original ores taken from the mines. These slags may therefore, be sent to a concentrator and the concentrate is returned to the smelter. This slag processing operation results in slag tailings. Slag resulting from converting and fire refining is also normally returned to the process to capture any remaining mineral values. Blowdown slurry/sludge that results from the sulfur recovery process is regulated by the RCRA as hazardous waste.

Secondary copper processing

Secondary copper processing produces the same types of wastes as primary pyrometallurgical copper processing. One type of secondary processing pollutant that differs from primary processing is the air emissions. Air pollutants are generated during the drying of chips and borings to remove excess oils and cuttings fluids and causes discharges of large amounts of dense smoke containing soot and unburned hydrocarbons. These emissions can be controlled by baghouses or direct-flame afterburners.

3.2.3 LEAD

The information on lead is collected from U.S. EPA,[1,2,17,18] the U.S. Department of Commerce,[12] and the U.S. Department of the Interior, Bureau of Mines.[14]

3.2.3.1 Industry Size and Geographic Distribution

The U.S. is the world's third largest primary lead producer, with one-seventh of all production reserves. Over 80% of the lead ore mined domestically comes from Missouri. The majority of lead ore mined in the U.S. is smelted in conventional blast furnaces and refined using pyrometallurgical methods.

The U.S. is the world's largest recycler of lead scrap and is able to meet about 72% of its total refined lead production needs from scrap recycling. The secondary lead industry consists of 16 companies that operate 23 battery breakers–smelters with capacities of between 10,000 and 120,000 t/yr; five smaller operations with capacities between 6000 and 10,000 t/yr; and 15 smaller plants that produce mainly specialty alloys for solders, brass and bronze ingots, and miscellaneous uses.

Treatment of Nonferrous Metal Manufacturing Wastes

3.2.3.2 Product Characterization

Within the U.S., the power storage battery industry is the largest end-user of lead, accounting for 83% of domestically consumed lead. Industrial demand for batteries is rising due both to the growth in demand for stationary batteries used in telecommunications and back-up power systems for computers, lighting, and security systems, as well as an increased need for mobile batteries used in fork lifts and other battery-powered vehicles. Additional lead end-uses and users of consequence are ammunition, consumers of lead oxides used in television glass and computers, construction (including radiation shielding) and protective coatings, and miscellaneous uses such as ballasts, ceramics, and crystal glass.

3.2.3.3 Industrial Process Description

This section specifically contains a description of commonly used production processes, associated raw materials, the byproducts produced or released, and the materials either recycled or transferred offsite. This discussion, coupled with schematic drawings of the identified processes, provide a concise description of where wastes may be produced in the process.

Primary lead processing

Primary lead production consists of four steps: sintering, smelting, drossing, and pyrometallurgical refining (Figure 3.4). To begin, a feedstock comprised mainly of lead concentrate is fed into a sintering machine. Other raw materials may be added, including iron, silica, limestone flux, coke, soda, ash, pyrite, zinc, caustic, and particulates gathered from pollution control devices. In the sintering machine the lead feedstock is subjected to blasts of hot air, which burn off the sulfur, creating sulfur dioxide. The lead material existing after this process contains about 9% of its weight in carbon.[1] The sinter is then fed along with coke, various recycled and cleanup materials, limestone, and other fluxing agents into a blast furnace for reducing, where the carbon acts as a fuel and smelts or melts the lead material. The molten lead flows to the bottom of the furnace where four layers form: "speiss" (the lightest material, basically arsenic and antimony); "matte" (copper sulfide and other

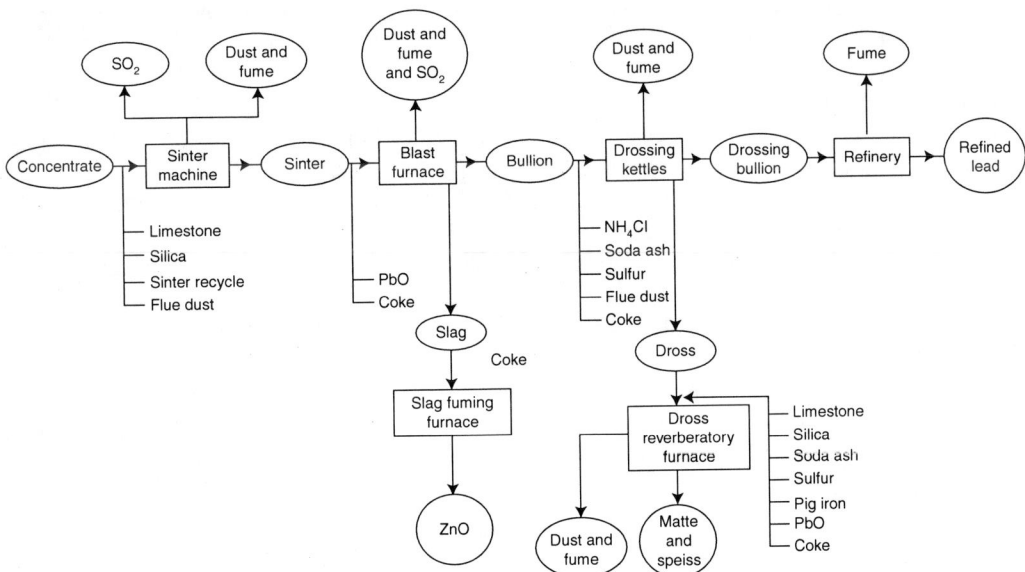

FIGURE 3.4 Primary lead production process. (From U.S. EPA, Profile of the Nonferrous Metals Industry, publication EPA/310-R-95-010, U.S. EPA, Washington, DC, September 1995.)

metal sulfides); blast furnace slag (primarily silicates); and lead bullion (98% weight lead). All layers are then drained off. The speiss and matte are sold to copper smelters for recovery of copper and precious metals. The blast furnace slag, which contains zinc, iron, silica, and lime, is stored in piles and is partially recycled. Sulfur oxide emissions are generated in blast furnaces from small quantities of residual lead sulfide and lead sulfates in the sinter feed.

Rough lead bullion from the blast furnace usually requires preliminary treatment in kettles before undergoing refining operations. During drossing the bullion is agitated in a drossing kettle and cooled to just above its freezing point (370 to 430°C). Dross, which is composed of lead oxide, along with copper, antimony, and other elements, floats to the top and solidifies above the molten lead.

The dross is removed and fed into a dross furnace for recovery of the nonlead mineral values. To enhance copper recovery, drossed lead bullion is treated by adding sulfur-bearing materials, zinc, and/or aluminum, lowering the copper content to approximately 0.01%.

During the fourth step the lead bullion is refined using pyrometallurgical methods to remove any remaining nonlead saleable materials (e.g., gold, silver, bismuth, zinc, and metal oxides such as antimony, arsenic, tin, and copper oxide). The lead is refined in a cast iron kettle over five stages. Antimony, tin, and arsenic are removed first. Then gold and silver are removed by adding zinc. Next, the lead is refined by vacuum removal of zinc. Refining continues with the addition of calcium and magnesium. These two materials combine with bismuth to form an insoluble compound that is skimmed from the kettle. In the final step caustic soda and/or nitrates may be added to the lead to remove any remaining traces of metal impurities. The refined lead will have a purity of 99.90 to 99.99%, and may be mixed with other metals to form alloys, or may be directly cast into shapes.

The processes used in the primary production of lead produce several wastestreams of concern under different regulatory scenarios. The listed RCRA hazardous wastes include smelting plant wastes that are sent to surface impoundments to settle. The impoundments are used to collect solids from miscellaneous slurries, such as acid plant blowdown, slag granulation water, and plant washings. Acid plant blowdown is generated during the production of lead in the same way it is produced at a copper plant—during the recovery of sulfur dioxide emissions. Slag granulation water is produced when hot slag from the process is sprayed with water to be cooled and granulated before transport to a slag pile. Plant washing is a housekeeping process and the washdown normally contains a substantial amount of lead and other process materials. When these materials accumulate in a surface impoundment or are dredged from the surface impoundment they are regulated as hazardous waste.

Secondary lead processing

The secondary production of lead begins with the recovery of old scrap from worn-out, damaged, or obsolete products and new scrap that is made of product wastes and smelter-refinery drosses, residues, and slags. The chief source of old scrap in the U.S. is lead-acid batteries, although cable coverings, pipe, sheet, and terne-bearing metals also serve as a source of scrap. Solder, a tin-based alloy, may also be recovered from the processing of circuit boards for use as lead charge.

Although some secondary lead is recovered directly for specialty products like babbitt metal, solder, re-melt, and copper-base alloys, about 97% of secondary lead is recovered at secondary lead smelters and refineries as either soft (unalloyed) or antimonial lead, most of which is recycled directly back into the manufacture of new batteries.[1] Unlike copper and zinc, where scrap processing varies tremendously by scrap type and ultimate use, the dominance of lead battery scrap allows for a more standard secondary recovery process. Before smelting, batteries must be broken by one of several techniques and then classified into their constituent products. The modern battery-breaking process classifies the lead into metallics, oxides, and sulfate fragments, and organics into separate casing and plate separator fractions. Cleaned polypropylene case fragments are recycled back into battery cases or other products. The dilute sulfuric acid is either neutralized for disposal or is recycled into the local acid market. One of three main smelting processes is then used to reduce the lead fractions to produce lead bullion.

The majority of domestic battery scrap is processed in blast furnaces or rotary reverberatory furnaces. Used to produce a semisoft lead, a reverberatory furnace is more suitable for processing fine particles and may be operated in conjunction with a blast furnace. The reverberatory furnace is a rectangular shell lined with refractory brick, and is fired directly with oil or gas to a temperature of 1260°C. The material is heated by direct contact with combustion gases. The average furnace can process about 45 t/d. About 47% of the charge is recovered as lead product and is periodically tapped into mold or holding pots. A total of 46% of the charge is removed as slag and later processed in blast furnaces. The remaining 7% of the furnace charge escapes as dust or fume. Short (batch) or long (continuous) rotary furnaces may be used. Slags from reverberatory furnaces are processed through the blast furnace for recovery of alloying elements.

Blast furnaces produce hard lead from charges containing siliceous slag from previous runs (~4.5% of the charge), scrap iron (~4.5%), limestone (~3%), and coke (~5.5%). The remaining 82.5% of the charge comprises oxides, pot furnace refining drosses, and reverberatory slag. The proportions of rerun slags, limestone, and coke, respectively, vary to as high as 8%, 10%, and 8% of the charge.[1] The processing capacity of the blast furnace ranges from 20 to 70 t/d. Similar to iron cupolas, the blast furnace is a vertical steel cylinder lined with refractory brick. Combustion air at 350 to 530 kg m^{-2} (0.5 to 0.75 psi) is introduced through tuyeres (pipes) at the bottom of the furnace. Some of the coke combusts to melt the charge, and the remainder reduces lead oxides to elemental lead.

As the lead charge melts, limestone and iron float to the top of the molten bath and form a flux that retards oxidation of the product lead. The molten lead flows from the furnace into a holding pot at a nearly continuous rate. The product lead constitutes roughly 70% of the charge. From the holding pot, the lead is usually cast into large ingots, called pigs or sows. About 18% of the charge is recovered as slag, with ~60% of this being matte. Roughly 5% of the charge is retained for reuse, and the remaining 7% of the charge escapes as dust or fume.

Refining/casting is the use of kettle-type furnaces for remelting, alloying, refining, and oxidizing processes. Materials charged for remelting are usually lead alloy ingots that require no further processing before casting. Alloying furnaces simply melt and mix ingots of lead and alloy materials. Antimony, tin, arsenic, copper, and nickel are the most common alloying materials. Refining furnaces, as in primary lead production, are used either to remove copper and antimony to produce soft lead, or to remove arsenic, copper, and nickel for hard lead production.

Newer secondary recovery plants use lead paste desulfurization to reduce sulfur dioxide emissions and waste sludge generation during smelting. Battery paste containing lead sulfate and lead oxide is desulfurized with soda ash to produce market-grade sodium sulfate solution. The desulfurized paste is processed in a reverberatory furnace. The lead carbonate product may then be treated in a short rotary furnace. The battery grids and posts are processed separately in a rotary smelter.

3.2.3.4 Raw Material Inputs and Pollution Outputs

The material inputs and pollution outputs resulting from primary and secondary lead processing are presented in Table 3.6.

Primary lead processing
Primary lead processing activities usually result in air emissions, process wastes, and other solid-phase wastes. The primary air emissions from lead processing are substantial quantities of SO_2 and/or particulates. Nearly 85% of the sulfur present in the lead ore concentrate is eliminated in the sintering operation. The off-gas containing a strong stream of SO_2 (5 to 7% SO_2) is sent to a sulfuric acid plant, and the weak stream (less than 0.5% SO_2) is vented to the atmosphere after removal of particulates. Particulate emissions from sinter machines range from 5 to 20% of the concentrated ore feed. Approximately 15% of the sulfur in the ore concentrate fed to the sinter machine is eliminated in the blast furnace. However, only half of this amount, about 7% of the total sulfur in the ore,

TABLE 3.6
Process Materials Inputs/Pollution Outputs for Lead

Process	Material Input	Air Emissions	Process Wastes	Other Wastes
Lead sintering	Lead ore, iron, silica, limestone flux, coke, soda, ash, pyrite, zinc, caustic, and baghouse dust	Sulfur dioxide, particulate matter containing cadmium and lead		
Lead smelting	Lead sinter, coke	Sulfur dioxide, particulate matter containing cadmium and lead	Plant washdown wastewater, slag granulation water	Slag containing impurities such as zinc, iron, silica, and lime, surface impoundment solids
Lead drossing	Lead bullion, soda ash, sulfur, baghouse dust, coke			Slag containing such impurities as copper, surface impoundment solids
Lead refining	Lead drossing bullion			
Lead-acid battery breaking	Lead-acid batteries			Polypropylene case fragments, dilute sulfuric acid
Secondary lead smelting	Battery scrap, rerun slag, drosses, oxides, iron, limestone, and coke	Sulfur dioxide, particulate matter containing cadmium and lead		Slag, emission control dust

Source: From U.S. EPA, Profile of the Nonferrous Metals Industry, publication EPA/310-R-95-010, U.S. EPA, Washington, DC, September 1995.

is emitted as SO_2. Particulate emissions from blast furnaces contain many different kinds of material, including a range of lead oxides, quartz, limestone, iron pyrites, iron-limestone-silicate slag, arsenic, and other metallic compounds associated with lead ores. The emission controls most commonly used are fabric filters and electrostatic precipitators.

As mentioned above, approximately 7% of the total sulfur present in lead ore is emitted as SO_2. The remainder is captured by the blast furnace slag. The blast furnace slag is composed primarily of iron and silicon oxides, as well as aluminum and calcium oxides. Other metals may also be present in smaller amounts, including antimony, arsenic, beryllium, cadmium, chromium, cobalt, copper, lead, manganese, mercury, molybdenum, silver, and zinc. This blast furnace slag is either recycled back into the process or disposed of in piles on site. About 50 to 60% of the recovery furnace output is slag and residual lead, which are both returned to the blast furnace. The remainder of this dross furnace output is sold to copper smelters for recovery of the copper and other precious metals.

The smelting of primary lead produces a number of wastewaters and slurries, including acid plant blowdown, slag granulation water, and plant washdown water. Slag granulation water is generated when slag is disposed. It can either be sent directly to a slag pile or granulated in a water jet before being transported to the slag pile. The granulation process cools newly generated hot slag with a water spray. Slag granulation water is often transported to surface impoundments for settling. Plant washdown water results from plant housekeeping and normally contains a substantial amount of lead and other process materials. Acid plant blowdown results from the conversion of SO_2 to sulfuric acid. All of these materials are included in the definition of hazardous waste.

Secondary lead processing

Secondary lead processing results in the generation of air emissions and solid-phase wastes. As with primary lead processing, reverberatory and blast furnaces used in smelting account for the vast majority of the total lead emissions. Other emissions from secondary smelting include oxides of sulfur and nitrogen, antimony, arsenic, copper, and tin. Smelting emissions are generally controlled with a settling and cooling chamber, followed by a baghouse. Other air emissions are generated during battery breaking. Emissions from battery breaking are mainly sulfuric acid and dusts containing dirt, battery case material, and lead compounds. Emissions from crushing are also mainly dusts.

The solid-phase wastes generated by secondary processing are emission control dust and slag. Slag is generated from smelting, and the emission control dust, when captured and disposed of, is considered to be hazardous waste.

3.2.4 Zinc

The information on zinc is collected from U.S. EPA,[1,2] the U.S. Department of Commerce,[12] and the U.S. Department of the Interior, Bureau of Mines.[14]

3.2.4.1 Industry Size and Geographic Distribution

Zinc is the fourth most widely used metal after iron, aluminum, and copper (lead is fifth). In abundant supply world-wide, zinc is mined and produced mainly in Canada, the former Soviet Union, Australia, Peru, Mexico, and the U.S. Historically, in the U.S. recoverable zinc has been mined in 19 states: Alaska, Arizona, Colorado, Idaho, Illinois, Kansas, Missouri, Montana, Nevada, New Jersey, New Mexico, New York, Oklahoma, Pennsylvania, Tennessee, Utah, Virginia, Washington, and Wisconsin. Nearly 50% of all domestic zinc is produced in Alaska. Other top producing states in order of output are Tennessee, New York, and Missouri.

The zinc industry employs 2200 workers at mines and mills and 1400 at primary smelters.[14] The four primary zinc smelters in the U.S. are located in Illinois, Oklahoma, Tennessee, and Pennsylvania. There are ten secondary zinc recovery plants in the U.S.[17]

3.2.4.2 Product Characterization

The U.S. accounts for almost one-quarter of worldwide slab zinc consumption and is the world's single largest market. About 80% of zinc is used in metal form, and the rest is used in compound form. In total, 90% of zinc metal is used for galvanizing steel (a form of corrosion protection) and for alloys, and is used in a wide variety of materials in the automotive, construction, electrical, and machinery sectors of the economy. Zinc compound use also varies widely, but is mainly found in the agricultural, chemical, paint, pharmaceutical, and rubber sectors of the economy.

3.2.4.3 Industrial Process Description

This section specifically contains a description of commonly used production processes, associated raw materials, the byproducts produced or released, and the materials either recycled or transferred offsite. This discussion, coupled with schematic drawings of the identified processes, provide a concise description of where wastes may be produced in the process.[1,2,14,18]

Primary zinc processing

The primary production of zinc begins with the reduction of zinc concentrates to metal. The zinc concentration process consists of separating the ore (which may be as little as 2% zinc) from waste rock by crushing and flotation, a process normally performed at the mining site and is discussed in the chapter on metal mining. Zinc reduction is accomplished in one of two ways: either pyrometallurgically by distillation (retorting in a furnace) or hydrometallurgically by electrowinning. Because hydrometallurgical refining accounts for more than 80% of total zinc refining, pyrometallurgical zinc refining will not be discussed in detail.

Four processing stages are generally used in hydrometallurgical zinc refining: calcining, leaching, purification, and electrowinning. Calcining, or roasting, is common to both pyrometallic and electrolytic (a form of hydrometallurgy) zinc refining, and is performed to eliminate sulfur and form leachable zinc oxide. Roasting is a high-temperature process that converts zinc sulfide concentrate to an impure zinc oxide called calcine. Roaster types include multiple-hearth, suspension, or fluidized-bed. In general, calcining begins with the mixing of zinc-containing materials with coal. This mixture is then heated, or roasted, to vaporize the zinc oxide, which is then moved out of the reaction chamber with the resulting gas stream. The gas stream is directed to the baghouse (filter) area, where the zinc oxide is captured in baghouse dust.

In a multiple-hearth roaster, the concentrate drops through a series of nine or more hearths stacked inside a brick-lined cylindrical column. As the feed concentrate drops through the furnace, it is first dried by the hot gases passing through the hearths and then oxidized to produce calcine. Multiple hearth roasters are unpressurized and operate at ~700°C (1300°F).

In a suspension roaster, the concentrates are blown into a combustion chamber. The roaster consists of a refractory-lined cylindrical shell, with a large combustion space at the top and two to four hearths in the lower portion. Additional grinding, beyond that required for a multiple-hearth furnace, is normally required to ensure that heat transfer to the material is sufficiently rapid for the desulfurization and oxidation reaction to occur in the furnace chamber. Suspension roasters are also unpressurized and operate at about 980°C (1800°F).

Fluidized-bed roasters require that the sulfide concentrates be finely ground. The concentrates are then suspended and oxidized on a feedstock bed supported on an air column. As in the suspension bed roaster, the reduction rates for desulfurization are more rapid than in the older multiple-hearth processes. Fluidized-bed roasters operate under a pressure slightly lower than atmospheric and at temperatures averaging 980°C (1800°F). In the fluidized-bed process, no additional fuel is required after ignition has been achieved. The major advantages of this roaster are greater throughput capacities and greater sulfur removal capabilities. All of the above calcining processes generate sulfur dioxide, which is controlled and converted to sulfuric acid as a marketable process byproduct.

Electrolytic processing of desulfurized calcine consists of three basic steps: leaching, purification, and electrolysis. Leaching refers to the dissolving of the captured calcine in a solution of sulfuric acid to form a zinc sulfate solution. The calcine may be leached once or twice. In the double-leach method, the calcine is dissolved in a slightly acidic solution to remove the sulfates. The calcine is then leached a second time in a stronger solution that dissolves the zinc. This second leaching step is in fact the beginning of the third step of purification, because many of the iron impurities (such as goethite and hematite) drop out of the solution as well as the zinc.

After leaching, the solution is purified in two or more stages by adding zinc dust. The solution is purified as the dust forces deleterious elements to precipitate so that they can be filtered out. Purification is usually conducted in large agitation tanks. The process takes place at temperatures ranging from 40 to 85°C (104 to 185°F), and pressures ranging from atmospheric to 2.4 atm. The elements recovered during purification include copper as a cake and cadmium as a metal. After purification the solution is ready for the final step—electrowinning.

Zinc electrowinning takes place in an electrolytic cell and involves running an electric current from a lead–silver alloy anode through the aqueous zinc solution. This process charges the suspended zinc and forces it to deposit onto an aluminum cathode (a plate with an opposite charge) that is immersed in the solution. Every 24 to 48 h, each cell is shut down, the zinc-coated cathodes removed and rinsed, and the zinc mechanically stripped from the aluminum plates. The zinc concentrate is then melted and cast into ingots, and is often as high as 99.995% pure.

Electrolytic zinc smelters contain up to several hundred cells. A portion of the electrical energy is converted into heat, which increases the temperature of the electrolyte. Electrolytic cells operate at temperature ranges from 30 to 35°C (86 to 95°F) at atmospheric pressure. During electrowinning a portion of the electrolyte passes through cooling towers to decrease its temperature and to evaporate the water it collects during the process.

Sulfur dioxide is generated in large quantities during the primary zinc refining process and sulfur fixation is carried out concurrently with the primary production process in order to meet CAA emission standards. Concentrations of sulfur dioxide in the off-gas vary with the type of roaster operation. Typical concentrations for multiple-hearth, suspension, and fluidized-bed roasters are 4.5 to 6.5%, 10 to 13%, and 7 to 12%, respectively. This sulfur dioxide is then converted into sulfuric acid.

The sulfur recovery process requires that the emissions from the zinc calcining or roasting process, where over 90% of potential sulfur dioxide is generated during primary zinc refining, flow through a filtering material in the air emissions scrubber to capture the sulfur. Blowdown slurry is formed from the mixture of the filtering material and sulfur emissions. This slurry contains not only sulfur, but cadmium and lead, materials that are always present in zinc ore. The acid plant blowdown slurry/sludge that results from thickening of blowdown slurry at primary zinc facilities is regulated by RCRA as hazardous waste.

During the electrolytic refining of zinc, solid materials in the electrolytic solution that have not been captured previously during purification may precipitate out in the electrolytic cell. When the cells undergo their periodic shutdown to recover zinc, this precipitated waste (known as anode slimes/sludges) is collected during cell cleaning. Once collected it is sent to a wastewater treatment plant. The resulting sludges are also regulated by RCRA as hazardous waste.

Secondary zinc processing

The secondary zinc industry processes scrap metals for the recovery of zinc in the form of zinc slabs, zinc oxide, or zinc dust. Zinc recovery involves three general operations: pretreatment, melting, and refining. Secondary recovery begins with the separation of zinc-containing metals from other materials, usually by magnetics, sink-float, or hand sorting. In situations where nonferrous metals have been mixed, molten zinc collects at the bottom of the sweat furnace and is subsequently recovered. The remaining scrap is cooled and removed to be sold to other secondary processors. In the case of zinc-galvanized steel, the zinc will be recovered largely in furnace dust after the scrap is charged into a steel-making furnace and melted. Almost all of the zinc in electric arc furnace (EAF) dust is first recovered in an upgraded, impure zinc oxide product and is then shipped to a primary pyrometallurgical zinc smelter for refinement to metal.

Clean new scrap, mainly brass and rolled zinc clippings and reject diecastings, generally requires only remelting before reuse. During melting, the zinc-containing material is heated in kettle, crucible, and reverberatory and electric induction furnaces. Flux is used to trap impurities from the molten zinc. Facilitated by agitation, flux and impurities float to the surface of the melt as dross, which is skimmed from the surface. The remaining molten zinc may be poured into molds or transferred to the refining operation in a molten state. Drosses, fragmentized diecastings, and mixed high-grade scrap are typically remelted, followed by zinc distillation with recovery as metal, dust, or oxide. Sometimes, high-purity drosses are simply melted and reacted with various fluxes to release the metallic content; often the recovered metal can be used directly as a galvanizing brightener or master alloy. Zinc alloys are produced from pretreated scrap during sweating and melting processes. The alloys may contain small amounts of copper, aluminum, magnesium, iron, lead, cadmium, and tin. Alloys containing 0.65 to 1.25% copper are significantly stronger than unalloyed zinc.

Medium- and low-grade skims, oxidic dust, ash, and residues generally undergo an intermediate reduction-distillation pyrometallurgical step to upgrade the zinc product before further treatment, or they are leached with acid, alkaline, or ammoniacal solutions to extract zinc. For leaching, the zinc-containing material is crushed and washed with water, separating contaminants from the zinc-containing material. The contaminated aqueous stream is treated with sodium carbonate to convert zinc chloride into sodium chloride and insoluble zinc hydroxide. The sodium chloride is separated from the insoluble residues by filtration and settling. The precipitate zinc hydroxide is dried and calcined (dehydrated into a powder at high temperature) to convert it into crude zinc oxide. The zinc oxide product is usually refined to zinc at primary zinc smelters. The washed zinc-containing metal portion becomes the raw material for the melting process.

Distillation retorts and furnaces are used either to reclaim zinc from alloys or to refine crude zinc. Bottle retort furnaces consist of a pear-shaped ceramic retort (a long-necked vessel used for distillation). Bottle retorts are filled with zinc alloys and heated until most of the zinc is vaporized, sometimes for as long as 24 h. Distillation involves vaporization of zinc at temperatures from 980 to 1250°C (1800 to 2280°F), and condensation as zinc dust or liquid zinc. Zinc dust is produced by vaporization and rapid cooling, and liquid zinc results when the vaporous product is condensed slowly at moderate temperatures.

Air pollution control can be an area of concern when pyrometallurgical processes are used in the secondary recovery of zinc. When the recovery process used is simply an iron pot remelt operation to produce zinc metal, fumes will not normally be generated. If slab zinc is needed and a rotary furnace is used, any air emissions are captured directly from the venting system (a rotating furnace sweats, or melts, the zinc separating it from drosses with different melting points, which allows it to be poured off separately). Air emissions become more of a concern when more complicated processes are used to produce zinc powder. Retort and muffle furnaces used to produce zinc powder heat the zinc and other charges to such a high temperature that the zinc vaporizes and is captured in the pollution control equipment. It is this zinc oxide dust that is the process' marketable product. Hoods are utilized around the furnace openings that are used to add additional charge. The fumes collected from the hoods are not normally of high quality and are used for products such as fertilizer and animal feed.

For the most part, the zinc materials recovered from secondary materials such as slab zinc, alloys, dusts, and compounds are comparable in quality to primary products. Zinc in brass is the principal form of secondary recovery, although secondary slab zinc has risen substantially over the last few years because it has been the principal zinc product of electric arc furnace (EAF) dust recycling. Impure zinc oxide products and zinc-bearing slags are sometimes used as trace element additives in fertilizers and animal feeds. About 10% of the domestic requirement for zinc is satisfied by old scrap.

As a result of environmental concerns, both domestic and worldwide secondary recovery of zinc (versus disposal) is expected to increase. However, the prospect for gains higher than 35 to 40% of zinc consumption is relatively poor because of the dissipative nature of zinc vapor.

3.2.4.4 Material Inputs and Pollution Outputs

The material inputs and pollution outputs resulting from primary and secondary zinc processing are presented in Table 3.7.

Primary zinc processing
Primary zinc processing activities generate air emissions, process wastes, and other solid-phase wastes. Air emissions are generated during roasting, which is responsible for more than 90% of the potential SO_2 emissions. Approximately 93 to 97% of the sulfur in the feed is emitted as sulfur oxides. Sulfur dioxide emissions from the roasting process at all four primary zinc processing facilities are recovered at onsite sulfuric acid plants. Much of the particulate matter emitted from primary zinc facilities is also attributable to roasters. Although the amount and composition of particulates varies with operating parameters, the particulates are likely to contain zinc and lead.

Wastewaters may be generated during the leaching, purification, and electrowinning stages of primary zinc processing when electrolyte and acid solutions become too contaminated to be reused again. This wastewater needs to be treated before discharge.

TABLE 3.7
Process Materials Inputs/Pollution Outputs for Zinc

Process	Material Input	Air Emissions	Process Wastes	Other Wastes
Zinc calcining	Zinc ore, coke	Sulfur dioxide, particulate matter containing zinc and lead		Acid plant blowdown slurry
Zinc leaching	Zinc calcine, sulfuric acid, limestone, spent electrolyte		Wastewaters containing sulfuric acid	
Zinc purification	Zinc-acid solution, zinc dust		Wastewaters containing sulfuric acid, iron	Copper cake, cadmium
Zinc electrowinning	Zinc in a sulfuric acid/aqueous solution, lead-silver alloy anodes, aluminum cathodes, barium carbonate, or strontium, colloidal additives		Dilute sulfuric acid	Electrolytic cell slimes/sludges
Secondary zinc smelting	Zinc scrap, electric arc furnace dust, drosses, diecastings, fluxes	Particulates		Slags containing copper, aluminum, iron, lead, and other impurities
Secondary zinc reduction distillation	Medium-grade zinc drosses, oxidic dust, acids, alkalines, or ammoniacal solutions	Zinc oxide fumes		Slags containing copper, aluminum, iron, lead, and other impurities

Source: From U.S. EPA, Profile of the Nonferrous Metals Industry, publication EPA/310-R-95-010, U.S. EPA, Washington, DC, September 1995.

Solid wastes, some of which are hazardous, are generated at various stages in primary zinc processing. Slurry generated during the operation of sulfuric acid plants is regulated as hazardous waste, as is the sludge removed from the bottom of electrolytic cells. The solid copper cake generated during purification is generally sent offsite to recover the copper.

Secondary zinc processing

Secondary zinc processing generates air emissions and solid-phase wastes. Air emissions result from sweating and melting and consist of particulates, zinc fumes, other volatile metals, flux fumes, and smoke generated by the incomplete combustion of grease, rubber, and plastics in the zinc scrap. Zinc fumes are negligible at low furnace temperatures. Substantial emissions may arise from incomplete combustion of carbonaceous material in the zinc scrap. These contaminants are usually controlled by afterburners, and particulate emissions are most commonly recovered by fabric filters. Emissions from refining operations are mainly metallic fumes. Distillation/oxidations operations emit their entire zinc oxide product in the exhaust dust. Zinc oxide is usually recovered in fabric filters with collection efficiencies of 9 to 99%.

The secondary zinc recovery process generates slags that contain metals such as copper, aluminum, iron, and lead. Although slag generated during primary pyrometallurgical processes is exempt from regulation as a hazardous waste under RCRA, slag resulting from secondary processing is not automatically exempt. Therefore if secondary processing slag exhibits a characteristic (e.g., toxicity for lead) it would need to be managed as a hazardous waste.

3.2.5 COLUMBIUM AND TANTALUM

Columbium (also known as niobium) and tantalum metals are produced from purified salts, which are prepared from ore concentrates and slags resulting from foreign tin production. The concentrates and slags are leached with hydrofluoric acid to dissolve the metal salts. Solvent extraction or ion exchange is used to purify the columbium and tantalum. The salts of these metals are then reduced by means of one of several techniques, including aluminothermic reduction, sodium reduction, carbon reduction, and electrolysis.[19-21] Owing to the reactivity of these metals, special techniques are used to purify and work the metal produced.

3.2.6 SILVER

There are four primary silver production facilities in the U.S. Of these, two discharge wastewaters. Wastes containing silver include materials from photography, the arts, electrical components, industry, and miscellaneous sources. These wastes are processed by a wide variety of techniques to recover the silver.[2] Because the process is highly specific for the type of waste, no attempt to discuss the various processes will be made in this chapter.

3.2.7 TUNGSTEN

There are several variations in the processes of this industry depending on the ore. In each process, one of the intermediate products is tungstic acid. The tungstic acid is converted to ammonium tungstate, which is dried and heated to form ammonium paratungstate. This intermediate is converted to oxides in a nitrogen–hydrogen atmosphere. Finally, the oxides are reduced to tungsten metal powder at high temperature in a hydrogen atmosphere.[2]

3.2.8 BERYLLIUM

Primary beryllium production occurs at two plants within the U.S. One of these plants discharges its wastewater to the environment. Because of the limited number of facilities, beryllium production will not be discussed in this chapter.

3.2.9 SELENIUM

Primary selenium recovery occurs at a single site that does not discharge to the environment. Consequently, this subcategory is not discussed further in this chapter.

3.3 WASTEWATER CHARACTERIZATION

Each metal subcategory uses different processes and emits different pollutant concentrations and types in the process wastewater. The following paragraphs and tables present information on the wastewater streams for each of the subcategories.[2,3]

Raw waste characteristics for the industry generally reflect the products and the methods used to manufacture them. Because there is such diversity in products, processing, raw materials, and process control, there is a wide range of characteristics. The variations exist among different streams within each subcategory, as well as among similar streams (such as casting wastewater) in different subcategories. Discharge of nonprocess wastes (sanitary, boiler blowdown, noncontact cooling water, and so on) with process wastestreams and other nonprocess-related variables such as raw water quality can contribute to this lack of uniformity.

3.3.1 Primary Aluminum

Process wastewater sources for this subcategory are primarily related to air pollution control. Wet air pollution controls on anode bake furnaces generate wastewater in plants utilizing prebaked anodes. Suspended solids, oil and grease, sulfur compounds, and fuel combustion products characterize this stream. Some organics may also be present as a result of the release of coal tar products during anode baking. Degassing with chlorine requires wet air pollution control methods and results in a wastewater stream. Cryolite recovery also produces a wastewater stream that has significant amounts of fluoride, suspended solids, and TOC. Other wastestreams may also be produced by cooling water, in cathode making, and from storm water runoff. Tables 3.8 and 3.9 present classical and toxic data for the primary aluminum subcategory.

TABLE 3.8
Classical Pollutants in the Raw Wastewater of the Primary Aluminum Subcategory

Pollutant	Number of Samples	Number of Detections	Concentration (mg/L)		
			Range	Median	Mean
COD	2	2	3.1–5700		2900
TOC	2	2	140–440		290
TSS	2	2	2100–11,000		6600
Total phenol	3	3	0.11–0.27	0.13	0.17
Oil and grease	2	2	4.2–5.5		4.9
Ammonia	1	1	25		
Fluoride	3	3	0.46–2700	170	960

Source: From Wang, L.K., Hung, Y.T., and Shammas, N.K., Eds., *Advanced Physicochemical Treatment Processes*, Humana Press, Totowa, NJ, 2006.

TABLE 3.9
Concentrations of Toxic Pollutants Found in Primary Aluminum Wastewater

Toxic Pollutant	Number of Samples	Number of Detections >10 µg/L	Concentration (µg/L)		
			Range	Median	Mean
Metals and Inorganics					
Antimony	3	2	ND–770	100	290
Arsenic	3	2	ND–260	130	130
Asbestos (fibers/L)	1	1	2.3×10^{10}	—	—
Beryllium	3	2	ND–75	33	36
Cadmium	3	2	2.3–<200	<24	<75
Chromium	3	2	ND–2200	84	760
Copper	3	3	13–140	77	77
Cyanide	3	2	<4–28,000	22	9300
Lead	3	2	0.56–770	650	470
Mercury	3	0	<0.1–1.3	<0.38	<0.59

Continued

TABLE 3.9 (continued)

Toxic Pollutant	Number of Samples	Number of Detections >10 µg/L	Concentration (µg/L)		
			Range	Median	Mean
Nickel	3	3	500–730	640	620
Selenium	3	1	ND–450	<0.19	150
Silver	3	1	ND–<250	<0.38	<83
Thallium	3	1	ND–<50	ND	<17
Zinc	3	2	ND–540	25	190
Phthalates					
Bis(2-ethylhexyl) phthalate	7	5	ND–450	—[a]	82
Butyl benzyl phthalate	7	2	ND–86	—[a]	22
Di-n-butyl phthalate	7	1	ND–120	—[a]	19
Diethyl phthalate	7	0	ND–2.5	—[a]	0.4
Phenols					
Phenol	6	1	ND–70	—[a]	12
Aromatics					
Benzene	8	1	ND–6.0	—[a]	0.8
Toluene	8	0	ND–1.0	—[a]	0.2
2,4-Dinitrotoluene	7	0	ND	—	—
Polycyclic Aromatic Hydrocarbons					
Acenaphthene	7	1	ND–50	—[a]	8.4
Acenaphthylene	7	1	ND–30	—[a]	5.6
Anthracene	7	4	ND–150	8.6	40
Benz (a) anthracene	7	3	ND–180	—[a]	38
Benzo (a) pyrene	7	3	ND–570	—[a]	95
Benzo (b) fluoranthene	7	1	ND–260	—[a]	37
1,12-Benzoperylene	7	1	ND–150	—[a]	24
Benzo (k) fluoranthene	7	2	ND–210	—[a]	39
Chrysene	7	2	ND–230	—[a]	40
Dibenz (a,h) anthracene	7	1	ND–110	—[a]	16
Fluoranthene	7	4	ND–320	49	95
Fluorene	7	1	ND–50	—[a]	7.4
Indeno(1,2,3-cd)pyrene	7	2	ND–350	—[a]	53
Naphthalene	7	1	ND–20	—[a]	3.0
Pyrene	7	4	ND–220	39	70
Phenanthrene	7	3	ND–230	—[a]	50
Halogenated Aliphatics					
Chloroform	8	0	ND–6.0	—[a]	0.8
Methylene chloride	8	1	ND–15	—[a]	3.0
Pesticides and Metabolites					
Gamma-BHC	7	0	ND–0.01	—[a]	—[b]

ND, not detectable.

[a] No median concentration is available in the reference.

[b] No mean concentration is available in the reference.

Source: From U.S. EPA, Treatability Manual, Volume II Industrial Descriptions, report EPA-600/2-82-001b, U.S. EPA, Washington, DC, September 1981.

3.3.2 Secondary Aluminum

Sources of process wastewater in the secondary aluminum industry include demagging air pollution control, wet nulling of residues, and contact cooling water. Removal of magnesium (demagging) involves the passage of chlorine or aluminum fluoride through the melt, leading to the release of magnesium in heavy fuming. The wastestreams from the air pollution control devices contain significant levels of suspended solids and chlorides or fluorides, as well as moderate amounts of heavy metals. Milling streams also contain suspended solids, and contact cooling water contains oil and grease, chlorides, and suspended solids. Tables 3.10 and 3.11 present classical and toxic pollutant concentrations found in the wastewater streams of this subcategory.

TABLE 3.10
Classical Pollutants in the Raw Wastewater of the Secondary Aluminum Subcategory

Pollutant	Number of Samples	Number of Detections	Concentration (mg/L)		
			Range	Median	Mean
COD	4	4	9–580	35	160
TOC	4	3	ND–140	3.8	37
TSS	4	4	63–13,000	150	3300
Total phenol	4	4	0.003–0.025	0.01	0.012
Oil and grease	4	4	3.2–98	13	32
Ammonia	2	2	<0.10–140	—	70
Chloride	3	3	400–6000	1400	2600

ND, not detectable.

Source: From U.S. EPA, Treatability Manual, Volume II Industrial Descriptions, report EPA-600/2-82-001b, U.S. EPA, Washington, DC, September 1981.

TABLE 3.11
Concentrations of Toxic Pollutants Found in Secondary Aluminum Wastewater

Toxic Pollutant	Number of Samples	Number of Detections >10 µg/L	Concentration (µg/L)		
			Range	Median	Mean
Metals and Inorganics					
Antimony	4	2	ND–950	150	310
Arsenic	4	3	ND–4000	32	1000
Asbestos (fibers/L)	1	1	7.5×10^8	—	—
Beryllium	4	4	<7.0–310	69	<110
Cadmium	4	4	<35–2000	260	<640
Chromium	4	4	<5–1200	68	<340
Copper	4	4	<70–6100	440	<1800
Cyanide	4	4	<1–7.8	4.6	<4.5
Lead	4	4	<65–5600	1500	<2200

Continued

TABLE 3.11 (continued)

Toxic Pollutant	Number of Samples	Number of Detections >10 µg/L	Concentration (µg/L)		
			Range	Median	Mean
Mercury	4	3	ND–6.4	0.38	1.8
Nickel	4	3	ND–620	<28	<170
Selenium	4	1	ND–200	ND	50
Silver	4	2	ND–30	<13	<14
Thallium	3	1	ND–540	ND	180
Zinc	4	4	<2000–5900	2200	<3100
Phthalates					
Bis(2-ethylhexyl) phthalate	6	4	ND–2000	46	380
Butyl benzyl phthalate	6	2	ND–98	—[a]	19
Di-n-butyl phthalate	6	3	ND–44	—[a]	16
Dimethyl phthalate	6	1	ND–56	—[a]	9.5
Di-n-octyl phthalate	6	1	ND–25	—[a]	4.2
Nitrogen Compounds					
3,3'-Dichlorobenzidine	6	0	ND–2.0	—[a]	0.3
Aromatics					
Benzene	10	1	ND–94	—[a]	9.4
1,4-Dichlorobenzene	6	1	ND–26	—[a]	4.3
Polycyclic Aromatic Hydrocarbons					
Acenaphthylene	6	1	ND–17	—[a]	2.8
Anthracene	6	0	ND–4.0	—[a]	0.7
Benzo (a) pyrene	6	1	ND–12	—[a]	2.0
Chrysene	6	1	ND–190	—[a]	32
Fluoranthene	6	2	ND–12	—[a]	3.8
Naphthalene	6	0	ND–1.0	—[a]	0.2
Phenanthrene	6	0	ND–10	—[a]	1.7
Pyrene	6	1	ND–24	—[a]	4.0
Polychlorinated Biphenyls					
Aroclor 1248	6	0	ND–0.3	—[a]	0.1
Aroclor 1254	6	0	ND–0.9	—[a]	0.4
Halogenated Aliphatics					
Carbon tetrachloride	10	0	ND–10	—[a]	1.0
Chloroform	10	6	ND–31	—[a]	3.4
Dichlorobromomethane	10	1	ND–19	—[a]	1.9
1,2-Dicloroethane	10	0	ND–1.0	—[a]	0.1
1,2-*trans*-Dichloroethylene	10	5	ND–57	9.5	19
Methylene chloride	10	1	ND–93	—[a]	9.3
Tetrachloroethylene	10	1	ND–310	—[a]	31
Trichloroethylene	10	5	ND–530	—[a]	61
Pesticides and Metabolites					
Alpha-BHC	6	0	ND–0.1	—[a]	—[b]
Beta-BHC	6	0	ND–0.4	—[a]	—[b]
Gamma-BHC	6	0	ND–0.1	—[a]	—[b]
Chlordane	6	0	ND–0.3	—[a]	0.1

Continued

TABLE 3.11 (continued)

Toxic Pollutant	Number of Samples	Number of Detections >10 µg/L	Concentration (µg/L)		
			Range	Median	Mean
4,4'-DDE	6	0	ND–0.01	—[a]	—[b]
4,4'-DDT	6	0	ND–0.02	—[a]	—[b]
Dieldrin	6	0	ND–0.2	—[a]	—[b]
Endrin	6	0	ND–0.01	—[a]	—[b]
Endrin aldehyde	6	0	ND–0.04	—[a]	0.01
Heptachlor	6	0	ND–0.04	—[a]	0.01
Heptachlor epoxide	6	0	ND–0.2	—[a]	—[b]
Isophorone	6	0	ND–3.0	—[a]	0.5

ND, not detectable.
[a] No median concentration is available in the reference.
[b] No mean concentration is available in the reference.
Source: From U.S. EPA, Treatability Manual, Volume II Industrial Descriptions, report EPA-600/2-82-001b, U.S. EPA, Washington, DC, September 1981.

3.3.3 PRIMARY COLUMBIUM AND TANTALUM

The production of columbium and tantalum involves the processing of ore concentrates and slags to obtain columbium and tantalum salts, and the subsequent reduction of those salts to the respective metals. The ore concentrates are dissolved by hydrofluoric acid, and the insoluble gangue is removed by filtration. Waste gangue is generally settled in holding ponds. Overflow from this pond is extremely acidic and contains metals, fluorides, and suspended solids. After filtration, the digested solution is extracted with an organic solvent, and the raffinate is discharged as a wastestream with high concentrations of organics, fluorides, metals, and suspended solids. The organic stream is then stripped with water to yield aqueous solutions of columbium and tantalum. Precipitation of the salts is accomplished by the addition of ammonia and is followed by filtration. The filtrate typically contains high concentrations of ammonia as well as significant levels of fluoride, various metals, and suspended solids. Conversion of the salts to metals produces wastewater from air pollution control scrubbers and reduction leachates. These streams contain high levels of dissolved solids and significant concentrations of fluoride.[2] Tables 3.12 and 3.13 present classical and toxic pollutant concentration data for this subcategory.

3.3.4 PRIMARY COPPER

Both smelting and refining are practiced by the primary copper industry. Some plants engage in smelting only, others practice only refining, and some facilities practice both operations. Significant differences in the wastewater characteristics associated with smelting and refining are found.

Smelting process wastewater sources include acid plant blowdown, contact cooling, and slag granulation. Acid plant blowdown results from the recovery of sulfur from the smelting operation. Contact casting cooling water used by primary copper smelters is normally recycled after cooling in towers or ponds. Furnace slag is disposed of by either dumping or granulation. Molten slag is granulated by using high-pressure water jets. The wastewater from this granulation is typically high in both suspended and dissolved solids and contains some toxic metals.

TABLE 3.12
Classical Pollutants in the Raw Wastewater of the Primary Columbium and Tantalum Subcategory

Pollutant	Number of Samples	Number of Detections	Concentration (mg/L)		
			Range	Median	Mean
COD	3	3	140–6600	400	2400
TOC	3	3	45–1000	120	390
TSS	3	3	570–8600	3900	4400
Total phenol	3	3	0.016–0.10	0.02	0.04
Oil and grease	3	3	5.3–16	7.3	9.5
Ammonia	3	3	31–2400	380	940
Fluoride	3	3	2200–6400	3500	4000
Chloride	1	1	120	—	—

Source: From U.S. EPA, Treatability Manual, Volume II Industrial Descriptions, report EPA-600/2-82-001b, U.S. EPA, Washington, DC, September 1981.

TABLE 3.13
Concentrations of Toxic Pollutants Found in Primary Columbium and Tantalum Wastewater

Toxic Pollutant	Number of Samples	Number of Detections >10 µg/L	Concentration (µg/L)		
			Range	Median	Mean
Metals and Inorganics					
Antimony	3	2	ND–11,000	10	3700
Arsenic	3	3	180–14,000	380	4900
Asbestos (fibers/L)	1	1	8.9×10^7	—	—
Beryllium	3	3	20–190	89	100
Cadmium	3	2	8.0–20,000	48	6700
Chromium	3	3	3000–510,000	3000	170,000
Copper	3	3	400–270,000	500	90,000
Cyanide	3	4	2–12	4	6
Lead	3	3	$3000–2.8 \times 10^7$	3000	8.7×10^6
Mercury	3	1	<0.1–36	6.0	<14
Nickel	3	3	600–2700	2000	1800
Selenium	3	1	ND–24,000	<10	8000
Silver	3	3	<20–620	60	230
Thallium	3	2	ND–<100	25	<42
Zinc	3	3	<540–710,000	6000	240,000
Phthalates					
Bis(2-ethylhexyl)phthalate	15	12	ND–1100	22	150
Butyl benzyl phthalate	15	2	ND–47	—[a]	6.3
Di-n-butyl phthalate	15	5	ND–60	—[a]	12
Diethyl phthalate	15	1	ND–17	—[a]	1.7

Continued

Treatment of Nonferrous Metal Manufacturing Wastes

TABLE 3.13 (continued)

Toxic Pollutant	Number of Samples	Number of Detections >10 µg/L	Concentration (µg/L)		
			Range	Median	Mean
Dimethyl phthalate	15	2	ND–39	—a	4.1
Di-n-octyl phthalate	15	1	ND–95	—a	6.6
Phenols					
Pentachlorophenol	8	1	ND–17	—a	2.1
Aromatics					
Benzene	22	2	ND–44	—a	4.4
2,4-Dinitrotoluene	15	1	ND–16	—a	1.7
2,6-Dinitrotoluene	15	1	ND–16	—a	—b
Nitrobenzene	15	2	ND–160	—a	18
1,2,4-Trichlorobenzene	15	2	ND–260	—a	22
Polycyclic Aromatic Hydrocarbons					
Acenaphthene	15	1	ND–17	—a	1.1
Acenaphthylene	15	0	ND–2.0	—a	0.2
Anthracene	15	0	ND–2.0	—a	0.3
Benz (a) anthracene	15	0	ND–1.0	—a	0.1
Benzo (a) pyrene	15	0	ND–1.0	—a	0.1
Benzo (ghi) perylene	15	0	ND–2.0	—a	0.2
2-Chloronaphthalene	15	0	ND–3.0	—a	0.3
Chrysene	15	1	ND–45	—a	3.1
Dibenz (ah) anthracene	15	0	ND–4.0	—a	0.3
Fluoranthene	15	0	ND–7.2	—a	1.1
Fluorene	15	2	ND–20	—a	1.3
Indeno (1,2,3-cd) pyrene	15	0	ND–4.0	—a	0.3
Naphthalene	15	1	ND–84	—a	6.1
Phenanthrene	15	0	ND–2.0	—a	0.3
Pyrene	15	0	ND–3.0	—a	0.5
Polychlorinated Biphenyls					
Aroclor 1248	15	1	ND–32	—a	2.6
Aroclor 1254	15	1	ND–52	—a	4.1
Halogenated Aliphatics					
Bromoform	22	1	ND–21	—a	1.2
Carbon tetrachloride	22	2	ND–74	—a	5.1
Chlorodibromomethane	22	3	ND–81	—a	5.2
Chloroform	22	7	ND–140	—a	7.8
Dichlorobromomethane	22	1	ND–13	—a	0.6
1,2-Dicloroethane	22	6	ND–150	—a	13
1,1-Dichloroethylene	22	1	ND–22	—a	1.4
1,2-*trans*-Dichloroethylene	22	6	ND–480	—a	49
Hexachloroethane	15	1	ND–23	—a	1.5
Methylene chloride	22	1	ND–88,000	—a	4000
1,1,2,2-Tetrachloroethane	22	0	ND–6.0	—a	0.5
Tetrachloroethylene	22	1	ND–65	—a	3.6
1,1,1-Trichloroethane	22	2	ND–40	—a	2.5

Continued

TABLE 3.13 (continued)

Toxic Pollutant	Number of Samples	Number of Detections >10 µg/L	Concentration (µg/L)		
			Range	Median	Mean
1,1,2-Trichloroethane	22	2	ND–29	—a	2.1
Trichloroethylene	22	3	ND–230	—a	21
Pesticides and Metabolites					
Aldrin	15	0	ND–4.0	—a	0.3
Alpha-BHC	15	0	ND–0.04	—a	—b
Beta-BHC	15	0	ND–4.5	—a	0.4
Delta-BHC	15	0	ND–4.0	—a	0.3
Gamma-BHC	15	0	ND–0.03	—a	—b
Chlordane	15	0	ND–0.8	—a	0.1
4,4'-DDE	15	0	ND–0.4	—a	—b
4,4'-DDT	15	0	ND–1.0	—a	0.1
Dieldrin	15	0	ND–0.1	—a	—b
Alpha-endosulfan	15	0	ND–0.01	—a	—b
Endosulfan sulfate	15	0	ND–0.03	—a	—b
Endrin	15	0	ND–5.4	—a	0.4
Endrin aldehyde	15	0	ND–0.2	—a	—b
Heptachlor	15	0	ND–0.5	—a	—b
Heptachlor epoxide	15	0	ND–0.1	—a	—b
Isophorone	15	1	ND–29	—a	2.1
Toxaphene	15	0	ND–0.1	—a	—b

ND, not detectable.
[a] No median concentration is available in the reference.
[b] No mean concentration is available in the reference.
Source: From U.S. EPA, Treatability Manual, Volume II Industrial Descriptions, report EPA-600/2-82-001b, U.S. EPA, Washington, DC, September 1981.

Refining operations have two principal wastestreams, waste electrolyte and cathode and anode washwater. Spent electrolyte is normally recycled. A bleed stream is treated to reduce copper and impurity concentration. Varying degrees of treatment are necessary because of the differences in the anode copper. Anode impurities, including nickel, arsenic, and traces of antimony and bismuth, may be present in the effluent if the spent electrolyte bleed stream is discharged. Tables 3.14 and 3.15 present classical and toxic pollutant data for raw wastewater in this subcategory.

3.3.5 SECONDARY COPPER

Wastewater is generated by several processes in this subcategory. Slag milling and classification generates wastewater that is high in suspended solids, copper, lead, and zinc. Air pollution control at the site generates acidic wastewater that contains significant levels of copper. Other wastewater sources may include contact cooling, electrolyte disposal, and slag granulation. Tables 3.16 and 3.17 present classical and toxic pollutant data for the secondary copper recovery subcategory.

TABLE 3.14
Classical Pollutants in Raw Wastewater from the Primary Copper Subcategory

Pollutant	Number of Samples	Number of Detections	Concentration (mg/L)		
			Range	Median	Mean
COD	3	3	<2.0–810	25	280
TOC	3	3	3.5–7.0	4.9	5.1
TSS	3	3	5.4–4500	18	1500
Total phenol	2	2	0.0055–0.033	—	0.019
Oil and grease	1	1	6.1	—	—

Source: From U.S. EPA, Treatability Manual, Volume II Industrial Descriptions, report EPA-600/2-82-001b, U.S. EPA, Washington, DC, September 1981.

TABLE 3.15
Concentrations of Toxic Pollutants Found in Primary Copper Wastewater

Toxic Pollutant	Number of Samples	Number of Detections >10 µg/L	Concentration (µg/L)		
			Range	Median	Mean
Metals and Inorganics					
Antimony	3	3	<50–3500	100	1200
Arsenic	3	3	<2.0–340,000	9300	120,000
Beryllium	3	0	<2–<7.7	6.0	<5.2
Cadmium	3	1	<5–9500	7.0	3200
Chromium	3	2	<10–73	51	45
Copper	3	3	1600–450,000	2300	150,000
Cyanide	2	1	1–20	—[a]	11
Lead	3	3	<20–170,000	470	57,000
Mercury	3	1	<0.5–49	4.6	18
Nickel	3	3	<20–1000	340	450
Selenium	3	2	7.5–310	15	110
Silver	3	3	20–510	54	190
Thallium	3	3	27–<100	<100	<76
Zinc	3	3	30–140,000	400	47,000
Phthalates					
Bis(2-ethylhexyl)phthalate	11	5	ND–78	—[a]	17
Butyl benzyl phthalate	11	0	ND–1.0	—[a]	0.1
Di-n-butyl phthalate	11	1	ND–75	0.7	7.6
Di-n-octyl phthalate	11	0	ND–3.0	—[a]	0.3
Phenols					
2,4-Dimethylphenol	2	1	ND–14	—[a]	7.0
Aromatics					
Benzene	11	0	ND–3.0	—[a]	0.7
Toluene	11	0	ND–1.0	—[a]	0.2

Continued

TABLE 3.15 (continued)

Toxic Pollutant	Number of Samples	Number of Detections >10 µg/L	Concentration (µg/L)		
			Range	Median	Mean
Polycyclic Aromatic Hydrocarbons					
Acenaphthylene	11	0	ND–3.0	—a	0.3
Anthracene	11	4	ND–21	—a	6.1
Benz (a) anthracene	11	0	ND–1.0	—a	0.1
Fluoranthene	11	0	ND–1.0	—a	0.3
Fluorene	11	0	ND–1.0	—a	0.1
Phenanthrene	11	4	ND–21	7.0	7.1
Pyrene	11	0	ND–1.0	—a	0.4
Polychlorinated Biphenyls					
Aroclor 1248	9	0	ND–0.6	—a	0.1
Aroclor 1254	11	0	ND–0.7	—a	0.1
Halogenated Aliphatics					
Carbon tetrachloride	11	3	ND–40	—a	8.4
Chlorodibromomethane	11	1	ND–13	—a	1.2
Chloroform	11	3	ND–93	5.0	16
Dichlorobromomethane	11	1	ND–14	—a	1.3
1,2-Dichloroethane	11	0	ND–7.0	—a	0.6
Methylene chloride	11	0	ND–6.8	—a	0.6
1,1,2,2-Tetrachloroethane	11	3	ND–12	—a	1.9
Tetrachloroethylene	11	4	ND–15	4.0	5.4
1,1,2-Trichloroethane	11	0	ND–2.0	—a	0.2
Trichloroethylene	11	0	ND–9.0	—a	1.5
Pesticides and Metabolites					
Beta-BHC	11	0	ND–0.01	—a	—b
Gamma-BHC	11	0	ND–0.04	—a	—b
Chlordane	11	0	ND–0.2	—a	—b
4,4'-DDD	11	0	ND–0.01	—a	—b
4,4'-DDT	11	0	ND–0.02	—a	—b
Dieldrin	11	0	ND–0.02	—a	—b
Beta-endosulfan	11	0	ND–0.01	—a	—b
Endrin	11	0	ND–0.1	—a	—b
Endrin aldehyde	11	0	ND–0.4	—a	—b
Heptachlor	11	0	ND–0.01	—a	—b
Heptachlor epoxide	11	0	ND–0.01	—a	—b
Isophorone	11	0	ND–3.0	—a	—b

ND, not detectable.

a No median concentration is available in the reference.

b No mean concentration is available in the reference.

Source: From U.S. EPA, Treatability Manual, Volume II Industrial Descriptions, report EPA-600/2-82-001b, U.S. EPA, Washington, DC, September 1981.

TABLE 3.16
Concentrations of Classical Pollutants in the Raw Wastewater of the Secondary Copper Subcategory

Pollutant	Number of Samples	Number of Detections	Concentration (mg/L)		
			Range	Median	Mean
COD	5	5	9.7–900	75	230
TOC	5	5	6.0–99	30	40
TSS	5	5	4.0–11,000	65	2700
Total phenol	4	4	0.0063–0.22	0.045	0.079
Oil and grease	4	4	1.7–30	4.2	10
Fluoride	1	1	0.29	—	—

Source: From U.S. EPA, Treatability Manual, Volume II Industrial Descriptions, report EPA-600/2-82-001b, U.S. EPA, Washington, DC, September 1981.

TABLE 3.17
Concentrations of Toxic Pollutants in the Raw Wastewater of the Secondary Copper Subcategory

Toxic Pollutant	Number of Samples	Number of Detections >10 µg/L	Concentration (µg/L)		
			Range	Median	Mean
Metals and Inorganics					
Antimony	5	2	ND–11,000	ND	2200
Arsenic	5	3	ND–4200	100	940
Asbestos (fibers/L)	2	2	3.3×10^7–1.0×10^{11}	—[a]	5.5×10^{10}
Beryllium	5	4	ND–160	30	58
Cadmium	5	4	5.0–1200	50	390
Chromium	5	4	5.0–2100	<240	<640
Copper	5	5	620–2.1×10^6	40,000	450,000
Cyanide	4	1	<1–26	6	<9.8
Lead	5	5	450–53,000	10,000	17,000
Mercury	5	0	ND–0.6	0.53	0.35
Nickel	5	4	7.0–3.1×10^6	3000	620,000
Selenium	5	2	ND–270	ND	98
Silver	5	3	ND–1600	<10	370
Thallium	5	2	ND–53	ND	21
Zinc	5	5	1400–1.5×10^6	40,000	330,000
Phthalates					
Bis(2-ethylhexyl) phthalate	12	9	ND–7000	53	1100
Butyl benzyl phthalate	12	1	ND–56	—[a]	5.3
Di-n-butyl phthalate	12	5	ND–390	9.5	56
Diethyl phthalate	12	2	ND–83	—[a]	11
Di-n-octyl phthalate	12	1	ND–67	5.8	—[b]

Continued

TABLE 3.17 (continued)

Toxic Pollutant	Number of Samples	Number of Detections >10 µg/L	Concentration (µg/L)		
			Range	Median	Mean
Aromatics					
Benzene	10	1	ND–13	—[a]	1.3
Ethylbenzene	10	0	ND–4.0	—[a]	0.4
Hexachlorobenzene	12	1	ND–5000	—[a]	420
Toluene	10	0	ND–10	—[a]	1.7
Polycyclic Aromatic Hydrocarbons					
Acenaphthene	12	2	ND–36	—[a]	4.6
Acenaphthylene	12	3	ND–120	—[a]	23
Anthracene	12	2	ND–3000	—[a]	260
Benzo (a) pyrene	12	0	ND–1.0	—[a]	0.1
Chrysene	12	2	ND–10,000	—[a]	840
Fluoranthene	12	3	ND–3000	1.0	280
Fluorene	12	3	ND–94	—[a]	14
Naphthalene	12	3	ND–5000	—[a]	550
Phenanthrene	12	3	ND–3000	—[a]	260
Pyrene	12	3	ND–7000	—[a]	610
Polychlorinated Biphenyls					
Aroclor 1248	14	0	ND–2.0	—[a]	0.5
Aroclor 1254	14	0	ND–3.0	—[a]	0.5
Halogenated Aliphatics					
Carbon tetrachloride	10	1	ND–120	—[a]	12
Chloroform	10	5	ND–1000	7.0	130
1,2-Dichloroethane	10	1	ND–32	—[a]	3.2
1,1-Dichloroethylene	10	2	ND–530	—[a]	57
1,2-*trans*-Dichloroethylene	10	0	ND–5.0	—[a]	0.5
Methylene chloride	10	2	ND–510	—[a]	80
1,1,2,2-Tetrachloroethane	10	0	ND–4.0	—[a]	0.4
Tetrachloroethylene	10	2	ND–72	—[a]	8.8
Trichloroethylene	10	1	ND–70	—[a]	7.1
Pesticides and Metabolites					
Aldrin	14	0	ND–0.2	—[a]	—[b]
Alpha-BHC	14	0	ND–0.2	—[a]	—[b]
Beta-BHC	14	0	ND–0.02	—[a]	—[b]
Delta-BHC	14	0	ND–0.2	—[a]	—[b]
Gamma-BHC	14	0	ND–0.04	—[a]	—[b]
Chlordane	14	0	ND–0.7	—[a]	0.1
4,4'-DDE	14	0	ND–0.02	—[a]	—[b]
4,4'-DDD	14	0	ND–0.1	—[a]	—[b]
4,4'-DDT	14	0	ND–0.03	—[a]	—[b]
Dieldrin	14	0	ND–0.03	—[a]	—[b]
Alpha-endosulfan	14	0	ND–0.3	—[a]	—[b]
Beta-endosulfan	14	0	ND–0.3	—[a]	—[b]
Endrin	14	0	ND–0.4	—[a]	—[b]

Continued

TABLE 3.17 (continued)

Toxic Pollutant	Number of Samples	Number of Detections >10 µg/L	Concentration (µg/L)		
			Range	Median	Mean
Endrin aldehyde	14	0	ND–0.3	—[a]	—[b]
Heptachlor	14	0	ND–0.02	—[a]	—[b]
Toxaphene	14	0	ND–0.4	—[a]	—[b]

ND, not detectable.
[a] No median concentration is available in the reference.
[b] No mean concentration is available in the reference.
Source: From U.S. EPA, Treatability Manual, Volume II Industrial Descriptions, report EPA-600/2-82-001b, U.S. EPA, Washington, DC, September 1981.

3.3.6 PRIMARY LEAD

Primary lead facilities have two major processes associated with wastewater generation. The smelting process generates a major wastestream from the sintering operation. These wastewaters are typically high in dissolved solids and metals such as cadmium, lead, and zinc. Acid plant blowdown, slag granulation, and air pollution control methods are also associated with smelting operations. Refining operations also generate wastewater from air pollution equipment and from noncontact cooling water. Tables 3.18 and 3.19 present classical and toxic pollutant data of the raw wastewater generated in this subcategory.

3.3.7 SECONDARY LEAD

The principal raw material for the secondary lead industry is scrap batteries. Wastewater is generated from battery acid streams, washdown streams, and saw cooling for cracking the batteries. These streams contain significant levels of suspended solids, antimony, arsenic, cadmium, lead, and zinc.

TABLE 3.18
Concentrations of Classical Pollutants in the Raw Wastewater of the Primary Lead Subcategory

Pollutant	Number of Samples	Number of Detections	Concentration (mg/L)		
			Range	Median	Mean
COD	2	2	3.7–170		87
TOC	2	1	ND–3.3		1.6
TSS	2	1	ND–26		13
Total phenol	2	1	ND–0.050		0.025
Ammonia	2	2	ND–3.8		1.9
Oil and grease	2	0	ND		—

ND, not detectable.
Source: From U.S. EPA, Treatability Manual, Volume II Industrial Descriptions, report EPA-600/2-82-001b, U.S. EPA, Washington, DC, September 1981.

TABLE 3.19
Concentrations of Toxic Pollutants in the Raw Wastewater of the Primary Lead Subcategory

Toxic Pollutant	Number of Samples	Number of Detections 10 µg/L	Concentration (µg/L)		
			Range	Median	Mean
Metals and Inorganics					
Antimony	2	1	ND–<330	—	<170
Arsenic	2	2	58–96	—	76
Beryllium	2	0	ND–6.7	—	3.4
Cadmium	2	2	700–1300	—	1000
Chromium	2	2	14–30	—	22
Copper	2	2	100–620	—	360
Cyanide	2	0	<0.02–0.12	—	0.07
Lead	2	2	7900–24,000	—	16,000
Mercury	2	0	0.67–7.5	—	4.1
Nickel	2	2	50–130	—	90
Selenium	2	1	5.4–<13	—	<9.2
Silver	2	1	ND–<20	—	<10
Thallium	2	1	ND–<100	—	<50
Zinc	2	2	5300–20,000	—	13,000
Polycyclic Aromatic Hydrocarbons					
Pyrene	3	0	ND–7.0	—	2.3
Halogenated Aliphatics					
Methylene chloride	4	1	ND–25	3.0	7.8

ND, not detectable.

Source: From U.S. EPA, Treatability Manual, Volume II Industrial Descriptions, report EPA-600/2-82-001b, U.S. EPA, Washington, DC, September 1981.

Smelting operations for this subcategory generate wastewater from air pollution control devices and contact cooling streams. Tables 3.20 and 3.21 present classical and toxic pollutant data for the raw wastewater in this subcategory.

3.3.8 SECONDARY SILVER

Secondary silver is recovered from photographic and nonphotographic sources. Wastewater sources from photographic wastes include leaching and stripping, precipitation and filtration of silver, electrolysis, and pollution control. Nonphotographic scrap wastewater is generated by similar processes. These wastewater streams contain significant concentrations of chromium, copper, lead, and zinc as well as some organic priority pollutants. Tables 3.22 and 3.23 present pollutant data for this subcategory.

3.3.9 PRIMARY TUNGSTEN

Tungsten production involves processing ore concentrates to obtain the salt ammonium paratungstate (APT), and subsequent reduction of APT to metallic tungsten. Wastewater is generated during all three processes and results from the precipitation and filtration of the salt, leaching to

TABLE 3.20
Concentrations of Classical Pollutants in the Raw Wastewater of the Secondary Lead Subcategory

Pollutant	Number of Samples	Number of Detections	Concentration (mg/L)		
			Range	Median	Mean
COD	3	3	65–230	160	150
TOC	3	3	4–140	70	71
TSS	4	4	0.056–4000	770	1400
Total phenol	4	4	<0.004–0.012	0.0091	0.0086
Oil and grease	3	3	6.5–40	36	28
Ammonia	1	1	12	—	—
Chloride	1	1	53	—	—

Source: From U.S. EPA, Treatability Manual, Volume II Industrial Descriptions, report EPA-600/2-82-001b, U.S. EPA, Washington, DC, September 1981.

TABLE 3.21
Concentrations of Toxic Pollutants in the Raw Wastewater of the Secondary Lead Subcategory

Toxic Pollutant	Number of Samples	Number of Detections >10 µg/L	Concentration (µg/L)		
			Range	Median	Mean
Metals and Inorganics					
Antimony	4	4	1600–80,000	39,000	40,000
Arsenic	3	3	3000–13,000	7100	7700
Asbestos (fibers/L)	1	1	1.3×10^{11}	—	—
Beryllium	3	1	1.0–30	4.5	12
Cadmium	4	4	240–2000	800	960
Chromium	4	4	110–1000	490	520
Copper	4	4	230–8000	3200	3600
Lead	4	4	$7000–2.0 \times 10^6$	24,000	510,000
Mercury	4	1	0.6–12	0.84	3.6
Nickel	4	4	210–2000	<970	1000
Selenium	3	0	ND–<10	<2.0	<4.0
Silver	3	3	110–250	120	160
Thallium	3	3	50–750	370	390
Zinc	4	4	870–15,000	3600	5700
Phthalates					
Bis(2-ethylhexyl) phthalate	5	4	ND–580	30	180
Butyl benzyl phthalate	5	1	ND–85	—[a]	17
Di-n-butyl phthalate	5	3	ND–27	13	12
Dimethyl phthalate	5	2	ND–13	—[a]	2.6
Di-n-octyl phthalate	5	2	ND–27	2.0	9.0

Continued

TABLE 3.21 (continued)

Toxic Pollutant	Number of Samples	Number of Detections >10 µg/L	Concentration (µg/L)		
			Range	Median	Mean
Nitrogen Compounds					
Benzidine	5	0	ND–6.0	—[a]	1.2
Aromatics					
Benzene	10	0	ND–2.0	—[a]	0.2
Chlorobenzene	10	0	ND–5.0	—[a]	0.5
Ethylbenzene	10	0	ND–1.2	—[a]	0.3
Nitrobonzene	5	1	ND–16	—[a]	3.2
Polycyclic Aromatic Hydrocarbons					
Acenaphthylene	5	1	ND–35	3.0	8.6
Anthracene	5	1	ND–20	—[a]	4.0
Benzo (a) pyrene	5	1	ND–10	—[a]	2.0
Benzo (b) fluoranthene	5	0	ND–5.3	—[a]	1.6
Benzo (k) fluoranthene	5	0	ND–5.3	—[a]	1.6
Chrysene	5	1	ND–546	40	110
Fluoranthene	5	2	ND–27	—[a]	7.6
Fluorene	5	0	ND–2.0	1.0	0.4
Indeno(1,2,3-cd)pyrene	5	0	ND–1.0	—[a]	0.2
Naphthalene	5	0	ND–4.0	—[a]	0.8
Phenanthrene	5	1	ND–20	—[a]	4.6
Pyrene	5	2	ND–38	1.0	10
Polychlorinated Biphenyls					
Aroclor 1248	5	0	ND–3.1	1.3	1.4
Aroclor 1254	5	0	ND–2.6	1.8	1.3
Halogenated Aliphatics					
Bromoform	10	1	ND–49	—[a]	5.7
Chloroform	10	3	ND–31	3.0	6.9
1,2-Dicloroethane	10	1	ND–10	4.0	4.0
1,1-Dichloroethylene	10	1	ND–10	2.0	3.7
1,1,2,2-Tetrachloroethane	10	0	ND–4.0	—[a]	1.0
Tetrachloroethylene	10	0	ND–5.0	—[a]	1.1
Trichloroethylene	10	0	ND–6.0	—[a]	0.8
Pesticides and Metabolites					
Aldrin	5	0	ND–0.1	—[a]	—[b]
Alpha-BHC	5	0	ND–0.2	—[a]	—[b]
Beta-BHC	5	0	ND–0.3	0.1	0.1
Gamma-BHC	5	0	ND–0.1	—[a]	—[b]
Chlordane	5	0	ND–0.2	0.2	0.2
4,4'-DDE	5	0	ND–0.2	—[a]	—[b]
4,4'-DDT	5	0	ND–0.1	—[a]	—[b]
Dieldrin	5	0	ND–0.2	—[a]	—[b]
Alpha-endosulfan	5	0	ND–0.2	—[a]	—[b]
Endrin	5	0	ND–4.0	—[a]	1.0
Endrin aldehyde	5	0	ND–0.6	—[a]	0.1

Continued

TABLE 3.21 (continued)

Toxic Pollutant	Number of Samples	Number of Detections >10 µg/L	Concentration (µg/L)		
			Range	Median	Mean
Heptachlor	5	0	ND–0.3	0.1	0.1
Heptachlor epoxide	5	0	ND–0.2	0.1	0.1
Isophorone	5	0	ND–2.7	—[a]	1.8

ND, not detectable.

[a] No median concentration is available in the reference.

[b] No mean concentration is available in the reference.

Source: From U.S. EPA, Treatability Manual, Volume II Industrial Descriptions, report EPA-600/2-82-001b, U.S. EPA, Washington, DC, September 1981.

TABLE 3.22
Concentrations of Classical Pollutants in the Raw Wastewater of the Secondary Silver Subcategory

Pollutant	Number of Samples	Number of Detections	Concentration (mg/L)		
			Range	Median	Mean
COD	3	3	230–12,000	3000	5100
TOC	3	3	19–9100	430	3200
TSS	3	3	110–1100	110	440
Total phenol	3	3	0.02–28	0.04	9.4
Oil and grease	3	3	8.0–100	17	42
Ammonia	2	2	12–1500	—	760
Fluoride	1	1	1.2	—	—
Chloride	1	1	32,000	—	—

Source: From U.S. EPA, Treatability Manual, Volume II Industrial Descriptions, report EPA-600/2-82-001b, U.S. EPA, Washington, DC, September 1981.

TABLE 3.23
Concentrations of Toxic Pollutants in the Raw Wastewater of the Secondary Silver Subcategory

Toxic Pollutant	Number of Samples	Number of Detections >10 µg/L	Concentration (µg/L)		
			Range	Median	Mean
Metals and Inorganics					
Antimony	3	1	ND–25,000	ND	8300
Arsenic	3	3	40–920	40	330
Asbestos (fibers/L)	1	1	2×10^9	—	—
Beryllium	3	2	ND–20	19	13
Cadmium	3	3	1000–80,000	3200	28,000
Chromium	3	3	2000–27,000	20,000	16,000

Continued

TABLE 3.23 (continued)

Toxic Pollutant	Number of Samples	Number of Detections >10 µg/L	Concentration (µg/L)		
			Range	Median	Mean
Copper	3	3	7300–70,000	60,000	46,000
Cyanide	3	2	1–2100	50	720
Lead	3	3	4000–50,000	4200	19,000
Mercury	3	0	ND–5.5	ND	1.8
Nickel	3	3	1100–800,000	30,000	280,000
Selenium	3	1	ND–590	ND	200
Silver	3	3	250–4700	410	1800
Thallium	3	1	ND–510	ND	170
Zinc	3	3	8400–2.0×10^6	20,000	680,000
Phthalates					
Bis(2-ethylhexyl) phthalate	5	4	7.0–34	11	18
Butyl benzyl phthalate	5	1	ND–53	—a	11
Di-n-butyl phthalate	5	4	ND–300	15	75
Diethyl phthalate	5	1	ND–38	—a	7.6
Di-n-octyl phthalate	5	3	ND–58	33	30
Aromatics					
Benzene	6	4	3.0–160	66	75
Chlorobenzene	6	0	ND–9.0	0.5	2.8
Ethylbenzene	6	3	ND–21	—a	9.2
Toluene	6	4	3.0–55	18	21
Polycyclic Aromatic Hydrocarbons					
Acenaphthene	5	1	ND–10	—a	2.0
Anthracene	5	0	ND–4.0	—a	0.8
Naphthalene	5	0	ND–1.0	—a	0.2
Phenanthrene	5	0	ND–4.0	—a	0.8
Pyrene	5	1	ND–2100	—a	430
Polychlorinated Biphenyls					
Aroclor 1248	3	0	ND–0.5	—a	0.2
Aroclor 1254	3	0	ND–0.7	—a	0.2
Halogenated Aliphatics					
Bromoform	6	1	ND–65	—a	11
Carbon tetrachloride	6	1	ND–2300	—a	380
Chlorodibromomethane	6	1	ND–64	—a	11
Chloroform	6	3	ND–890	8.5	160
1,2-Dichloroethane	6	3	ND–560	21	120
1,1-Dichloroethylene	6	2	ND–6100	—a	1100
Methylene chloride	6	3	ND–3100	170	1000
1,1,2,2-Tetrachloroethane	4	1	ND–32	—a	8.0
Tetrachloroethylene	6	5	ND–110	36	43
1,1,1-Trichloroethane	6	2	ND–22	—a	7.3
Trichloroethylene	6	5	ND–900	230	360
Pesticides and Metabolites					
Aldrin	3	0	ND–1.1	—a	0.4
Beta-BHC	3	0	ND–0.02	—a	—b

Continued

TABLE 3.23 (continued)

Toxic Pollutant	Number of Samples	Number of Detections >10 µg/L	Concentration (µg/L)		
			Range	Median	Mean
Delta-BHC	3	0	ND–1.1	—[a]	0.4
Chlordane	3	0	ND–0.1	—[a]	—[b]
4,4'-DDE	3	0	ND–0.01	—[a]	—[b]
4,4'-DDD	3	0	ND–0.1	—[a]	—[b]
4,4'-DDT	3	0	ND–0.01	—[a]	—[b]
Dieldrin	3	0	ND–0.01	—[a]	—[b]
Endrin	3	0	ND–2.0	—[a]	0.7
Heptachlor	3	0	ND–0.02	—[a]	—[b]

ND, not detectable.
[a] No median concentration is available in the reference.
[b] No mean concentration is available in the reference.
Source: From U.S. EPA, Treatability Manual, Volume II Industrial Descriptions, report EPA-600/2-82-001b, U.S. EPA, Washington, DC, September 1981.

convert it to tungstic acid, and air pollution control methods associated with the processes. Wastewaters may be acidic and contain significant concentrations of chlorides, arsenic, lead, zinc, and ammonia. Tables 3.24 and 3.25 present classical and toxic pollutant data for the primary tungsten subcategory.

3.3.10 Primary Zinc and Cadmium

Wastewater is generated in the primary zinc and primary cadmium recovery subcategories by acid plant blowdown, which results from sulfuric acid recovery, air pollution control, leaching, anode/cathode washing, and contact cooling. The streams may contain significant concentrations of lead, arsenic, cadmium, and zinc. Tables 3.26 and 3.27 present classical and toxic pollutant data for the primary zinc and primary cadmium subcategories.

TABLE 3.24
Concentrations of Classical Pollutants in the Raw Wastewater of the Primary Tungsten Subcategory

Pollutant	Number of Samples	Number of Detections	Concentration (mg/L)		
			Range	Median	Mean
COD	3	3	120–880	320	440
TOC	3	3	6.0–270	27	100
TSS	3	3	42–6700	210	2300
Total phenol	3	3	0.038–0.089	0.039	0.055
Oil and grease	3	3	6.3–17	6.8	10
Ammonia	3	3	3.9–1600	900	830
Chloride	2	2	850–26,000	—	13,000

Source: From U.S. EPA, Treatability Manual, Volume II Industrial Descriptions, report EPA-600/2-82-001b, U.S. EPA, Washington, DC, September 1981.

TABLE 3.25
Concentrations of Toxic Pollutants in the Raw Wastewater of the Primary Tungsten Subcategory

Toxic Pollutant	Number of Samples	Number of Detections >10 µg/L	Concentration (µg/L)		
			Range	Median	Mean
Metals and Inorganics					
Antimony	3	1	ND–800	ND	270
Arsenic	3	3	20–7200	210	2500
Asbestos (fibers/L)	1	1	6.0×10^9	—	—
Beryllium	3	1	2.0–29	9	13
Cadmium	3	3	19–190	20	76
Chromium	3	3	44–2000	48	700
Copper	3	3	95–3000	120	1700
Cyanide	3	2	2–140	13	52
Lead	3	3	180–20,000	240	6800
Mercury	3	0	0.21–3	1.0	1.4
Nickel	3	3	45–1000	92	380
Selenium	3	2	ND–1000	20	340
Silver	3	3	76–270	86	140
Thallium	3	2	ND–600	200	270
Zinc	3	3	250–1900	520	890
Phthalates					
Bis(2-ethylhexyl) phthalate	5	4	ND–880	10	180
Di-n-butyl phthalate	5	2	ND–23	—a	—b
Di-n-octyl phthalate	5	1	ND–1.0	—a	0.2
Aromatics					
Benzene	9	0	ND–3.0	—a	0.7
Ethylbenzene	9	1	ND–11	—a	2.2
Toluene	9	2	ND–45	3.0	11
Polycyclic Aromatic Hydrocarbons					
Acenaphthene	5	1	ND–100	—a	21
Acenaphthylene	5	1	ND–110	—a	23
Anthracene	5	1	ND–150	—a	30
Benzo (a) pyrene	5	0	ND–1.0	—a	0.2
Chrysene	5	1	ND–240	—a	48
Fluoranthene	5	0	ND–1.0	—a	0.2
Fluorene	5	1	ND–55	—a	11
Naphthalene	5	1	ND–1100	—a	220
Polychlorinated Biphenyls					
Aroclor 1248	5	0	ND–1.0	0.2	0.3
Aroclor 1254	5	0	ND–5.4	0.4	1.4
Halogenated Aliphatics					
Bromoform	9	2	ND–48	—a	9.3
Chlorodibromomethane	9	1	ND–38	—a	4.2
Chloroform	9	2	ND–1800	—a	210
1,2-Dichloroethane	9	0	ND–8.0	—a	2.1

Continued

TABLE 3.25 (continued)

Toxic Pollutant	Number of Samples	Number of Detections >10 μg/L	Concentration (μg/L)		
			Range	Median	Mean
1,1-Dichloroethylene	9	2	ND–19	—[a]	4.3
1,2-*trans*-Dichloroethylene	9	0	ND–2.0	—[a]	0.2
1,1,2,2-Tetrachloroethane	9	1	ND–35	—[a]	5.2
Tetrachloroethylene	9	5	ND–69	—[a]	20
1,1,1-Trichloroethene	9	1	ND–10	—[a]	1.1
Trichloroethylene	9	2	ND–19	—[a]	2.9
Pesticides and Metabolites					
Aldrin	5	0	ND–7.0	—[a]	1.4
Alpha-BHC	5	0	ND–0.6	—[a]	0.1
Beta-BHC	5	0	ND–0.2	—[a]	0.1
Gamma-BHC	5	0	ND–0.2	—[a]	0.1
Chlordane	5	0	ND–1.2	—[a]	0.2
4,4′-DDT	5	0	ND–0.1	—[a]	—
Dieldrin	5	0	ND–0.1	—[a]	0.1
Alpha-Endosulfan	5	1	ND–15	0.1	3.2
Beta-Endosulfan	5	1	ND–15	—[a]	3.1
Endrin	5	0	ND–0.8	—[a]	0.2
Endrin aldehyde	5	0	ND–0.9	0.2	0.3
Heptachlor	5	0	ND–0.2	—[a]	0.1
Heptachlor epoxide	5	0	ND–0.2	—[a]	0.1

ND, not detectable.
[a] No median concentration is available in the reference.
[b] No mean concentration is available in the reference.
Source: From U.S. EPA, Treatability Manual, Volume II Industrial Descriptions, report EPA-600/2-82-001b, U.S. EPA, Washington, DC, September 1981.

TABLE 3.26
Concentrations of Classical Pollutants in the Raw Wastewater of the Primary Zinc and Cadmium Subcategories

Pollutant	Number of Samples	Number of Detections	Concentration (mg/L)		
			Range	Median	Mean
COD	2	2	20–59	—	40
TOC	2	2	7.3–9.3	—	8.3
TSS	2	2	13–15	—	14
Total phenol	4	4	0.002–0.025	0.007	0.010
Oil and grease	2	2	10–14	—	12

Source: From U.S. EPA, Treatability Manual, Volume II Industrial Descriptions, report EPA-600/2-82-001b, U.S. EPA, Washington, DC, September 1981.

TABLE 3.27
Concentrations of Toxic Pollutants in the Raw Wastewater of the Primary Zinc and Cadmium Subcategories

Toxic Pollutant	Number of Samples	Number of Detections >10 µg/L	Concentration (µg/L)		
			Range	Median	Mean
Metals and Inorganics					
Antimony	4	4	2.0–2100	59	550
Arsenic	4	4	3.0–3000	150	820
Asbestos (fibers/L)	2	2	3.2×10^7–4.3×10^7	—	3.8×10^7
Beryllium	4	1	2.0–20	7.5	9.3
Cadmium	4	4	350–44,000	3500	13,000
Chromium	4	4	24–610	65	190
Copper	4	4	37–26,000	1200	7100
Cyanide	4	2	2–380	6.7	99
Lead	4	4	280–18,000	4400	6700
Mercury	4	1	2.9–52	5.5	16
Nickel	4	4	50–4300	590	1400
Selenium	4	4	24–1200	360	490
Silver	4	4	25–740	58	220
Thallium	2	2	20–360	—	190
Zinc	4	4	8700–1.7×10^6	160,000	510,000
Phthalates					
Bis(2-ethylhexyl) phthalate	9	6	ND–98	15	28
Butyl benzyl phthalate	9	1	ND–30	—[a]	3.3
Di-n-butyl phthalate	9	1	ND–26	5.0	3.6
Diethyl phthalate	9	1	ND–18	—[a]	2.7
Dimethyl phthalate	9	1	ND–22	—[a]	2.4
Phenols					
Pentachlorophenol	9	0	ND–8.0	—[a]	0.9
Aromatics					
Benzene	9	1	ND–24	—[a]	2.7
Ethylbenzene	9	0	ND–2.0	—[a]	0.2
Hexachlorobenzene	9	1	ND–100	—[a]	11
Toluene	9	1	ND–54	7.0	7.5
Polycyclic Aromatic Hydrocarbons					
Acenaphthene	9	1	ND–18	—[a]	2.0
Anthracene	9	0	ND–0.4	—[a]	—[b]
Chrysene	9	1	ND–11	—[a]	2.2
Fluoranthene	9	1	ND–15	—[a]	1.7
Fluorene	9	2	ND–14	—[a]	1.6
Pyrene	9	2	ND–15	—[a]	3.2
Halogenated Aliphatics					
Chloroform	9	3	ND–71	53	16
1,1-Dichloroethane	9	1	ND–180	—[a]	20
1,2-Dichloroethane	9	1	ND–22	—[a]	2.9
1,1-Dichloroethylene	9	1	ND–23	—[a]	2.6

Continued

TABLE 3.27 (continued)

Toxic Pollutant	Number of Samples	Number of Detections >10 μg/L	Concentration (μg/L)		
			Range	Median	Mean
Methylene chloride	9	4	ND–2,600	—a	350
Tetrachloroethylene	9	0	ND–8.0	—a	0.9
Trichloroethylene	9	1	ND–160	7.2	19
Trichlorofluoromethane	9	1	ND–100	—a	12
Pesticides and Metabolites					
Isophorone	9	1	ND–18	—a	2.0

ND, not detectable.
a No median concentration is available in the reference.
b No mean concentration is available in the reference.
Source: From U.S. EPA, Treatability Manual, Volume II Industrial Descriptions, report EPA-600/2-82-001b, U.S. EPA, Washington, DC, September 1981.

3.4 POLLUTANT REMOVABILITY AND TREATMENT

There are several methods for pollutant removal currently used in this industry. Some are used industry-wide; others are used only in specific applications. Those used industry-wide include physical-chemical methods (precipitation,[22] coagulation and flocculation,[24] pH adjustment, and stripping and physical separation methods), filtration,[24] sedimentation,[25] and centrifugation.[26] Lime, caustic, soda ash, and calcium chloride are used as precipitants in the industry, especially for removal of soluble metals. In the coagulation-flocculation system, polymer, lime, and iron or aluminum salts are mixed into the wastestream to facilitate agglomeration of colloidal suspensions. Air/steam stripping are widely practiced techniques for the reduction of volatile compounds such as ammonia, hydrogen sulfide, and organics.

The physical separation methods find wide application in this industry because of the nature of the wastes. Centrifugation may be feasible in some applications, but it is not suitable for abrasive or very fine particles (less than 5 μm).

There are several potential treatment technologies that may be applicable, but are more expensive than the methods currently used. These potential treatments include: sulfide precipitation, ultrafiltration, reverse osmosis, deep-well disposal, activated carbon adsorption or activated alumina adsorption, solidification, or ion exchange.[19–21]

Pollutant removal data for toxic organic pollutants in the subcategories are presented in Tables 3.28 through 3.37. The average removal percentage was determined by comparing the average raw wastewater concentrations found in Section 3.3 with the average treated wastewater concentrations presented in these tables.

3.5 MANAGEMENT OF CHEMICALS IN THE WASTESTREAM

The Pollution Prevention Act of 1990[27] requires facilities to report information about the management of Toxic Relief Inventory (TRI) chemicals in waste and efforts made to eliminate or reduce those quantities. The data summarized in Table 3.38 cover a four-year period and is meant to provide a basic understanding of the quantities of waste handled by the industry, the methods typically used to manage this waste, and recent trends in these methods.[1] TRI waste management data can be used to assess trends in source reduction within individual industries and facilities, and for specific TRI chemicals. This information could then be used as a tool in identifying opportunities for pollution prevention compliance assistance activities.

TABLE 3.28
Removability of Toxic Organic Pollutants from Wastewater in the Primary Aluminum Subcategory

Toxic Pollutant	Number of Samples	Number of Detections >10 µg/L	Treated Effluent Concentration (µg/L)			Average Percent Removal
			Range	Median	Mean	
Phthalates						
Bis(2-ethylhexyl) phthalate	9	2	ND–120	—a	17	79
Butyl benzyl phthalate	9	1	ND–75	—a	9.6	56
Di-n-butyl phthalate	9	3	ND–30	—a	5	74
Diethyl phthalate	9	0	ND	—	—	NM
Dimethyl phthalate	9	0	ND–5.0	—a	1	NM
Di-n-octyl phthalate	9	1	ND–13	—a	1.8	NM
Phenols						
Phenol	4	0	ND	—	—	>99
Aromatics						
Benzene	14	2	ND–33	—a	4.0	NM
2,4-Dinitrotoluene	9	0	ND–7.0	—a	0.9	NM
2,6-Dinitrotoluene	9	0	ND–1.0	—a	0.1	NM
Ethylbenzene	14	1	ND–12	—a	0.8	NM
Toluene	14	0	ND–6.8	—a	0.5	NM
Polycyclic Aromatic Hydrocarbons						
Acenaphthene	9	4	ND–13	—a	5.0	40
Acenaphthylene	9	0	ND–7.0	—a	1.9	66
Anthracene	9	2	ND–11	2.6	4.7	88
Benz (a) anthracene	9	0	ND–6.0	—a	0.7	98
Benzo (a) pyrene	9	0	ND–8.0	—a	2.1	98
Benzo (b) fluoranthene	9	0	ND–6.0	—a	0.7	98
Benzo (ghi) perylene	9	1	ND–11	—a	0.1	>99
Benzo (k) fluoranthene	9	0	ND–6.0	—a	1.1	97
Chrysene	9	1	ND–140	—a	17	58
Dibenz (ah) anthracene	9	0	ND	—	—	>99
Fluoranthene	9	5	ND–79	11	22	77
Fluorene	9	0	ND–1.0	—a	0.2	97
Indeno (1,2,3-cd) pyrene	9	0	ND–1.0	—a	0.1	>99
Naphthalene	9	0	ND–1.0	—a	0.1	97
Phenanthrene	9	2	ND–11	—a	44	NM
Pyrene	9	3	ND–80	9.0	20	71
Halogenated Aliphatics						
Chloroform	14	2	ND–320	—a	23	NM
1,2-Dichloroethane	14	0	ND–5.5	—a	0.4	NM
1,1-Dichloroethylene	14	1	ND–4100	—a	290	NM
Methylene chloride	14	6	ND–4200	—a	360	NM
1,1,2,2-Tetrachloroethane	14	0	ND–1.0	—a	0.1	NM
Tetrachloroethylene	14	1	ND–61	—a	44	NM
Trichloroethylene	14	1	ND–120	—a	8.5	NM

Continued

TABLE 3.28 (continued)

Toxic Pollutant	Number of Samples	Number of Detections >10 µg/L	Treated Effluent Concentration (µg/L) Range	Median	Mean	Average Percent Removal
Pesticides and Metabolites						
Aldrin	8	0	ND–0.1	—[a]	—[b]	NM
Delta-BHC	8	0	ND–0.1	—[a]	—[b]	NM
Gamma-BHC	8	0	ND–0.01	—[a]	—[b]	NM
Chlordane	8	0	ND–0.1	—[a]	—[b]	NM
4,4'-DDT	8	0	ND–0.01	—[a]	—[b]	NM
Dieldrin	8	0	ND–0.1	—[a]	—[b]	NM
Endrin aldehyde	8	0	ND–0.2	—[a]	—[b]	NM
Heptachlor	8	0	ND–0.2	—[a]	—[b]	NM
Heptachlor epoxide	8	0	ND–0.2	—[a]	—[b]	NM
Isophorone	9	0	ND	—[a]	—[b]	>99
PCB 1248	8	0	ND–0.4	—[a]	—[b]	NM
PCB 1254	8	0	ND–0.2	—[a]	—[b]	NM

ND, not detectable; NM, not meaningful.

[a] No median concentration is available in the reference.

[b] No mean concentration is available in the reference.

Source: From U.S. EPA, Treatability Manual, Volume II Industrial Descriptions, report EPA-600/2-82-001b, U.S. EPA, Washington, DC, September 1981.

TABLE 3.29
Removability of Toxic Organic Pollutants from Wastewater in the Secondary Aluminum Subcategory

Toxic Pollutant	Number of Samples	Number of Detections >10 µg/L	Treated Effluent Concentration (µg/L) Range	Median	Mean	Average Percent Removal
Phthalates						
Bis(2-ethylhexyl) phthalate	7	6	ND–1200	5.3	290	24
Butyl benzyl phthalate	7	0	ND–2.0	—[a]	0.6	97
Di-n-butyl phthalate	7	3	ND–50	—[a]	13	19
Dimethyl phthalate	7	0	ND–3.0	—[a]	0.6	94
Di-n-octyl phthalate	7	1	ND–100	—[a]	15	NM
Nitrogen Compounds						
3,3'-Dichlorobenzidine	7	0	ND	—	—	>99
Aromatics						
Benzene	11	0	ND–5.0	—[a]	0.7	93
Chlorobenzene	11	0	ND–7.0	—[a]	1.5	NM
1,4-Dichlorobenzene	7	0	ND	—	—	>99
Ethylbenzene	11	0	ND–6.0	—[a]	0.5	NM
1,2,4-Trichlorobenzene	7	1	ND–2.0	—[a]	0.3	NM
Polycyclic Aromatic Hydrocarbons						
Acenaphthylene	7	0	ND	—	—	<99
Benzo (a) pyrene	7	0	ND–1.0	—[a]	0.1	95

Continued

TABLE 3.29 (continued)

Toxic Pollutant	Number of Samples	Number of Detections >10 µg/L	Treated Effluent Concentration (µg/L)			Average Percent Removal
			Range	Median	Mean	
Benzo (b) fluoranthene	7	0	ND–2.0	—a	0.3	NM
Benzo (k) fluoranthene	7	0	ND–2.0	—a	0.3	NM
Benzo (ghi) perylene	7	0	ND–2	—a	0.3	NM
Chrysene	7	0	ND–2.5	—a	0.4	99
Fluoranthene	7	0	ND	—	—	>99
Naphthalene	7	0	ND–1.0	—a	0.1	50
Pyrene	7	0	ND–1.0	—a	0.1	98
Anthranene/Phenanthrene	7	0	ND	—	—	>99
PCB 1254	7	0	ND–0.3	—a	—b	0
PCB 1248	7	0	ND–0.3	—a	—b	0
Halogenated Aliphatics						
Bromoform	11	0	ND–4.7	—a	1.0	NM
Carbon tetrachloride	11	0	ND–6.0	—a	0.5	50
Chlorodibromomethane	11	2	ND–29	—a	4.9	NM
Chloroform	11	6	ND–170	—a	32	NM
Dichlorobromomethane	11	3	ND–18	—a	3.0	NM
1,1-Dichloroethane	11	0	ND–7.0	—a	0.6	NM
1,2-Dichloroethane	11	2	ND–20	—a	2.3	NM
1,2-*trans*-Dichloroethylene	11	2	ND–75	1.0	9.2	52
Methylene chloride	11	1	ND–200	—a	16	NM
1,1,2,2-Tetrachloroethane	11	0	ND–1.0	—a	0.1	NM
1,1,1-Trichloroethane	11	0	ND–5.0	—a	0.5	NM
1,1,2-Trichloroethane	11	0	ND–8.5	—a	2.3	NM
Tetrachloroethylene	11	0	ND–4	—a	0.8	98
Trichloroethylene	11	0	ND–7	—a	1.0	98
Pesticides and Metabolites						
Isophorone	7	0	ND	—	—	>99
Chlordane	7	0	ND	—	—	>99
Aldrin	7	0	ND–0.1	—a	—b	NM
Dieldrin	7	0	ND	—	—	>99
4,4′-DDT	7	0	ND–0.1	—a	—b	NM
4,4′-DDE	7	0	ND–0.04	—a	—b	NM
Alpha-Endosulfan	7	0	ND–0.03	—a	—b	NM
Endrin	7	0	ND–0.4	—a	—b	NM
Endrin aldehyde	7	0	ND–0.6	—a	0.1	0
Heptachlor	7	0	ND	—a	—	>99
Heptachlor epoxide	7	0	ND–0.1	—a	—b	NM
Alpha-BHC	7	0	ND–0.2	—a	—b	NM
Beta-BHC	7	0	ND–0.1	—a	—b	65
Gamma-BHC	7	0	ND–0.02	—a	—b	80

ND, not detectable; NM, not meaningful.

a No median concentration is available in the reference.
b No mean concentration is available in the reference.

Source: From U.S. EPA, Development document for effluent limitations guidelines and standards for the nonferrous metals manufacturing point source category, report EPA-440/1-79/019-a, U.S. EPA, Washington, DC, 622 pp, 1979.

TABLE 3.30
Removability of Toxic Organic Pollutants from Wastewater in the Primary Columbium and Tantalum Subcategories

Toxic Pollutant	Number of Samples	Number of Detections >10 µg/L	Treated Effluent Concentration (µg/L)			Average Percent Removal
			Range	Median	Mean	
Phthalates						
Bis(2-ethylhexyl) phthalate	4	0	ND–9.5	2.8	3.8	97
Butyl benzyl phthalate	4	0	ND	—	—	>99
Di-n-butyl phthalate	4	0	ND–9.0	—[a]	2.2	82
Diethyl phthalate	4	0	ND–2.0	—[a]	0.5	71
Dimethyl phthalate	4	1	ND–20	—[a]	5.0	NM
Di-n-octyl phthalate	4	0	ND–2.0	—[a]	0.5	92
Phenols						
Pentachlorophenol	2	0	ND	—	—	>99
Aromatics						
Benzene	7	1	ND–40	—[a]	6.9	NM
Chlorobenzene	7	1	ND–65	—[a]	13	NM
2,4-Dinitrotoluene	4	0	ND	—[a]	—	>99
2,6-Dinitrotoluene	4	0	ND	—	—	>99
Ethylbenzene	7	1	ND–49	—[a]	7.0	NM
Nitrobenzene	4	0	ND	—	—	>99
Toluene	7	2	ND–92	—[a]	15	NM
1,2,4-Trichlorobenzene	4	2	ND–17	7.5	8.0	64
Polycyclic Aromatic Hydrocarbons						
Acenaphthene	4	2	ND–16	6.9	7.4	NM
Acenaphthylene	4	0	ND–2.8	0.9	1.2	NM
Anthracene	4	0	ND–12	1.5	3.8	NM
Benz (a) anthracene	4	0	ND	—	—	>99
Benzo (a) pyrene	4	0	ND	—	—	>99
Benzo (b) fluoranthene	4	0	ND–2.0	—[a]	0.5	NM
Benzo (ghi) perylene	4	0	ND–1.0	—[a]	0.2	0
Benzo (k) fluoranthene	4	0	ND–2.0	—[a]	0.5	NM
2-Chloronaphthalene	4	0	ND	—	—	>99
Chrysene	4	0	ND	—	—	>99
Dibenz (ah) anthracene	4	0	ND	—	—	>99
Fluoranthene	4	0	ND	—	—	>99
Fluorene	4	1	ND–69	—[a]	17	NM
Indeno (1,2,3-cd) pyrene	4	0	ND	—	—	>99
Naphthalene	4	0	ND	—	—	>99
Phenanthrene	4	1	ND–12	1.5	3.8	NM
Pyrene	4	0	ND–4.9	0.4	1.4	NM
Polychlorinated Biphenyls						
Aroclor 1248	3	0	ND	—	—	>99
Aroclor 1254	3	0	ND	—	—	>99

Continued

TABLE 3.30 (continued)

Toxic Pollutant	Number of Samples	Number of Detections >10 µg/L	Treated Effluent Concentration (µg/L)			Average Percent Removal
			Range	Median	Mean	
Halogenated Aliphatics						
Bromoform	7	0	ND	—	—	>99
Carbon tetrachloride	7	2	ND–110	—ᵃ	21	NM
Chlorodibromomethane	7	0	ND–5.0	—ᵃ	0.7	87
Chloroform	7	3	ND–47	—ᵃ	9.0	NM
Dichlorobromomethane	7	1	ND–16	—ᵃ	2.3	NM
1,2-Dichloroethane	7	2	ND–18	3.0	5.9	55
1,1-Dichloroethylene	7	2	ND–140	—ᵃ	21	NM
1,2-*trans*-Dichloroethylene	7	0	ND	—	—	>99
Hexachloroethane	4	0	ND	—	—	>99
Methylene chloride	7	1	ND–600	—ᵃ	85	98
1,1,2,2-Tetrachloroethane	7	1	ND–49	—ᵃ	7.0	NM
Tetrachloroethylene	7	4	ND–190	10	54	NM
1,1,1-Trichloroethane	7	0	ND	—	—	>99
1,1,2-Trichloroethane	6	0	ND–5.0	—ᵃ	—ᵇ	83
Trichloroethylene	7	3	ND–190	—ᵃ	32	NM
Pesticides and Metabolites						
Aldrin	3	0	ND–0.5	—ᵃ	0.2	33
Alpha-BHC	3	0	ND–0.01	—ᵃ	—ᵇ	75
Beta-BHC	3	0	ND–0.3	0.1	0.1	75
Delta-BHC	3	0	ND–0.5	—ᵃ	0.2	33
Gamma-BHC	3	0	ND	—	—	>99
Chlordane	3	0	ND–1.0	—ᵃ	—ᵇ	NM
4,4′-DDE	3	0	ND	—	—	>99
4,4′-DDT	3	0	ND	—	—	>99
Dieldrin	3	0	ND–0.01	—ᵃ	—ᵇ	90
Alpha-endosulfan	3	0	ND	—	—	>99
Beta-endosulfan	3	0	ND	—	—	>99
Endosulfan sulfate	3	0	ND	—	—	>99
Endrin	3	0	ND–0.01	—ᵃ	—ᵇ	99
Endrin aldehyde	3	0	ND	—	—	>99
Heptachlor	3	0	ND–0.3	—ᵃ	0.1	40
Heptachlor epoxide	3	0	ND	—	—	>99
Isophorone	4	0	ND	—	—	>99
Toxaphene	3	0	ND	—	—	>99

ND, not detectable; NM, not meaningful.
ᵃ No median concentration is available in the reference.
ᵇ No mean concentration is available in the reference.
Source: From U.S. EPA, Development document for effluent limitations guidelines and standards for the nonferrous metals manufacturing point source category, report EPA-440/1-79/019-a, U.S. EPA, Washington, DC, 622 pp, 1979.

TABLE 3.31
Removability of Toxic Organic Pollutants from Wastewater in the Primary Copper Subcategory

Toxic Pollutant	Number of Samples	Number of Detections >10 µg/L	Treated Effluent Concentration (µg/L) Range	Median	Mean	Average Percent Removal
Phthalates						
Bis(2-ethylhexyl) phthalate	5	4	ND–480	17	110	NM
Butyl benzyl phthalate	5	1	ND–48	—a	9.6	NM
Di-n-butyl phthalate	5	2	ND–73	—a	25	NM
Di-n-octyl phthalate	5	1	ND–190	—a	38	NM
Phenols						
2,4-Dimethylphenol	2	0	ND	—	—	>99
Aromatics						
Benzene	5	0	ND–1.0	—a	0.4	43
Chlorobenzene	5	0	ND–6.0	—a	1.2	NM
Toluene	5	0	ND	—	—	NM
Polycyclic Aromatic Hydrocarbons						
Acenaphthylene	5	0	ND	—	—	>99
Anthracene	5	3	ND–17	—a	6.2	NM
Benz (a) anthracene	5	0	ND	—	—	>99
Chrysene	5	0	ND–2.0	—a	0.4	NM
Fluoranthene	5	0	ND–2.0	—a	0.4	NM
Fluorene	5	1	ND–14	—a	2.8	NM
Phenanthrene	5	1	ND–17	—a	3.4	52
Pyrene	5	0	ND	—	—	>99
Polychlorinated Biphenyls						
Aroclor 1248	5	0	ND–1.0	1.0	0.8	NM
Aroclor 1254	5	0	ND–1.5	—a	0.5	NM
Halogenated Aliphatics						
Carbon tetrachloride	5	0	ND	—	—	>99
Chlorodibromomethane	5	0	ND	—	—	>99
Chloroform	5	0	ND	—	—	>99
Dichlorobromomethane	5	0	ND	—	—	>99
1,2-Dichloroethane	5	0	ND	—	—	>99
1,1-Dichloroethylene	5	0	ND–10	—	3.8	NM
Methylene chloride	5	0	ND	—	—	>99
1,1,2,2-Tetrachloroethane	5	0	ND–9.0	—	3.2	NM
Tetrachloroethylene	5	0	ND–3.0	—	1.0	81
1,1,1-Trichloroethane	5	0	ND–10	—	3.4	NM
1,1,2-Trichloroethane	5	0	ND	—	—	>99
Trichloroethylene	5	0	ND–3.0	—	0.6	60
Pesticides and Metabolites						
Beta-BHC	5	0	ND–0.2	—a	0.1	NM
Gamma-BHC	5	0	ND–0.01	—a	—b	75
Chlordane	5	0	ND–0.9	—a	0.2	0

Continued

TABLE 3.31 (continued)

Toxic Pollutant	Number of Samples	Number of Detections >10 µg/L	Treated Effluent Concentration (µg/L)			Average Percent Removal
			Range	Median	Mean	
4,4′-DDE	5	0	ND–0.1	—a	—b	NM
4,4′-DDT	5	0	ND–0.1	—a	—b	NM
Dieldrin	5	0	ND	—a	—b	>99
Alpha-endosulfan	5	0	ND–0.04	—a	—b	NM
Beta-endosulfan	5	0	ND	—	—	>99
Endosulfan sulfate	5	0	ND–0.2	—a	0.1	NM
Endrin	5	0	ND–0.1	—a	—b	0
Endrin aldehyde	5	0	ND–0.4	—a	0.1	0
Heptachlor	5	0	ND–0.2	—a	—b	NM
Heptachlor epoxide	5	0	ND–0.1	—a	—b	NM
Isophorone	5	0	ND	—	—	>99
4,4′-DDD	5	0	ND	—	—	>99

ND, not detectable; NM, not meaningful.
a No median concentration is available in the reference.
b No mean concentration is available in the reference.
Source: From U.S. EPA, Development document for effluent limitations guidelines and standards for the nonferrous metals manufacturing point source category, report EPA-440/l-79/019-a, U.S. EPA, Washington, DC, 622 pp, 1979.

TABLE 3.32
Removability of Toxic Organic Pollutants from Wastewater in the Secondary Copper Subcategory

Toxic Pollutant	Number of Samples	Number of Detections >10 µg/L	Treated Effluent Concentration (µg/L)			Average Percent Removal
			Range	Median	Mean	
Phthalates						
Bis(2-ethylhexyl) phthalate	13	10	ND–590	34.0	84	92
Butyl benzyl phthalate	13	2	ND–23	—a	3.3	38
Di-n-butyl phthalate	13	7	ND–110	16.0	32	43
Diethyl phthalate	13	3	ND–82	—a	15	NM
Dimethyl phthalate	13	4	ND–1.3×10^3	1.0	210	NM
Di-n-octyl phthalate	13	2	ND–170	—a	15	NM
Aromatics						
Benzene	13	0	ND–3.0	—a	0.2	85
Ethylbenzene	13	0	ND–2.0	—a	0.2	50
Hexachlorobenzene	13	2	ND–220	—a	30	93
Nitrobenzene	13	0	ND–1.0	—a	—b	NM
Toluene	13	1	ND–69	—a	5.6	NM
Polycyclic Aromatic Hydrocarbons						
Acenaphthene	13	1	ND–36	—a	2.8	39
Acenaphthylene	13	0	ND	—	—	>99
Anthracene	13	5	ND–140	5.0	19	93
Benzo (a) pyrene	13	0	ND–9.0	—a	1.5	NM

Continued

TABLE 3.32 (continued)

Toxic Pollutant	Number of Samples	Number of Detections >10 µg/L	Treated Effluent Concentration (µg/L) Range	Median	Mean	Average Percent Removal
Benzo (b) fluoranthene	13	1	ND–12	—a	—b	NM
Benzo (k) fluoranthene	13	1	ND–12	—a	—b	NM
Chrysene	13	0	ND–8.0	—a	0.8	>99
Dibenz (ah) anthracene	13	0	ND–8.0	—a	0.6	NM
Fluoranthene	13	1	ND–17	2.0	3.9	99
Fluorene	13	3	ND–100	—a	23	NM
Indeno (1,2,3-cd) pyrene	13	0	ND–8.0	—a	0.6	NM
Naphthalene	13	2	ND–930	—a	87	84
Phenanthrene	13	5	ND–140	5.0	19	93
Pyrene	13	3	ND–38	3.0	7.8	99
Polychlorinated Biphenyls						
Aroclor 1248	13	0	ND–2.2	—a	0.2	60
Aroclor 1254	13	0	ND–1.7	—a	0.2	60
Halogenated Aliphatics						
Carbon tetrachloride	13	1	ND–260	—a	20	NM
Chloroform	13	6	ND–320	—a	43	67
Dichlorobromomethane	13	0	ND–7.0	—a	0.5	NM
1,2-Dichloroethane	13	0	ND–1.0	—a	—b	97
1,1-Dichloroethylene	13	0	ND	—	—	>99
1,2-*trans*-Dichloroethylene	13	0	ND	—	—	>99
Methylene chloride	13	0	ND	—	—	>99
1,1,2,2-Tetrachloroethane	13	1	ND–14	—a	2.6	NM
Tetrachloroethylene	13	1	ND–12	—a	1.7	81
Trichloroethylene	13	0	ND–2.0	—a	0.2	97
Pesticides and Metabolites						
Aldrin	13	0	ND	—	—	>99
Alpha-BHC	13	0	ND–0.2	—a	—b	0
Beta-BHC	13	0	ND–0.2	—a	—b	95
Gamma-BHC	13	0	ND–0.1	—a	—b	NM
Chlordane	13	0	ND–0.5	—a	0.1	0
4,4'-DDE	13	0	ND–0.1	—a	0.1	NM
4,4'-DDD	13	0	ND–0.04	—a	—b	60
4,4'-DDT	13	0	ND–0.1	—a	0.1	NM
Dieldrin	13	0	ND–0.2	—a	—b	NM
Alpha-endosulfan	13	0	ND–0.6	—a	0.1	NM
Beta-endosulfan	13	0	ND–0.1	—a	—b	67
Endrin	13	0	ND–0.1	—a	—b	75
Endrin aldehyde	13	0	ND–0.4	—a	0.1	NM
Heptachlor	13	0	ND–0.2	—a	—b	NM
Heptachlor epoxide	13	0	ND–0.1	—a	—b	NM
Toxaphene	13	0	ND	—	—	>99

ND, not detectable; NM, not meaningful.

a No median concentration is available in the reference.

b No mean concentration is available in the reference.

Source: From U.S. EPA, Development document for effluent limitations guidelines and standards for the nonferrous metals manufacturing point source category, report EPA-440/1-79/019-a, U.S. EPA, Washington, DC, 622 pp, 1979.

TABLE 3.33
Removability of Toxic Organic Pollutants from Wastewater in the Primary Lead Subcategory

Toxic Pollutant	Number of Samples	Number of Detections >10 µg/L	Treated Effluent Concentration (µg/L)			Average Percent Removal
			Range	Median	Mean	
Pyrene	1	0	ND	—	—	>99
Methylene chloride	1	1	54	—	—	NM

ND, not detectable; NM, not meaningful.
a No median concentration is available in the reference.
b No mean concentration is available in the reference.
Source: From U.S. EPA, Development document for effluent limitations guidelines and standards for the nonferrous metals manufacturing point source category, report EPA-440/1-79/019-a, U.S. EPA, Washington, DC, 622 pp, 1979.

TABLE 3.34
Removability of Toxic Organic Pollutants from Wastewater in the Secondary Lead Subcategory

Toxic Pollutant	Number of Samples	Number of Detections >10 µg/L	Treated Effluent Concentration (µg/L)			Average Percent Removal
			Range	Median	Mean	
Phthalates						
Bis(2-ethylhexyl) phthalate	4	2	ND–22	9.5	12	97
Butyl benzyl phthalate	4	0	ND–4.0	—a	1.0	94
Di-n-butyl phthalate	4	1	ND–35	1.5	9.5	21
Dimethyl phthalate	4	0	ND	—	—	>99
Di-n-octyl phthalate	4	0	ND–2.0	—a	0.5	94
Nitrogen Compounds						
Benzidine	4	0	ND	—	—	>99
Aromatics						
Benzene	7	0	ND–7.0	—a	1.0	NM
Chlorobenzene	7	0	ND	—	—	>99
Ethylbenzene	7	0	ND–4.0	—a	0.6	NM
Nitrobenzene	4	0	ND	—	—	NM
Toluene	7	0	ND–1.0	—a	0.3	NM
Polycyclic Aromatic Hydrocarbons						
Acenaphthylene	4	0	ND	—	—	>99
Anthracene	4	0	ND–2.0	—a	0.5	88
Benzo (a) pyrene	4	0	ND	—	—	>99
Benzo (b) fluoranthene	4	0	ND	—	—	>99
Benzo (ghi) perylene	4	0	ND–1.0	—a	0.3	NM
Benzo (k) fluoranthene	4	0	ND	—	—	>99

Continued

TABLE 3.34 (continued)

Toxic Pollutant	Number of Samples	Number of Detections >10 µg/L	Treated Effluent Concentration (µg/L)			Average Percent Removal
			Range	Median	Mean	
Chrysene	4	0	ND–2.0	—ª	0.5	>99
Fluoranthene	4	0	ND	—	—	>99
Fluorene	4	0	ND	—	—	>99
Indeno (1,2,3-cd) pyrene	4	0	ND	—	—	>99
Naphthalene	3	0	ND–3.0	—ª	0.8	0
Phenanthrene	4	0	ND–2.0	—ª	0.5	89
Pyrene	4	0	ND	—	—	<99
Polychlorinated Biphenyls						
Aroclor 1248	4	0	ND–1.6	1.0	0.9	36
Aroclor 1254	4	0	ND–1.9	1.3	1.1	15
Halogenated Aliphatics						
Bromoform	7	0	ND	—	—	>99
Chloroform	7	4	ND–32	—ª	4.6	33
1,2-Dichloroethane	7	0	ND–2.0	—ª	0.3	93
1,1-Dichloroethylene	7	1	ND–17	—ª	2.4	35
1,2-*trans*-Dichloroethylene	7	1	ND–22	—ª	3.1	NM
1,1,2,2-Tetrachloroethane	7	0	ND	—	—	>99
Tetrachloroethylene	7	0	ND–3.0	—ª	0.6	45
1,1,2-Trichloroethane	7	0	ND–7.2	—ª	1.0	NM
Trichloroethylene	7	1	ND–28	1.0	4.7	NM
Pesticides and Metabolites						
Aldrin	4	0	ND	—	—	>99
Alpha-BHC	4	0	ND–0.04	—ª	—ᵇ	80
Beta-BHC	4	0	ND–0.3	—ª	0.1	0
Gamma-BHC	4	0	ND–0.02	—ª	—ᵇ	80
Chlordane	4	1	ND–31	9.0	15	NM
4,4′-DDE	4	0	ND–0.02	—ª	—ᵇ	90
4,4′-DDT	4	0	ND–0.1	—ª	—ᵇ	0
Dieldrin	4	0	ND–0.4	0.2	0.1	NM
Alpha-endosulfan	4	0	ND	—ª	—ᵇ	>99
Beta-endosulfan	4	0	ND–0.1	—ª	—ᵇ	NM
Endrin	4	0	ND	—	—	>99
Endrin aldehyde	4	0	ND	—	—	>99
Heptachlor	4	0	ND–0.3	—ª	0.1	0
Heptachlor epoxide	4	0	ND–0.1	—ª	—ᵇ	>99
Isophorone	4	0	ND	—	—	>99

ND, not detected; NM, not meaningful.

ª No median concentration is available in the reference.

ᵇ No mean concentration is available in the reference.

Source: From U.S. EPA, Development document for effluent limitations guidelines and standards for the nonferrous metals manufacturing point source category, report EPA-440/1-79/019-a, U.S. EPA, Washington, DC, 1979.

TABLE 3.35
Removability of Toxic Organic Pollutants from Wastewater in the Secondary Silver Subcategory

Toxic Pollutant	Number of Samples	Number of Detections >10 µg/L	Treated Effluent Concentration (µg/L)			Average Percent Removal
			Range	Median	Mean	
Phthalates						
Bis(2-ethylhexyl) phthalate	5	3	3.4–120	17	37	NM
Butyl benzyl phthalate	5	2	ND–52	1.0	18	NM
Di-n-butyl phthalate	5	1	ND–79	7.0	19	75
Diethyl phthalate	5	0	ND	—	—	>99
Di-n-octyl phthalate	5	2	ND–69	—[a]	16	47
Aromatics						
Benzene	9	4	ND–59	—[a]	14	81
Chlorobenzene	9	0	ND–4.0	—[a]	0.4	86
Ethylbenzene	9	2	ND–14	—[a]	3.9	58
Toluene	9	1	ND–19	—[a]	2.7	87
Polycyclic Aromatic Hydrocarbons						
Acenaphthene	5	0	ND	—	—	>99
Anthracene	5	0	ND	—	—	>99
Fluoranthene	5	1	ND–200	—[a]	40	NM
Naphthalene	5	0	ND	—	—	>99
Phenanthrene	5	0	ND	—	—	>99
Pyrene	5	1	ND–180	—[a]	36	92
Polychlorinated Biphenyls						
Aroclor 1248	2	0	0.3–1.9	1.1	1.1	NM
Aroclor 1254	2	0	0.2–2.6	1.4	1.4	NM
Halogenated Aliphatics						
Bromoform	9	1	ND–13	—[a]	1.4	87
Carbon tetrachloride	9	5	ND–1,700	19	310	18
Chlorodibromomethane	9	4	ND–2,800	—[a]	750	NM
Chloroform	9	6	ND–2,900	130	440	NM
1,2-Dichloroethane	9	3	ND–240	2.0	48	60
1,1-Dichloroethylene	9	3	ND–3,400	—[a]	390	65
1,3-Dichloropropylene	9	0	ND	—	—	—
1,2-*trans*-Dichloroethylene	9	1	ND–44	—[a]	4.9	NM
Methylene chloride	9	2	ND–790	—[a]	160	84
1,1,2,2-Tetrachloroethane	8	2	ND–25	—[a]	5.9	26
Tetrachloroethylene	9	4	ND–35	—[a]	8.3	81
1,1,1-Trichloroethane	9	0	ND–5.0	—[a]	0.6	92
Trichloroethylene	9	3	ND–330	—[a]	51	86
Pesticides and Metabolites						
Aldrin	2	0	ND	—	—	>99
Alpha-BHC	2	0	ND–0.1		0.05	NM
Beta-BHC	2	0	0.01–0.04		0.025	NM
Delta-BHC	2	0	ND		—	>99

Continued

Treatment of Nonferrous Metal Manufacturing Wastes

TABLE 3.35 (continued)

Toxic Pollutant	Number of Samples	Number of Detections >10 µg/L	Treated Effluent Concentration (µg/L) Range	Median	Mean	Average Percent Removal
Gamma-BHC	2	0	ND–0.03		0.015	NM
Chlordane	2	0	ND–0.1		0.05	0
4,4'-DDE	2	0	ND–0.01		0.005	NM
4,4'-DDD	2	0	ND–0.01		0.005	NM
4,4'-DDT	2	0	0.02–0.03		0.025	NM
Dieldrin	2	0	ND–0.1		0.05	NM
Endrin	2	0	ND–0.2		0.1	86
Endrin aldehyde	2	0	ND–0.5		0.25	NM
Heptachlor	2	0	0.01–0.04		0.025	NM

ND, not detectable; NM, not meaningful.
a No median concentration is available in the reference.
b No mean concentration is available in the reference.
Source: From U.S. EPA, Development document for effluent limitations guidelines and standards for the nonferrous metals manufacturing point source category, report EPA-440/1-79/019-a, U.S. EPA, Washington, DC, 1979.

TABLE 3.36
Removability of Toxic Organic Pollutants from Wastewater in the Primary Tungsten Subcategory

Toxic Pollutant	Number of Samples	Number of Detections >10 µg/L	Treated Effluent Concentration (µg/L) Range	Median	Mean	Average Percent Removal
Phthalates						
Bis(2-ethylhexyl) phthalate	2	2	32–730	—	380	NM
Di-n-butyl phthalate	2	2	22–66	—	44	NM
Diethyl phthalate	2	1	ND–16	—	8.0	NM
Dimethyl phthalate	2	1	ND–230	—	120	NM
Di-n-octyl phthalate	2	1	ND–43	—	22	NM
Aromatics						
Benzene	4	1	ND–17	7.5	8.0	NM
Chlorobenzene	4	0	ND–1.0	—a	—b	NM
Ethylbenzene	4	0	ND–1.0	—a	0.3	86
Nitrobenzene	2	0	ND–5.5	—a	2.8	NM
Toluene	4	0	ND–1.0	—a	0.3	97
1,2,4-Trichlorobenzene	2	0	4.0–5.5	—a	4.8	NM
Polycyclic Aromatic Hydrocarbons						
Acenaphthene	2	0	ND	—	—	>99
Acenaphthylene	2	0	ND	—	—	>99
Anthracene	2	0	ND–8.0	—	4.0	87
Benzo (a) pyrene	2	0	ND–1.0	—	0.5	NM

Continued

TABLE 3.36 (continued)

Toxic Pollutant	Number of Samples	Number of Detections >10 µg/L	Treated Effluent Concentration (µg/L)			Average Percent Removal
			Range	Median	Mean	
Chrysene	2	0	ND	—	—	>99
Fluoranthene	2	0	ND–1.0	—	0.5	NM
Fluorene	2	0	ND	—	—	>99
Naphthalene	2	1	ND–32	—	16	93
Phenanthrene	2	0	ND–8.0	—	4.0	NM
Pyrene	2	1	ND–15	—	7.5	NM
Polychlorinated Biphenyls						
Aroclor 1248	2	0	ND–2.4	—	1.2	NM
Aroclor 1254	2	0	ND–1.9	—	1.0	29
Halogenated Aliphatics						
Bromoform	4	0	ND	—	—	>99
Chlorodibromomethane	4	0	ND	—	—	>99
Chloroform	4	2	ND–870	29	230	NM
Dichlorobromomethane	4	1	ND–12	6.0	6.0	NM
1,2-Dichloroethane	4	2	ND–29	7.5	11	NM
1,1-Dichloroethylene	4	2	ND–29	10	12	NM
1,2-*trans*-Dichloroethylene	4	0	ND–2.0	—	0.5	NM
1,1,2,2-Tetrachloroethane	4	0	ND–9.0	5.3	5.0	4
Tetrachloroethylene	4	1	3.0–20	7.0	9.3	54
1,1,1-Trichloroethane	4	0	ND	—	—	>99
Trichloroethylene	4	3	ND–86	38	41	NM
Pesticides and Metabolites						
Aldrin	2	0	ND	—	—	>99
Alpha-BHC	2	0	ND	—	—	>99
Beta-BHC	2	0	ND	—	—	>99
Gamma-BHC	2	0	ND–0.1	—	—	50
Chlordane	2	0	ND–0.5	—	0.3	NM
4,4′-DDD	2	0	ND–0.2	—	0.1	NM
4,4′-DDT	2	0	ND	—	—	>99
Dieldrin	2	0	ND	—	—	>99
Alpha-endosulfan	2	0	ND–0.6	—	0.3	91
Beta-endosulfan	2	0	ND–0.2	—	0.1	97
Endrin	2	0	ND	—	—	>99
Endrin aldehyde	2	0	ND	—	—	>99
Heptachlor	2	0	ND	—	—	>99
Heptachlor epoxide	2	0	ND	—	—	>99
Isophorone	2	0	ND–6.0	—	3.0	NM

ND, not detectable; NM, not meaningful.

[a] No median concentration is available in the reference.

[b] No mean concentration is available in the reference.

Source: From U.S. EPA, Development document for effluent limitations guidelines and standards for the nonferrous metals manufacturing point source category, report EPA-440/1-79/019-a, U.S. EPA, Washington, DC, 1979.

TABLE 3.37
Removability of Toxic Organic Pollutants from Wastewater in the Primary Zinc Subcategory

Toxic Pollutant	Number of Samples	Number of Detections >10 µg/L	Treated Effluent Concentration (µg/L)			Average Percent Removal
			Range	Median	Mean	
Phthalates						
Bis(2-ethylhexyl) phthalate	11	4	ND–170	14	22	21
Butyl benzyl phthalate	11	0	ND–0.1	—ª	—ᵇ	99
Di-n-butyl phthalate	11	1	ND–12	4.0	1.6	56
Diethyl phthalate	11	0	ND–0.9	—ª	0.1	96
Dimethyl phthalate	11	1	ND–22	—	2.4	96
Di-n-octyl phthalate	11	0	ND–1.0	—	0.1	NM
Nitrogen Compounds						
3,3′-Dichlorobenzidine	11	0	ND–2.0	—ª	0.2	NM
Phenols						
Pentachlorophenol	11	0	ND	—	—	>99
Aromatics						
Benzene	11	0	ND–3.0	—ª	0.4	85
Ethylbenzene	11	0	ND–6.0	—ª	0.5	NM
Hexachlorobenzene	11	0	ND	—	—	>99
Toluene	11	0	ND–5.3	3.0	0.8	89
1,2,4-Trichlorobenzene	11	1	ND–47	—ª	4.3	NM
Polycyclic Aromatic Hydrocarbons						
Acenaphthylene	11	0	ND–8.0	—ª	0.7	65
Anthracene	11	0	ND–9.0	7.0	1.6	NM
Chrysene	11	0	ND–0.7	—ª	—ᵇ	94
Fluoranthene	11	0	ND	—	—	>99
Fluorene	11	0	ND–3.0	—ª	0.3	81
Naphthalene	11	0	ND–6.0	—ª	0.5	NM
Phenanthrene	11	0	ND–9.0	—ª	1.4	NM
Pyrene	11	0	ND–8.0	—ª	0.9	72
Polychlorinated Biphenyls						
Aroclor 1248	11	0	ND–7.0	—ª	0.6	NM
Aroclor 1254	11	0	ND–9.8	—ª	0.9	NM
Halogenated Aliphatics						
Bromoform	11	1	ND–44	—ª	4.0	NM
Chloroform	11	1	ND–54	—ª	5.4	66
1,1-Dichloroethane	11	0	ND	—	—	>99
1,2-Dichloroethane	11	0	ND	—	—	>99
1,1-Dichloroethylene	11	0	ND	—	—	>99
Methylene chloride	11	0	ND–7.0	—ª	0.8	>99
Tetrachloroethylene	11	1	ND–22	—ª	2.6	NM
Trichloroethylene	11	1	ND–19	—ª	2.0	89
Trichlorofluoromethane	11	0	ND	—	—	>99

Continued

TABLE 3.37 (continued)

Toxic Pollutant	Number of Samples	Number of Detections >10 μg/L	Treated Effluent Concentration (μg/L)			Average Percent Removal
			Range	Median	Mean	
Pesticides and Metabolites						
Alpha-BHC	11	0	ND–0.7	—a	0.1	NM
Beta-BHC	11	0	ND–0.03	—a	—	NM
Chlordane	11	0	ND–1.6	—a	0.2	NM
4,4′-DDE	11	0	ND–0.2	—a	0.01	NM
4,4′-DDT	11	0	ND–0.4	—a	0.03	NM
Dieldrin	11	0	ND–0.03	—a	—	NM
Heptachlor	11	0	ND–0.7	—a	0.1	NM
Heptachlor epoxide	11	0	ND–0.7	—a	0.1	NM
Isophorone	11	0	ND	—	—	>99
Nitrogen Compounds						
3,3′-Dichlorobenzidene	11	0	ND–2.0	—a	0.2	NM

ND, not detectable; NM, not meaningful.
a No median concentration is available in the reference.
b No mean concentration is available in the reference.
Source: From U.S. EPA, Development document for effluent limitations guidelines and standards for the nonferrous metals manufacturing point source category, report EPA-440/l-79/0l9-a, U.S. EPA, Washington, DC, 1979.

TABLE 3.38
Source Reduction and Recycling Activity for Nonferrous Metals Manufacturing Wastes

Year	Production Related Waste (metric ton)	% Reported as Released and Transferred	% Recycled	Onsite % Energy Recovery	% Treated	% Recycled	Offsite % Energy Recovery	% Treated	% Remaining Releases and Disposal
1	852,000	28	42.98	1.05	23.93	17.38	0.15	0.89	12.68
2	905,000	35	44.77	0.99	23.75	17.17	0.16	0.33	12.85
3	915,000	—	46.79	0.88	23.12	16.60	0.14	0.35	12.11
4	920,000	—	48.42	1.01	21.16	16.39	0.18	0.39	12.45

Source: From U.S. EPA, Profile of the Nonferrous Metals Industry, publication EPA/310-R-95-010, U.S. EPA, Washington, DC, September 1995.

Although the quantities reported for the first two years are estimates of quantities already managed, the quantities reported for the third and fourth years are projections only. U.S. EPA requires these projections to encourage facilities to consider future waste generation and source reduction of those quantities as well as movement up the waste management hierarchy. Future-year estimates are not commitments that facilities reporting under TRI are required to meet.

Table 3.38 shows that the primary and secondary metals industry managed 905,000 t of production-related waste (total quantity of TRI chemicals in the waste from routine production operations) in Year 2 (column B). Column C reveals that of this production-related waste, 35% was either transferred offsite or was released to the environment. Column C is calculated by dividing the total TRI transfers and releases by the total quantity of production-related waste. In other words, about 70% of the industry's TRI wastes were managed onsite through recycling, energy recovery, or treatment, as shown in columns D, E, and F, respectively. The majority of waste that is released or transferred offsite can be divided into portions that are recycled offsite, recovered for energy offsite, or treated offsite as shown in columns G, H, and I, respectively. The remaining portion of the production-related wastes (12.85%), shown in column J, is either released to the environment through direct discharges to air, land, water, and underground injection, or it is disposed offsite.

From the presented yearly data it is apparent that the portion of TRI wastes reported as recycled onsite has increased and the portions treated or managed through energy recovery onsite have remained steady, but are projected to decrease between the first and fourth years.

3.5.1 Chemical Release and Transfer Profile

This section is designed to provide background information on the pollutant releases that are reported by this industry. The best source of comparative pollutant release information is the TRI. Pursuant to the Emergency Planning and Community Right-to-Know Act (EPCRA),[28] TRI includes self-reported facility release and transfer data for over 600 toxic chemicals. Facilities within manufacturing industries that have more than ten employees and that are above weight-based reporting thresholds are required to report TRI onsite releases and offsite transfers. The information presented in here focuses primarily on the onsite releases reported by each sector. Because TRI requires consistent reporting regardless of sector, it is an excellent tool for drawing comparisons across industries.

Although this section does not present historical information regarding TRI chemical releases over time, note that, in general, toxic chemical releases have been declining.[13] Although onsite releases have decreased, the total amount of reported toxic waste has not declined because the amount of toxic chemicals transferred offsite has increased. Better management practices have led to increases in offsite transfers of toxic chemicals for recycling. More detailed information can be obtained from U.S. EPA's annual Toxics Release Inventory Public Data Release book, or directly from the Toxic Release Inventory System database.

Tables 3.39 and 3.40 illustrate TRI releases and transfers for the primary nonferrous metals smelting and refining industry. For this industry as a whole, chlorine comprises the largest number of TRI releases. This is reflected in the fact that chlorine is a byproduct of the magnesium industry and the largest reporter is a magnesium facility. The other top releases are copper compounds, zinc compounds, lead compounds, and sulfuric acid.

Tables 3.41 and 3.42 illustrate the TRI releases and transfers for the secondary nonferrous metals smelting and refining industry. For the industry as a whole, the largest releases were the various metals: aluminum (fume or dust), zinc compounds, lead compounds, copper, and zinc (fume or dust).

3.5.2 Summary of the Toxicity of Top Chemicals

The following is a synopsis of current scientific toxicity and fate information for the top chemicals (by weight) that facilities within this sector self-reported as released to the environment based upon TRI data. The information contained below is based upon exposure assumptions that have been conducted using standard scientific procedures. The effects listed must be taken in context of these exposure assumptions that are more fully explained within the full chemical profiles in the Hazardous Substances Data Bank (HSDB) and the Integrated Risk Information System (IRIS), both accessed via the Internet.

TABLE 3.39
Transfers for Primary Smelting and Refining in lb/yr (1 kg/yr = 0.454 lb/yr)

Chemical Name	No. of Facilities Reporting	Fugitive Air	Point Air	Water Discharges	Under-Ground Injection	Land Disposal	Total Releases	Average Releases per Facility
Copper	20	9412	248,340	508	0	500,254	758,514	37,926
Chlorine	19	153,751	67,037,082	2803	0	11	67,193,647	3,536,508
Sulfuric acid	15	24,527	1,013,009	0	5,700,000	100,920	6,838,456	455,897
Hydrogen fluoride	14	1,565,588	1,520,212	5	0	0	3,085,805	220,415
Manganese	11	15	5130	0	0	5	5150	468
Zinc compounds	10	47,545	102,940	8505	5	42,345,637	42,504,632	4,250,463
Chromium	8	10	398	5	0	0	413	52
Copper compounds	8	559,987	408,015	1502	65,000	27,574,267	28,608,771	3,576,096
Hydrochloric acid	8	3853	6,155,294	0	5	5	6,159,157	769,895
Lead compounds	8	68,834	274,504	7263	730	7,713,452	8,064,783	1,008,098
Arsenic compounds	7	7147	30,181	3005	52,000	2,190,652	2,282,985	326,141
Antimony compounds	6	6319	4398	3143	2100	661,740	677,700	112,950
Cadmium compounds	6	1286	18,912	311	0	39,734	60,243	10,041
Nickel compounds	6	1323	8956	225	4200	1,149,028	1,163,732	193,955
Nitric acid	6	15	23,670	0	5	15	23,705	3951
Aluminum (fume or dust)	5	5760	32,472	44	0	5	38,281	7656
Lead	5	138,589	96,836	18	0	2,352,628	2,588,071	517,614
Nickel	5	345	781	4	0	29,052	30,182	6036
Silver compounds	5	848	2210	270	100	19,633	23,061	4612
Barium compounds	4	5	1850	0	890	456,308	459,053	114,763
Arsenic	3	270	28,264	9	0	7114	35,657	11,886
Cadmium	3	981	6181	11	0	4824	11,997	3999
Chromium compounds	3	250	592	250	0	190,005	191,097	63,699

Treatment of Nonferrous Metal Manufacturing Wastes

Manganese compounds	3	620	823	0	0	2,400,643	2,402,086	800,695
Selenium compounds	3	1350	38,000	250	2300	120,265	162,165	54,055
Zinc (fume or dust)	3	10,190	25,682	46	0	4,010,295	4,046,213	1,348,738
1,1,1-Trichloroethane	3	75,031	0	0	0	0	75,031	25,010
Anthracene	2	250	25,487	0	0	0	25,737	12,869
Antimony	2	500	10,915	5	0	0	11,420	5710
Cobalt	2	250	5	0	0	0	255	128
Cobalt compounds	2	669	262	255	0	5	1191	596
Cyanide compounds	2	0	0	500	0	0	500	250
Ethylene glycol	2	0	0	0	0	0	0	0
Phosphoric acid	2	0	0	0	0	0	0	0
Thiourea	2	60	0	0	5300	255	5615	2808
Ammonia	1	250	0	0	0	0	250	250
Beryllium compounds	1	0	0	0	0	0	0	0
Cresol (mixed isomers)	1	250	0	250	0	750	1250	1250
Decabromodiphenyl oxide	1	0	250	0	0	0	250	250
Dichlorodifluoromethane	1	18,000	0	0	0	0	18,000	18,000
M-Xylene	1	14,000	0	0	0	0	14,000	14,000
Naphthalene	1	0	467	0	0	0	467	467
Phenol	1	0	0	1	0	0	1	1
Styrene	1	1900	0	0	0	5	1905	1905
Thallium	1	5	0	0	0	755	1010	1010
Titanium tetrachloride	1	250	250	0	0	0	500	500
1,2,4-Trimethylbenzene	1	18,000	0	0	0	0	18,000	18,000
Total	—	2,738,235	77,122,618	29,188	5,832,635	91,868,262	177,590,938	—

Source: From U.S. EPA, Profile of the Nonferrous Metals Industry, publication EPA/310-R-95-010, U.S. EPA, Washington, DC, September 1995.

TABLE 3.40
Transfers for Primary Smelting and Refining in lb/yr (1 kg/yr = 0.454 lb/yr)

Chemical Name	No. of Reporting Facilities	POTW Discharge	Disposal	Recycling	Treatment	Energy Recovery	Total Transfers	Average per Facility
Copper	20	5	17,596	124,723	0	0	142,324	7116
Chlorine	19	0	0	9991	0	0	9991	526
Sulfuric acid	15	1	600	6,454,346	0	0	6,454,947	430,330
Hydrogen fluoride	14	0	0	0	0	0	0	0
Manganese	11	0	14	46,752	0	0	46,766	4251
Zinc compounds	10	760	2,692,570	750,680	833,231	0	4,277,241	427,724
Chromium	8	0	0	2361	0	0	2361	295
Copper compounds	8	459	2,900,850	3,882,069	93,989	0	6,877,367	859,671
Hydrochloric acid	8	0	0	0	0	0	0	0
Lead compounds	8	2401	2,253,086	2,289,461	11,239	0	4,556,187	569,523
Arsenic compounds	7	386	1,649,205	174,013	634,487	0	2,458,091	351,156
Antimony compounds	6	1749	345,100	29,836	15,262	0	391,947	65,325
Cadmium compounds	6	346	26,097	420,187	62,987	0	509,617	84,936
Nickel compounds	6	260	5	237,910	3931	0	242,106	40,351
Nitric acid	6	0	5	0	11,000	0	11,005	1834
Aluminum	5	0	317,650	3,826,700	0	0	4,144,350	828,870
Lead	5	5	5	640,899	0	0	640,909	128,182
Nickel	5	5		633	0	0	638	128
Silver compounds	5	174	5765	8756	255	0	14,950	2990
Barium compounds	4	0	0	0	0	0	0	0
Arsenic	3	5	250	55,713	0	0	55,968	18,656
Cadmium	3	5		212,387	0	0	212,392	70,797
Chromium compounds	3	0	1200	15,000	0	0	16,200	5400

Treatment of Nonferrous Metal Manufacturing Wastes

Chemical	# Facilities	Fugitive Air	Stack Air	Water	Underground Injection	Land	Total On-Site	Total Off-Site
Manganese compounds	3	41	0	5639	0	0	5680	1893
Selenium compounds	3	0	19,005	0	0	0	19,005	6335
Zinc (fume or dust)	3	250	0	412,568	0	0	412,818	137,606
1,1,1-Trichloroethane	3	0	0	0	250	0	250	83
Anthracene	2	0	14,032	0	0	0	14,032	7016
Antimony	2	0	4110	1,911,550	0	0	1,915,660	957,830
Cobalt	2	0	0	0	0	0	0	0
Cobalt compounds	2	250	0	77,640	0	0	77,890	38,945
Cyanide compounds	2	0	53,213	0	1813	0	55,026	27,513
Ethylene glycol	2	0	0	0	8673	0	8673	4337
Phosphoric acid	2	0	0	0	160	0	160	80
Thiourea	1	0	0	0	0	0	0	0
Ammonia	1	0	0	0	0	0	0	0
Beryllium compounds	1	0	0	0	0	0	0	0
Cresol (mixed isomers)	1	0	4374	0	0	0	4374	4374
Decabromodiphenyl oxide	1	0	0	0	0	0	0	0
Dichlorodifluoromethane	1	0	0	0	0	0	0	0
M-Xylene	1	0	0	0	0	0	0	0
Naphthalene	1	0	0	0	0	0	0	0
Phenol	1	0	0	0	0	0	0	0
Styrene	1	5	750	0	0	0	755	755
Thallium	1	0	0	0	0	0	0	0
Titanium tetrachloride	1	0	0	0	0	0	0	0
4-Trimethylbenzene	1	0	0	0	0	0	0	0
Total	225	7107	10,304,732	21,590.56	1,677.277	33,579,680		108187.82

Source: From U.S. EPA, Profile of the Nonferrous Metals Industry, publication EPA/310-R-95-010, U.S. EPA, Washington, DC, September 1995.

TABLE 3.41
Transfers for Secondary Smelting and Refining in lb/yr (1 kg/yr = 0.454 lb/yr)

Chemical Name	No. of Facilities Reporting Chemical	Fugitive Air	Point Air	Water Discharges	Under-ground Injection	Land Disposal	Total Releases	Average Releases per Facility
Copper	74	17,235	56,198	2720	0	221,287	297,440	4019
Nickel	38	5646	5873	262	0	12,934	24,715	650
Chlorine	32	5103	6304	0	0	0	11,407	356
Lead	30	13,964	29,230	571	0	750	44,515	1484
Copper compounds	29	11,921	35,205	358	0	1500	48,984	1689
Lead compounds	25	11,211	115,573	404	0	147,930	275,118	11,005
Manganese	25	7848	3547	10	0	74,536	85,941	3,438
Aluminum (fume or dust)	24	34,297	196,604	922	11	641,760	873,594	36,400
Zinc compounds	24	41,195	263,420	3049	0	0	307,664	12,819
Sulfuric acid	21	6917	1730	0	0	0	8647	412
Chromium	19	1465	1937	255	0	2005	5662	298
Zinc (fume or dust)	19	57,759	79,392	331	0	0	137,482	7236
Hydrochloric acid	14	17,116	604,670	0	0	0	621,786	44,413
Nickel compounds	13	1113	1492	297	0	0	2902	223
Chromium compounds	10	276	617	0	0	0	893	89
Ammonia	9	1,343,335	168,094	53,229	57,053	353,800	1,975,511	219,501
Antimony	9	364	373	586	0	5	1328	148
Antimony compounds	9	115	1294	44	0	67,760	69,213	7690
Silver	9	21	517	251	0	0	789	88
Silver compounds	9	1033	823	5	0	0	1861	207
Manganese compounds	8	1074	3426	570	0	0	5070	634
Nitric acid	8	1008	2628	0	0	0	3636	455

Arsenic	7	310	308	36	0	5	94
Arsenic compounds	7	10	573	16	0	27,104	3958
Barium compounds	6	298	2011	0	0	2309	385
Cadmium compounds	6	545	5409	20	0	5974	996
Cobalt	6	905	680	5	0	1610	268
Cadmium	3	250	874	281	0	1405	468
Hexachloroethane	3	0	11,536	0	0	11536	3845
Aluminum oxide (fibrous form)	2	0	53	0	0	53	27
Barium	2	20	45	0	0	65	33
Beryllium	2	0	5	0	0	5	3
Methanol	2	1000	0	0	0	1000	500
Molybdenum trioxide	2	500	4205	18,750	0	23,455	11,728
Ammonium sulfate (solution)	1	250	0	0	0	250	250
Cobalt compounds	1	0	0	0	0	0	0
Mercury compounds	1	250	5	5	5	265	265
Phosphoric acid	1	0	0	0	0	0	0
Phosphorus (yellow or white)	1	0	0	0	0	0	0
Polychlorinated Biphenyls	1	0	1	0	0	1	1
Selenium	1	250	0	0	0	250	250
Xylene (mixed isomers)	1	250	0	0	0	250	250
1,1,1-Trichloroethane	—						
Totals		1,584,854	1,604,652	82,977	57,064	1,551,401	4,880,948

Source: From U.S. EPA, Profile of the Nonferrous Metals Industry, publication EPA/310-R-95-010, U.S. EPA, Washington, DC, September 1995.

TABLE 3.42
Transfers for Secondary Smelting and Refining in lb/yr (1 kg/yr = 0.454 lb/yr)

Chemical Name	No. of Facilities Reporting Chemical	POTW Discharge	Disposal	Recycling	Treatment	Energy Recovery	Total Transfers	Average per Facility
Copper	74	7024	139,130	20,126,255	20,233	0	20,292,642	274,225
Nickel	38	282	9366	78,143	3984	0	91,775	2415
Chlorine	32	2545	0	0	0	0	2545	80
Lead	30	1106	675,459	1,749,221	16,055	0	2,441,841	81,395
Copper compounds	29	82	658,756	806,437	537,038	0	2,002,313	69,045
Lead compounds	25	810	5,543,943	11,216,399	1,020,276	0	17,781,428	711,257
Manganese	25	501	108,806	67,048	1236	0	177,591	7104
Aluminum (fume or dust)	24	500	966,226	15,417	0	0	982,143	40,923
Zinc compounds	24	1661	129,752	5,571,000	229,930	0	5,932,343	247,181
Sulfuric acid	21	5	0	7,332,842	0	0	7,332,847	349,183
Chromium	19	51	11,812	43,378	83	0	55,324	2912
Zinc (fume or dust)	19	5	164,242	1,048,567	8180	0	1,220,994	64,263
Hydrochloric acid	14	0	750	56,965	27,557	0	85,272	6091
Nickel compounds	13	23	34,996	1,531,600	4777	0	1,571,396	120,877
Chromium compounds	10	251	165,015	214,000	4664	0	383,930	38,393
Ammonia	9	0	621,718	0	0	0	621,718	69,080
Antimony	9	927	127,443	8180	880	0	137,430	15,270
Antimony compounds	9	614	935,418	641,800	10,710	0	1,588,542	176,505
Silver	9	755	0	8680	0	0	9435	1048
Silver compounds	9	20	835	485,550	186	0	486,591	54,066
Manganese compounds	8	75	29,005	128,500	0	0	157,580	19,698
Nitric acid	8	5	1500	11,299	750	0	13,554	1694
Arsenic	7	67	51,353	0	1784	0	53,204	7601

Treatment of Nonferrous Metal Manufacturing Wastes

Arsenic compounds	7	110	196,876	55,734	0	252,720	36,103
Barium compounds	6	4448	115,647	82,700	31,094	233,889	38,982
Cadmium compounds	6	257	0	393,000	0	393,257	65,543
Cobalt	6	5	905	35,045	15	35,970	5995
Cadmium	3	0	12,930	23,795	900	37,625	12,542
Hexachloroethane	3	0	0	0	0	0	0
Aluminum oxide (fibrous form)	2	0	0	0	0	0	0
Barium	2	5	62,710	0	250	62,965	31,483
Beryllium	2	0	0	7,930	—	7930	3965
Methanol	2	0	0	0	0	0	0
Molybdenum trioxide	2	0	—	165,100	17,150	182,250	91,125
Ammonium sulfate (solution)	1	0	0	0	0	0	0
Cobalt compounds	1	0	0	0	0	0	0
Mercury compounds	1	0	33,200	0	10	33,210	33,210
Phosphoric acid	1	0	0	0	0	0	0
Phosphorus (yellow or white)	1	250	0	0	0	250	250
Polychlorinated Biphenyls	1	0	255	0	0	255	255
Selenium	1	0	2673	0	510	3183	3183
Xylene (mixed isomers)	1	0	0	0	0	0	0
1,1,1-Trichloroethane	1	0	0	0	0	0	0
Totals	—	22,384	10,800,721	51,904,585	1,938,252	64,665,942	—

Source: From U.S. EPA, Profile of the Nonferrous Metals Industry, publication EPA/310-R-95-010, U.S. EPA, Washington, DC, September 1995.

3.5.2.1 Chlorine

Chlorine is a highly reactive gas. Most of the chlorine released to the environment will quickly evaporate. Breathing small amounts of chlorine for short periods of time can affect the respiratory tract in humans, causing symptoms such as coughing and chest pain. It is irritating to the skin, eyes, and respiratory tract. Repeated long-term exposure to chlorine can cause adverse effects on the blood and respiratory systems. There is currently no evidence to suggest that this chemical is carcinogenic. Ecologically, chlorine is highly toxic to aquatic organisms at low doses.

3.5.2.2 Copper

Metallic copper probably has little or no toxicity, although copper salts are more toxic. Inhalation of copper oxide fumes and dust has been shown to cause metal fume fever: irritation of the upper respiratory tract, nausea, sneezing, coughing, chills, aching muscles, gastric pain, and diarrhea. However, the respiratory symptoms may be due to a nonspecific reaction to the inhaled dust as a foreign body in the lung, and the gastrointestinal symptoms may be attributed to the conversion of copper to copper salts in the body.

It is unclear whether long-term copper poisoning exists in humans. Some have related certain central nervous system disorders, such as giddiness, loss of appetite, excessive perspiration, and drowsiness to copper poisoning. Long-term exposure to copper may also cause hair, skin, and teeth discoloration, apparently without other adverse effects. There is currently no evidence to suggest that this chemical is carcinogenic.

People at special risk from exposure to copper include those with impaired pulmonary function, especially those with obstructive airway diseases, because the breathing of copper fumes might cause exacerbation of pre-existing symptoms due to its irritant properties.

Ecologically, copper is a trace element essential to many plants and animals. However, high levels of copper in soil can be directly toxic to certain soil microorganisms and can disrupt important microbial processes in soil, such as nitrogen and phosphorus cycling. Copper is typically found in the environment as a solid metal in soils and soil sediment in surface water. There is no evidence that biotransformation processes have a significant bearing on the fate and transport of copper in water.

3.5.2.3 Hydrochloric Acid

Concentrated hydrochloric acid is highly corrosive. Hydrochloric acid is primarily a concern in its aerosol form. Acid aerosols have been implicated in causing and exacerbating a variety of respiratory ailments. Dermal exposure and ingestion of highly concentrated hydrochloric acid can result in corrosivity. There is currently no evidence to suggest that this chemical is carcinogenic.

Ecologically, accidental releases of solution forms of hydrochloric acid may adversely affect aquatic life by including a transient lowering of the pH (i.e., increasing the acidity) of surface waters. Releases of hydrochloric acid to surface waters and soils will be neutralized to an extent due to the buffering capacities of both systems. The extent of these reactions will depend on the characteristics of the specific environment.

3.5.2.4 Lead

Short-term lead poisoning is relatively infrequent and occurs from ingestion of acid-soluble lead compounds or inhalation of lead vapors. Symptoms include nausea, severe abdominal pain, vomiting, diarrhea or constipation, shock, tingling, pain, muscle weakness, and kidney damage. Death may occur in one to two days. If the patient survives the acute episode, characteristic signs and symptoms of chronic lead poisoning are likely to appear. Chronic lead poisoning affects the gastrointestinal, neuromuscular, blood, kidney, and central nervous systems. Individuals with chronic lead poisoning appear ashen, with an appearance of "premature aging," with stooped posture, poor muscle tone, and emaciation. Neuromuscular syndrome (muscle weakness, easy fatigue, localized paralysis) and central nervous system syndrome (progressive mental deterioration, decreased intelligence, loss of motor skills and speech, hyperkinetic and aggressive behavior disorders, poorly controlled convulsive disorder, severe learning impairment) usually result from intense exposure, while the abdominal syndrome (anorexia, muscle discomfort, malaise, headache, constipation, severe abdominal pain, persistent metallic taste) is a more common manifestation of a very slowly and insidiously developing intoxication.

In the U.S., the central nervous system syndrome is usually more common among children, and the gastrointestinal syndrome is more prevalent in adults. Exposure to lead is also linked to decreased fertility in men. Lead is a probable human carcinogen, based on sufficient animal evidence. Populations at increased risk of toxicity from exposure to lead include developing fetuses and young children, individuals with decreased kidney function, and children with sickle-cell anemia.

If released or deposited on soil, lead will be retained in the upper 2 to 5 cm of soil. Leaching is not important under normal conditions, nor generally is the uptake of lead from soil into plants. Lead enters water from atmospheric fallout, runoff, or wastewater; it is effectively removed from the water column to the sediment predominantly by adsorption to organic matter and clay minerals. Some lead re-enters the water column through methylation by microorganisms. Volatilization is negligible. Lead does not appear to bioconcentrate significantly in fish but does in some shellfish such as mussels. When released to the atmosphere, lead will generally be in dust or adsorbed to particulate matter and subject to gravitational settling.

3.5.2.5 Zinc and Zinc Compounds

Zinc is a nutritional trace element; toxicity from ingestion is low. Severe exposure to zinc might give rise to gastritis with vomiting due to swallowing of zinc dust. Short-term exposure to very high levels of zinc is linked to lethargy, dizziness, nausea, fever, diarrhea, and reversible pancreatic and neurological damage. Long-term zinc poisoning causes irritability, muscular stiffness and pain, loss of appetite, and nausea. There is currently no evidence to suggest that this chemical is carcinogenic.

Zinc chloride fumes cause injury to mucous membranes and to the skin. Ingestion of soluble zinc salts may cause nausea, vomiting, and purging.

Significant zinc contamination of soil is only seen in the vicinity of industrial point sources. Zinc is a relatively stable soft metal, although it burns in air. Zinc bioconcentrates in aquatic organisms.

3.5.3 Comparison of Toxic Release Inventory between Industries

Figure 3.5 is a graphical representation of a summary of the TRI data for the nonferrous metals industry and the other sectors. The bar graph presents the total TRI releases and total transfers on the left axis and the triangle points show the average releases per facility on the right axis. Industry sectors are presented in the order of increasing total TRI releases. The graph is meant to facilitate comparisons between the relative amounts of releases, transfers, and releases per facility both within and between these sectors.

3.5.4 Pollution Prevention Opportunities

The best way to reduce pollution is to prevent it in the first place. Some companies have creatively implemented pollution prevention techniques that improve efficiency and increase profits while at the same time minimizing environmental impacts. This can be done in many ways, such as reducing material inputs, reengineering processes to reuse byproducts, improving management practices, and using substitution of toxic chemicals. Some smaller facilities are able to get below regulatory thresholds just by reducing pollutant releases through aggressive pollution prevention including a discussion of associated costs, time frames, and policies.

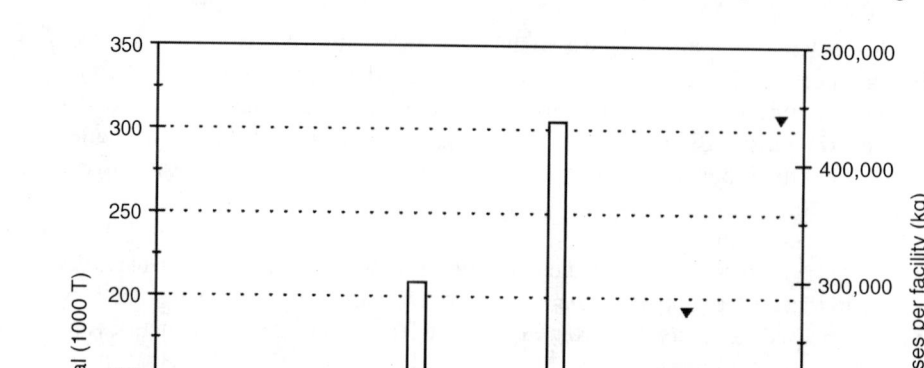

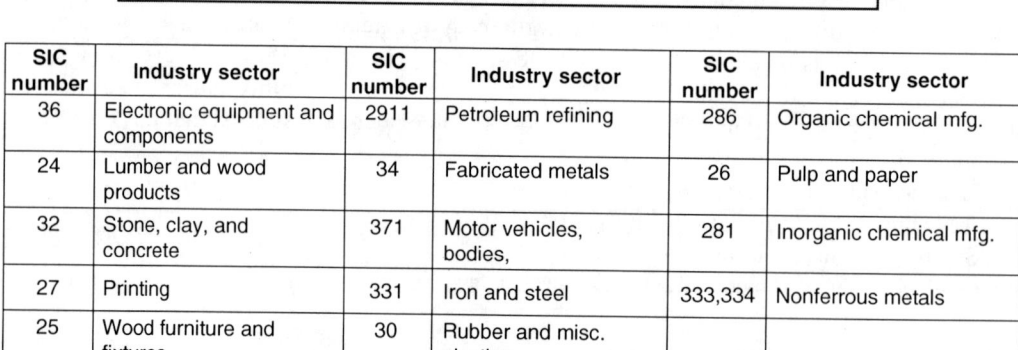

FIGURE 3.5 Releases and transfers by industry. (From U.S. EPA, Profile of the Nonferrous Metals Industry, publication EPA/310-R-95-010, U.S. EPA, Washington, DC, September 1995.)

There are great efforts all around the world for improving recycling and waste reduction, including in Hong Kong[29] and Japan.[30] In Japan,[30] the following targets were set in 1998 for recycling and the reduction of final quantities of nonferrous metal industrial wastes by 2010:

1. Japan Mining Industry Association: reduction by 41%
2. Japan Brass Maker's Association: reduction by 13%
3. Japan Aluminum Association: reduction by 14%
4. Japan Electric Wire and Cable Maker's Association: reduction by 25%

Pollution prevention in the U.S., whether through source material reduction/reuse, or waste recycling, is practiced in various sectors of the nonferrous metals industry. Pollution prevention

Treatment of Nonferrous Metal Manufacturing Wastes

techniques and processes currently used by the nonferrous metals industry can be grouped into the following general categories[1]:

1. Process equipment modification
2. Raw materials substitution or elimination
3. Solvent recycling
4. Precious metals recovery

3.5.4.1 Process Equipment Modification

Process equipment modification is used to reduce the amount of waste generated. Many copper, lead, and zinc refiners have modified their production processes by installing sulfur fixation equipment. This equipment not only captures the sulfur before it enters the atmosphere (helping the refining plant meet CAA sulfur standards), but processes it so that a marketable sulfuric acid is produced. Another example is the use of prebaked anodes in primary aluminum refining. When a prebaked anode is used, the electrolytic cell, or pot, can be closed, thereby increasing the efficiency of the collection of fluoride emissions. In addition, new carbon liners have been developed that significantly increase the life of the aluminum reduction cell. This has resulted in large reductions in the amount of spent potliner material generated by the aluminum industry.

3.5.4.2 Raw Materials Substitution or Elimination

Raw material substitution or elimination is the replacement of raw materials with other materials that produce less waste, or a nontoxic waste. Material substitution is inherent in the secondary nonferrous metals industry primarily by substituting scrap metal, slag, and baghouse dust for ore feedstock. All of these materials, whether in the form of aluminum beverage cans, copper scrap, or lead-acid batteries, are commonly added to other feedstock or charges (usually slag containing residual metals) to produce marketable grades of metal. Primary nonferrous metals refining also uses previously refined metals as feedstock, especially zinc-containing electric arc furnace dust (a byproduct of the iron and steel industry).

3.5.4.3 Precious Metals Recovery

Precious metals recovery is the modification of a refining process to allow the capture of marketable precious metals such as gold and silver. Like sulfur fixation, precious metals recovery is a common waste minimization practice. During primary copper smelting, appreciable amounts of silver and gold present in copper ore will be concentrated into the anode copper and can be recovered as a byproduct in the electrorefining process (as the copper anode is electrochemically dissolved and the copper attaches itself to the cathode, silver and gold drop out and are captured in the slime at the bottom of the tank). In the lead refining process the copper often present in lead ore is removed during the initial lead bullion smelting process as a constituent of dross. Silver and gold are removed from the lead bullion later in the process by adding certain fluxes that cause them to form an impure alloy. The alloy is then refined electrolytically and separated into gold and silver. Precious metals recovery also takes place during zinc refining to separate out copper, a frequent impurity in zinc ore. Copper is removed from the zinc ore during the zinc purification process (after zinc undergoes leaching, zinc dust is added, which forces many of the deleterious elements to drop out; copper is recovered in a cake form and sent for refining).

3.5.5 Important Pollution Prevention Case Studies

Various pollution prevention case histories have been documented for nonferrous metals refining industries. In particular, the actions of the AMPCO Metal Manufacturing Company, Inc., typify industry efforts to simultaneously lessen the impact of the industrial process on the environment, reduce energy consumption, and lower production costs.[1]

3.5.5.1 The Use of Electric Induction to Replace Fossil Fuel Combustion

AMPCO Metal Manufacturing Company, Inc. (Ohio) participated in the development of pollution prevention technologies. The project, sponsored by the U.S. DOE and U.S. EPA, consists of researching and developing the use of electric induction to replace fossil fuel combustion as is currently used to heat tundishes. Tundishes are used to contain the heated reservoir of molten alloy in the barstock casting process. The fossil fuel combustion process currently used requires huge amounts of energy and produces tremendous amounts of waste gases, including combustion bases and lead and nickel emissions.

Heating the tundish by electric induction instead of fossil fuel combustion will substantially improve the current process, saving energy and reducing pollution. Energy efficiency will jump to an estimated 98%, saving 16 billion chu (28.9 billion Btu)/yr/unit. Industry-wide energy savings in 2010 are estimated to be 114 billion chu (206 billion Btu)/yr, assuming a 70% adoption at U.S. foundries.

In addition to the energy savings, the new process also has substantial environmental benefits. Along with the elimination of lead and nickel gases, carbon dioxide, carbon monoxide, and nitrogen oxide emissions from combustion will decrease. The consumption of refractory (a heat-resisting ceramic material) will decline by 80%, resulting in a similar reduction of refractory waste disposal. In all, the prevention of various forms of pollution is estimated to be 66.7 million kg (147 million lb/yr) by 2010.

Economically, the elimination of lead and nickel emissions will result in an improved product because exposure of the metal to combustion gases in the current process results in porosity and entrainment of hydrogen gas in the metal. Overall, AMPCO estimates an annual savings in operations and maintenance expenses of USD 1.2 million with the use of this technology. Assuming the same 70% industry adoption, economic savings by 2010 could reach USD 5.8 million. Without the new electric induction heating process, the capital costs required for compliance could be USD 3 million.

3.5.5.2 Processing Nonferrous Metal Hydroxide Sludge Wastes

Nonferrous metal hydroxide sludge wastes contain a large quantity of water, and the content of valuable metals is too small to allow economical smelting of these wastes. However, the wastes are a burden on the environment and they can only be deposited in special garbage dumps, which are very costly. Therefore a method for processing and recovery of the valuable materials and metal in the waste is highly desirable.

Such a method has been patented in the U.S.[31] for processing the nonferrous metal hydroxide sludge wastes containing chromium, copper, zinc, and nickel for simultaneous recovery and separation of the individual nonferrous metals sequentially. In this method, the individual nonferrous metals, such as, chromium, copper, zinc, and nickel can be individually and economically separated from the collected nonferrous metal hydroxide sludge wastes. This is achieved by the combination of the following steps performed in sequence[31]:

1. Chlorinating the aqueous waste sludge suspension (to oxidize the chromium) at temperatures of 20 to 80°C and pH values between 4 and 13. The chlorinated sludge is then acidified with sulfuric acid to a pH of 1.0 to 3.0. The insoluble components are then separated, followed by the separation of the chromium(VI) from the solution using a fixed-bed anion exchanger (at pH values of <3).
2. Copper is separated from the remaining solution by means of known and conventional liquid–liquid-extraction procedures.
3. Zinc is separated from the remaining solution, which also contains chloride and sulfate by means of liquid–liquid extraction.

4. Aluminum is precipitated and separated from the remaining solution in the form of the hydroxide.
5. Nickel is separated from the filtrate by means of liquid–liquid-extraction.
6. The individual nonferrous metal fractions thus obtained are then processed in conventional ways.[32]

REFERENCES

1. U.S. EPA, Profile of the Nonferrous Metals Industry, publication EPA/310-R-95-010, U.S. EPA, Washington, DC, September 1995.
2. U.S. EPA, Treatability Manual, Volume II Industrial Descriptions, report EPA-600/2-82-001b, U.S. EPA, Washington, DC, September 1981.
3. U.S. EPA, Development document for effluent limitations guidelines and standards for the nonferrous metals manufacturing point source category, report EPA-440/1-79/019-a, U.S. EPA, Washington, DC, 1979.
4. ITC, International Trade Statistics by Product Group and Country, International Trade Centre, available at http://www.intracen.org/tradstat/sitc3-3d/ep288.htm, 2007.
5. 2-36, U.S. EPA, Effluent guidelines and standards for nonferrous metals, 4OCFR421; 39FR12822, April 8, 1974; amended by 40FR8514, February 27, 1975; 40FR48348, October 15, 1975; 41FR 54850, December 15, 1976, U.S. EPA, Washington, DC, 1974–1976.
6. Federal Register, Resource Conservation and Recovery Act (RCRA), 42 U.S. Code s/s 6901 et seq. 1976, United States Government, Public Laws, available at www.access.gpo.gov/uscode/title42/chapter82_.html, January 2004.
7. U.S. EPA, Resource Conservation and Recovery Act (RCRA)—Orientation Manual, U. S. EPA, report EPA530-R-02-016, Washington, DC, January 2003.
8. Federal Register, Clean Water Act (CWA), 33 U.S.C. ss/1251 et seq., 1977, U.S. Government, Public Laws, available at www.access.gpo.gov/uscode/title33/chapter26_.html, May 2002.
9. Federal Register, Clean Air Act (CAA), 42 U.S.C. s/s 7401 et seq., 1970, U.S. Government, Public Laws, available at www.epa.gov/airprogm/oar/caa/index.html, March 2006.
10. Aluminum Association, Aluminum Know the Facts, The Aluminum Association Inc., Arlington, VA, available at http://www.aluminum.org, 2007.
11. U.S. TC, Industry & Trade Summary—Aluminum, The U.S. Trade Commission, Washington, DC, 1994.
12. U.S. DOC, U.S. Industrial Outlook 1994—Metals, U.S. Department of Commerce, Washington, DC, 1994.
13. U.S. EPA, Guidelines and Standards for the Nonferrous Metals Forming and Iron and Steel Copper Forming Aluminum Metal Powder Production and Powder Metallurgy Point Source Category, EPA 440184019b, U.S. EPA, Washington, DC, NTIS PB95-154357, February 1984.
14. DOI, Recycled Metals in The United States, A Sustainable Resource, U.S. Department of the Interior, Bureau of Mines, Washington, DC, 1995.
15. ICA, Copper Connects Life, The International Copper Association, Ltd., New York, NY, available at http://www.copperinfo.org/aboutica/index.html, 2007.
16. U.S. Congress, Copper Technology and Competitiveness, Congress of the United States, Office of Technology Assessment, Washington, DC, 1994.
17. U.S. EPA, Compilation of Air Pollutant Emission Factors (AP42), U.S. EPA, Washington, DC, 1995.
18. U.S. EPA, 1990 Report to Congress on Special Wastes from Mineral Processing, U.S. EPA, Washington, DC, 1990.
19. Wang, L.K., Hung, Y.T. and Shammas, N.K., Eds., *Physicochemical Treatment Processes*, Humana Press, Totowa, NJ, 2005.
20. Wang, L.K., Hung, Y.T., and Shammas, N.K., Eds., *Advanced Physicochemical Treatment Processes*, Humana Press, Totowa, NJ, 2006.
21. Wang, L.K., Hung, Y.T., and Shammas, N.K., Eds., *Advanced Physicochemical Treatment Technologies*, Humana Press, Totowa, NJ, 2007.
22. Wang, L.K., Vaccari, D.A., Li,Y., and Shammas, N.K., Chemical precipitation, in *Physicochemical Treatment Processes*, Wang, L.K., Hung, Y.T., and Shammas, N.K., Eds., Humana Press, Totowa, NJ, pp. 141–198, 2005.
23. Shammas, N.K., Coagulation and flocculation, in *Physicochemical Treatment Processes*, Wang, L.K., Hung, Y.T., and Shammas, N.K., Eds., Humana Press, Totowa, NJ, pp. 103–140, 2005.

24. Chen, J.P., Chang, S.Y., and Hung, Y.T., Gravity filtration, in *Physicochemical Treatment Processes*, Wang, L.K., Hung, Y.T., and Shammas, N.K., Eds., Humana Press, Totowa, NJ, pp. 501–544, 2005.
25. Shammas, N.K., Kumar, I.J., Chang, S.Y., and Shammas, N.K., Sedimentation, in *Physicochemical Treatment Processes*, Wang, L.K., Hung, Y.T., and Shammas, N.K., Eds., Humana Press, Totowa, NJ, pp. 379–430, 2005.
26. Wang, L.K., Chang, S.Y., Hung, Y.T., Muralidhara, H.S., and Chauhan, S.P., Centrifugation, clarification and thickening, in *Biosolids Treatment Processes*, Wang, L.K., Shammas, N.K., and Hung, Y.T., Eds., Humana Press, Totowa, NJ, pp. 101–134, 2007.
27. U.S. EPA, Pollution Prevention Act of 1990, United States Code Title 42, U.S. EPA, Washington, DC, available at http://www.epa.gov/p2/pubs/p2policy/act1990.htm, October 2006.
28. Federal Register, The Emergency Planning & Community Right-To-Know Act (EPCRA), 42 U.S.C. 11011 et seq., 1986, United States Government, Public Laws, available at www.access.gpo.gov/uscode/title42/chapter116_.html, January 2004.
29. HKEPA, Recovery and Recycling of Metal Waste in Hong Kong, Waste Reduction and Recovery Factsheet No 3., Waste Reduction Group, Hong Kong Environmental Protection Agency, June 2007, available at http://www.epd.gov.hk/epd/english/environmentinhk/waste/guide_ref/files/wr_metal.pdf, 2007.
30. ISC, Guideline for Waste Treatment/Recycling by Businesses, Waste Recycling Subcommittee, Industrial Structure Council, July 2001, available at http://www.meti.go.jp/policy/recycle/main/english/pamphlets/pdf/ceBusie_e.pdf, 2007.
31. Muller, W. and Witzke, L., Processing Nonferrous Metal Hydroxide Sludge Wastes. U.S. patent 4,151,257. Free patent available at http://www.freepatentsonline.com/4151257.html, 2007.
32. Wang, L.K., Shammas, N.K., and Hung, Y.T., Eds., Biosolids Engineering and Management, Humana Press, Totowa, NJ, 2008.

4 Management, Minimization, and Recycling of Metal Casting Wastes

An Deng, Yung-Tse Hung, and Lawrence K. Wang

CONTENTS

4.1	Industry Description		152
	4.1.1	Casting Flow	153
		4.1.1.1 Overview	153
		4.1.1.2 Core Making	153
		4.1.1.3 Molding	154
		4.1.1.4 Melting and Pouring	155
		4.1.1.5 Casting Cleaning and Inspection	156
		4.1.1.6 Reclamation of Molds and Cores	157
	4.1.2	Casting Processes	157
		4.1.2.1 Sand Casting	157
		4.1.2.2 Shell Casting	157
		4.1.2.3 Investment Casting	157
		4.1.2.4 Die Casting	157
		4.1.2.5 Permanent Mold Casting	158
		4.1.2.6 Centrifugal Casting	158
	4.1.3	Sand Casting Systems	158
		4.1.3.1 Green Sand System	158
		4.1.3.2 Chemically Bonded Sand Systems	159
	4.1.4	Casting Metals	161
		4.1.4.1 Iron Castings	161
		4.1.4.2 Steel Castings	162
		4.1.4.3 Aluminum Castings	162
		4.1.4.4 Copper Castings	162
4.2	Characterization of Wastes		162
	4.2.1	General	162
	4.2.2	Air Emission	163
		4.2.2.1 Origin	163
		4.2.2.2 Characterization	163
	4.2.3	Wastewater	163
		4.2.3.1 Origin	163
		4.2.3.2 Characterization	164
	4.2.4	Spent Foundry Sand	164
		4.2.4.1 Origin	164
		4.2.4.2 Physical Properties	165

        4.2.4.3 Chemical Compositions 166
        4.2.4.4 Trace Element Characterization 166
        4.2.4.5 Mechanical Properties 167
     4.2.5 Baghouse Dust 168
        4.2.5.1 Origin 168
        4.2.5.2 Physical Properties 168
        4.2.5.3 Chemical Composition 168
     4.2.6 Furnace Slag 168
        4.2.6.1 Origin 168
        4.2.6.2 Physical Properties 170
        4.2.6.3 Chemical Compositions 171
        4.2.6.4 Mechanical Properties 174
        4.2.6.5 Thermal Properties 175
4.3 Source Reduction 175
     4.3.1 Chemical Substitution or Minimization 175
     4.3.2 In-Plant Reclamation 176
     4.3.3 Waste Segregation 177
     4.3.4 Process Modifications to Reduce Emission 177
4.4 Solid Wastes Reuse Technologies 178
     4.4.1 General 178
     4.4.2 Reuse Evaluation Framework 178
        4.4.2.1 Technical Implementability 178
        4.4.2.2 Environmental Issues 180
     4.4.3 Reuse in Civil Engineering 180
        4.4.3.1 Asphalt Concrete 181
        4.4.3.2 Portland Cement Concrete 183
        4.4.3.3 Portland Cement 186
        4.4.3.4 Embankment or Fill Material 187
        4.4.3.5 Flowable Fill 189
        4.4.3.6 Landfill Liner, Cover, and Hydraulic Barrier 190
     4.4.4 Agricultural Applications 191
        4.4.4.1 Topsoil 191
        4.4.4.2 Growing Amendments 192
     4.4.5 General Processes 192
        4.4.5.1 Crushing, Screening, and Storage 192
        4.4.5.2 Design and Construction 192
     4.4.6 Unresolved Issues 193
4.5 Barriers to Solid Waste Reuse 193
     4.5.1 Education 193
     4.5.2 Environmental Regulation 193
     4.5.3 Guidelines, Procedures, and Specifications 194
     4.5.4 Economics 194
     4.5.5 Market Potential 194
References 195

4.1 INDUSTRY DESCRIPTION

The metal casting industry, also known as the foundry industry, is one of the largest recyclers in the world. For centuries, this industry has been converting a huge volume (e.g., 15 to 20 million tons in the U.S.) of scrap metal that would otherwise be disposed in landfills, into manufactured useful

Metal Casting Wastes

products. This scrap metal forms the raw material charged into furnaces of the foundry facility and converted into usable castings. The casting categories include many general ferrous and nonferrous metals and their alloys, including iron, steel, aluminum, copper, magnesium, and zinc. Major end-use markets cross all sectors of the global economy, examples being the automotive industry, transportation equipment, construction, mining and oil field machinery, and industrial machinery.

4.1.1 Casting Flow

4.1.1.1 Overview

Metal casting is a process in which molten metal is poured into a mold to produce metal products. In the mold, the molten metal cools and shapes into castings by filling the preset mold internal space. The most common metal casting process is sand casting, which uses sand as the major molding and core-making material. Besides sands, other materials can also be used as molding materials, such as ceramic mold for investment casting and metal mold for die casting.[1,2] A general metal casting flow diagram is shown in Figure 4.1. A schematic process is shown in Figure 4.2. The casting begins with the customer demands and material preparations, including metal product specification, sands, binders, and scraps. Manufactured molds and cores are assembled in the assembly area, and made ready for pouring.

As the assembled molds are being placed on the pour-off lines, the scrap metal is melted in the furnace. Molten metal from the furnace is brought to the molds on the pouring lines in a refractory lined pouring ladle. Once poured, the molds are allowed to cool before being sent to the shakeout processes. At the shakeout, the castings are separated from the sand mold. The sand is sent to a reclamation system so that it can be reused in the molding process.

The materials comprising the core and mold in the casting processes have the properties of porosity, cohesion, and refractoriness. Sand has globally been selected as one of the materials meeting the property requirements. Its aggregate porosity, connected as passages, allows air and steam to escape from the mold during casting. The sand particles can adhere together into all kinds of molding shapes. In particular, sands have the ability to withstand severe heat and resist penetration of the molten metal, and impart a smooth and desirable appearance to the casting.

4.1.1.2 Core Making

Cores are separate shapes that are placed in the mold to provide castings with contours, cavities, and passages that are not achievable by the mold alone. A core has to be fixed tightly in place while the metal flows around it. Cores are made by mixing sand with binders and catalysts, which are

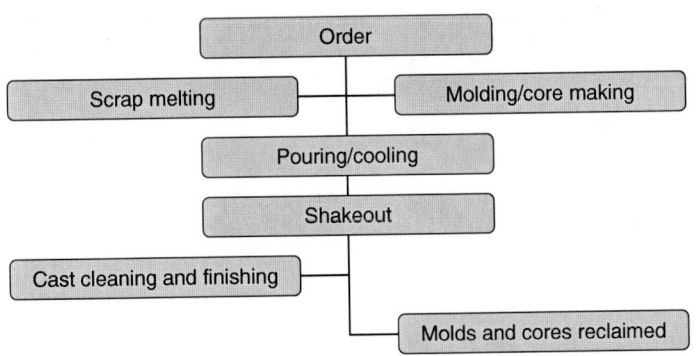

FIGURE 4.1 General metal casting flow chart. (From U.S. EPA, Summary of Factors Affecting Compliance by Ferrous Foundries, Vol. 1, EPA-340/1-80-020, U.S. EPA, Washington, DC, January 1981.)

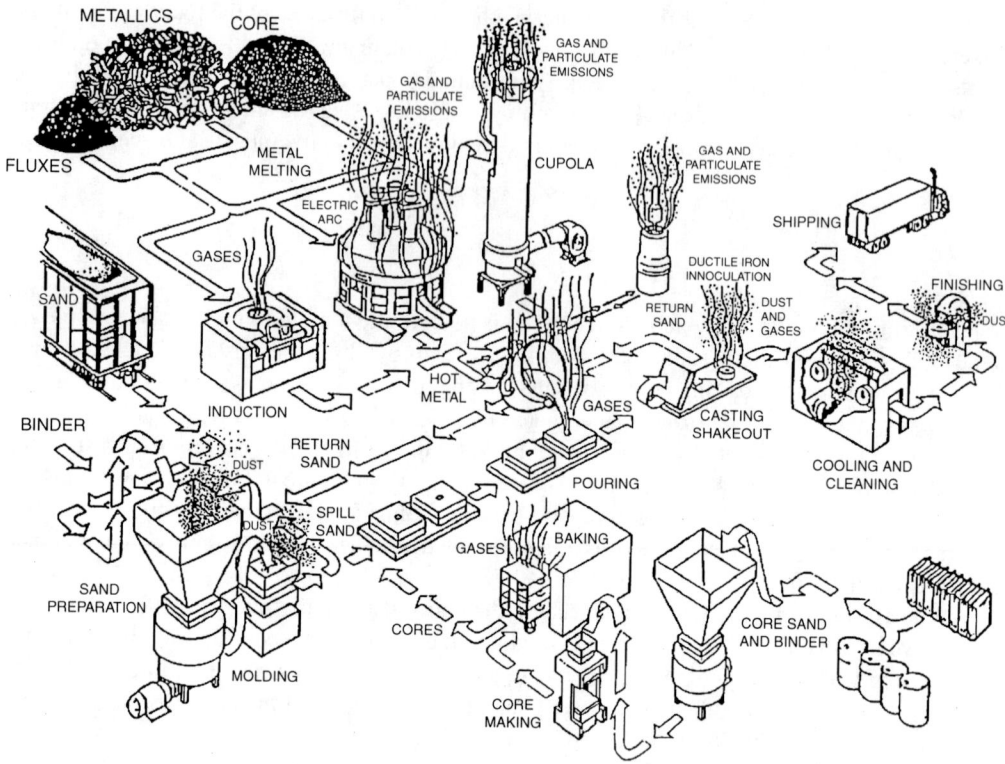

FIGURE 4.2 Metal casting process.

activated to bond sand into various shapes. Figure 4.3 demonstrates the typical core-making process. The sand and binders are blended uniformly in a mixer. The mixture is then discharged into a core machine, where continuous curing with a catalyst is applied. After the core is cured, it is removed and sent to a molding assembly area.

4.1.1.3 Molding

Molding is the process where a pattern is pressed or embedded into special sand to the desired shape or form. Alternatively, the pattern can be placed on a molding board, and the sand rammed or compressed around the pattern. Figure 4.4 shows the typical molding flow. The sand and binder are first

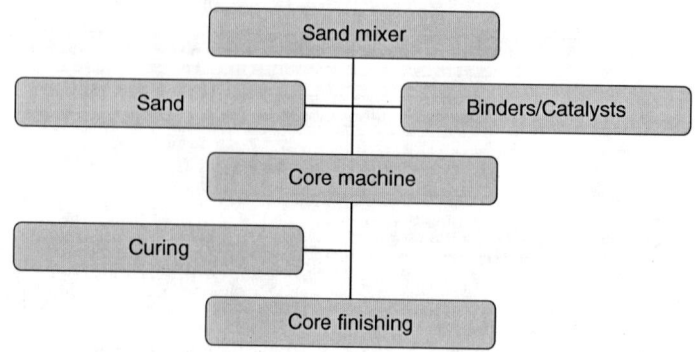

FIGURE 4.3 Core making flow chart.

Metal Casting Wastes

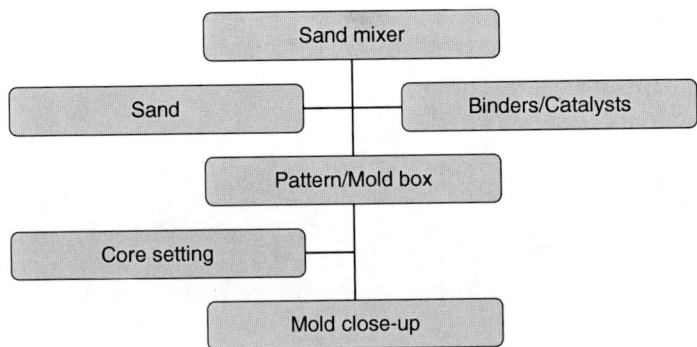

FIGURE 4.4 Typical molding flow chart.

mixed in a mixer. Then the mixed sand is discharged onto a pattern face mounted in a mold box. The sand in the box is compacted to its maximum density. Once the sand is set, the pattern is removed from the sand, and cores are placed in the mold. The mold is then closed up and moved to the pouring lines.

4.1.1.4 Melting and Pouring

Figure 4.5 describes the typical metal melting flow chart. At first, the customer's metal specification for the casting will determine what type of scrap metal will be used to feed the furnace. Once charged, the furnace uses multiple melting powers or burners (that is, electricity, kerosene, gas, coke, charcoal, and used engine oil) to melt the scrap metal. Scrap metals may be fully melted in tens of minutes, depending upon the size of the vessel used. Alloys are added according to the metal specification. The addition is determined by spectro analysis in the melting process. When the melting is complete, the molten metal is placed in a pouring shank and sent to the pouring line.

Furnace types include cupolas, electric arc, induction, hearth or reverberatory, and crucible. Because of the different characteristics of metals, different inputs are required and different pollution is released from each type. Table 4.1 summarizes the types of furnaces depending on the type of metal being used.

The molds are cast and allowed to cool for a suitable time, often 30 to 40 min, before shaking out the castings. The shakeout work may be supplied by a vibrating conveyor or a rotating drum, which cause the molds to be broken up by the vibration, exposing the casting for removal. The sand

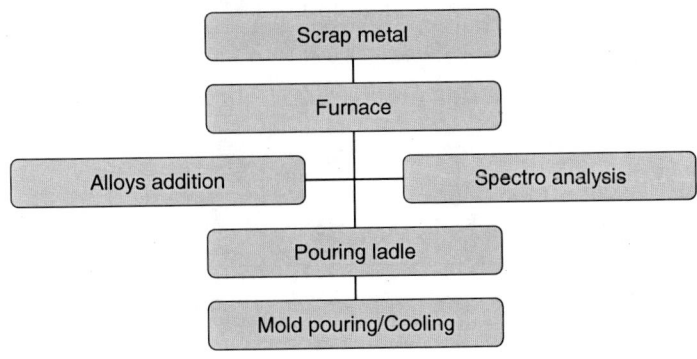

FIGURE 4.5 Typical melting flow chart.

TABLE 4.1
Common Types of Metal-Melting Furnaces

Furnace Type	Raw Materials	Outputs	Process
Cupola furnace	Iron ore, scrap iron, lime, coke	Molten iron	Alternative layers of metal and coke are fed into the top of the furnace. The metal is melted by hot gasses from the coke combustion. Impurities react with the lime and are separated.
Electric arc furnace	Scrap iron, flux	Molten iron and steel	Electric arcs from carbon electrodes melt the scrap metal. The flux reacts with impurities.
Induction furnace	Scrap iron or nonferrous metals	Molten iron or nonferrous metals	Induction furnaces are the most common type used by both ferrous and nonferrous foundries. Copper coils heat the metal using alternating currents. The flux reacts with impurities.
Reverberatory, hearth, or crucible furnace	Nonferrous metals, flux	Molten nonferrous metals	Reverberatory furnaces melt metals in batches using a pot-shaped crucible that holds the metal over an electric heater or fuel-free burner. The flux reacts with impurities.

Source: From WMRC, Primary Metal, Illinois Waste Management and Research Centre, available at http://www.wmrc.uiuc.edu/info/library_docs/manuals/primmetals/chapter3.htm.

from the mold is separated and processed through a reclamation system for reuse in molding and core making.

4.1.1.5 Casting Cleaning and Inspection

The foundry cleaning room is a collection area where castings are finished to meet the casting specifications. A sample flow chart is shown Figure 4.6. When castings are removed from the shake-out, they are run through the shot blast to remove sand and expose the surface for inspection and further work. Castings are inspected for defects such as cracks, flashing, and inclusions. If none are found the castings are sent to the heat-treating department. If defects are present that require welding or grinding the castings are sent to the appropriate area to have the defect corrected. Once rework is completed, the castings are sent to the heat-treating department. After being heat treated, the castings are again sent through the shot blast before being sent to the final inspection area.

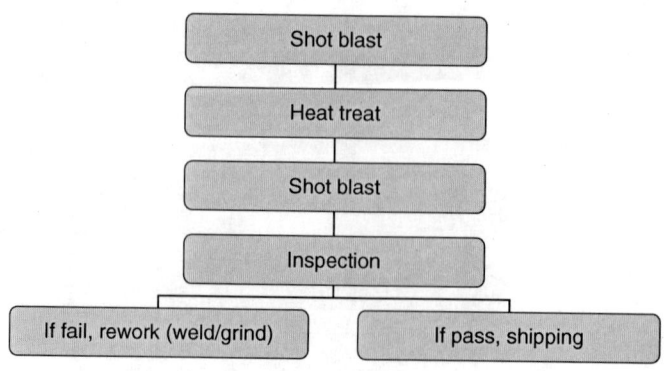

FIGURE 4.6 Casting, cleaning and inspection flow chart.

Metal Casting Wastes

4.1.1.6 Reclamation of Molds and Cores

After shakeout, the return sand is reclaimed by a crushing process and by screening out core lumps, nonmagnetic metallics, and other unwanted material. Burnt binders (such as clay, resin, and other foreign fines) will excessively build up in the matrix of reclaimed molds and cores, and may reduce the gas/heat permeability within the molds and cores. This used molding sand will be put back through the sieve to have the correct amount of water added. Sometimes, the reclaim process may not be sufficient to recondition the technically acceptable or functional refractory materials, which eventually become excess foundry sand and is removed from the system. New sand, additional water, make-up binder, and additional catalysts are added to ensure molding and core-making quality.

4.1.2 CASTING PROCESSES

Casting processes can be divided according to the refractory materials used, that is, sand, ceramics, and metals. The principal sand casting processes used in the metal casting industry are sand mold casting, expendable pattern (lost foam) casting, shell mold casting, plaster mold casting, and vacuum (V-process) casting. Processes that use some form of disposable ceramic molds include the ceramic mold process and investment mold casting. Processes that use a reusable metal mold include die casting, permanent mold casting, and centrifugal casting.

4.1.2.1 Sand Casting

This is the earliest and the most commonly used casting process. It has the advantages of wide metal suitability, low cost, and simple operation. It uses sand as a refractory material. Many types of sand are utilized by the foundry industry. However, because of its wide availability and relatively low cost, silica sand is the one that makes most metal castings. Silica sand is composed of the mineral quartz (SiO_2), which has a fusion point of approximately 1670°C (3090°F), which is often lowered by the presence of appreciable quantities of minerals with lower fusion points.

4.1.2.2 Shell Casting

In this process, the mold cavity is formed by a shell of resin-bonded sand. The shell is built up in layers, starting with a very fine-grained dip-coat, which is then dusted with a fine powder (molochite or zircon). Once the first coat is set hard, the wax is dipped a second time. The third and successive coats are dusted with coarse stucco. This coarse aggregate builds up the strength of the shell. Shell mold castings surpass ordinary sand castings in surface finish and dimensional accuracy, and cool at slightly higher rates. In addition, equipment costs are higher, and the size and complexity of castings that can be produced are more limited.

4.1.2.3 Investment Casting

This process is used to produce intricate, thin-section parts with great dimensional accuracy, fine detail, and very smooth surfaces. All ferrous and nonferrous alloys can be cast in investment molds. Investment casting begins with expendable wax patterns that are assembled into clusters, then coated with a series of successively coarser ceramic slurries. The assembly is then fired in a furnace to dry and harden the ceramic shell and to melt out the wax, leaving a cavity into which molten metal is poured to form the casting.

4.1.2.4 Die Casting

A die is a reusable mold, usually made of steel, for the mass production of small parts in low-melting-point alloys—usually zinc or aluminum alloys. For the mass production of small parts that have no undercuts, the durability and excellent surface quality of the die, in addition to the saving in labor costs, make die casting a competitive and worthwhile process.

4.1.2.5 Permanent Mold Casting

Metal molds and cores are used in permanent mold casting. The process works best in continuous operation so that the mold temperature can be maintained within a fixed operating range. The operating temperature of the mold is one of the most important factors in successful permanent mold casting. Mold cavities are machined from solid blocks of graphite. Mold life is the major cost factor in permanent mold casting.

4.1.2.6 Centrifugal Casting

Centrifugal force is used to introduce molten metal into a mold cavity that is spinning around its axis. Cast iron pipe is produced in centrifugal molds, and copper-base alloy bearings are also commonly produced this way. Permanent metal molds are usually coated or lined to extend operating life.

4.1.3 SAND CASTING SYSTEMS

Of the many casting processes, sand casting is principally addressed in this section as this process not only prevails in the casting industry, but also generates a vast volume of solid wastes. Sand casting systems are possibly the most versatile foundry method, and are largely divided according to the binder and bonding manners into two categories, green sand and chemically bonded sand systems. The green sand system uses sand, clay, water, and additives as components, and bonds sand particles together by relying upon mechanical forces generated by mixing the clay and water. Chemically bonded sand systems use sand, resins, or inorganic binders, and sometimes water and catalysts, as components. The bonding forces are generated by chemical reactions (polymerization) between the resins/inorganic binders and catalysts.

4.1.3.1 Green Sand System

Green sand molding is the most widely used molding process in the world, accounting for up to 90% ferrous sand casting materials.[4] It is low in cost, high in performance, and the materials are reusable. Green, in this sense, does not refer to color, but is a technical point, indicating a natural bonding effect (with water, but without artificial binders, additives, or catalysts). Green sand consists of 85 to 95% high-quality silica sand, 4 to 10% bentonite clay (as a binder), 2 to 5% water, and 2 to 10% sea coal (a carbonaceous mold additive to improve casting surface finish). A machine, known as a muller, is used to coat the sand quickly and uniformly with a clay and water mixture (glue). A muller is capable of producing more than several tons of prepared molding sand in a few minutes.

Sand is composed of grains ranging from 0.05 to 2 mm in diameter. The physical properties of sand that can affect green sand system performance include grain shape, grain size distribution, grain fineness, permeability, density, and coefficient of thermal expansion. The chemical properties of sand that can affect green sand system performance include chemical composition, loss on ignition, pH value, and fusion point.

Clays used in foundries include hydrous alumina silicates, known as bentonites. Their properties provide cohesion and plasticity in the green state and also high strength when dried. There are three clays that are commonly used in foundries: western bentonite [sodium bentonite, burnout point 1290°C (2350°F)], southern bentonite [calcium bentonite, burnout point 1065°C (1950°F)], and fire clay [kaolinite, burnout point 1425°C (2600°F)].

The water used for a green sand system should be clean and it should be consistent. It should have a pH that is neutral to alkaline, not acidic, because acids prevent bentonite from swelling. In

Metal Casting Wastes

addition to sand, clay, and water, there are a number of other materials, or additives, that can enhance, control, and optimize the performance of green sand systems.

4.1.3.2 Chemically Bonded Sand Systems

Foundry cores and molds may be made using resin-coated sand prepared by a number of different bonding processes (e.g., no-bake, cold-box, warm-box, hot-box, shell) that use all sorts of different binders (resins) with unique chemistries. These binders can be triggered, based upon the processes, in two ways—self-setting and triggered setting. In the self-setting system (also known as a self-set, cold setting, cold-box, or no-bake process), sand, binder, and a hardening chemical are mixed together; the binder and hardener start to react immediately, but sufficiently slowly to allow the sand to be formed into a mold or core, which continues to harden further until strong enough to allow casting. In triggered setting system, sand and binder are mixed and blown or rammed into a core box. Little or no hardening reaction occurs until triggered by applying heat or a catalyst gas. Setting then takes place in seconds.

Self-setting systems
These include the following:

1. *Furanes.* Furane sands use a furane resin and an acid catalyst. The resins are urea-formaldehyde (UF), phenol-formaldehyde (PF), or UFPF resins with additions of furfuryl alcohol (FA). The speed of setting is controlled by the percentage of acid catalyst used and the strength of the acid. Ratios such as resin at 0.8 to 1.5% of base sand, catalyst at 40 to 60% of resin are normally used, depending on the sand temperature and the speed of setting required. The optimum ambient temperature is 20 to 30°C, resulting in a compressive strength of typically 4000 kPa (600 psi).
2. *Phenolic-isocyanates (phenolic-urethanes).* The binder is supplied in three parts: a phenolic resin in an organic solvent (0.8%), methylene diphenyl diisocyanate (MDI) (0.5%), and a liquid amine catalyst. When mixed with sand, the amine causes a reaction between the resin and the MDI, forming urethane bonds, which rapidly set the mixture. The speed of setting is controlled by the type of catalyst. The optimum cure temperature is 25 to 30°C. Compression strength is typically over 4000 kPa (600 psi).
3. *Alkaline phenolic resin, ester hardened.* The binder is a low viscosity, highly alkaline phenolic resole resin (1.2 to 1.7%). The hardener is a liquid organic ester (18 to 25%). Sand is mixed with hardener and resin, usually in a continuous mixer. The speed of setting is controlled by the type of ester used. Low sand temperature slows the cure rate, but special hardeners are available for cold and warm sand. In 24 hr compression strength can reach 4000 kPa (600 psi).
4. *Ester silicate process.* Sand is mixed with a suitable grade of sodium silicate (2.2 to 2.8% of sand weight), often incorporating a breakdown agent, together with 10 to 12% (based on silicate) of liquid organic ester hardener. The acid ester reacts with the sodium silicate, hardening the sand. The speed of hardening is controlled by the type of ester used. The final compression and tensile strength achieved are 2000 to 5000 kPa (300 to 700 psi) and 700 kPa (100 psi), respectively.
5. *Portland cement process.* Sand is mixed with portland cement (10% of sand weight) and water (5% of sand weight). Molds are air dried for 24 hr and may then be dried out more rapidly.

Triggered setting systems
Triggered setting systems are used to make cores for repetition foundries. After the mixed sand is blown into the core box, the cores must be cured in the box until sufficient strength has been achieved to allow stripping without damage or distortion. Usually the core continues to harden after stripping. Tensile strengths of 1000 to 2000 kPa (150 to 300 psi) are typical, equating

roughly to transverse strengths of 1500 to 3000 kPa. Final strengths may be higher. Triggered setting systems are categorized into two groups, that is, heat triggered processes and gas triggered processes.

Heat triggered processes include the following bonding systems:

1. *Shell process.* Sand is precoated with a solid phenolic novolak resin and a catalyst to form a dry, free-flowing material. The coated sand is blown into a heated core box or dumped onto a heated pattern plate, causing the resin to melt and then harden. Shell molds are normally 20 to 25 mm thick. Resin additions are 2.5 to 4.5% of sand weight, and the catalyst hexamine is added at 11 to 14% of the resin content. The minimum curing time is 90 sec but 2 min is common. A 3.5% resin content will give a tensile strength of 1400 kPa (200 psi).
2. *Hot-box process.* The binder is an aqueous PFUF or UFFA resin, and the catalyst is an aqueous solution of ammonium salts, usually chloride and bromide. Sand is mixed with the liquid resin (2.0 to 2.5% of sand weight) and catalyst (20 to 25% of resin weight) and blown into a heated core box. The heat liberates acid vapor from the catalyst, which triggers the hardening reaction. Hardening continues after removal of the core from the box. Thin section cores cure in 5 to 10 sec. As cores increase in section size, curing time must be extended up to about 1 min for a 50-mm section. The final tensile strength is 1400 to 2800 kPa (200 to 400 psi).
3. *Warm-box process.* The binder, 1.3 to 1.5% of sand weight, is a reactive, high FA binder. The catalyst, 20% of sand weight, is usually a copper salt of sulfonic acid. Sand, binder, and catalyst are mixed and blown into a heated core box. The heat activates the catalyst, which causes the binder to cure. Curing time is 10 to 30 sec depending on thickness. The final tensile strength can be 3000 to 4000 kPa (400 to 600 psi).
4. *Oil sand.* Certain natural oils, such as linseed oil, known as "drying oils," polymerize and harden when exposed to air and heat. Silica sand is mixed with the drying oil (1 to 2% of sand weight), a cereal binder (1 to 2% of sand weight), and water (2 to 2.5% of sand weight). The resulting mixture is either manually packed or blown into a cold core box. Applied backing will harden the oil and the core becomes rigid. A recirculation air oven is needed because oxygen is necessary to harden the oil. The temperature is normally 230°C, allowing 1 hr for each 25-mm section thickness. Correctly baked cores develop a tensile strength of 1340 kPa (200 psi).

Gas triggered processes include the following bonding systems:

1. *Phenolic-urethane-amine gassed (cold-box) process.* The binder is supplied in two parts: a solvent-based phenolic resin (0.8 to 1.5% of sand weight), a polyisocyanate (0.8 to 1.5% of sand weight), MDI (methylene diphenyl diisocyanate) in a solvent. The resins are mixed with sand and the mixture blown into a core box. An amine gas [TEA (triethylamine) or DMEA (dimethyl ethyl amine)] is blown into the core, catalyzing the reaction and causing almost instant hardening. The tensile strength immediately after curing is high at 2000 kPa (300 psi), and the transverse strength is 2700 kPa (400 psi).
2. *Alkaline phenolic resin gassed process.* Alkaline phenol-formaldehyde resin containing a coupling agent is used. The resin (2.0 to 2.5% of sand weight) is mixed with sand, and the mixture is blown into a core box. Carbon dioxide is passed through the mixture, lowering the pH and activating the coupling agent, which causes crosslinking and hardening of the resin. Strength continues to develop after the core is ejected as further crosslinking occurs and moisture dries out. The compression strength is 2000 to 3000 kPa (300 to 400 psi), and the tensile strength is 500 to 800 kPa (70 to 110 psi).

Metal Casting Wastes

3. *The SO_2 process.* Sand is mixed with a furane polymer resin (1.2 to 1.4% of sand weight) and an organic peroxide (such as methyl ethyl ketone peroxide at 25 to 60% of resin weight). The mixture is blown into the core box and hardened by passing sulfur dioxide gas through the compacted sand. The gas reacts with the peroxide-forming carbon trioxide and then H_2SO_4, which hardens the resin binder. The tensile strength is 1250 kPa (180 psi) after 6 hr.
4. *SO_2-cured epoxy resin.* Modified epoxy/acrylic resins (1.2 to 1.4% of sand weight) are mixed with organic peroxide (26 to 60% of resin weight), the mixture is blown into the core box and a hardening mechanism similar to the SO_2 process takes place.
5. *Ester-cured alkaline phenolic system.* The resin is an alkaline phenolic resin (essentially the same as the self-hardening resins of this type). Sand is mixed with the resin and blown or manually packed into a core box. A vaporized ester, methyl formate, is passed through the sand, hardening the binder. The total resin and peroxide addition is 1.5%. Compression strengths of 5000 kPa (700 psi) are possible.
6. *Carbon dioxide–silica process.* Sand is mixed with sodium silicate (3.0 to 3.5% of sand volume), and the mixture is blown or hand-rammed into a core box or around a pattern. Carbon dioxide gas is passed through the compacted sand to harden the binder. The bonding strength eliminates the need for drying or baking the mold and metal can be poured into the mold immediately. Over-gassing should be avoided because it makes the mixture friable.

4.1.4 Casting Metals

The metal casting industry conventionally divides casting products into ferrous and nonferrous metals, in particular, iron-based, steel-based, aluminum-based, and copper-based castings. The other castings of low fractions include magnesium, lead, zinc, and their alloys. In the U.S., the foundry industry currently produces 11 million tons of metal product per year, with a shipment value of $19 billion. Of them, iron and steel accounted for 84% of metals cast.[5] The remaining 15% of foundry operations are concerned with aluminum, copper, zinc, and lead production. Table 4.2 summarizes critical physical and thermal properties of aluminum, iron/steel, and cast iron.

4.1.4.1 Iron Castings

Iron is the world's most widely used metal. Iron castings encompass a family of ferrous alloys: gray iron, alloy iron, white iron, malleable iron, ductile iron, and compacted graphite iron. Wide variations in properties can be achieved by varying the balance between carbon and silicon, by alloying,

TABLE 4.2
Physical and Mechanical Properties of Aluminum, Iron/Steel, and Cast Iron

Properties	Al	Fe	Gray Iron	Ductile Iron
Content (%)	Pure	Pure	10–40	40–70
Atomic weight (g/cm³)	26.98	55.85	48.44	48.44
Density, solid (g/cm³)	2.70	7.86	7.1–7.35	7.06–7.44
Density, liquid (g/cm³)	2.38	7.01	7.1	7.1
Melting point (°C)	660.37	1536	1120–1180	1120–1180
Thermal expansion coefficient, 0°C (K^{-1} 10^{-6})	23.9	11.7	11.7	11.4–12.8
Heat conductivity, 25°C (W/m K)	237	74.4	45–52	25–42

and by applying various types of heat treatment. Iron castings have good fluidity and mold filling during the casting process, with low shrinkage on cooling.

4.1.4.2 Steel Castings

The most common types of steels used in castings are carbon steels, which contain only carbon as the major alloying element. Carbon steels are classified by their carbon content into three groups: low-carbon steel (C < 0.20%), medium-carbon steel (C = 0.20 to 0.50%), and high-carbon steel (C > 0.50%). Steel's hardness also depends upon the carbon content.

4.1.4.3 Aluminum Castings

Aluminum is a light metal with good tensile strength. It is easily cast, extruded, or pressed. At present, aluminum is the second most widely used metal after iron. Aluminum castings can be cast by virtually all of the common casting processes. It is common to add the alloying constituents as solids to molten aluminum: Al–Cu, Al–Mg, Al–Zn, Al–Sn. The potential for the use of aluminum in automotive applications is considerable, including engine blocks, heads, pistons, rocker covers, inlet manifolds, differential casings, steering boxes, brackets, wheels, and so on. Castings may also be used for household and hospital utensils, and machinery.

4.1.4.4 Copper Castings

Copper is a soft metal that is resistant to corrosion and is a good conductor of heat and electricity. It is most commonly used for electrical wiring and hot water pipes. Copper is second only to aluminum in importance among the nonferrous metals. Products include bushings and bearings, propellers, and other cast products. Copper-base alloys are grouped according to composition: pure copper, high-copper alloys, brasses, leaded brasses, bronzes, aluminum bronzes, silicon bronzes, copper–nickel alloys, and copper–nickel–zinc alloys. In brasses, zinc is the principal alloying element. Tin is the principal alloying element in cast bronze alloys. Copper castings are produced by several methods, including centrifugal molds, green sand molds, and die casting.

4.2 CHARACTERIZATION OF WASTES

Three major solid wastes—spent foundry sand, furnace slag, and baghouse dust—are discharged from metal casting facilities. In the U.S., the annual generation of foundry solid waste is believed to range from 9 to 13.6 million metric tons (10 to 15 million tons).[6] Of them, spent foundry sand can account for nearly 70% of a foundry's total wastestream.[1] In addition to solid waste, wastewater and air emissions are also discharged from a metal casting facility. Reliable quantification of physical properties and chemical characterization of the byproduct is important for the marketability of the materials. This section focuses on the characterization of the solid wastes of the metal casting industry. Characterization of air emission and wastewater shall also be addressed according to some limited documental data. Mainly four aspects of characterization for a solid wastestream are included: origin, physical properties, chemical properties, and mechanical properties.

4.2.1 GENERAL

Prior to their acceptance for beneficial treatment or reuse, foundry wastes discharged from casting processes are characterized and must comply with environmental protection laws and regulations. Countries vary significantly in constituting environmental protection laws. In the U.S., numerous federal environmental laws (or acts) and regulations have been promulgated to protect human health and the environment. Table 4.3 lists most of the federal laws or regulations involved in managing wastes of the metal casting industry. These acts are the unique measures assessing the environmental impact and reuse acceptance of foundry solid waste. Thus, detailed physical and chemical characterization of foundry waste materials is necessary in order to obtain permits for reusing foundry byproducts.

TABLE 4.3
Federal Legislation Related to Solid Waste Management

Title	Year of Promulgation or Amendment
Solid Waste Disposal Act (SWDA)	1965
National Environmental Policy Act (NEPA)	1969
Occupational Safety and Health Act (OSHA)	1970
Clean Air Act (CAA)	1970, 1977, 1990
Clean Water Act (CWA)	1977, 1981, 1987
Safe Drinking Water Act (SDWA)	1974, 1977, 1986
Toxic Substances Control Act (TSCA)	1976
Resource Conservation and Recovery Act (RCRA)	1976, 1980
Comprehensive Environmental Response, Compensation and Liabilities Act (CERCLA or Superfund)	1980
Hazardous and Solid Waste Amendments (HSWA)	1984
Superfund Amendments	1986
Pollution Prevention Act (PPA)	1990

4.2.2 Air Emission

4.2.2.1 Origin

Air emission, known as a gaseous waste, is the largest waste source from foundries.[2] Emission sources include the binder systems used in mold making, vapors from metal melting, and airborne sand used in the pouring and shakeout steps. Very limited quantified data are available about the characterization of air emissions. They are thought containing metals dust, semivolatile and volatile organic compounds. They mainly come from the melting procedures. Pouring and cooling steps contribute about 16% of the total organic and semivolatile wastes from foundries.[7]

4.2.2.2 Characterization

Air emission composition is closely related to its form of generation or collection. Cupola furnaces produce more metallic air emissions than other furnace types. Lower metal emissions are released from induction furnaces and core- and mold-making processes. Emissions from the pouring process depend on the metal temperature. The hotter the metals, the higher the metal emissions.[7] Organic air emissions arise largely from vaporized resins, solvents, and catalysts, which are used extensively in core- and mold-making steps. With the promulgation of the Clean Air Act and its amendments, as well as increasingly stringent regulations from U.S. EPA, more air emissions studies are being conducted.

The principal gases produced were found to be hydrogen, carbon monoxide, carbon dioxide, methane, nitrogen, oxygen, and water vapor. Volatile hydrocarbons, including ethane, ethylene, propane, propylene, acetylene, FA, methanol, and ethanol, constitute up to 5% of the gas volume. Benzene, toluene, nitrous oxide, and hydrogen cyanide were identified in the atmosphere near a pouring line in a foundry using alkyd isocyanate resin bonded molds. Concentrations detected in the foundry atmosphere were generally low.

4.2.3 Wastewater

4.2.3.1 Origin

Wastewater discharge, known as liquid pollution in a facility, makes up a small portion of the total wastestream from foundries.[2] Wastewater mainly comes from the noncontact cooling water used to cool metal and other work pieces or from wet scrubber air emission systems. Water runoff from floor cleaning and other maintenance procedures may also contribute to wastewater. However, the volumes of liquid waste are relatively small and do not pose a large pollution problem for foundries. Some plants have water treatment facilities to remove contaminants for water reuse.

U.S. EPA promulgated wastewater discharge regulations for the foundry industry in October 1985, which are published in the Code of Federal Regulations at 40 CFR Part 464.[8] The regulations cover 28 process segments (processes such as casting quench, grinding scrubber, mold cooling) in four subcategories: aluminum casting, copper casting, ferrous casting, and zinc casting. It is noted that the cast metals have unique properties that influence the way they are melted and processed and, thus, affect the process wastewater characteristics.

4.2.3.2 Characterization

Table 4.4 presents wastewater flow characterization for the foundry industry by casting metals. Also presented in this table is the level of process water recycle, and the number of plants surveyed with central wastewater treatment facilities for all of the processes at that plant. The discharge flow represents all processes within the specific metal casting facilities.

Many toxic pollutants were detected in the process wastewaters from metal molding and casting processes. The toxic pollutants detected most frequently in concentrations at or above 0.1 mg/L were phenolic compounds and heavy metals. The pollutants include 2,4,6-trichlorophenol, 2,4-dimethylphenol, phenol, 2-ethylhexyl, cadmium, chromium, copper, lead, nickel, and zinc. Each type of operation in the foundry industry can produce different types of pollutants in the wastewater stream. Also, because each subcategory operation often involves different processes, pollutant concentrations per casting metals may vary.

4.2.4 SPENT FOUNDRY SAND

4.2.4.1 Origin

Foundries purchase new, virgin sand to make casting molds, and the sand is reused numerous times within the foundry. However, heat and mechanical abrasion eventually render the sand unsuitable for use in casting molds, and a portion of the sand is continuously removed and replaced with virgin sand. The removed sand becomes spent foundry sand, which is discarded from the foundry facility.

In the U.S., the foundry industry produces roughly seven to eight million tons of spent sand each year,[1] which are available to be recycled into nonfoundry applications. However, less than 15% of

TABLE 4.4
Wastewater Flow Characterization by Casting Metals

	Iron/Steel Casting	Aluminum Casting	Copper and Alloy Casting	Magnesium Casting	Zinc Casting
Applied flow (ML/yr)	397,000	14,500	34,900	8.18	4040
Recycle flow (ML/yr)	317,000	7530	25,300	0	3430
Direct discharge flow (ML/yr)	69,300	5700	9610	8.18	5050
Indirect discharge flow (ML/yr)	11,600	1260	48	0	100
100% recycle flow (ML/yr)	189,000	408	3340	0	1010
Central treatment facilities (no. of plants)	109	12	10	0	13
Operation treatment facilities (no. of plants)	205	20	14	3	12

Note: ML/yr, million liters per year.

Metal Casting Wastes

the spent foundry sand is recycled. It is believed that a greater percentage of spent foundry sand can be safely and economically recycled, as has been encouraged by many successful case studies. Concentrating energies on the largest volume stream first will have the greatest economic impact for the industry as a whole.

Spent material often contains casting residues, such as degraded binders, metals, and oversized mold/core materials. Spent foundry sand may also contain some leachable contaminants, including heavy metals and phenols that are absorbed by the sand during the molding process and casting operations.[9] The detection of heavy metals is of greater concern in nonferrous foundry sands generated from nonferrous foundries.[10] Spent foundry sand from the brass or bronze foundries, in particular, may contain high concentrations of cadmium, lead, copper, nickel, and zinc.[11]

4.2.4.2 Physical Properties

Spent foundry sand can be divided, based upon bonding processes, into two categories—spent green sand and spent chemically bonded sand. Spent green sand is black in color due to its carbon content, and has clay contents that result in a fraction of the material passing a No. 200 sieve (0.075 mm). Chemically bonded sands are generally yellowish in color and coarser in texture than clay bonded sands.

Physical properties involve tests of the physical index parameters of the materials. For spent foundry sand, these parameters include particle gradation, unit weight, specific density, moisture content, adsorption, hydraulic conductivity, clay content, plastic limit, and plastic index. These parameters determine the suitability of spent foundry sand for uses in potential applications. Typical physical properties of spent green foundry sand are listed in Table 4.5.

The grain size distribution of spent foundry sand is very uniform, with approximately 85 to 95% of the material between 0.6 mm and 0.15 mm (No. 30 and No. 100) sieve sizes. Five to twelve percent of foundry sand can be expected to be smaller than 0.075 mm (No. 200 sieve). The particle shape is typically subangular to round. Spent foundry sand gradations are too fine to satisfy the fine aggregate standard specified in specification ASTM C33 Standard Specification for Concrete Aggregates.

Spent foundry sand has low absorption and is nonplastic. Reported values of absorption were found to vary widely (0 to 5%), which can also be attributed to the presence of binders and

TABLE 4.5
Typical Physical Properties of Spent Green Foundry Sand

Property	Results	Test Methods
Specific gravity	2.39–2.55	ASTM D854
Bulk relative density, kg/m^3 (lb/ft^3)	2590 (160)	ASTM C48/AASHTO T84
Absorption (%)	0.45	ASTM C128
Moisture content (%)	0.1–10.1	ASTM D2216
Clay lumps and friable particles	1–44	ASTM C142/AASHTO T112
Coefficient of permeability (cm/sec)	10^{-3}–10^{-6}	AASHTO T215/ASTM D2434
Plastic limit/plastic index	Nonplastic	AASHTO T90/ASTM D4318

Source: From AFS, Alternative Utilization of Foundry Waste Sand, final report (Phase I) for Illinois Department of Commerce and Community Affairs, American Foundrymen's Society, Des Plaines, IL, July 1991. Javed, S. and Lovell, C.W., Use of Foundry Sand in Highway Construction, Joint Highway Research Project No. C-36-50N, Purdue University, West Lafayette, IN, July 1994. Javed, S., Lovell, C. W., and Wood, L.E., Waste Foundry Sand in Asphalt Concrete, in *Transportation Research Record*, No 1437, Transportation Research Board, Washington, DC, 1994.

additives.[11,12] The content of organic impurities (particularly from sea coal binder systems) can vary widely and can be quite high. This may preclude its use in applications where organic impurities could be important (e.g., portland cement concrete aggregate).[9] The specific gravity of foundry sand has been found to vary from 2.39 to 2.55. This variability has been attributed to the variability in fines and additive contents in different samples.[11] In general, foundry sands are dry, with moisture contents less than 2%. A large fraction of clay lumps and friable particles have been reported, which are attributed to the lumps associated with the molded sand, which are easily disintegrated in the test procedure.[11] The variation in permeability listed in Table 4.5 is a direct result of the fraction of fines in the samples collected.

4.2.4.3 Chemical Compositions

The chemical compositions of materials are usually expressed in terms of simple oxides calculated from elemental analysis determined by x-ray fluorescence. For spent foundry sand, the chemical parameters include bulk oxides mass composition, loss on ignition, and total oxygen demand. Table 4.6 lists the general chemical properties of spend foundry sand. It is shown that spent foundry sand consists primarily of silica dioxide.

Depending on the binder and type of metal cast, the pH of spent foundry sand[12] can vary from approximately 4 to 8. As such, it has been reported that some spent foundry sands can be corrosive to metals.[14] Spent foundry sand must be monitored to assess the need to establish controls for potential phenol discharges.[9,15,16]

4.2.4.4 Trace Element Characterization

Trace element characterization represents concentrations of elements that are contaminated in materials or their leachates in a trace level, generally reported in units of mg/kg or mg/L. Although in minimum quantities, trace elements need to be characterized to assess the hazardous impact of the

TABLE 4.6
Foundry Sand Sample Chemical Oxide Composition

Constituent	Value (%)
SiO_2	87.91
Al_2O_3	4.70
Fe_2O_3	0.94
CaO	0.14
MgO	0.30
SO_3	0.09
Na_2O	0.19
K_2O	0.25
TiO_2	0.15
P_2O_5	0.00
Mn_2O_3	0.02
SrO	0.03
Loss on ignition	5.15, on average
Total	99.87

Source: From AFS, Alternative Utilization of Foundry Waste Sand, final report (Phase I) for Illinois Department of Commerce and Community Affairs, American Foundrymen's Society, Des Plaines, IL, July 1991.

materials and their compliance with environmental protection laws. Total analyses and leaching analyses are generally used. The former quantifies as a dry-basis the mass percentage of trace elements by following U.S. EPA standard environmental analytical methods or approved analytical chemistry methods, relying upon techniques of inductively coupled plasma atomic emission mass spectrometry (ICP-AES, ICP-MS) and gas chromatography interfaced with a mass spectrometer (GC-MS). Leaching analysis is often run as a simulation of the field extraction effect, in which materials are extracted into aqueous media by leachate fluid, groundwater, rainfall, or other fluids. Currently, three leaching protocols are frequently documented to simulate field extraction variation: the Toxicity Characteristic Leaching Procedure (TCLP, U.S. EPA Method 1311), the Synthetic Precipitation Leaching Procedure (SPLP, U.S. EPA Method 1312), and the Standard Test Method for Shake Extraction of Solid Waste with Water (ASTM D3987). TCLP and SPLP are acidic toxicity tests, whereas ASTM D3987 is a neutral leaching procedure.

Many studies have been conducted on metal contaminants in foundry sands. Spent foundry sand segregated from the other wastestreams leaches regulated metals well below the toxicity characteristic levels.[12] It is also found that spent foundry sands produced by iron, steel, and aluminum foundries are rarely hazardous, whereas spent foundry sand collected from copper-based facilities may render leachate with regulated elements exceeding regulation threshold values.[12,14] Only iron and manganese, which are not regulated under RCRA, were recorded at increased leaching potentials on a number of occasions. Lead, chromium, copper, and zinc are reported to be of concern for mixed foundry wastes. There is no direct correlation between the total metal content and the leachability under TCLP. Quantities of total metal content in spent and virgin sands and in sandy soils are typically of the same order of magnitude, which suggests an opportunity for spent foundry sand replacing conventional sand and natural soil in many applications, without posing environmental threats.

Few studies have been conducted to determine organic residues in spent foundry sand and leachates from disposal sites. It is reported that several organic compounds are present in the spent foundry sand but have concentrations below the regulated toxicity characteristic limits. Organic compounds of concern include benzoic acid, naphthalene, methylnaphthalenes, phenol, methylenebisphenol, diethylphenol, and 3-methylbutanoic acids.[12] These compounds are thought to be derived from the decomposition of organic binders such as phenolic urethane, furan, and alkyd isocyanate.

4.2.4.5 Mechanical Properties

Typical mechanical properties of spent foundry sand are listed in Table 4.7. Spent foundry sand has good durability characteristics as measured by low microdeval abrasion and magnesium sulfate soundness loss tests.[17] Recent studies have reported relatively high soundness loss, which is

TABLE 4.7
Typical Mechanical Properties of Spent Foundry Sand

Property	Results	Test Method
Microdeval abrasion loss (%)	<2	—
Magnesium sulfate soundness loss (%)	5–47	ASTM C88
Friction angle (°)	33–40	
California bearing ratio (%)	4–20	ASTM D1883

Source: From MNR, Mineral Aggregate Conservation, Reuse and Recycling, report for Ontario Ministry of Natural Resources, Ontario, Canada, February 1992. Javed, S. and Lovell, C.W., Use of Foundry Sand in Highway Construction, Joint Highway Research Project No. C-36-50N, Purdue University, West Lafayette, IN, July 1994.

attributed to samples of bound sand loss and not a breakdown of individual sand particles.[11] The angle of shearing resistance (friction angle) of foundry sand has been reported to be in the range of 33 to 40°, which is comparable to that of conventional sands.[11]

4.2.5 BAGHOUSE DUST

4.2.5.1 Origin

Baghouse dusts are fine particles that are captured from the gas collection and cleaning system (baghouse) installed in a metal casting facility. Baghouses consist of several rows or compartments of fabric filters that collect the dust during the operation of a metal casting facility. Most of these systems are preceded by cyclones, which are primary collection devices used to capture the coarser particles emitted from the casting processes. Metal casting facilities that do not have baghouse collection systems are equipped with wet scrubbers to control air emissions.

The primary sources of airborne dust are the vibrating shakeout and the sand carryover, as well as the sand that remains attached to castings after shakeout. When a green sand mold is placed onto a conventional vibrating shakeout, a large amount of fine silica dust is released into the plant. The amount is directly related to the sand temperature at shakeout. The higher the sand temperature at shakeout, the more dust is carried into the atmosphere and captured by the baghouse. The second source of dust is sand still attached to castings after shakeout (in pockets, cavities, and corners); this may be released into the atmosphere in the shot blasting process.

4.2.5.2 Physical Properties

As baghouse dust accounts for a minimum of the total foundry solid waste, less attention is placed on characterizing this wastestream. Few data are available giving its physical properties and chemical composition. Visually, it is a very fine powder, dark gray in color. The dust may demonstrate physical properties that are similar to clay soils.

4.2.5.3 Chemical Composition

The main mineral components of baghouse dust are silica, clay, some resin evaporation residue, and metal fines. Its composition is related to the way it is collected. Shake-out dust mainly contains silica and clay. Metal fines may be present in the dust collected from areas used for cleaning, grinding, and melting processes.

4.2.6 FURNACE SLAG

4.2.6.1 Origin

In the melting process, metal scraps and fluxes (limestone or dolomite) are charged into a furnace, sometimes along with coke for fuel if a blast furnace is used. Upon heating using electricity (arc furnace) or burning (blast furnace), scraps are melted into a molten phase. The metal is subsequently gravimetrically separated from the composite flux, leaving the residual slag. Flux is used to adequately render the slag fluid so that it can be separated from the molten iron, and it then flows freely from the cupola.

Furnace slag is a nonmetallic byproduct produced in the melting process. It consists primarily of silicates, alumina silicates, and calcium–alumina–silicates. The molten slag, which absorbs much of the sulfur from the charge, comprises ~20% by mass of iron production. As a byproduct of the melting process, furnace slags vary considerably in form depending on the melted metals used, furnace types, and slag cooling method. Figure 4.7 demonstrates the major types of furnace slags.

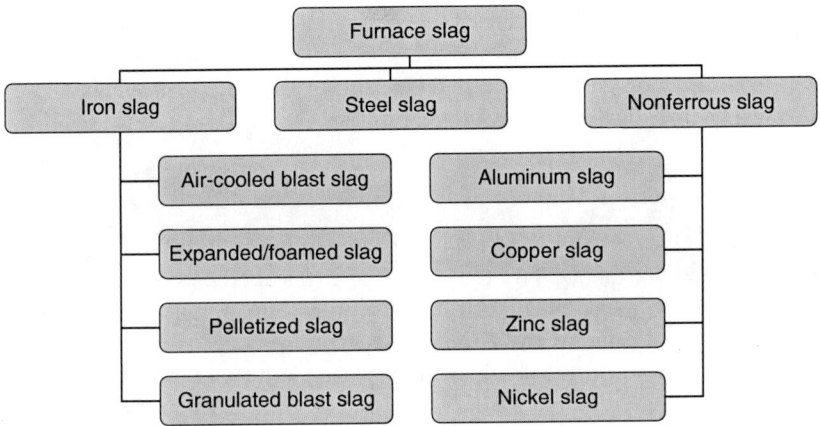

FIGURE 4.7 Furnace slag types.

For iron slag, subcategories include air-cooled blast furnace slag, expanded or foamed slag, pelletized slag, and granulated blast furnace slag. The generation of each slag is described below:

1. *Air-cooled blast furnace slag.* If the liquid slag is poured into beds and slowly cooled under ambient conditions, a crystalline structure is formed, and a hard, lump slag is generated, which can subsequently be crushed and screened.
2. *Expanded or foamed blast furnace slag.* If the molten slag is cooled and solidified by adding a controlled volume of water, air, or steam, the process of cooling and solidification can be accelerated, increasing the cellular nature of the slag and generating a lightweight expanded or foamed product. Foamed slag is distinguishable from air-cooled blast furnace slag by its relatively high porosity and low bulk density.
3. *Pelletized blast furnace slag.* If the molten slag is cooled and solidified with water and air quenched in a spinning drum, pellets, rather than a solid mass, can be produced. By controlling the process, the pellets can be made more crystalline, which is beneficial for aggregate use, or more vitrified (glassy), which is more desirable in cementitious applications. More rapid quenching results in greater vitrification and less crystallization.
4. *Granulated blast furnace slag.* If the molten slag is cooled and solidified by rapid water quenching to a glassy state, little or no crystallization occurs. This process results in the formation of sand-sized (or frit-like) fragments, usually with some friable clinker-like material. The physical structure and gradation of granulated slag depend on the chemical composition of the slag, its temperature at the time of water quenching, and the method of production. When crushed or milled to very fine cement-sized particles, ground granulated blast furnace slag has cementitious properties, which makes it a suitable partial replacement for or additive to portland cement.

Steel slag, a byproduct of steel making, is a complex solution of silicates and oxides that solidifies upon cooling. The main components are consist of carbon, silicon, manganese, phosphorus, some iron as liquid oxides, lime, and dolime. There are several different types of steel slag produced during the steel-making process. These different types are referred to as furnace or tap slag, synthetic or ladle slags, and pit or cleanout slag.

Nonferrous slag is mostly formed by dumping it into a pit and simply allowing it to air cool, solidifying under ambient conditions. A small proportion is granulated, and by using rapid water and air quenching results in the production of a vitrified product. Similar to the generation of iron slag, the cooling rate has a strong influence on the mineralogy and, consequently, the physical and cementitious properties of the nonferrous slag. Slag generation is highly dependent on specific processes and sources. Consequently, slag properties can vary between plants and different ore sources, and must be investigated on a case-by-case basis.

4.2.6.2 Physical Properties

There can be considerable variability in the physical properties of blast furnace slag, depending on the slag generation method. Table 4.8 lists some typical physical properties of air-cooled, expanded, and pelletized blast furnace iron slags. Crushed air-cooled blast furnace slag is angular, roughly cubical, and has textures ranging from rough, vesicular (porous) surfaces to glassy (smooth) surfaces with fractures. Some air-cooled blast furnace slag has been reported to have a compacted unit weight as high as $1940 kg/m^3$ ($120 lb/ft^3$). The water absorption of air-cooled blast furnace slag can be as high as 6%.

Crushed expanded slag is angular, roughly cubical in shape, and has a texture that is rougher than that of air-cooled slag. Its porosity is higher than that of air-cooled blast furnace slag aggregates. The bulk relative density of expanded slag is difficult to determine accurately, but it is ~70% that of air-cooled slag. Typical compacted unit weights for expanded blast furnace slag aggregates range from $800 kg/m^3$ ($50 lb/ft^3$) to $1040 kg/m^3$ ($65 lb/ft^3$).[19]

Unlike air-cooled and expanded blast furnace slag, pelletized blast furnace slag has a smooth texture and rounded shape. Consequently, its porosity and water absorption are much lower than those of air-cooled blast furnace slag or expanded blast furnace slag. Pellet sizes range from 13 mm (1/2 in.) to 0.1 mm (No. 140 sieve size), with the bulk of the product in the 9.5 mm (3/8 in.) to +1.0 mm (No. 18 sieve size) range. Pelletized blast furnace slag has a unit weight of about $840 kg/m^3$ ($52 lb/ft^3$).[20]

Granulated blast furnace slag is a glassy granular material that varies, depending upon its chemical composition and mode of generation, from a coarse, popcorn-like friable structure greater than 4.75 mm (No. 4 sieve) in diameter to dense, sand-sized grains passing a 4.75 mm (No. 4) sieve. Grinding reduces the particle size to cement fineness, allowing its use as a supplementary cementitious material in portland cement concrete.

Steel slag aggregates are highly angular in shape and have a rough surface texture. They have high bulk specific gravity and moderate water absorption (<3%). Table 4.9 lists some typical physical properties of steel slag.

Table 4.10 lists some typical physical properties for nonferrous slags. Because they have similar properties, lead, lead–zinc, and zinc slags are grouped together.

TABLE 4.8
Typical Physical Properties of Blast Furnace Slag

	Slag Types		
Property	Air-Cooled	Expanded	Pelletized
Specific gravity	2.0–2.5	—	—
Compacted unit weight	1120–1360	(800–1040)	840
(kg/m^3) (lb/ft^3)	(70–85)	(50–65)	(52)
Absorption (%)	1–6	—	—

Source: From AASHTO, AASHTO Designation M240: Blended hydraulic cements, in *Standard Specification for Materials*, American Association of State Highway and Transportation Officials, 1986.

TABLE 4.9
Typical Physical Properties of Steel Slag

Property	Value
Specific gravity	3.2–3.6
Unit weight (kg/m^3) (lb/ft^3)	1600–1920 (100–120)
Absorption (%)	up to 3

Air-cooled copper slag has a black color and glassy appearance. As a general rule, its specific gravity will vary with iron content, from a low of 2.8 to as high as 3.8.[21] The unit weight of copper slag is somewhat higher than that of conventional aggregate. The absorption of the material is typically very low (0.13%).[22]

Granulated copper slag is more porous and therefore has lower specific gravity and higher absorption than air-cooled copper slag. The granulated copper slag is made up of regularly shaped, angular particles, mostly between 4.75 mm (3/4 in.) and 0.075 mm (No. 200 sieve) in size.

Granulated nickel slag is essentially an angular, black, glassy slag, with most particles in the size range of −2 mm (No. 10 sieve) to +0.15 mm (No. 100 sieve).[21] It is more porous, with lower specific gravity and higher absorption than air-cooled nickel slag.

Slags from lead, lead–zinc, and zinc groups are often black to red in color and glassy. They have sharp, angular particles that are cubical in shape. The unit weights of lead, lead–zinc, and zinc slags are somewhat higher than conventional aggregate materials. Granulated lead, lead–zinc, and zinc slags tend to be porous, with up to 5% absorption.[23] The specific gravity can vary from less than 2.5 to as high as 3.6.[21,23] These slags are made up of regularly shaped, angular particles, mostly between 4.75 mm (3/4 in.) and 0.075 mm (No. 200 sieve) in size.

4.2.6.3 Chemical Compositions

Table 4.11 depicts the typical chemical composition of blast furnace iron slag. It is suggested that the chemical composition of blast furnace slags produced in North America has remained relatively consistent over the years.

TABLE 4.10
Typical Physical Properties of Nonferrous Slags

Property	Copper Slag	Nickel Slag	Lead, Lead–Zinc, and Zinc Slags
Appearance	Black, glassy, more vesicular when granulated	Reddish brown to brown-black, massive, angular, amorphous texture	Black to red, glassy, sharp angular (cubical) particles
Unit weight (kg/m^3)	2800–3800	3500	2500 or 3600
Absorption (%)	0.13	0.37	5.0

Source: From MNR, Mineral Aggregate Conservation, Reuse and Recycling, report for Ontario Ministry of Natural Resources, Ontario, Canada, February 1992. JEGEL, Manitoba Slags, Deposits, Characterization, Modifications, Potential Utilization, report, John Emery Geotechnical Engineering Limited, Toronto, Ontario, 1986. Hughes, M.L. and Haliburton, T.A., Use of zinc smelter waste as highway construction material, *Highway Research Record*, 430, 16–25, 1973. Mantell, C.L., *Solid Wastes: Origin, Collection, Processing and Disposal*, John Wiley & Sons, New York, 1975. With permission.

TABLE 4.11
Typical Composition of Blast Furnace Slag

Constituent	Percent							
	1949[a]		1957[a]		1968[a]		1985[a]	
	Mean	Range	Mean	Range	Mean	Range	Mean	Range
CaO	41	34–48	41	31–47	39	32–44	39	34–43
SiO_2	36	31–45	36	31–44	36	32–40	36	27–38
Al_2O_3	13	10–17	13	8–18	12	8–20	10	7–12
MgO	7	1–15	7	2–16	11	2–19	12	7–15
FeO or Fe_2O_3	0.5	0.1–1.0	0.5	0.2–0.9	0.4	0.2–0.9	0.5	0.2–1.6
MnO	0.8	0.1–1.4	0.8	0.2–2.3	0.5	0.2–2.0	0.44	0.15–0.76
Sulfur	1.5	0.9–2.3	1.6	0.7–2.3	1.4	0.6–2.3	1.4	1.0–1.9

[a] Data source is National Slag Association data: 1949 (22 sources); 1957 (29 sources); 1968 (30 sources), and 1985 (18 sources).

Source: From MNR, Mineral Aggregate Conservation, Reuse and Recycling, report for Ontario Ministry of Natural Resources, Ontario, Canada, February 1992.

When ground to the proper fineness, the chemical composition and glassy (noncrystalline) nature of vitrified slags are such that when combined with water, these vitrified slags react to form cementitious hydration products. The magnitude of these cementitious reactions depends on the chemical composition, glass content, and fineness of the slag. The chemical reaction between ground granulated blast furnace slag and water is slow, but it is greatly enhanced by the presence of calcium hydroxide, alkalis, and gypsum ($CaSO_4$).

Because of these cementitious properties, ground granulated blast furnace slag can be used as a supplementary cementitious material either by premixing the slag with portland cement or hydrated lime to produce a blended cement (during the cement production process), or by adding the slag to portland cement concrete as a mineral admixture.

Blast furnace slag is mildly alkaline and exhibits a pH range of 8 to 10 in solution. Although blast furnace slag contains a small component of elemental sulfur (1 to 2%), the leachate tends to be slightly alkaline and does not present a corrosion risk to steel in pilings,[24] or to steel embedded in concrete made with blast furnace slag cement or aggregates.[25]

Table 4.12 lists the range of compounds present in steel slag from a typical base oxygen furnace. The predominant compounds are dicalcium silicate, tricalcium silicate, dicalcium ferrite, merwinite, calcium aluminate, calcium–magnesium iron oxide, and some free lime and free magnesia. The relative proportions of these compounds depend on the steel-making practice and the steel slag cooling rate. If the cooling rate of the steel slag is sufficiently low, crystalline compounds are generally formed. As a result, not all steel slags are suitable as aggregates.

Steel slag may expand when in contact with moisture. Free calcium and magnesium oxides are generally not completely consumed in the steel slag, and there is general agreement in the technical literature that the hydration of lime and magnesia is largely responsible for the expansive nature of most steel slags.[6,27] The free lime hydrates rapidly and can cause large volume changes over a relatively short period of time (weeks), whereas magnesia hydrates much more slowly and contributes to long-term expansion that may take years to develop.

Steel slag is mildly alkaline, with a solution pH value of 8 to 10. However, the pH of leachate from steel slag can exceed 11, a level that can be corrosive to aluminum or galvanized steel pipes placed in direct contact with the slag.

The chemical properties of copper, lead, lead–zinc, and zinc slags are essentially as ferrous silicates, whereas nickel slags are primarily calcium/magnesium silicates. Table 4.13 lists typical chemical compositions of these slags.

TABLE 4.12
Typical Steel Slag Chemical Composition

Constituent	Composition (%)
CaO	40–52
SiO_2	10–19
FeO	10–40 (70–80% FeO, 20–30% Fe_2O_3)
MnO	5–8
MgO	5–10
Al_2O_3	1–3
P_2O_5	0.5–1
S	<0.1
Metallic Fe	0.5–10

Source: From Emery, J.J., Slag Utilization in Pavement Construction, *Extending Aggregate Resources*, ASTM Special Technical Publication 774, American Society for Testing and Materials, 1982, pp. 95–118. With permission.

During slag production, the sudden cooling that results in the vitrification of nonferrous slags (typically in the granulating process) prevents the molecules from being locked up in crystals. In the presence of an activator (such as calcium hydroxide from hydrating portland cement), vitrified nonferrous slags react with water to form stable, cementitious, hydrated calcium silicates. The reactivity depends on the fineness to which the slag is ground (reactivity increases with fineness)[30] and the

TABLE 4.13
Typical Chemical Compositions of Nonferrous Slag

Element	Copper Slag (%)	Nickel Slag (%)	Lead Slag (%)	Lead–Zinc Slag (%)
SiO_2	36.6	29.0	35.0	17.6
Al_2O_3	8.1	Trace	—	6.1
Fe_2O_3	—	53.06	—	—
CaO	2.0	3.96	22.2	19.5
MgO	—	1.56	—	1.3
FeO	35.3	—	28.7	—
K_2O	—	—	—	—
F	—	—	—	—
MnO	—	Trace	—	2.0
P_2O_5	—	—	—	—
Cu	0.37	—	—	—
BaO	—	—	—	2.0
SO_3	—	0.36	—	—
Free CaO	—	—	—	—
S	0.7	—	1.1	2.8
PbO	—	—	—	0.8

Source: From OECD, *Use of Waste Materials and Byproducts in Road Construction*, Organization for Economic Co-operation and Development, Paris, 1977.

chemical composition of the slag and its glass content. These vitrified slags can be of such composition that when ground to proper fineness, they may also react directly with water to form hydration products that provide the slag with cementitious properties.

There is some evidence that nickel slag can be involved in the corrosion of iron and steel in the presence of moisture (probably galvanic corrosion). In Canada, where nickel slag is used in fill applications, it is common practice to provide a layer [typically 150 mm (6 in.) thick] of natural aggregate between ferrous materials and the slag.[21]

4.2.6.4 Mechanical Properties

Of all the slag types generated, air-cooled blast furnace is the type that is most commonly used as an aggregate material. Processed air-cooled blast furnace slag exhibits favorable mechanical properties for aggregate use, including good abrasion resistance, good soundness characteristics, and high bearing strength. Table 4.14 lists typical mechanical properties of air-cooled blast furnace slag aggregates.

Table 4.15 lists some typical mechanical properties of steel slag. Processed steel slag has favorable mechanical properties for aggregate use, including good abrasion resistance, good soundness characteristics, and high bearing strength.

TABLE 4.14
Typical Mechanical Properties of Air-Cooled Blast Furnace Slag

Property	Value
Los Angeles abrasion (%)	35–45
Sodium sulfate soundness loss (%)	12
Angle of internal friction (°)	40–45
Hardness[a]	5–6
California bearing ratio (CBR) (%), top size 19 mm (3/4 in.)[b]	Up to 250

[a] Hardness of dolomite measured on same scale is 3 to 4.
[b] Typical CBR value for crushed limestone is 100%.

Source: From Noureldin, A.S. and McDaniel, R.S., Evaluation of Steel Slag Asphalt Surface Mixtures, presented at the 69th annual meeting, Transportation Research Board, Washington, January, 1990.

TABLE 4.15
Typical Mechanical Properties of Steel Slag

Property	Value
Los Angeles abrasion (%)	20–25
Sodium sulfate soundness loss (%)	<12
Angle of internal friction (°)	40–50
Hardness[a]	6–7
California bearing ratio (CBR), (%), top size 19 mm (3/4 in.)[b]	Up to 300

[a] Hardness of dolomite measured on same scale is 3 to 4.
[b] Typical CBR value for crushed limestone is 100%.

Source: From Noureldin, A.S. and McDaniel, R.S., Evaluation of Steel Slag Asphalt Surface Mixtures, presented at the 69th annual meeting, Transportation Research Board, Washington, January, 1990.

TABLE 4.16
Typical Mechanical Properties of Nonferrous Slags

Test	Nickel Slag	Copper Slag	Lead, Lead–Zinc, and Zinc Slags
Los Angeles abrasion loss (%)	22.1	24.1	No data
Sodium sulfate soundness loss (%)	0.40	0.90	No data
Angle of internal friction (°)	~40	40–53	No data
Hardness	6–7	6–7	No data

Source: From Hughes, M.L. and Haliburton, T.A., Use of zinc smelter waste as highway construction material, Highway Research Record, 430, 16–25, 1973. Das, B.M., Tarquin A.J., and Jones, A.Q., Geotechnical properties of copper slag, *Transportation Research Record*, 941, National Research Board, Washington, DC, 1993.

Table 4.16 presents typical mechanical properties for nonferrous slags. Processed air-cooled and granulated copper and nickel slags have a number of favorable mechanical properties for aggregate use, including excellent soundness characteristics, good abrasion resistance, and good stability (high friction angle due to sharp, angular shape). However, nonferrous slags tend to be vitreous, or glassy, which adversely affects their frictional properties (skid resistance), a potential problem if used in pavement surfaces.

4.2.6.5 Thermal Properties

Thermal property is another critical property for furnace slag. Because of their more porous structure, blast furnace slag aggregates have lower thermal conductivities than conventional aggregates. Their insulating value is of particular advantage in applications such as frost tapers (transition treatments in pavement subgrades between frost-susceptible and nonfrost-susceptible soils) or pavement base courses over frost-susceptible soils.

Owing to their high heat capacity, steel slag aggregates have been observed to retain heat considerably longer than conventional natural aggregates. The heat retention characteristics of steel slag aggregates can be advantageous in hot mix asphalt repair work in cold weather.

4.3 SOURCE REDUCTION

The Pollution Prevention Act (PPA, 1990) set a priority for reducing the amount of manufacturing waste through "source reduction"—preventing the generation of waste on the factory floor. The second-best option is to recycle wastes for other uses. The next option in the priority list is to recover the energy content of any wastes that are generated. The last resort is to treat the wastestream. This solid waste management hierarchy also applies to foundry solid waste. In this section, following waste characterization and preceding waste reuse, source reduction regarding solid waste generation and pollution emission is addressed. Effective measures include chemical substitution, in-plant reclamation, waste segregation, and process modifications to reduce emission.

4.3.1 CHEMICAL SUBSTITUTION OR MINIMIZATION

Regulated chemicals are of particular concern to foundrymen, waste recyclers, and decision-makers. An increasingly applied measure is to substitute or at least minimize the use of these chemicals in the plant, and basically eliminate the source of the environmental threat. For instance, all resin manufacturers are reducing the free-phenol content of their products to mitigate the discharge of phenol. Water-based refractory coatings are replacing solvent-based products, leading to casting improvements, such as a reduction in the number of scrap pieces and improved cycle times, and most importantly, a lighter environmental impact.[32]

Targeted chemicals can also be heavy metals. Nonleaded brass castings (also described as very-low-lead alloys, because no lead is intentionally added to them) are an important new approach to reducing lead in drinking water. A variety of other approaches to meeting lead release requirements have been tried or are currently being used, including the use of organic and inorganic coatings, the chemical removal of interior surface lead, and the reduction of internal surface areas of devices by implementing design changes. Most of the lead-free alloys contain bismuth as the major alloying element. Bismuth replaces lead in copper alloys and contributes to the machinability and pressure tightness of the alloys. Bismuth, like lead, is almost completely insoluble in copper and has a low melting point. It is not known to be toxic to humans and is used as a chemical compound in a popular remedy for upset stomachs.

Chemical substitution or minimization may bring great benefits through a managed scrap charge process. Metal scraps are carefully charged, screening out heavy metal-rich scraps and avoiding a mixed metal scrap charge. Scraps containing toxic polymer materials shall be treated before being charged.

4.3.2 IN-PLANT RECLAMATION

In-plant reclamation refers to the sand reclamation process in a foundry facility, which directly minimizes the generation of spent foundry sand. Sand reclamation includes physical, chemical, or thermal treatment of foundry sands so they may be safely substituted for new sand in molding and core-making mixes.

Mechanical attrition is used to remove most of the spent binder. First, dry attrition or abrasion processes crush lumps to grain size. Mechanical abrasion is then used to separate the binder from the sand grains. Sometimes, sand is pneumatically propelled against a metal target plate. The impact of the sand on the plate scrubs off the clay and resin coating from the sand grains. Fines are separated and removed by dry classification.

Depending on the binder system used, 60 to 80% of the mechanically reclaimed sand can be reconditioned satisfactorily for molding, with the addition of clean sand. The remaining 20 to 40% of the mechanically treated sand may then be thermally treated to remove the residual organic binder, restoring the sand to a clean condition. Mechanical attrition has the lowest cost. It allows lump breaking, removes and segregate metal scraps, mechanically scrubs as much binder as possible while avoiding breakage of grains, and removes dust, fines and binder residue by air classification.

In some cases, particularly for resin bonded sand, thermal treatment is used to burn the resin binder and carbonaceous residues. Thermal treatments are usually gas heated, but electric or oil heating can also be used. Sand is heated to approximately 500 to 800°C (930 to 1475°F), at which temperature sand bonded with an entirely organic binder system can be reclaimed up to 100%. The sand is then cooled and crushed to grain size by mechanical scrubbing. Binder systems containing inorganic chemicals, for example, silicate-based systems, cement systems, and phosphoric acid systems, are difficult to reclaim at high percentages because no burnout of the inorganic material occurs. Thermal reclamation is costly because of the large amount of heat and relatively expensive equipment needed. The ensured reclamation quality (sand thermal stabilization and clean up) and the need to remove resin residue, however, has led to its increasing use.

Wet reclamation, although being phased out in the U.S., has been used for silicate bonded sand. After the sand is crushed to grain size, water scrubbing using mechanical agitation is used to wash off the silicate residues, then dried. This process requires a large amount of water and also the treatment and clarification of the water before its recirculation and disposal. In addition, the capital cost of the equipment is high and it requires a large amount of floor space.

Whatever method of reclamation is used, there is always some loss of sand so that 100% reclamation can never be achieved. Sand losses include burn-on, spillage, and inefficiency in the sand system and the need to remove fines. Total sand losses of up to 10% may be expected.

Metal Casting Wastes

4.3.3 Waste Segregation

Waste segregation helps separate hazardous materials from nonhazardous materials, divide recyclable materials from nonrecyclable materials, make materials largely "pure" and then consistent in physical and chemical property, and leads to managed waste disposal. There are up to 40 wastestreams covering spent sand, slag, and dust, inclusive of spent molding sand, core sand waste, cupola slag, scrubber sludge, baghouse dusts, shotblast fines, buffing wastes, and others.[32] In a facility, workers tend to group several wastestreams and discard them as a composite.[12] As a result, complex properties with wide variation are identified, either rendering an assessment that the materials are hazardous, although only a minimum stream deserves the classification, or impeding the recycling program by worsening the materials' consistency.

Besides the segregation of generated wastestreams, in-plant reclamation also considers material division. In typical foundry processes, sand from collapsed molds or cores are subjected to reclamation. However, it is well known that reclamation of sand is easiest when only one type of chemical binder is used. If more than one binder is used, care must be taken to ensure that the binder systems are compatible. Shaken-out green sand and chemical bonded sand are better separated from each other to ensure their rebonding and casting quality. Waste segregation, such as separating fresh casting mixtures and core sand that have not been in contact with hot metal from the other wastestreams, also mitigates the organic compounds identified in the wastestreams.

4.3.4 Process Modifications to Reduce Emission

Toxins, such as benzene, naphthalene, phenol, toluene, xylene, formaldehyde, and mercury, are found in resins and scraps and released as a result of evaporation and solvent processes and during combustion. Respiration of these emissions affects the brain and central nervous system, causes irritation to the skin, eyes, nose, and throat, breathing difficulties, lung problems, impaired memory, stomach discomfort, liver, and kidney changes. Clean-air regulations as well as workplace safety and health standards, however, have forced operators to address the issue. To meet the challenge of providing environmentally friendly core binders and melting processes, a number of suppliers have introduced technologies that are a promising step towards a new generation of binders that both reduce the amount of volatile organic compounds (VOCs) and hazardous air pollutants (HAPs) released and yet do not compromise casting quality.

There is a large amount of development work going on worldwide to improve the performance of core binders and to make them more environmentally friendly. The inorganic binder system is a green and relatively environmentally safe bonding process. For example, for silicate-based processes, there is very little in the way of fumes or smells during core manufacture, storage, or casting. A recently improved resin/CO_2 process involves a water-based alkaline phenolic resin. Both the binder and sand are gas-cured with CO_2 to activate the coupling agent, with low emission. Protein-based foundry sand binders are an entirely new class of sand core binders. They are made from high strength collagens with an additive to promote rapid thermal breakdown of the binder coating. This binder essentially has zero odor, contains no hazardous chemicals, and offers excellent sand reclamation. More improvements come from renewed sodium silicate, modifications of the PUCB and alkaline phenolic resins, and the introduction of new binders. The ultimate industry goal is to develop bonding systems that provide an equally good casting surface, improve the shakeout behavior, and eliminate the use of noxious gaseous catalysts and scrubbers.

The dry ice blaster is an effective and mess-free method for in-place cleaning that eliminates the need to disassemble machinery before it is cleaned. Compressed air propels tiny dry ice pellets at supersonic speeds so they flash freeze and then lift grime, paint, rust, mold, and other contaminants from metal surfaces. Pellets vaporize quickly into the air, leaving no wastewater or solvents, only the soiled contaminant to be swept up.

Special baghouse filters are designed for high-efficiency filtration with a unique three-layer construction. The dust filtration is effective for a wide range of particle sizes. The layered design includes a polypropylene prefilter layer, a melt-blown polypropylene microfiber final filter layer, and a polypropylene outer migration barrier layer, resulting in a cost-effective filter bag.

The volume of sand carried over and adhering to the castings is the biggest factor relating to shot blast costs. By stopping the sand from going into the casting cleaning department, and keeping it in the sand system where it belongs, the benefits go right to the bottom line. In addition to saving on all the shot blasting costs, other savings include less wear on the dust collectors, reduced wastestreams, no airborne silica dust, and reduced cleanup time.

4.4 SOLID WASTES REUSE TECHNOLOGIES

As the volume of solid wastes generated out of the metal casting industry and the cost of waste disposal continue to increase, there is increased pressure and incentive to divert valuable materials from the wastestreams, recover and recycle these materials for use in secondary applications, which in turn reduces the burden on landfills and minimizes the need for virgin materials. Many examples show that it is not only better for the environment but is profitable for the metal casting industry to deliver or even sell waste materials, for instance, spent foundry sand, to an alternative user. These foundries have significantly reduced the volume of byproduct materials going to landfill and actually offset the total cost of transporting the byproduct "in" and "out." This section summarizes reuse technologies of solid wastes from the metal casting industry.

4.4.1 General

The beneficial reuse of foundry solid waste has long been carried out informally, particularly in the U.S. Foundry solid waste has always been used as fills around the foundry or nearby neighborhood. With the promulgation of strict environmental protection laws, foundry solid waste is now required to be landfilled. Later, spent foundry sand was selected as a daily cover for landfills that are "cover short." However, many recyclers believe that foundry solid waste should not necessarily be disposed of in landfills where other hazardous industrial waste belongs, simply because the main fraction of foundry solid waste is nonhazardous and has value in fully or partially substituting for currently in-use materials, for example, construction aggregates, soils, and minerals. Thus, reuse of foundry solid waste is marketable.

All reuse options of foundry solid waste are largely categorized into two domains: uses in civil engineering and uses in agricultural applications. In the civil engineering domain, the solid wastes can be used as aggregate materials in asphalt concrete, portlant cement concrete, flowable fill, and highway embankment fills, as raw feed for portland cement production, and as barrier materials in hydraulic cutoff wall or permeable reactive wall. In the agricultural domain, spent foundry sand has alternative uses, such as in manufactured soil and agricultural amendments. In addition to reuse technologies, requisite qualification inspections of waste materials, mainly technical evaluation, environmental concerns, and economic consideration, should be addressed.

4.4.2 Reuse Evaluation Framework

Any proposal to incorporate an unconventional material, and particularly a waste or byproduct material, into a functional product, requires the rendered product to provide reliable, safe, and cost-effective service during its useful life. Such requirements necessitate qualification evaluations to be performed on wastes before their acceptance as alternative materials. At least three evaluation aspects shall be included in an evaluation framework, that is, technical implementability, environmental safety, and economic benefits.

4.4.2.1 Technical Implementability

As they are unconventional materials, foundry solid waste lacks documented procedures qualifying its substitution for conventional materials, which is a primary barrier in the reuse program. Necessary

Metal Casting Wastes

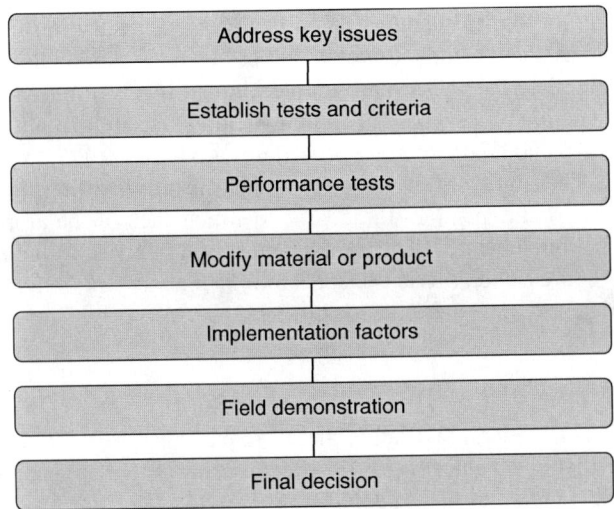

FIGURE 4.8 Technical evaluation of foundry solid waste reuse program.

procedures therefore include evaluation tests, assessment procedures, and criteria to address the technical performance and characteristics that a functional material shall present. A procedural framework needs to be outlined with which one can decide if a waste or byproduct material can be used fully or partially in replacing a conventional material.

There are seven major steps, as shown in Figure 4.8, in an unconventional material technical evaluation process that should be considered:

1. *Address key issues.* Identify all relevant engineering, environmental, occupational health and safety, recyclability, and economic issues that will arise when assessing the use of unconventional materials in functional products. Efforts should be concentrated on compiling and assessing existing data, which include previous laboratory testing, field demonstrations, and the performance history from previous projects that have made use of the proposed material in the proposed application. Incorporating existing data into this process can be of great assistance in the task of defining all relevant technical issues and avoiding any unnecessary duplication of prior efforts.
2. *Establish tests and criteria.* Establish laboratory testing and assessment procedures and criteria that the material and the product should meet prior to accepting solid waste incorporation. Although at the present time, there is an absence of generally accepted test methods and criteria to address all of the proposed key issues, solid waste has generally undergone significant laboratory and field demonstration testing, by referring to suitable specifications, to ensure an equivalent or better performance criteria defined so that the material and application is met. In some cases, such as blast furnace slag, formal specifications have eventually been developed.
3. *Performance tests.* This step is to implement testing and assessment procedures, at a bench-scale, to determine whether the material will meet the criteria established.
4. *Modify material or product.* If a material is not capable of meeting established material or product criteria, it is useful to consider whether additional or alternative material processing or product modification could achieve the desired results.
5. *Implementation factors.* There are always some nontechnical issues that could prevent widespread implementation of the unconventional materials. These nontechnical issues involve institutional acceptability, political acceptability, and public acceptability.

6. *Field demonstration.* A field demonstration is always necessary to supplement a bench-scale evaluation, as some technical issues cannot be undertaken in a laboratory environment. In addition, field data help address nontechnical issues. Proper planning is critical for the implementation of a successful demonstration to ensure that all monitoring equipment, construction, and quality control procedures are in place.
7. *Final decision.* The ability to arrive at a final decision regarding the acceptability of a material for use will depend on the degree to which each of the aforementioned steps were planned and implemented. The establishment of a stepwise framework with specific objectives, procedures, and criteria is critical to implementing an evaluation approach that will address all necessary issues.

4.4.2.2 Environmental Issues

In the U.S., three pieces of federal legislation that were passed from 1969 to 1980, and the implementing rules and regulations that followed, initiated a series of fundamental changes in the management of waste and byproduct materials. They presently affect the way in which regulatory agencies address waste and byproduct material use. These acts include the National Environmental Policy Act (NEPA, 1969), the Resource Conservation and Recovery Act (RCRA, 1976, 1980), and the Comprehensive Environmental Response, Compensation, and Liabilities Act (CERCLA) or Superfund (1980).

NEPA was the major environmental legislation representing the nation's commitment to protect and maintain environmental quality. This act introduced the requirement that environmental impact statements be prepared on all federal actions.

RCRA was passed to manage nonhazardous and hazardous wastes and underground storage tanks, with an emphasis placed on the recovery of reusable materials as an alternative to their disposal. This act introduced the concept of the separate management of hazardous and nonhazardous wastes, and defined procedures to identify whether a waste is hazardous or nonhazardous. A waste exhibits the characteristic of toxicity, classified as a hazardous material, if the concentration of any of 39 selected analytes in the Toxicity Characteristic Leaching Procedure (TCLP) extract exceed regulatory action levels.

CERCLA was promulgated to address the release or imminent release of hazardous substances into the environment and established the mechanisms for responding to those releases and assessing liability. Regulations and procedures that evolved from CERCLA introduced the concept of human health risk assessments. CERCLA also provided the legal framework for assigning liability and assessing monetary damages for environmental impairment.

Although none of the three laws or their implementing regulations directly addressed the reuse of waste materials, they necessitate a series of evaluations in the reuse program of solid waste, which include the preparation of an environmental assessment, a human health risk assessment, or an ecosystem risk assessment.

4.4.3 Reuse in Civil Engineering

One of the largest opportunities to recycle foundry solid waste lies in the construction industry. Eight of the most researched civil applications are described in this section, where vast foundry solid waste was and is being consumed. These applications or products include asphalt concrete, portland cement concrete, granular base, embankment or fill, stabilized base, cement, flowable fill, and landfill liner and cover. A waste material (such as spent foundry sand, dust or slag) may partially or fully suit these applications, depending their generation form and characteristics. A general overview of each application or product is provided, which includes a description of the application or product (components, material properties, test standards), and documented research work and case studies on reusing foundry solid wastes.

Metal Casting Wastes

4.4.3.1 Asphalt Concrete

Description

Asphalt concrete is primarily used as a structural pavement surface constructed over a subgrade and a subbase. It is designed to support the traffic load and distribute the load over the roadbed. Asphalt concrete pavements can be constructed using hot mix or cold mix asphalt. Hot mix asphalt is a mixture of fine and coarse aggregate with asphalt cement binder that is mixed, placed, and compacted in a heated condition. Cold mix asphalt is a mixture of emulsified asphalt and aggregate, produced, placed, and compacted at ambient air temperature. Cold mix asphalt pavement usually requires an overlay of hot mix asphalt or surface treatment to resist traffic action.

Aggregates used in asphalt concrete mixtures comprise ~95% of the mix by mass and ~80% by volume. Thus, the aggregate material(s) used in asphalt concrete have a profound influence on the properties and performance of the mixture. Proper aggregate grading, strength, toughness, and shape are needed for improving final product uses. Table 4.17 provides a list of standard test methods that are used to assess the suitability of aggregates for use in asphalt paving applications.

TABLE 4.17
Asphalt Concrete Aggregate Test Procedures

	Test Method	Specification
General specifications	Coarse aggregate for bituminous paving mixtures	ASTM D692
	Fine aggregates for bituminous paving mixtures	ASTM D1073/AASHTO M29
	Steel slag aggregates for bituminous paving mixtures	ASTM D5106
	Aggregate for single or multiple surface treatments	ASTM D1139
	Crushed aggregate for macadam pavements	ASTM D693
Gradation	Sieve analysis of fine and coarse aggregates	ASTM C136/AASHTO T27
	Sizes of aggregate for road and bridge construction	ASTM D448/AASHTO M43
Particle shape	Index of aggregate particle shape and texture	ASTM D3398
	Flat and elongated particles in coarse aggregate	ASTM D4791
	Uncompacted void content of fine aggregate (as influenced by particle shape, surface texture, and grading)	ASTM C1252/AASHTO TP33
Particle texture	Accelerated polishing of aggregates using the British wheel	ASTM D3319/T279
	Insoluble residue in carbonate aggregates	ASTM D3042
	Centrifuge kerosene equivalent	ASTM D5148
Particle strength	Resistance to degradation of large-size coarse aggregate by abrasion and impact in the Los Angeles machine	ASTM C535
	Resistance to degradation of small-size coarse aggregate by abrasion and impact in the Los Angeles machine	ASTM C131/AASHTO T96
	Degradation of fine aggregate due to attrition	ASTM C1137
Durability	Aggregate durability index	ASTM D3744/AASHTO T210
	Soundness of aggregates by use of sodium sulfate or magnesium sulfate	ASTM C88/AASHTO T104
	Soundness of aggregates by freezing and thawing	AASHTO T103
Specific gravity and adsorption	Specific gravity and absorption of coarse aggregate	ASTM C127/AASHTO T85
	Specific gravity and absorption of fine aggregate	ASTM C128/AASHTO T84
Unit weight	Unit weight and voids in aggregate	ASTM C29/C29M/AASHTO T19
Volume stability	Potential expansion of aggregates from hydration reactions	ASTM D4792
Deleterious components	Sand equivalent value of soils and fine aggregate	ASTM D2419
	Clay lumps and friable particles in aggregates	ASTM C142

Source: From Federal Highway Administration, available at http://www.tfhrc.gov/hnr20/recycle/waste/app.htm.

TABLE 4.18
Asphalt Concrete Test Procedures

Property	Test Method	Specification
Stability and flow characteristics	Marshall method	AASHTO T245
	Hveem method	AASHTO T246, T247
	Asphalt Institute recommended cold mix method	Asphalt Institute Cold Mix Manual
	Resistance to plastic flow of bituminous mixtures using Marshall Apparatus	ASTM D1559
Stripping resistance	Immersion—Marshall method	ASTM D4867
	Immersion—Marshall method	AASHTO T283
Resilient modulus	Superpave mix design	Asphalt Institute Superpave Series No. 1 (SP-1), No. 2 (SP-2)
Unit weight	Theoretical maximum specific gravity and density of bituminous paving mixtures	ASTM D2041
Compacted density	In-place density of compacted bituminous paving mixtures	ASTM D2950

Source: From Federal Highway Administration, available at http://www.tfhrc.gov/hnr20/recycle/waste/app.htm.

Asphalt concrete is properly proportioned to resist the potentially damaging effects in the road. Asphalt concrete paving mixtures should be evaluated for the following properties: stability, flow, air voids, stripping resistance, resilient modulus, compacted density, and unit weight. Table 4.18 provides a list of standard laboratory tests that are presently used to evaluate the mix design or expected performance of fresh and hardened asphalt concrete.

Use of spent foundry sands

Spent foundry sand has been used as a substitute for fine aggregate in asphalt paving materials.[6,11,13] As spent foundry sand is a poorly graded fine sand [largely sized between 0.3 mm and 0.15 mm, 5 to 15% fines content passing 0.075 mm openings (No. 200 sieve)], which essentially excludes it as a solid substitution for conventional fine aggregate. To satisfy the gradation requirements for hot mix asphalt fine aggregates specified in AASHTO M29, the spent foundry sand must be blended with natural sand at selected percentages. Satisfactory performance has been obtained from hot mix pavements incorporating up to 15% spent foundry sand.

The properties of spent foundry sand that are of particular interest when foundry sand is used in asphalt paving applications include particle shape, gradation, durability, and plasticity. With the exception of gradation, clean, processed foundry sands can generally satisfy the physical requirements for hot mix asphalt fine aggregate (AASHTO M29). Round to subangular shapes facilitate the work consistency of asphalt mixture. The hydrophilic nature of the (primarily silica) foundry sand, however, can result in stripping of the asphalt cement coating surrounding the aggregate grains, when over 15% spent foundry sand of total aggregate is blended with the conventional fine sand. This problem can be mitigated by using an antistripping additive.[13]

Use of furnace slag

Air-cooled furnace (ferrous and nonferrous) slag is considered to be a conventional aggregate and can replace both coarse and fine aggregates in asphalt paving applications. Surface-treated pavements incorporating air-cooled furnace slag aggregate demonstrate a number of favorable mechanical properties for use as aggregate, including good friction resistance, good resistance to stripping, fair wear resistance, good soundness characteristics, or good resistance to freeze–thaw weathering.[35]

Air-cooled blast furnace iron slag, however, is more absorptive than conventional aggregate and therefore has a higher asphalt cement demand. It also has a lower compacted unit weight than conventional mineral aggregates, which results in a higher asphalt pavement yield (greater volume for the same weight). This is offset somewhat by the higher yield (volume per mass) of air-cooled furnace slag paving mixtures.[19]

The resistance of air-cooled blast furnace slag to impact is not very high and the material can break down under heavy traffic conditions. Such aggregate is better suited to surface treatment applications on light traffic pavements. Some nonferrous slags are vitreous or "glassy," which can adversely affect their frictional resistance properties. Some glassy nonferrous slags may also be susceptible to moisture-related damage (stripping).

Variability in the production process can result in poor consistency in the physical properties (gradation, specific gravity, absorption, and angularity) of air-cooled furnace slag. This lack of consistency has occasionally contributed to hot mix asphalt performance problems, such as flushing due to high binder content (too rich), raveling due to low binder content, and high fines-to-asphalt ratios (too lean).[10] To minimize problems associated with the variable properties of some air-cooled furnace slag aggregates, a comprehensive scout testing program may be necessary to monitor the gradation, specific gravity, absorption, and angularity of air-cooled blast furnace slag used in asphalt concrete.[26]

The potential for expansion because of free lime or magnesia in the steel slag is of particular concern, which could result in pavement cracking if ignored. It is recommended that no detectable soft lime particles or lime-oxide agglomerations be present.[36]

Some of the engineering properties of air-cooled furnace slag that are of particular interest when air-cooled furnace slag is used as an aggregate in asphalt concrete include gradation, grain shape and texture, bulk and compacted density, absorption, abrasion, stability, friction properties, and freeze–thaw resistance. Blast furnace slag should be crushed and screened to produce aggregate that satisfies the gradation requirements for hot mix asphalt as specified in ASTM D692 for coarse aggregate and AASHTO M29 for fine aggregate. For surface treatments, air-cooled furnace slag aggregate should satisfy gradation specification requirements in ASTM D1139.

Specific physical, chemical, and mineralogical properties of furnace slags depend in great part on the type of slag, method of production, type of furnace, and cooling procedures associated with their respective production processes. Consequently, each slag aggregate must be considered by mineralogical type on a source-specific and cooling (air-cooled or granulated) basis, with recognition of the inherent variability of the slag composition and the presence of potentially foreign materials.

Conventional asphalt mix design methods (e.g., Marshall, Hveem, SHRP) are applicable for the design of hot mix asphalt containing furnace slag aggregates. No special procedures are required for aggregate gradations. Both coarse and fine slag aggregates can be incorporated in hot mix asphalt, provided that the physical requirements of ASTM D692 and/or AASHTO M29 are satisfied. No special provisions are required for furnace slag, and conventional hot mix gradations specifications may be used. Blending with other suitable hot mix asphalt aggregates may be necessary to achieve gradation specifications compliance. Owing to the difference in unit weights, mix designs are usually calculated on a volumetric basis.

4.4.3.2 Portland Cement Concrete

Description
Portland cement concrete is perhaps the most popular and highest use volume construction material. It has the technical advantages of high strength, long durability, solid hardness, and reliable bearing capacity, making it globally favored for various structures. Basic components of portland cement concrete include coarse aggregate (gravel at 40 to 50% mass percentage), fine aggregate (sand at 20 to 25% mass percentage), portland cement (at 10 to 15% mass percentage), and water (at 15 to 20% mass percentage). Concrete can be either cast-in-place, or precast into concrete products such as bricks, pipes, and blocks. The aggregate functions as a filler material, which is bound together by hardened portland cement paste formed by chemical reactions (hydration) between the portland cement and water. In addition to these basic components, supplementary

cementitious materials and chemical admixtures are often used to enhance or modify the properties of the fresh or hardened concrete. The coarse and fine aggregates comprise about 80 to 85% of the mix by mass (60 to 75% of the mix by volume). Proper aggregate grading, strength, durability, toughness, shape, and chemical properties are needed for concrete mixture strength and performance.

As aggregates used in concrete mixtures comprise the major components in the mixture, the aggregate materials used have a profound influence on the properties and performance of the mixture in both the plastic and hardened states. Important properties for aggregates that are used in concrete paving mixtures include gradation, absorption, particle shape and surface texture, abrasion resistance, durability, deleterious materials, and particle strength. Table 4.19 provides a list of standard test methods that are used to assess the suitability of conventional mineral aggregates in portland cement concrete paving applications.

The mix proportions for concrete paving mixtures are determined by attaining optimum characteristics of the mix in both the plastic and hardened states. The designed mixture can be properly placed and consolidated, finished to the required texture and smoothness, and will have the desired properties necessary for pavement performance. Concrete paving mixtures should be

TABLE 4.19
Concrete Aggregate Test Procedures

Property	Test Method	Specification
General	Concrete aggregates	ASTM C33
	Ready-mixed concrete	ASTM C94/AASHTO M157M
	Concrete made by volumetric batching and continuous mixing	ASTM C685/AASHTO M241
	Terminology related to concrete and concrete aggregates	ASTM C125
Gradation	Sizes of aggregate for road and bridge construction	ASTM D448/AASHTO M43
	Sieve analysis of fine and coarse aggregate	ASTM C136/AASHTO T27
Absorption	Specific gravity and absorption of coarse aggregate	ASTM C127/AASHTO T85
	Specific gravity and absorption of fine aggregate	ASTM C128/AASHTO T84
Particle shape and surface texture	Flat and elongated particles in coarse aggregate	ASTM D4791
	Uncompacted voids content of fine aggregate	ASTM C1252/AASHTO TP33
	Index of aggregate particle shape and texture	ASTM D3398
Abrasion resistance	Resistance to degradation of large-size coarse aggregate by abrasion and impact in the Los Angeles machine	ASTM C535
	Resistance to degradation of small-size coarse aggregate by abrasion and impact in the Los Angeles machine	ASTM C131/AASHTO T96
Durability	Aggregate durability index	ASTM D3744/AASHTO T210
	Soundness of aggregates by use of sodium sulfate or magnesium sulfate	ASTM C88/AASHTO T104
	Soundness of aggregates by freezing and thawing	AASHTO T103
Deleterious components	Petrographic examination of aggregates for concrete	ASTM C295
	Organic impurities in fine aggregate for concrete	ASTM C40
	Clay lumps and friable particles in aggregates	ASTM C142
	Plastic fines in graded aggregates and soils by use of the sand equivalent test	ASTM D2419
Volume stability	Potential volume change of cement–aggregate combinations	ASTM C342
	Accelerated detection of potentially deleterious expansion of mortar bars due to alkali–silica reaction	ASTM C227

Source: From Federal Highway Administration, available at http://www.tfhrc.gov/hnr20/recycle/waste/app2.htm.

Metal Casting Wastes

evaluated for the following properties: slump, workability, setting time and air content at the fresh state, strength, density, durability, air content, frictional resistance, and volume stability at the hardened state. Table 4.20 provides a list of standard laboratory tests that are presently used to evaluate these properties.

Use of spent foundry sand

Spent foundry sand is thought of as a beneficial substitute for fine sand for use in portland cement concrete. Prior to acceptance of inclusion, test standards applied on conventional fine sand shall be referred to as the standards for spent foundry sand to compare the physical properties of conventional sand and spent foundry sand. The most important parameters are particle size distribution, fineness modulus, dust content, density, organics content, deleterious materials content, and grain shape. Although no spent foundry sand satisfies all of the specifications, foundry sand can be blended with conventional sand to be incorporated into the concrete matrix. The replacing ratio normally starts at one-third.

In the production of foundry sand concrete, conventional mixing, placing, and curing are easily referable.

TABLE 4.20
Concrete Paving Materials Test Procedures

Property	Test Method	Specification
General	Ready-mixed concrete	ASTM C94/AASHTO M157
	Concrete made by volumetric batching and continuous mixing	ASTM C685/AASHTO M241
	Concrete aggregates	ASTM C33
	Terminology related to concrete and concrete aggregates	ASTM C125
	Pozzolan use as a mineral admixture	ASTM C618
	Ground blast furnace slag specifications	ASTM C989
	Chemical admixtures for concrete	ASTM C494
	Air entraining agents	ASTM C260
	Silica fume specifications	ASTM C1240
Slump	Slump of hydraulic cement concrete	ASTM C143/AASHTO T119
Workability	Bleeding of concrete	ASTM C232/AASHTO T158
Hydration and setting	Time of setting of concrete mixtures by penetration resistance	ASTM C403
Strength	Compressive strength of cylindrical concrete specimens	ASTM C39/AASHTO T22
	Flexural strength of concrete (using simple beam with third-point loading)	ASTM C78/AASHTO T96
	Splitting tensile strength of cylindrical concrete specimens	ASTM C496/AASHTO T198
Air content	Microscopic determination of parameters of the air-void system in hardened concrete	ASTM C457
	Air content of freshly mixed concrete by the pressure method	ASTM C231/AASHTO T152
	Air content of freshly mixed concrete by the volumetric method	ASTM C173/AASHTO T196
	Unit weight, yield, and air content of concrete	ASTM C138
Density	Specific gravity, absorption, and voids in hardened concrete	ASTM C642
Durability	Resistance of concrete to rapid freezing and thawing	ASTM C666
	Scaling resistance of concrete surfaces exposed to deicing chemicals	ASTM C131/AASHTO T96
Volume stability	Length change of hardened hydraulic-cement mortar and concrete	ASTM C157
	Length change of concrete due to alkali–carbonate rock reaction	ASTM C1105

Source: From Federal Highway Administration, available at http://www.tfhrc.gov/hnr20/recycle/waste/app2.htm.

Use of furnace slag
Ground granulated blast furnace slag has been used for many years as a supplementary cementitious material in portland cement concrete, as a mineral admixture. The use of ground granulated blast furnace slag as a partial portland cement replacement takes advantage of the energy invested in the slag-making process and its corresponding benefits with respect to the enhanced cementitious properties of the slag. Rapid quenching is important if cementitious properties are to be achieved. The chemical composition of ground granulated blast furnace slag use in portland cement concrete must also conform to sulfur and sulfate content limitations outlined in AASHTO M302.

Granulated blast furnace slag is a glassy granular material, and its particle distribution, shape, and grain size vary, depending on the chemical composition and method of production, from popcorn-like friable particles to dense, sand-sized grains. Processing for use as a supplementary cementitious material requires grinding of the slag, typically using the same or similar plant and equipment as for portland cement production.

The properties of concrete mixes containing ground granulated blast furnace slag that are of particular interest when it is used as partial cement replacement include strength development, workability, heat of hydration, resistance to alkali–aggregate reactivity, resistance to sulfate attack, and salt scaling. Special production characteristics and performance may be exhibited when ground granulated blast furnace slag is incorporated, such as a slower strength development, longer-lasting workability, low slump loss, lower heat of hydration, reduced alkali–aggregate reaction, improved resistance to sulfate attack, and susceptibility to salt scaling.[38,39]

The most frequently used proportioning recommendations for ground granulated blast furnace slag use in concrete mix designs are covered in ACI 226.1R, Ground Granulated Blast-Furnace Slag as a Cementitious Constituent in Concrete. Some agencies require that a salt scaling test also be completed for selected concrete mixes subjected to deicing salts.[38] The same equipment and procedures as used for conventional portland cement concrete may be used to batch, mix, transport, place, and finish concrete containing ground granulated blast furnace slag.

4.4.3.3 Portland Cement

Description
Portland cement is a fine, soft, powdery substance that acts as a critical component in producing portland cement concrete. When mixed in contact with water, the cement will hydrate and generate complex chemicals that eventually bind the sand and gravel into a hard, solid mass, known as concrete.

There are eight types of portland cement as specified in standard ASTM C150 Standard Specification for Portland Cement. Each cement is manufactured with special use and chemical requirements. Their manufacturing process, however, is basically the same. Portland cement is a chemical product of a kiln process, where blended ground raw materials undergo chemical transformation. Raw materials must comprise a selected ratio of calcium oxide, silica, alumina, and iron oxide. Most of these ingredients are usually contained in shale, dolomite, and limestone, known as the prevailing raw materials. As sand is a good source of silica, alumina, and iron oxides, it is often used in as one of the raw minerals in manufacturing cement.

Use of spent foundry sand
Spent foundry sand can be used as a good source of silica in manufacturing portland cement. Also, the clay fraction of foundry sand is an additional source of iron and aluminum oxides. According to the portland cement industry, spent foundry sand can be beneficially used in the manufacture of portland cement, as sand possesses the following characteristics:

1. Silica content >80%
2. Low alkali level
3. Uniform particle size

Manufacturers also request that spent foundry sand should be cleaned from other foundry byproducts. Core butts should be ground to a uniform grain size to improve the kiln process. In addition, adequate supplies of spent foundry sand are viable in the manufacturing process.

Use of furnace slag
Furnace slag can be used as a source of aluminum oxide and magnesia in manufacturing portland cement. Furnace slag is used as a component of blended cement. The use of ground granulated blast furnace slag in portland cement is governed by AASHTO M302. Three types of ground granulated slag cements are typically manufactured: portland cement is covered by AASHTO M85, portland blast furnace slag cement and slag cement by AASHTO M240.

Use of baghouse dust
Although low in volume compared with other foundry solid waste, baghouse dust may still be used beneficially in the production of portland cement. This opportunity arises from its attractive mineral composition: silica, clay, and metal fines, which are needed in the cement kiln. Also, special efforts may be undertaken to characterize its chemical composition and purity.

4.4.3.4 Embankment or Fill Material

Description
An embankment refers to a volume of longitudinal earthen material that is placed and compacted for the purpose of raising the level of a roadway (or railway) above the level of the existing ground surface. A fill refers to a volume of earthen material that is placed and compacted for the purpose of filling in a hole, cavity, or excavation. Embankments or fills are constructed of materials that usually consist of soil, but may also include aggregate, rock, or crushed paving material.

Soils range from granular soils (sand and gravel), which are highly desirable, to the more finely sized soils (silt and clay). Well-graded soils are preferred as they are readily compacted and give firm bearing capacity. Concerns in selecting soils are the presence of unsuitable or deleterious materials, such as tree roots, branches, stumps, sludge, metal, or trash. Other oversized materials, such as rocks, large stones, reclaimed paving materials, or air-cooled slags, can be used for the construction of embankment bases. Although the use of oversized materials can result in a stable embankment base, the oversized materials should have strong particles that do not readily break down under the action of construction machinery, but have a range of sizes so that void spaces are at least partially filled.

Some of the more important properties of materials that are used for the construction of embankments or fills include gradation, unit weight, specific gravity, moisture–density characteristics, shear strength, compressibility, bearing capacity, permeability, and corrosion resistance. Table 4.21 provides a list of the standard test methods usually used to assess the suitability of conventional earthen fill materials for use in embankment or fill construction.

Use of spent foundry sand
Embankment and fill applications are the biggest end-user of spent foundry sand. Natural soils are often composed primarily of sand, clay, and water. Most spent foundry sands have these same constituents, which suggests spent foundry sand as a good fill material. The immediate benefits include saving virgin soil materials and reduce the bottom line of the foundry industry. It is also reported that foundry sand as a fill material may present better performance then conventional materials, including better resistance to freeze–thaw distress.

The physical properties of concern for construction fill applications are the relationship between moisture and density, plasticity, the liquid limit, and particle size distribution. The same set of construction machinery for conventional fills, such as bulldozers, compactors, and grabbers, is suitable for fill earth works containing spent foundry sand.

Use of furnace slag
Both air-cooled blast furnace slag and expanded blast furnace slag can be used as a conventional aggregate in embankment or fill. They are generally considered by many specifying agencies to be

TABLE 4.21
Embankment or Fill Material Test Procedures

Property	Test Method	Specification
Gradation	Particle size analysis of soils	ASTM D422
	Sieve analysis of fine and coarse aggregate	ASTM D136
Unit weight and specific gravity	Unit weight and voids in aggregate	ASTM D29
	Specific gravity of soils	ASTM D854
	Relative density of cohesionless soils	ASTM D2049
	Maximum index density of soils using a vibratory table	ASTM D4253
	Minimum index density of soils and calculation of relative density	ASTM D4254
Moisture–density characteristics	Moisture–density relations of soils and soil–aggregate mixtures using 5.5 lb (2.49 kg) rammer and 12 in. (305 mm) drop	ASTM D698 (Standard)
	Moisture–density relations of soils and soil–aggregate mixtures using 10 lb (4.54 kg) rammer and 18 in. (457 mm) drop	ASTM D1557 (Modified)
Compacted density (in-place density)	Density of soil in place by the sand-cone method	ASTM D1556
	Density and unit weight of soil in place by the rubber balloon method	ASTM D2167
	Density of soil and soil–aggregate in place by nuclear methods (shallow-depth)	ASTM D2922
	Density of soil in place by the sleeve method	ASTM D4564
Shear strength	Unconsolidated undrained compressive strength of cohesive soils in triaxial compression	ASTM D2850
	Direct shear test of soils under consolidated drained conditions	ASTM D3080
	Consolidated-undrained triaxial compression test on cohesive soils	ASTM D4767
Compressibility	One-dimensional consolidation properties of soils	ASTM D2435
	One-dimensional consolidation properties of soils using controlled-strain loading	ASTM D4186
	One-dimensional swell or settlement potential of cohesive soils	ASTM D4546
Bearing capacity	California Bearing Ratio (cbr) of laboratory-compacted soils	ASTM D1883
	Bearing ratio of soils in place	ASTM D4429
Permeability	Permeability of granular soils by constant head	ASTM D2434
Corrosion resistance	pH of soil for use in corrosion testing	ASTM G51
	Field measurement of soil resistivity using the Wenner four-electrode method	ASTM G57
	Pore water extraction and determination of the soluble salt content of soils by refractometer	ASTM D4542

Source: From Federal Highway Administration, available at http://www.tfhrc.gov/hnr20/recycle/waste/app4.htm.

conventional aggregates and require minimal processing to satisfy conventional soil and aggregate engineering requirements. Although there is little documented use of nonferrous slags as aggregate in embankments or fill, both air-cooled and granulated nonferrous slags are potentially useful for these applications. Nonferrous slag that is suitable for use as a granular base will generally exceed specifications for embankment and fill construction. The high stability of nonferrous slag aggregates can be used advantageously to provide good load transfer to weaker subgrades.

Critical properties qualifying slag use in embankment and fills include gradation, stability, compacted density, drainage characteristics, and corrosivity. Blast furnace slag requires crushing processes to satisfy the physical requirements for use in embankments. Air-cooled nonferrous slags are fine enough to save crushing processes. If necessary, nonferrous slag aggregates can be blended with conventional embankment or fill materials (rock, soil, aggregates) to meet required gradation specifications. In addition, if used as materials for embankments and fills, air-cooled blast furnace takes advantages of high shear strength, reduced post-compaction settlement, slightly lighter weight, good drainage characteristics, no frost susceptibility, and no corrosion risk to steel.

Design procedures for embankments or fill containing blast furnace slag are the same as design procedures for embankments or fills using conventional materials. The same equipment and procedures used for handling, stockpiling, placing, and compacting conventional aggregates may be used for air-cooled blast furnace slag.

4.4.3.5 Flowable Fill

Description

Flowable fill refers to cementitious slurry created by blending fine aggregate or filler, water, and cementitious material(s), and is used primarily as a backfill in lieu of compacted soil. This mixture is capable of filling all voids in irregular excavations (such as utility trenches) and inaccessible places (such as narrow cavities), is self-leveling, self-setting, and hardens in a matter of a few hours without the need for compaction in layers. Flowable fill is sometimes referred to as controlled density fill (CDF), controlled low strength material (CLSM), lean concrete slurry, and unshrinkable fill. The applications of flowable fill are numerous and include restoration of utility cuts in county roads, backfilling structures, filling abandoned wells, filling voids under existing pavements, and pipe embedments.[41–46]

Flowable fill is defined by the American Concrete Institute[47] as a self-compacting cementitious material that is in a flowable state at placement and has a compressive strength of 8.3 MPa (1200 psi) or less at 28 days. Most current applications for flowable fill involve unconfined compressive strengths of 2.1 MPa (300 psi) or less, which makes possible its excavation at a later date.

Fine aggregate or filler material are important components in the mixture, which provides the solids to develop compressive strength. The aggregates must be sufficiently finely graded to enhance the flowability of the mix, but also granular enough to be able to drain some of the excess water from the mix prior to initial hardening. Sand is the most commonly used flowable fill material, although other materials (such as coal bottom ash, fly ash, and quarry fines) have also been used. Important properties include gradation and unit weight, which have a direct effect on the flow characteristics and yield of fresh flowable fills.

The most important physical characteristics of fresh and hardened flowable fill mixtures are its strength development, flowability, hardening time, bleeding and shrinkage, unit weight, bearing capacity, shear strength, and corrosion resistance. Table 4.22 lists the standard test methods usually used to evaluate flowable fill materials.

Use of spent foundry sand

Natural sand is a major component of most flowable fill mixes. Ferrous spent foundry sand can be used as substitute for natural sand (fine aggregate) in flowable fill.[48,49] Spent sands from nonferrous foundries and foundry baghouse dust can contain high concentrations of heavy metals that may preclude their use in flowable fill applications. Some of the engineering properties of spent foundry sand that are of particular interest when foundry sand is used in flowable fill applications include particle shape, gradation, strength characteristics, soundness, deleterious substances, and corrosivity.

Flowable fill mixes are usually designed on the basis of compressive strength, generally after 28 days of ambient temperature curing, but sometimes on the basis of longer term (90 days or more) strength. They are designed to have high fluidity during placement [typical slump of 150 to 200 mm (6 to 8 in.)] and to develop limited strength [typically between 340 and 1400 kPa (50 and 200 psi)], which is sufficient to support traffic without settling, yet can be readily excavated. The mix design shown in Table 4.23 could be referred to as a starting point for mixing formulation for a flowable fill containing spent foundry sand.

Construction procedures for flowable fill materials are no different than those for conventional earth backfill materials. The same methods and equipment used to mix, transport, and place flowable

TABLE 4.22
Flowable Fill Test Procedures

Property	Test Method	Specification
Strength development	Unconfined compressive strength of cohesive soil	ASTM D2166
	Unconfined compressive strength index of chemical-grouted soils	ASTM D4219
Flowability	Slump of portland cement concrete	ASTM C143
	Flow of grout for preplaced aggregate (flow cone method)	ASTM C939
Hardening time	Time of setting of concrete mixtures by penetration resistance	ASTM C403
Bleeding and shrinkage	Change in height at early ages of cylindrical specimens from cementitious mixtures	ASTM C827
Unit weight	Unit weight, yield, and air content of concrete	ASTM C138
Bearing strength	California Bearing Ratio (CBR) of laboratory-compactive soils	ASTM D1883
Shear strength	Unconsolidated undrained compressive strength of cohesive soils in triaxial compression	ASTM D2850
	Direct shear test of soils under consolidated drained conditions	ASTM D3080
Corrosion resistance	pH of soil for use in corrosion testing	ASTM G51
	Field measurement of soil resistivity using the Wenner four-electrode method	ASTM G57
	Optimum SO_3 in portland cement	ASTM C563

fill made with conventional aggregates may be used for flowable fill incorporating spent foundry sand. Special measures may be required to control the early contact water leachate (containing phenols) from spent foundry sand stockpiles. The construction of an impervious pad (to collect surface moisture or precipitation passing through the stockpile) and subsequent filtration (through an activated carbon filter) of the leachate has reportedly been effective in limiting the phenol concentration of the discharge.

4.4.3.6 Landfill Liner, Cover, and Hydraulic Barrier

Description

The landfill liner, cover, and hydraulic barrier all belong to the subsurface pollutant engineered containment system. The liner is designed at the bottom of a landfill to contain downward leachate. The cover is designed at the top of a landfill to prevent precipitation from infiltrating into the landfill. The hydraulic barrier, or cutoff walls, is a vertical compacted earthen system to contain horizontal flow of plume. The ultimate purpose of these barriers is to isolate contaminants from the environment and, therefore, to protect the soil and groundwater from pollution originating in the landfill or polluted site.

TABLE 4.23
Mixing Proportions of Flowable Fill with Spend Foundry Sand (kg/m³)

	Cement	Fly Ash	Spend Foundry Sand	Water
Minimum	25	334	818	291
Maximum	94	463	1264	504
Average	57	383	1075	399

Source: From Deng, A., *Excess Foundry Sand Characterization and Experimental Investigation in Controlled Low-Strength Material*, PhD Dissertation of the Pennsylvania State University, University Park, PA, August 2004.

Metal Casting Wastes

The primary characteristic necessary for a liner, cover, or cutoff wall is low permeability, which essentially enables them to slow down the seepage or diffusion of chemicals. Clay is therefore the main material used to construct these containment systems. The thickness and chemical compatibility of containment systems are of concern in assessing the performance of a system. For example, clay liners are constructed as a simple liner that is 2 to 5 ft thick. In composite and double liners, the compacted clay layers are usually between 2 and 5 ft thick, depending on the characteristics of the underlying geology and the type of liner to be installed. Regulations specify that the clay used can only allow water to penetrate at a rate of less than 1.2 in./yr. However, the effectiveness of clay liners can be reduced by fractures induced by freeze–thaw cycles, drying out, and the presence of some chemicals.

Use of spent foundry sand
Most spent foundry sand discarded is green sand. The primary components of green sand are silica and bentonite. Thus, green sands are essentially a sand–bentonite mixture, which makes them potentially useful as a liner and cover materials, that is, for hydraulic barrier layers.

The critical properties of green sand affecting its performance as a hydraulic barrier material include grain size distribution, compaction curves, and hydraulic conductivity when compacted. In general, hydraulic conductivity in use should be less than 1×10^{-7} cm/s, which is the criterion for a conventional clay barrier. Sometimes, if aggressive leachate might penetrate through the barrier, the chemical compatibility and durability of a green sand barrier must be researched. As zero-valence iron, clay, and carbonaceous materials help the containment of chemicals, green sand is thought to be an active containment media in the subsurface cleaning domain.

4.4.4 Agricultural Applications

An emerging domain for some spent foundry sand reuse is as a component in the manufacture of topsoil and growing amendments or composites. In many parts of the globe, high-quality topsoils for landscaping are not available in urban areas. Commercial landscapers and nursery growers frequently manufacture topsoil and composite by blending composted materials and low-quality soils, which not only exhausts natural resource, but also increases manufacture cost. Spent foundry sand has been reported to be amended into a product matching the characteristics of topsoil and amendments; this could be another competitive and vast market for both the metal casting industry and horticultural professionals.

4.4.4.1 Topsoil

Description
Topsoil is the uppermost layer of the Earth's surface, ranging in depth from a few inches to many feet. Topsoil has been created over long time by the physical and chemical action of climate, weather on the Earth's parent rock materials, and decay of plants. As a result, a considerable accumulation of decaying organic matter is found in topsoil. Topsoil is the base for gardening and landscaping activities, where plants are grown by gardening efforts. Unfortunately, genuine topsoil created by natural forces is often unavailable because it is so scarce, and when it can be obtained it is often very expensive.

Topsoil should have a loose and open structure so that it drains fast to keep the ground surface dry. At the same time, it must be able to retain enough moisture in order that plants growing in it are not constantly subjected to drought stress. The properties of interest include particle gradation, clay content, nutrient content, and retention capacity.

Use of spent foundry sand
Spent foundry green sand is of particular interest to soil blending companies that produce topsoil, because of its dark color, clay content, moisture retention, and consistency. A high sand content is required in topsoil, so spent foundry sand could be a major component. Spent sand reduces the

formation of clumps and prevents the mix from compacting, which allows air to circulate within the topsoil and to stimulate decomposition. The U.S. Agricultural Research Service is leading some pilot studies conducted to investigate the feasibility of beneficially using foundry sand as a topsoil replacement.

4.4.4.2 Growing Amendments

Description

Very similar in growing function to topsoil, growing amendments are also manufactured for agronomic purposes. They are specifically designed as a composite for serving gardening, greenhouse, nursery, or horticultural industries.

The growing amendments should be porous and well drained, yet retentive of sufficient moisture to supply adequate water volume for plants. A relatively low level of soluble salts is necessary to maintain a mild environment for plants. However, an adequate exchange capacity is preferred to retain and supply the elements necessary for plant growth. The media should be free from harmful pests, pathogenic organisms, insects, nematodes, and weed seeds. In use, it should present stable biological and chemical characteristics. For sand-based growing amendments, the components generally include peat, bark, and sand. The former two components account for 60 to 80% in weight.

Use of spent foundry sand

Replacing natural soil/sand with spent foundry sand in agronomic applications represents an excellent market for the beneficial reuse of foundry byproduct. The presence of clay in foundry sand is beneficial because clay increases the capacity of soils to retain nutrients and therefore reduce the amount of additional nutrient required for plant growth.[50] A pilot study[51] has indicated that spent foundry sand can be incorporated into growing mixes for the nursery industry. The pilot study[51] used the mixing formulation of 50% manufactured growing mix and 50% foundry sand. The materials should avoid the complexity of a wide carbon: nitrogen ratio, high pH, and high water holding capacity, which leads to the easy development of successful growing amendments containing spend foundry sand. The nursery industry and the greenhouse industry represent an excellent market for local small and large foundries.

4.4.5 GENERAL PROCESSES

4.4.5.1 Crushing, Screening, and Storage

For solid wastes to be suitable as a full or partial replacement for components in other applications, it should be free of objectionable material such as wood, garbage, and metal that can be introduced at the foundry. It should be free of foreign material and thick coatings of burnt carbon, binders, and mold additives that could inhibit product manufacture, such as cement hydration. It may be necessary to crush the solid waste to reduce the size of oversized core butts or unclasped molds. Magnetic separation is a good solution to producing a suitable coarse or fine aggregate product.

Aggregates should exhibit consistent physical and chemical characteristics to quality for most of the aforementioned applications. However, current practices often lead to a composite of various foundry wastestreams. Special attention is required to set up a rigorous quality control system with waste supply on a source-specific basis.

Stockpiles of a sufficient volume of solid wastes should be supplied for reuse requirements.

4.4.5.2 Design and Construction

Conventional structural design and construction procedures for a construction are generally applicable to a construction incorporating foundry solid wastes. The same production methods and equipment used for conventional manufacture can be used for production of manufacture using foundry solid waste.

Metal Casting Wastes

4.4.6 Unresolved Issues

From an engineering perspective, recycled materials should be used in such a manner that the expected performance of the product will not be compromised. Waste and byproduct materials, however, differ vastly in their types and properties and, as a result, in the end-use applications for which they may be suited. Experience and knowledge regarding the use of these materials vary from material to material as well as from facility to facility. To recover these materials for potential use, engineers, researchers, generators, and regulators need to be aware of the properties of the materials, how they can be used, and what limitations may be associated with their use.

Most foundries have installed sand reclamation systems that screen the metal and debris out of the sand so that a good, clean product is available for reuse in a variety of applications and industries. This is a good start in the strategy of reusing foundry solid waste. Depending on the projected end-use, it may be important to segregate wastestreams at the foundry, as each stream can have different characteristics. Additionally, some waste materials, such as bulk spent sand, are typically unrecoverable during the "shakeout" and finishing processes. These sands may be contaminated with metal or very large chunks of burned cores (referred to as core "butts") and will need to undergo some type of segregation, crushing, and screening before recycling. Some hard chunks may not even be crushable, and have to be landfilled.

4.5 BARRIERS TO SOLID WASTE REUSE

Besides technical aspects, wide acceptance of reusing foundry solid wastes as marketable materials will only be achieved by removing barriers or limitations arising from public perception (education or training), environmental regulation, engineering guidelines and procedures, economics, and market potentials. These barriers are basically nontechnical but take considerable efforts to address. Unlike the technical aspects of a reuse program, many parties are involved, such as the government, the public, academics, and industrial and commercial departments. A coordinated and consistent framework needs to be constructed among these parties, aiming at eliminating barriers to the foundry solid waste reuse program.

4.5.1 Education

Public acceptance of foundry waste reuse significantly depends upon their understanding of nature and the performance of foundry waste materials and generated products. In general, negative descriptions, such as its black appearance, the presence of casting byproducts and heavy metals, high melting temperature and sometimes odor, may automatically bring objections into the public mind. It is unfair. The public should be well educated to understand the generation and characterization of foundry solid waste. Documented technical data and environmental regulations are to be presented to convince people that foundry solid waste (at least not all) is not hazardous or as bad as they thought.

Communication channels shall be set up between industry and academics. There has been inconsistency with regards to the characterization of foundry solid waste between industry and academics. The former cares about the workability and efficiency of materials in generating products. The latter concentrate on the technical behavior of materials if reused. The way that metal casters define the characteristics of their sands is completely different from what the contractor wants to know. For example, metal casters talk about ground fineness number, whereas contractors want to know fine and clay contents. At the point of reusing their solid waste, metal casters should divert their attention from regulators and customers to researchers, working within a well channeled system.

4.5.2 Environmental Regulation

Solid waste regulations are frequently cited as barriers for metal industrial byproduct recycling. Research indicates that most ferrous spent foundry sand meets nonhazardous standards under the

RCRA. Competing granular materials, such as sands, gravels, and native soils, are not regulated materials although their environmental profiles may be similar to spent sands.

In some case, experts may debate the reuse of nonhazardous materials, which, they insist, should still be dumped to general landfill sites where nonhazardous materials belong, like municipal solid waste. It is also insisted that there is no documented regulation requiring the reuse of nonhazardous materials. Therefore, to defend the beneficial reuse program of foundry solid waste, regulations should specifically permit their marketing.

Environmental regulation should be complied with to legally validate a reuse program. It is critical that recyclers become familiar with the federal and state regulations relative to their materials. Before reuse starts, materials should be tested according to these environmental regulations to determine whether they are hazardous or nonhazardous. Knowing and understanding the rules and regulations will lead to a better reuse program.

4.5.3 Guidelines, Procedures, and Specifications

Conventional materials have been approved to enter the market, supported by many standalone guidelines, procedures, and specifications. As such, suppliers and users favor the selection of conventional materials. Foundry solid waste is being put to a competitive disadvantage against conventional materials, just because no standalone guidelines and procedures are universally documented for their potential markets. This barrier could be eliminated by showing data demonstrating that foundry solid waste is at least as good as, if not better than, conventional materials for target end-uses. A trial and error procedure is normally used in bench-scale tests, where guidelines, procedures, and specifications are developed by referring to documented ones. Successful experimental and field demonstration then further modifies and finalizes guidelines, procedures, and specifications.

4.5.4 Economics

Economical factors, such as disposal costs, the availability of conventional materials, and transportation costs, are critical considerations. As with any material, transportation costs are generally the highest cost factor in recycling solid waste. The most economically sustainable options for recycling foundry solid waste will generally match the volume and characteristics of the materials with nearby businesses and construction projects. Small foundries may not generate enough material on a weekly or monthly basis to satisfy the need for construction sands. In this case, it may be necessary to collect similar wastestreams from multiple sources or to partially substitute for conventional materials in order to meet volume requirements.

Some end-use applications may prefer the characteristics of foundry solid waste. For instance, spent foundry sand is a uniformly graded fine aggregate containing chemically active iron and organics. Spent foundry sand can be superior to other types of granular materials, such as compacted soils or clays, for hydraulic barriers. In this case, spent foundry sand provides better performance at lower cost.

4.5.5 Market Potential

One particular mistake that foundries make is improperly defining the market potential of their byproducts. Competitive material availability and transportation costs will dictate market acceptance in most cases. They must study the landscape before attempting to enter the market. Active marketing efforts will always get paid back. Keep in mind that many potential customers are cost-conscious, and that is an advantage to the foundry byproducts process. Aiming low to establish a market is a great strategy for getting in the door. Foundries need to value market sustainability and cost reduction over the best short-term deal. Build partnerships with end-users and long-term progress will be established.

Before entering the market, the following questions should be addressed in order to have a good start. The ultimate goal is that the bottom line of reuse is well understood, making sure the materials are characterized properly, and then marketing them according to the appropriate regulations.

1. Is the volume of material supplies adequate for the quantity expectations of a potential customer?
2. Will the properties and variability of the materials satisfy the quality expectations of a potential customer?
3. Will any processing be required to consistently guarantee the expected quality level?
4. Is all cost taken into account?
5. What is the cost of the material the byproduct is to replace?
6. Are all permits obtained for the actions?

REFERENCES

1. Wright, R.J., Take a new look at sand reclamation, *Foundry Management and Technology*, 3, 22–24, 2001.
2. Leidel, D.S., Pollution Prevention and Foundries, in *Industrial Pollution Prevention Handbook*, Freedman, H. M., Eds., McGraw-Hill, New York, 1984.
3. WMRC, Primary Metal, Illinois Waste Management and Research Centre, available at http://www.wmrc.uiuc.edu/info/library_docs/manuals/primmetals/chapter3.htm.
4. AFS, Alternative Utilization of Foundry Waste Sand, final report (Phase I) for Illinois Department of Commerce and Community Affairs, American Foundrymen's Society, Des Plaines, IL, July 1991.
5. McKinley, M.D. et al., Waste Management Study of Foundries Major Wastestreams, Phase II, HWRIC TR-016, April 1994.
6. Collins, R.J. and Ciesielski, S.K., Recycling and Use of Waste Materials and Byproducts in Highway Construction, National Cooperative Highway Research Program Synthesis of Highway Practice 199, Transportation Research Board, Washington, DC, 1994.
7. Shah, D.B. and Phadke, A.V., Lead removal of foundry waste by solvent extraction, *Journal of Air and Waste Management*, 45, 150–155, 1995.
8. CFR, Metal Molding and Casting Industry Point Source Category Effluent Limitations Guidelines, Pretreatment Standards and New Source Performance Standards, Part 464, Code of Federal Register, Washington, DC, 1985.
9. MOEE, Spent Foundry Sand—Alternative Uses Study, report for Ontario Ministry of the Environment and Energy and the Canadian Foundry Association, Ontario, Canada, July 1993.
10. MNR, Mineral Aggregate Conservation, Reuse and Recycling, report for Ontario Ministry of Natural Resources, Ontario, Canada, February 1992.
11. Javed, S. and Lovell, C.W., Use of Foundry Sand in Highway Construction, Joint Highway Research Project No. C-36-50 N, Purdue University, West Lafayette, IN, July 1994.
12. Deng, A., Excess Foundry Sand Characterization and Experimental Investigation in Controlled Low-Strength Material, PhD Dissertation of the Pennsylvania State University, University Park, PA, August 2004.
13. Javed, S., Lovell, C. W., and Wood, L.E., Waste foundry sand in asphalt concrete, in *Transportation Research Record*, No 1437, Transportation Research Board, Washington, DC, 1994.
14. U.S. EPA, Foundry Sands Recycling, report EPA/530/F-07/018, U.S. EPA, Washington, DC, 2007, available at http://www.epa.gov/epaoswer/osw/conserve/foundry/foundry-st.pdf), 2007.
15. Ham, R.K., Boyle, W.C., Engroff, E.C., and Fero, R.L., Determining the presence of organic compounds in foundry waste leachates, in *Modern Casting*, American Foundrymen's Society, August 1989.
16. Johnson, C.K., Phenols in foundry waste sand, in *Modern Casting*, American Foundrymen's Society, January 1981.
17. Ontario Ministry of Transportation, Resistance of Fine Aggregate to Degradation by Abrasion in the MicroDuval Apparatus, LS-619, Ontario Ministry of Transportation, Ontario, Canada, 1996.
18. AASHTO, AASHTO Designation M240: Blended hydraulic cements, in *Standard Specification for Materials*, American Association of State Highway and Transportation Officials, 1986.
19. NSA, *Processed Blast Furnace Slag: The All Purpose Construction Aggregate*, 188.1, National Slag Association, Alexandria, VA, 1988.

20. Emery, J.J., Pelletized lightweight slag aggregate, in *Proceedings of Concrete International 1980*, Concrete Society, April, 1980.
21. JEGEL, *Manitoba Slags, Deposits, Characterization, Modifications, Potential Utilization*, report, John Emery Geotechnical Engineering Limited, Toronto, Ontario, 1986.
22. Hughes, M.L. and Haliburton, T.A., Use of zinc smelter waste as highway construction material, *Highway Research Record*, 430, 16–25, 1973.
23. Mantell, C.L., *Solid Wastes: Origin, Collection, Processing and Disposal*, John Wiley & Sons, New York, 1975.
24. NSA, Blast Furnace Slag, Ideal Backfill Material for Steel Sheet Piling, 166.2, National Slag Association, Alexandria, VA, 1966.
25. Short, A., The use of lightweight concrete in reinforced concrete construction, *The Reinforced Concrete Review*, 5 (3), September 1959.
26. Emery, J.J., Slag utilization in pavement construction, *Extending Aggregate Resources*, ASTM Special Technical Publication 774, American Society for Testing and Materials, 1982, pp. 95–118.
27. JEGEL, *Steel Slag Aggregates Use in Hot Mix Asphalt Concrete*, final report for the Steelmaking Slag Technical Committee, April, 1993.
28. OECD, *Use of Waste Materials and Byproducts in Road Construction*, Organization for Economic Co-operation and Development, Paris, 1977.
29. Tennessee Department of Transportation, Test Reports on Samples of Coarse and Fine Aggregates, July, 1995.
30. Noureldin, A.S. and McDaniel, R.S., Evaluation of Steel Slag Asphalt Surface Mixtures, presented at the 69th annual meeting, Transportation Research Board, Washington, January, 1990.
31. Das, B.M., Tarquin A.J., and Jones, A.Q., Geotechnical Properties of Copper Slag, *Transportation Research Record*, 941, National Research Board, Washington, DC, 1993.
32. Lee, H., Coatings go beyond appearance to provide quality control, *Foundry Management and Technology*, 1, 2008.
33. Winkler, E.S. and Bol'shakov, A.A., Characterization of Foundry Sand Waste, technical report 31, University of Massachusetts at Amherst, MA, 2000.
34. Federal Highway Administration, available at http://www.tfhrc.gov/hnr20/recycle/waste/app.htm.
35. ASA/RTA, A Guide to the Use of Slag in Roads, Australian Slag Association and Roads and Traffic Authority of NSW, New South Wales, Australia, 1993.
36. Farrand, B. and Emery, J., Recent Improvements in the Quality of Steel Slag Aggregate, paper prepared for presentation at 1995 annual meeting of the Transportation Research Board, Washington, DC, January, 1995.
37. Federal Highway Administration, available at http://www.tfhrc.gov/hnr20/recycle/waste/app2.htm, 2008.
38. Hogan, F.J., The effect of blast furnace slag cement on alkali aggregate reactivity: a literature review, *Cement Concrete and Aggregates*, 7, 2, 1985.
39. ASTM, Standard Specification C672: Scaling Resistance of Concrete Surfaces Exposed to Deicing Chemicals, in *Annual Book of ASTM Standards*, Vol. 04.02, American Society for Testing and Materials, West Conshohocken, PA, 1993.
40. Federal Highway Administration, available at http://www.tfhrc.gov/hnr20/recycle/waste/app4.htm.
41. Adaska, W.S. and Krell, W.C., Bibliography on controlled low-strength materials (CLSM), *Concrete International*, 14 (10), 42–43, December 1992.
42. Naik, T.R., Ramme, B.W., and Kolbeck, H.J., Filling abandoned underground facilities with CLSM fly ash slurry, *Concrete International*, 12 (7), 19–25, July 1990.
43. Larsen, R.L., Sound uses of CLSM in the environment, *Concrete International*, 12 (7), 26–29, July 1990.
44. Ambroise, J., Amoura, A., and Péra, J., Development of flowable high volume – fly ash mortars, in *Proceedings of the 11th International Symposium on the Use and Management of Coal Combustion Byproducts (CCBs)*, Vol. 2, American Coal Ash Association, 1995.
45. Newman, F.B., Di Gioia, A.M., and Rojas-Gonzalez, L.F., CLSM backfills for bridge abutments, in *Proceedings of the 11th International Symposium on the Use and Management of Coal Combustion Byproducts (CCBs)*, Vol. 2, American Coal Ash Association, 1995.
46. ASTM, Standard specification C33: concrete aggregate, in *Annual Book of ASTM Standards*, Vol. 04.02, American Society for Testing and Materials, West Conshohocken, PA, 1996.
47. ACI Committee 229, Controlled Low Strength Materials, ACI 229R, American Concrete Institute, Farmington Hills, MI, 1999.

48. Bhat, S.T. and Lovell, C.W., Design of Flowable Fill: Waste Foundry Sand as a Fine Aggregate, Paper 961066, Transportation Research Board, 75th Annual Meeting, Washington, DC, 1996.
49. Naik, T.R., Foundry Industry Byproducts Utilization, report CBU-1989-01, University of Wisconsin-Milwaukee, WI, February 1989.
50. Regan, R.W. and Voigt, R.C., Working with the regulators for the beneficial use of foundry residuals, in *Proceedings of the 28th Mid-Atlantic Industrial Waste Conference*, Buffalo, NY, 1996.
51. Dunkelberger, J. and Regan, R., Evaluation of spent foundry sand as growing mix amendment: potential beneficial use option, in *Proceedings of AFS 101st Casting Congress*, Seattle, WA, April 1997.
52. U.S. EPA, Summary of Factors Affecting Compliance by Ferrous Foundries, Vol. 1, EPA-340/1-80-020, U.S. EPA, Washington, DC, January 1981.

5 Waste Treatment in the Aluminum Forming Industry

Lawrence K. Wang and Nazih K. Shammas

CONTENTS

5.1	Industry and Process Description	200
	5.1.1 Casting	200
	5.1.1.1 Direct Chill Casting	200
	5.1.1.2 Continuous Casting	201
	5.1.1.3 Stationary Casting	201
	5.1.2 Rolling	201
	5.1.3 Extrusion	201
	5.1.4 Forging	202
	5.1.5 Drawing	202
	5.1.6 Heat Treatment	203
	5.1.7 Surface Treatment	203
5.2	Aluminum Forming Industry Subcategory Description	203
	5.2.1 Rolling with Neat Oils	205
	5.2.2 Rolling with Emulsions	206
	5.2.3 Extrusion	206
	5.2.4 Forging	206
	5.2.5 Drawing with Neat Oils	206
	5.2.6 Drawing with Emulsions or Soaps	206
5.3	Aluminum Forming Industry Waste Characterization	206
	5.3.1 Direct Chill Casting	206
	5.3.2 Rolling with Emulsions	207
	5.3.3 Extrusion	207
	5.3.4 Forging	207
	5.3.5 Drawing with Emulsions or Soaps	209
	5.3.6 Heat Treatment	209
	5.3.7 Etch or Cleaning	216
5.4	Plant-Specific Description of Case Histories	216
	5.4.1 Plant A: Case History	216
	5.4.2 Plant B: Case History	218
	5.4.3 U.S. EPA Data on Full-Scale Treatment of Aluminum Forming Wastewaters	218
5.5	Pollutant Removability	221
5.6	Treatment Technology Costs	222
5.7	Aluminum Forming Industrial Effluent Limitations	223
5.8	Technical Terminologies of Aluminum Forming Operations and Pollution Control	224
References		226

5.1 INDUSTRY AND PROCESS DESCRIPTION

The aluminum forming industry is a manufacturing industry in which aluminum or aluminum alloys are made into semifinished aluminum products using hot or cold working processes. The aluminum forming manufacturing operations include the rolling, drawing, extruding, and forging of aluminum. In the U.S., the industry consists of about 300 plants owned by about 150 companies. The industry employs about 30,000 workers.

As well as the aluminum forming manufacturing operations of rolling, drawing, extruding, and forging, there are associated processes, such as the casting of aluminum alloys for subsequent forming, heat treatment, cleaning, etching, and solvent degreasing.[1-6]

Surface treatment of aluminum (such as cleaning, etching, and solvent degreasing) is any chemical or electrochemical treatment applied to the surface of aluminum. Such surface treatment is considered to be an important part of aluminum forming. For the purposes of government regulation, surface treatment of aluminum is considered to be an integral part of aluminum forming whenever it is performed at the same plant site at which aluminum is formed, and such operations are not considered for government regulation under the Electroplating and Metal Finishing provisions of U.S. 40 CFR parts 413 and 433.

Casting aluminum when performed as an integral part of aluminum forming and located on site at an aluminum forming plant is considered an aluminum forming operation and hence is covered under these government guidelines.

When aluminum forming is performed on the same site as primary aluminum reduction the casting shall be regulated by the nonferrous metals guidelines if there is no cooling of the aluminum prior to casting. If the aluminum is cooled prior to casting then the casting shall be regulated by the aluminum forming guidelines. The major aluminum forming processes are briefly described in the narrative below.

5.1.1 Casting

Before aluminum alloys can be used for rolling or extrusion and subsequently for other aluminum forming operations, they are usually cast into ingots of suitable size and shape.

The aluminum alloys used as the raw materials for casting operations are sometimes purchased from nearby smelters and transported to the forming plants in a molten state. Usually, however, purchased aluminum ingots are charged together with alloying elements into melting furnaces at the casting plants. Several types of furnaces can be used, but reverberatory furnaces are the most common.

At many plants, fluxes are added to the metal to reduce hydrogen contamination, remove oxides, and eliminate undesirable trace elements. Solid fluxes such as hexachloroethane, aluminum chloride, and anhydrous magnesium chloride may be used, but it is more common to bubble gases such as chlorine, nitrogen, argon, helium, and mixtures of chlorine and inert gases through the molten metal.

The casting methods used in aluminum forming can be divided into three classes: direct chill casting, continuous casting, and stationary casting.

5.1.1.1 Direct Chill Casting

Vertical direct chill casting is the most widely used method of casting aluminum for subsequent forming. Direct chill casting is characterized by continuous solidification of the metal while it is being poured. The length of an ingot cast using this method is determined by the vertical distance it is allowed to drop rather than by mold dimensions. Molten aluminum is tapped from the melting furnace and flows through a distributor channel into a shallow mold. Noncontact cooling water circulates within this mold, causing solidification of the aluminum. The base of the mold is attached to a hydraulic cylinder that is gradually lowered as pouring continues. As the solidified aluminum leaves the mold it is sprayed with contact cooling water, reducing the temperature of the forming ingot. The cylinder continues to lower into a tank of water, causing cooling of the ingot as it is immersed. When the cylinder has reached its lowest position, pouring stops and the ingot is lifted from the pit. The hydraulic cylinder is then raised and positioned for another casting cycle. Lubrication of the mold is required to ensure proper ingot quality. Lard or castor oil is usually applied before casting begins and may be reapplied during the drop.

5.1.1.2 Continuous Casting

Unlike direct chill casting, no restrictions are placed on the length of the casting as it is not necessary to interrupt production to remove the cast product. Continuous casting eliminates or reduces the degree of subsequent rolling required. Because continuous casting affects the mechanical properties of the aluminum cast, the use of continuous casting is limited by the alloys used, the nature of subsequent forming operations, and the desired properties of the finished product. Continuous casting techniques have been found to significantly reduce or eliminate the use of contact cooling water and oil lubricants.

5.1.1.3 Stationary Casting

Molten aluminum is poured into cast-iron molds and allowed to air cool. Lubricants and cooling water are not required. Melting and casting procedures are dictated by the intended use of the ingots produced. Frequently the ingots are used as raw material for subsequent aluminum forming operations at the plant. Other plants sell these ingots for reprocessing.

5.1.2 ROLLING

The rolling process is used to transform cast aluminum ingot into any one of a number of intermediate or final products. Pressure exerted by the rollers as aluminum is passed between them flattens the metal and may cause work hardening.

Heat treatment is usually required before and between stages of the rolling process. Ingots are usually made homogeneous in grain structure prior to hot rolling in order to remove the effects of casting on the aluminum's mechanical properties. Annealing is typically required during cold rolling to keep the metal ductile and remove the effects of work hardening. The kind and degree of heat treatment applied depends on the alloy involved, the nature of the rolling operation, and the properties desired in the product.

It is necessary to use a cooling and lubricating compound during rolling to prevent excessive wear on the rolls, to prevent adhesion of aluminum to the rolls, and to maintain a suitable and uniform rolling temperature. Oil-in-water emulsions, stabilized with emulsifying agents such as soaps and other polar organic materials, are used for this purpose in hot rolling operations.

5.1.3 EXTRUSION

In the extrusion process, high pressures are applied to a cast billet of aluminum, forcing the metal to flow through a die orifice. The resulting product is an elongated shape or tube of uniform cross-sectional area.

Extrusions are manufactured using either a mechanical or a hydraulic extrusion press. A heated cylindrical billet is placed into the ingot chamber and the dummy block and ram are placed into position behind it. Pressure is exerted on the ram by hydraulic or mechanical means, forcing the metal to flow through the die opening. The extrusion is sawed off next to the die, and the dummy block and ingot butt are released. Hollow shapes are produced with the use of a mandrel positioned in the die opening so that the aluminum is forced to flow around it. A less common technique, indirect extrusion, is similar except that in this method the die is forced against the billet, extruding the metal in the opposite direction through the tam stem. A dummy block is not used in indirect extrusion.

Although aluminum can be extruded cold, it is usually first heated to a temperature ranging from 375 to 525°C, so that little work hardening will be imposed on the product. Heat treatment is frequently used after extrusion to achieve the desired mechanical properties.

The extrusion process requires the use of a lubricant to prevent adhesion of the aluminum to the die and ingot container walls. In hot extrusion, limited amounts of lubricant are applied to the ram and die face or to the billet ends. For cold extrusion, the container walls, billet surfaces, and die orifice must be lubricated with a thin film of viscous or solid lubricant. The lubricant most commonly used in extrusion is graphite in an oil or water base. A less common technique, spraying liquid nitrogen on the billet prior to extrusion, is also used. The nitrogen vaporizes during the extrusion process and acts as a lubricant.

5.1.4 Forging

Closed die forging, the most prevalent method, is accomplished by hammering or squeezing the aluminum between two steel dies, one fixed to the hammer or press ram and the other to the anvil. Forging hammers, mechanical presses, and hydraulic presses can be used for the closed die forging of aluminum alloys. The heated stock is placed in the lower die and, by one or more blows of the ram, forced to take the shape of the die set. In closed die forging, aluminum is shaped entirely within the cavity created by these two dies. The die set comes together to completely enclose the forging, giving lateral restraint to the flow of the metal.

The process of open die forging is similar to that described above but in this method the shape of the forging is determined by manually turning the stock and regulating the blows of the hammer or strokes of the press. Open die forging requires a great deal of skill and only simple, roughly shaped forgings can be produced. Its use is usually restricted to items produced in small quantities and to development work where the cost of making closed-type dies is prohibitive.

The process of rolled ring forging is used in the manufacture of seamless rings. A hollow cylindrical billet is rotated between a mandrel and pressure roll to reduce its thickness and increase its diameter.

Proper lubrication of the dies is essential in forging aluminum alloys. Colloidal graphite in either water or an oil medium is usually sprayed onto the dies for this purpose.

5.1.5 Drawing

The term drawing, when it applies to the manufacture of tube, rod, bar, or wire, refers to the pulling of metal through a die or succession of dies to reduce its diameter, alter the cross-sectional shape, or increases its hardness. In the drawing of aluminum tubing, one end of the extruded tube is swaged to form a solid point and then passed through the die. A clamp, known as a bogie, grips the swaged end of tubing. A mandrel is inserted into the die orifice, and the tubing is pulled between the mandrel and die, reducing the outside diameter and the wall thickness of the tubing. Wire, rod, and bar drawing is accomplished in a similar manner but the aluminum is drawn through a simple die orifice without using a mandrel.

In order to ensure uniform drawing temperatures and avoid excessive wear on the dies and mandrels used, it is essential that a suitable lubricant be applied during drawing. A wide variety of lubricants are used for this purpose. Heavier draws may require oil-based lubricants, but oil-in-water emulsions are used for many applications. Soap solutions may also be used for some of the lighter draws. Drawing oils are usually recycled until their lubricating properties are exhausted.

Intermediate annealing is frequently required between draws in order to restore the ductility lost by cold working of the drawn product. Degreasing of the aluminum may be required to prevent burning of heavy lubricating oils in the annealing furnaces.

Aluminum Forming Industry

5.1.6 Heat Treatment

Heat treatment is an integral part of aluminum forming and is practiced at nearly every plant in the category. It is frequently used both in-process and as a final step in forming to give the aluminum alloy the desired mechanical properties. The general types of heat treatment applied are as follows:

1. *Homogenizing.* This increases the workability and helps control recrystallization and grain growth following casting.
2. *Annealing.* This softens work-hardened and heat-treated alloys, relieves stress, and stabilizes properties and dimensions.
3. *Solution heat treatment.* This improves mechanical properties by maximizing the concentration of hardening constituents in the solid solution.
4. *Artificial aging.* This provides hardening by precipitation of constituents from the solid solution.

Homogenizing, annealing, and aging are dry processes whereas solution heat treatment typically involves significant quantities of contact cooling water.

5.1.7 Surface Treatment

A number of chemical or electrochemical treatments may be applied after the forming of aluminum or aluminum alloy products. Solvent, acid and alkaline solutions, and detergents can be used to clean soils such as oil and grease from the aluminum surface. Acid and alkaline solutions can be used to etch the product or brighten its surface. Acid solutions are also used for deoxidizing and desmutting.

Surface treatments and their associated rinses are usually combined in a single line of successive tanks. Wastewater discharges from these lines are typically commingled before treatment or discharge. In some cases, rinsewater from one treatment is reused in the rinse of another. These treatments may be used for cleaning purposes, to provide the desired finish for an aluminum formed product, or to prepare the aluminum surface for subsequent coating by processes such as anodizing, conversion coating, electroplating, painting, and porcelain enameling.

A number of different terms are commonly used in referring to sequences of surface treatments, for example, pickling lines, cleaning lines, etch lines, preparation lines, and pretreatment lines. The terminology depends, to some degree, on the purpose of the lines, but usage varies within the industry. In addition, the characteristics of wastewater generated by surface treatment are determined by the unit components of the treatment lines rather than the specific purpose of its application. Cleaning and etch line is used in this section to refer to any surface treatment processes other than solvent cleaning.

5.2 ALUMINUM FORMING INDUSTRY SUBCATEGORY DESCRIPTION

Division of the industry into subcategories provides a mechanism for addressing processes, products, and other variations that result in distinct wastewater characteristics. The aluminum forming industry is comprised of separate and distinct processes with enough variability in products and wastes to require categorization into a number of discrete subcategories. The individual processes, wastewater characteristics, and treatability comprise the most significant factors in the subcategorization of this complex industry. Other factors either served to support and substantiate the subcategorization or were shown to be inappropriate bases for subcategorization. From this evaluation, the following are the established subcategories:

1. Rolling with neat oils
2. Rolling with emulsions
3. Extrusion

4. Forging
5. Drawing with neat oils
6. Drawing with emulsions

Each industrial subcategory is broken into "core" and "additional allocation" operations. The core is defined as those operations that always occur in the subcategory or do not affect the wastewater characteristics from the subcategory facilities (e.g., dry operations, zero-pollutant-allocation operations, or operations that contribute insignificant pollutants and wastewater volume in comparison with other streams). These operations that do not contribute to the wastewater characteristics will not occur at every plant, which should not affect wastewater treatment.

Operations that may affect wastewater characteristics but are not included in the core are classified as additional allocation operations. These are ancillary operations involving discharged wastewater streams of significant pollutant concentrations and flows that may or may not be present at any one facility. If an additional allocation operation occurs at a facility, the wastewater from the operation would occur in addition to the core wastewater, with a subsequent modification to the performance expected from a treatment facility. The most common additional allocation operations are as follows:

1. Cooling water from direct chill casting
2. Quench water from heat treatment
3. Rinsewater from cleaning and etch lines

The designation of core and additional allocation operations is listed by subcategory in Table 5.1. More than one subcategory may be associated with a specific facility. A brief description of the subcategories follows.

TABLE 5.1
Summary of Core and Additional Allocation Operations Associated with Aluminum Forming Industry Subcategories

Core	Additional Allocation
Subcategory 1: Rolling with Neat Oils	
Rolling using neat oils	Solution heat treating
Roll grinding	Cleaning or etching
Degassing	Annealing
Stationary casting	
Continuous sheet casting	
Homogenizing	
Artificial aging	
Degreasing	
Cleaning or etching	
Sawing	
Stamping	
Subcategory 2: Rolling with Emulsions	
Rolling with emulsified lubricants	Direct chill casting or
Roll grinding	continuous rod casting
Degassing	Solution heat treatment
Stationary casting	Cleaning or etching

Continued

TABLE 5.1 (continued)

Core	Additional Allocation
Homogenizing	
Artificial aging	
Cleaning or etching	

Subcategory 3: Extrusion

Core	Additional Allocation
Extrusion die cleaning	Direct chill or continuous rod casting
Extrusion dummy block cooling	Press and solution heat treatment
Degassing	Cleaning or etching
Stationary casting	Extrusion die cleaning
Artificial aging	Annealing
Annealing	
Degreasing	
Cleaning or etching	

Subcategory 4: Forging

Core	Additional Allocation
Artificial aging	Forging
Annealing	Solution heat treatment
Degreasing	Cleaning or etching
Cleaning or etching	
Sawing	

Subcategory 5: Drawing with Neat Oils

Core	Additional Allocation
Drawing with neat oils	Continuous rod casting
Continuous rod casting	Solution heat treatment
Stationary casting	Cleaning or etching
Artificial aging	
Annealing	
Degreasing	
Cleaning or etching	
Sawing	
Stamping	
Swaging	

Subcategory 6: Drawing with Emulsions or Soaps

Core	Additional Allocation
Drawing with emulsions or soaps	Continuous rod casting
Continuous sheet casting	Solution heat treatment
Stationary casting	Cleaning or etching
Artificial aging	
Annealing	
Degreasing	
Cleaning or etching	

Source: U.S. EPA, *Development Document for Effluent Limitations Guidelines and Standards for the Aluminum Forming Point Source Category*, Vols. 1 & 2, U.S. EPA, Washington, DC, 1984; U.S. EPA, *Aluminum Forming Point Source Category*, available at http://www.access.gpo.gov/nara/cfr/waisidx_03/40cfr467_03.html, 2008.

5.2.1 Rolling with Neat Oils

This subcategory is applicable to all wastewater discharges resulting from or associated with aluminum rolling operations in which neat oils are used as a lubricant. The rolling with neat oils subcategory consists of approximately 45 plants, 22 of which use only this process. Half of the plants (23 of 45) associated with this subcategory were also associated with one or more additional subcategories.

5.2.2 Rolling with Emulsions

This subcategory is applicable to all wastewater discharges resulting from or associated with aluminum rolling operations in which oil-in-water emulsions are used as lubricants. The rolling with emulsions subcategory consists of approximately 23 plants, of which only one uses this process exclusively. Thus, 96% of the plants in this subcategory were also included in one or more other subcategories.

5.2.3 Extrusion

All wastewater discharges resulting from or associated with extrusion are applicable to this subcategory. The extrusion subcategory consists of approximately 157 plants, more than in any other subcategory. Of these plants, 140 use the extrusion process exclusively. Although most of the plants in this subcategory (89%) are not associated with any other subcategories, some overlap does occur.

5.2.4 Forging

This subcategory is applicable to all wastewater discharges resulting from or associated with forging of aluminum or aluminum alloy products. The forging subcategory consists of approximately 15 plants, 12 of which use only this process. Thus, only 20% of the plants have operations that overlap with one or more other subcategories.

5.2.5 Drawing with Neat Oils

All wastewater discharges resulting from or associated with drawing operations that use neat oil lubricants are applicable to this subcategory. Fifty of the sixty-two plants that comprise the drawing with neat oils subcategory use this process exclusively. The remaining 12 plants in this subcategory were also associated with one or more additional subcategories.

5.2.6 Drawing with Emulsions or Soaps

This subcategory is applicable to all wastewater discharges resulting from or associated with the drawing of aluminum products using oil-in-water emulsion or soap solution lubricants. Eight of the eleven plants that comprise this subcategory use the drawing with emulsions or soaps process exclusively. Overlap with other subcategories occurs at the remaining three plants.

5.3 ALUMINUM FORMING INDUSTRY WASTE CHARACTERIZATION

Wastewater characterization for the aluminum forming industry has been developed on a wastestream basis, rather than on a subcategory basis. Table 5.2 summarizes the wastewater sources reported for this industry. Wastewater flow rates identified for these sources are presented in Table 5.3.

The pollutants characteristic of the industry wastewaters are summarized in Table 5.4 through Table 5.11, for both classical and toxic pollutants. The toxic pollutant data have been developed using a verification protocol established by U.S. EPA, with the exception of the following: selenium, silver, thallium, and 2,3,7,8-tetrachlorodibenzo-p-dioxin (TCCD). Table 5.12 presents the minimum detection limit for the toxic pollutants. Any value below the minimum limit is listed in the summary tables as below detection limit (BDL).

5.3.1 Direct Chill Casting

Of the approximately 266 plants in the aluminum forming industry, 57 cast aluminum or aluminum alloys using the direct chill method. Because the ingot or billet produced by direct chill casting is used as stock for subsequent rolling or extrusion, this wastewater stream is associated with both rolling with emulsions and extrusion categories. Table 5.4 summarizes the classical and toxic pollutant data associated with the contact cooling water wastestream from direct chill casting.

TABLE 5.2
Wastewater Sources Reported in Aluminum Forming Industry Processes

Wastewater Source	Plants Known to Have Process Wastewater
Direct chill cooling	29
Continuous rod casting cooling	3
Continuous rod casting lubricant	2
Continuous sheet casting	3
Stationary mold casting	0
Air pollution control for metal treatment	5
Rolling with neat oils	45
Rolling with emulsions	27
Roll grinding emulsions	4
Extrusion die cleaning bath	11
Extrusion die cleaning rinse	5
Air pollution control for extrusion die cleaning	2
Extrusion dummy block cooling	3
Air pollution control for forging	3
Drawing with neat oils	55
Drawing with emulsions or soaps	5
Heat treatment quench	43
Air pollution control for annealing furnace	1
Annealing furnace seal	1
Degreasing solvents	2
Cleaning and etch line baths	12
Cleaning and etch line rinses	20
Air pollution control for etch lines	4
Saw oil	3
Swaging and stamping	0

Source. U.S. EPA, *Development Document for Effluent Limitations Guidelines and Standards for the Aluminum Forming Point Source Category*, Vols. 1 & 2, U.S. EPA, Washington, DC, 1984; U.S. EPA, *Aluminum Forming Point Source Category*, available at http://www.access.gpo.gov/nara/cfr/waisidx_03/40cfr467_03.html, 2008.

5.3.2 ROLLING WITH EMULSIONS

Rolling operations that use oil-in-water emulsions as coolants and lubricants are found in 27 plants of the aluminum forming industry. Table 5.5 summarizes the classical and toxic pollutant data for the rolling with emulsions subcategory.

5.3.3 EXTRUSION

The wastewater characterization data for the extrusion die cleaning rinse are summarized by classical and toxic pollutants in Table 5.6.

5.3.4 FORGING

Of the approximately 15 aluminum forging plants, three use wet scrubbers to control particulates and smoke generated from the partial combustion of oil-based lubricants in the forging process.

TABLE 5.3
Summary of Wastewater Flows Reported for Aluminum Forming Industry Processes

Operation	Number of Plants	Number with Zero Discharges	Minimum (cu m/Mg)[a]	Minimum (gal/ton)	Mean (cu m/Mg)	Mean (gal/ton)	Median (cu m/Mg)	Median (gal/ton)	Maximum (cu m/Mg)	Maximum (gal/ton)
Direct chill cooling (no recycle)	16	0	0.0003	0.08	7.9	1900	0.96	230	92	22,000
Direct chill cooling (recycle)	24	6	0.0003	0.08	0.96	230	0.43	104	5.8	1400
Continuous rod casting cooling	1	0	—	—	1.0	250	—	—	—	—
Continuous sheet casting	4	2	0.001	0.24	0.0009	0.22	0.0005	0.12	0.003	0.64
Rolling with emulsions	20	0	0.0003	0.08	0.035	8.4	0.005	1.2	0.3	73
Extrusion die cleaning										
Caustic bath	11	0	0.0002	0.06	0.014	3.3	0.008	1.9	0.054	13
Rinse	5	0	0.001	0.31	0.018	4.4	0.012	2.8	0.054	13
Extrusion dummy block cooling	2	0	2.1	500	2.1	510	—	—	2.2	520
Drawing with emulsions or soaps	6	1	0.003	0.81	1700	400,000	0.67	160	10,000	2,400,000
Heat treatment quench	52	9	0.021	5	5.8	1400	2.3	560	32	7700
Annealing furnace, air pollution control	1	0	—	—	0.026	6.3	—	—	—	—
Cleaning and etch line, rinse	20	0	0.001	0.34	22	5300	5.0	1200	150	36,000
Cleaning and etch line, air pollution control	3	0	0.54	130	2.0	490	1.0	240	4.6	1100
Saw oil lubricants	6	1	0.0004	0.10	0.0018	0.42	0.001	0.25	0.006	1.5

[a] cu m/Mg = m³/10⁶ g.

Source: U.S. EPA, *Development Document for Effluent Limitations Guidelines and Standards for the Aluminum Forming Point Source Category*, Vols. 1 & 2, U.S. EPA, Washington, DC, 1984; U.S. EPA, *Aluminum Forming Point Source Category*, available at http://www.access.gpo.gov/nara/cfr/waisidx_03/40cfr467_03.html, 2008.

TABLE 5.4
Summary of Pollutant Data for the Direct Chill Casting Subcategory Verification Data

Pollutant	Number of Samples/ Number of Detections	Range of Detections	Median of Detections	Mean of Detections
Classical Pollutants (mg/L)				
COD	12/12	<5–420	72	<170
Suspended solids	12/12	<1–220	26	<39
Oil and grease	13/13	<5–210	68	<74
TOC	12/12	1–150	16	47
pH (pH units)	12/12	6.0–8.4	7.4	7.4
Phenols, total	12/12	<0.001–0.12	0.01	0.024
Toxic Pollutants (µg/L)				
Toxic metals				
Lead	12/12	BDL–100	<20	27
Mercury	12/12	BDL–20	1.2	3.2
Zinc	12/12	BDL–1000	100	200
Toxic Organics				
Bis(2-ethylhexyl) phthalate	12/9	BDL–280	46	70
Butyl benzyl phthalate	12/4	BDL–360	130	160
Di-n-butyl phthalate	12/7	BDL–43	20	20
Di-n-octyl phthalate	12/3	BDL–94	40	46
Phenol	12/5	BDL–56	50	33
2-Chlorophenol	12/2	BDL–12	—	BDL
Benzene	12/8	BDL–13	BDL	BDL
Chloroform	12/11	BDL–96	14	22
Methylene chloride	12/12	BDL–<240	95	<98
Polychlorinated biphenyls				
PCB 1232, 1248, 1260, 1016	12/5	BDL–32	BDL	12

BDL, below detection limit.
Source: U.S. EPA, *Development Document for Effluent Limitations Guidelines and Standards for the Aluminum Forming Point Source Category*, Vols. 1 & 2, 1, U.S. EPA, Washington, DC, 1984; U.S. EPA, *Aluminum Forming Point Source Category*, available at http://www.access.gpo.gov/nara/cfr/waisidx_03/40cfr467_03.html, 2008.

The summaries of the classical and toxic pollutant data on the air pollution controls for the forging subcategory are contained in Table 5.7.

5.3.5 DRAWING WITH EMULSIONS OR SOAPS

Eight of the 266 plants in the aluminum forming industry draw aluminum products using oil-in-water emulsions and three use soap solutions as drawing lubricants. These solutions are frequently recycled and discharged periodically after their lubrication properties are exhausted. Table 5.8 summarizes the classical and toxic pollutant data for the drawing with emulsions subcategory.

5.3.6 HEAT TREATMENT

Heat treatment of aluminum products frequently involves the use of a water quench in order to achieve the desired metallic properties. Of the 266 aluminum forming plants, 84 use heat treatment processes that involve water quenching. The sampling data for classical and toxic pollutants from

TABLE 5.5
Summary of Pollutant Data for the Rolling with Emulsions Subcategory Verification Data

Pollutant	Number of Samples/ Number of Detections	Range of Detections	Median of Detections	Mean of Detections
Classical Pollutants (mg/L)				
Dissolved solids	2/2	27,000–34,000	—	30,000
Suspended solids	3/3	890–3900	2400	2400
TOC	3/3	1800–23,000	6800	11,000
Phenols, total	1/1	0.24	—	—
Oil and grease	4/4	1300–31,000	19,000	11,000
Aluminum	3/3	44–210,000	20,000	77,000
Calcium	3/3	22–27,000	18,000	15,000
Magnesium	3/3	11–17,000	12,000	9400
pH (pH units)	1/1	7.0	—	—
Toxic Pollutants (µg/L)				
Metals and inorganics				
Arsenic	3/3	BDL–16	BDL	BDL
Cadmium	3/3	15–180	65	87
Chromium	3/3	41–120	120	93
Copper	3/3	630–7400	4100	4100
Cyanide	3/3	BDL–940	BDL	350
Lead	3/3	2000–57,000	12,000	24,000
Nickel	3/3	86–210	130	140
Zinc	3/3	1400–4200	2200	2600
Toxic organics				
Bis(2-ethylhexyl) phthalate	4/1	1900	—	—
Butyl benzyl phthalate	4/1	190	—	—
Di-n-butyl phthalate	4/1	19,000	—	—
2,4,6-Trichlorophenol	4/1	22	—	—
Phenol	4/2	60–9900	—	5000
Toluene	3/3	BDL–130	40	58
Ethylbenzene	3/2	40–40	—	—
Acenaphthene	4/1	95	—	—
Naphthalene	4/3	10–380	150	180
Chrysene	4/2	<10–360	—	<180
Anthracene	4/3	90–<1100	200	<460
Fluorene	4/3	40–450	70	190
Phenanthrene	4/3	90–<1100	200	<460
Pyrene	4/2	20–98	—	59
Methylene chloride	3/3	BDL–1200	BDL	400
Tetrachloroethylene	3/3	BDL–3600	10	1200
Polychlorinated biphenyls				
PCB 1242, 1254, 1221, total	3/1	63	—	—
PCB 1232, 1248, 1260, 1016, total	3/1	65	—	—
Pesticides				
4,4-DDE	3/1	BDL	—	—
Alpha-endosulfan	3/1	BDL	—	—
Endrin aldehyde	3/1	58	—	—
Alpha-BHC	3/1	BDL	—	—
Beta-BHC	3/1	18	—	—

BDL, below detection limit.

Source: U.S. EPA, *Development Document for Effluent Limitations Guidelines and Standards for the Aluminum Forming Point Source Category*, Vols. 1 & 2, U.S. EPA, Washington, DC, 1984; U.S. EPA, *Aluminum Forming Point Source Category*, available at http://www.access.gpo.gov/nara/cfr/waisidx_03/40cfr467_03.html, 2008.

Aluminum Forming Industry

TABLE 5.6
Summary of Pollutant Data for the Extrusion Subcategory Verification Data

Pollutant	Number of Samples/ Number of Detections	Range of Detections
Classical Pollutants (mg/L)		
COD	1/1	12
TOC	1/1	19
Dissolved solids	1/1	3200
Suspended solids	1/1	28
Oil and grease	1/1	8
Phenol, total	1/1	0.005
Aluminum	1/1	400
Calcium	1/1	<1
Magnesium	1/1	<1
Sulfate	1/1	60
pH (pH units)	1/1	11
Toxic Pollutants (µg/L)		
Metals and inorganics		
Cadmium	1/1	20
Chromium	1/1	90
Copper	1/1	200
Lead	1/1	600
Mercury	1/1	0.7
Zinc	1/1	100
Toxic organics		
Bis(2-ethylhexyl) phthalate	1/1	27
Methylene chloride	1/1	36

BDL, below detection limit.

Source: U.S. EPA, *Development Document for Effluent Limitations Guidelines and Standards for the Aluminum Forming Point Source Category*, Vols. 1 & 2, U.S. EPA, Washington, DC, 1984; U.S. EPA, *Aluminum Forming Point Source Category*, available at http://www.access.gpo.gov/nara/cfr/waisidx_03/40cfr467_03.html, 2008.

TABLE 5.7
Summary of Pollutant Data for the Forging Subcategory Verification Data

Pollutant	Number of Samples/ Number of Detections	Range of Detections
Classical Pollutants (mg/L)		
COD	1/1	350
TOC	1/1	98
Dissolved solids	1/1	390
Suspended solids	1/1	2
Oil and grease	1/1	160
Phenols, total	1/1	0.07
Aluminum	1/1	0.5
Calcium	1/1	59
Magnesium	1/1	10
Sulfate	1/1	95

Continued

TABLE 5.7 (continued)

Pollutant	Number of Samples/ Number of Detections	Range of Detections
Toxic Pollutants (µg/L)		
Toxic metals		
Lead	1/1	2000
Zinc	1/1	300
Toxic organics		
2,4-Dichlorophenol	1/1	38
Fluoranthene	1/1	18
Methylene chloride	1/1	950
2,4-Dinitrophenol	1/1	23
4,6-Dinitro-o-cresol	1/1	24
N-Nitrosodiphenylamine	1/1	17
Benzo(a)anthracene	1/1	19
Chrysene	1/1	19
Anthracene/phenanthrene	1/1	28
Pyrene	1/1	21
Polychlorinated biphenyls		
PCB 1242, 1254, 1221	1/1	1.3

Source: U.S. EPA, *Development Document for Effluent Limitations Guidelines and Standards for the Aluminum Forming Point Source Category*, Vols. 1 & 2, U.S. EPA, Washington, DC, 1984; U.S. EPA, *Aluminum Forming Point Source Category*, available at http://www.access.gpo.gov/nara/cfr/waisidx_03/40cfr467_03.html, 2008.

TABLE 5.8
Summary of Pollutant Data for fhe Drawing Oil Emulsion-Soap Subcategory Verification Data

Pollutant	Number of Samples/ Number of Detections	Range of Detections
Classical Pollutants (mg/L)		
Oil and grease	1/1	1500
pH (pH units)	1/1	7.2
Aluminum	1/1	340
Calcium	1/1	130
Magnesium	1/1	37
Toxic Pollutants (µg/L)		
Toxic organics		
1,1,1-Trichloroethane	1/1	530
1,1-Dichloroethane	1/1	97
p-Chloro-m-cresol	2/1	28
2-Chlorophenol	2/1	130
2,4-Dinitrotoluene	2/1	77
Ethylbenzene	1/1	15
Methylene chloride	1/1	3

Continued

TABLE 5.8 (continued)

Pollutant	Number of Samples/Number of Detections	Range of Detections
Isophorone	2/1	39
Bis(2-ethylhexyl) phthalate	1/1	34
Di-n-butyl phthalate	1/1	23
Di-n-octyl phthalate	1/1	23
Toluene	1/1	200
Pesticides		
Alpha-endosulfan	2/1	BDL
Polychlorinated biphenyls		
PCB 1254	2/1	BDL
PCB 1248	2/1	BDL
Metals and inorganics		
Arsenic	1/1	37
Cadmium	1/1	11
Chromium	1/1	8000
Copper	1/1	480
Lead	1/1	140
Nickel	1/1	34
Zinc	1/1	46,000

BDL, below detection limit.

Source: U.S. EPA, *Development Document for Effluent Limitations Guidelines and Standards for the Aluminum Forming Point Source Category*, Vols. 1 & 2, U.S. EPA, Washington, DC, 1984; U.S. EPA, *Aluminum Forming Point Source Category*, available at http://www.access.gpo.gov/nara/cfr/waisidx_03/40cfr467_03.html, 2008.

TABLE 5.9
Summary of Pollutant Data for the Rolling Heat Treatment Quench Subcategory Verification Data

Pollutant	Number of Samples/Number of Detections	Range of Detections	Mean of Detections
Classical Pollutants (mg/L)			
Oil and grease	2/2	12–13	12
Suspended solids	2/1	3	—
pH (pH units)	2/2	7.1–7.9	7.5
Aluminum	2/2	<0.2–0.4	<0.3
Calcium	2/2	41–51	46
Iron	2/2	<0.1–<0.1	—
Magnesium	2/2	11–20	16
COD	2/2	<5–7	<6
Dissolved solids	2/2	110–410	260
Sulfate	2/2	<3–70	<36
TOC	2/2	<1–2	<2
Phenols, total	2/2	0.01–0.01	—

Continued

TABLE 5.9 (continued)

Pollutant	Number of Samples/ Number of Detections	Range of Detections	Mean of Detections
Toxic Pollutants (µg/L)			
Toxic organics			
Chloroform	2/2	BDL–20	12
Methylene chloride	2/2	38–<40	<39
Metals and inorganics			
Nickel	2/2	BDL–20	12

BDL, below detection limit.

Source: U.S. EPA, *Development Document for Effluent Limitations Guidelines and Standards for the Aluminum Forming Point Source Category*, Vols. 1 & 2, U.S. EPA, Washington, DC, 1984; U.S. EPA, *Aluminum Forming Point Source Category*, available at http://www.access.gpo.gov/nara/cfr/waisidx_03/40cfr467_03.html, 2008.

TABLE 5.10
Summary of Pollutant Data for the Heat Treatment Quench Forging Subcategory Verification Data

Pollutant	Number of Samples/ Number of Detections	Range of Detections	Median of Detections	Mean of Detections
	Wastestream: Forging			
Classical Pollutants (mg/L)				
Oil and grease	3/3	4–87	14	35
Suspended solids	3/3	4–240	22	88
pH (pH units)	3/3	7.7–8.2	—	8.0
Aluminum	4/4	<1–9	<1.1	<3
Calcium	4/4	38–80	58	58
Magnesium	4/4	8–35	21	21
COD	3/3	<5–56	18	<26
Dissolved solids	4/4	190–1400	690	740
Sulfate	4/4	30–330	130	160
TOC	3/3	<2–14	3	<6.3
Phenols, total	3/3	0.003–0.8	0.02	0.27
Toxic Pollutants (µg/L)				
Toxic organics				
Bis (2-ethylhexyl) phthalate	4/4	BDL–890	BDL	240
Metals and inorganics				
Cadmium	4/4	BDL–12	<10	<6
Chromium	4/4	7–72,000	23,000	30,000
Copper	4/4	<50–380	85	<150
Lead	4/3	<50–17,000	60	<5700
Mercury	4/4	BDL–0.5	<0.2	0.24
Nickel	4/4	BDL–<20	<7	<9.1
Zinc	4/4	50–5200	120	1400

Continued

Aluminum Forming Industry

TABLE 5.10 (continued)

Pollutant	Number of Samples/ Number of Detections	Range of Detections	Median of Detections	Mean of Detections
Wastestream: Drawing				
Classical Pollutants (mg/L)				
Oil and grease	1/1	20	—	—
Suspended solids	1/1	19	—	—
pH (pH units)	1/1	8.2	—	—
COD	1/1	92,000	—	—
Phenols, total	1/1	0.005	—	—
TOC	1/1	19,000	—	—
Toxic Pollutants (µg/L)				
Toxic organics				
Benzene	1/1	2100	—	—
Chloroform	1/1	12,000	—	—
Methylene chloride	1/1	31,000	—	—
Bis(2-ethylhexyl) phthalate	1/1	310	—	—
Di-n-butyl phthalate	1/1	330	—	—
Diethyl phthalate	1/1	160	—	—
Dimethyl phthalate	1/1	20	—	—
Tetrachloroethylene	1/1	<4000	—	—
Toluene	1/1	320	—	—
Trichloroethylene	1/1	430	—	—
Polychlorinated biphenyls				
PCB 1242, 1254, 1221	1/1	4.5	—	—
PCB 1232, 1248, 1260, 1016	1/1	3.2	—	—
Metals and inorganics				
Antimony	1/1	<100	—	—
Copper	1/1	<16	—	—
Cyanide	1/1	1300	—	—
Mercury	1/1	10	—	—
Wastestream: Extrusion Press				
Classical Pollutants (mg/L)				
Oil and grease	5/5	8–130	17	37
Suspended solids	5/5	<1–59	3	<19
pH (pH units)	5/5	7.4–9.2	7.8	8.1
COD	5/5	<5–210	74	<74
TOC	5/5	<1–88	27	<33
Phenols, total	4/4	0.001–0.015	0.012	0.01
Toxic Pollutants (µg/L)				
Toxic organics				
2-Chlorophenol	5/1	20	—	—
1,2-*trans*-Dichloroethylene	5/2	BDL–13	—	9
Methylene chloride	5/5	49–210,000	100	42,000
Bis(2-ethylhexyl)phthalate	5/4	BDL–100	18	36
Butyl benzyl phthalate	5/5	BDL–67	BDL	20

Continued

TABLE 5.10 (continued)

Pollutant	Number of Samples/ Number of Detections	Range of Detections	Median of Detections	Mean of Detections
Metals and inorganics				
Copper	4/4	BDL–100	36	44
Nickel	4/4	BDL–<17	BDL	<6.1
Wastestream: Extrusion Solution				
Classical Pollutants (mg/L)				
Oil and grease	1/1	41	—	—
Suspended solids	2/2	<2–2	—	<2
pH	2/2	7.3–7.3	—	—
Aluminum	2/2	<0.5–0.54	—	<0.5
Calcium	2/2	38–58	—	48
Magnesium	2/2	5.3–25	—	15
COD	2/2	7–20	—	14
Dissolved solids	2/2	160–580	—	370
Sulfate	2/2	7–120	—	64
TOC	2/2	1.8–2.7	—	2.2
Phenols, total	2/2	0.001–0.01	—	0.005
Toxic Pollutants (µg/L)				
Toxic organics				
Methylene chloride	2/2	10–630	—	320
Metals and inorganics				
Chromium	2/2	18–5100	—	2600
Nickel	2/2	BDL–18	—	10

BDL, below detection limit.

Source: U.S. EPA, *Development Document for Effluent Limitations Guidelines and Standards for the Aluminum Forming Point Source Category*, Vols. 1 & 2, U.S. EPA, Washington, DC, 1984; U.S. EPA, *Aluminum Forming Point Source Category*, 2008.

heat treatment quenching processes are presented in Tables 5.9 and 5.10 by the aluminum forming operation that it follows.

5.3.7 Etch or Cleaning

Thirty plants in the aluminum forming industry use etch or cleaning lines. Rinsing is usually required following successive chemical treatments within these etch or cleaning lines. Wastewater discharge values tend to increase as the number of rinses increase. Table 5.11 summarizes the classical and toxic pollutant data for etch line rinses.

5.4 PLANT-SPECIFIC DESCRIPTION OF CASE HISTORIES

5.4.1 Plant A: Case History

A very limited amount of individual plant specific data for the aluminum forming industry is available. Data available on the influent and effluent streams are discussed briefly in the following subsections for specific plants. This aluminum processing plant uses lime precipitation (pH adjustment) followed by coagulant addition and sedimentation as its treatment system. Data

TABLE 5.11
Summary of Pollutant Data for the Etch Line Rinses Subcategory Verification Data

Pollutant	Number of Samples/Number of Detections	Range of Detections	Median of Detections	Mean of Detections
Classical Pollutants (mg/L)				
Oil and grease	18/16	2–47	11	16
Suspended solids	18/18	<1–2700	95	<300
pH (pH units)	16/16	1.1–12	7.0	6.4
Aluminum	19/19	0.9–1200	100	300
Calcium	19/18	0.03–1200	26	90
Iron	18/15	0.1–200	1.9	20
Magnesium	1/1	10	—	—
COD	17/17	<5–280	35	<90
Dissolved solids	19/19	20–48,000	660	4600
Sulfate	19/19	1–9400	40	600
TOC	18/18	<1–180	10	<31
Phenols, total	18/18	0.003–0.04	0.008	0.01
Toxic Pollutants (μg/L)				
Toxic organics				
Acenaphthene	19/5	BDL–17	BDL	10
Benzene	19/9	BDL–34	BDL	12
Chloroform	19/18	BDL–75	16	24
1,2-*trans*-Dichloroethyl	19/5	BDL–110	BDL	22
2,4-Dimethyl phenol	19/2	BDL–19	—	12
Methylene chloride	19/18	BDL–2100	120	380
Isophorone	19/2	BDL–16	—	10
4-Nitrophenol	19/1	18	—	—
Phenol	19/9	BDL–63	BDL	12
Bis(2-ethylhexyl) phthalate	19/18	BDL–120	BDL	26
Butyl benzyl phthalate	19/9	BDL–66	BDL	11
Di-n-butyl phthalate	19/14	BDL–68	BDL	10
Di-n-octyl phthalate	19/5	BDL–38	BDL	12
Diethyl phthalate	19/9	BDL–22	BDL	BDL
Chlordane	19/12	BDL–BDL	—	—
Polychlorinated biphenyls				
PCB 1242, 1254, 1221	19/11	BDL–16	BDL	BDL
PCB 1232, 1248, 1260, 101	19/12	BDL–20	BDL	BDL
Metals and inorganics				
Arsenic	19/17	BDL–200	BDL	17
Beryllium	19/15	BDL–<20	BDL	<10
Cadmium	19/15	BDL–210	<10	<23
Chromium	19/15	7–860,000	110	52,000
Copper	19/15	9–2,400,000	200	160,000
Lead	19/16	BDL–10,000	250	1100
Mercury	19/15	BDL–21	0.5	2.0
Nickel	19/15	BDL–2800	<6	220
Zinc	19/15	BDL–2,100,000	300	140,000

BDL, below detection limit.

Source: U.S. EPA, *Development Document for Effluent Limitations Guidelines and Standards for the Aluminum Forming Point Source Category*, Vols. 1 & 2, U.S. EPA, Washington, DC, 1984; U.S. EPA, *Aluminum Forming Point Source Category*, 2008.

TABLE 5.12
Minimum Detection Limits for Toxic Pollutants

Toxic Pollutant	Concentration (µg/L)
Organic pollutants	10
Pesticides	5
Metals	
Antimony	100
Arsenic	10
Asbestos	1×10^7 fibers/L
Beryllium	10
Cadmium	2
Chromium	5
Copper	9
Cyanide	100
Lead	20
Mercury	0.1
Nickel	5
Selenium	10
Silver	20
Thallium	100
Zinc	50

Source: U.S. EPA, *Development Document for Effluent Limitations Guidelines and Standards for the Aluminum Forming Point Source Category*, Vols. 1 & 2, U.S. EPA, Washington, DC, 1984; U.S. EPA, *Aluminum Forming Point Source Category*, 2008.

on the pollutant removal efficiency at Plant A are summarized in Table 5.13. No production or water usage data are available for this plant.

5.4.2 Plant B: Case History

No plant-specific identification number was available for this facility. The wastewater from Plant B contains pollutants from both metals processing and finishing operations. It is treated by precipitation–settling followed by filtration with a rapid sand filter. A clarifier is used to remove much of the solids load. Table 5.14 summarizes the data on pollutant removal efficiency at Plant B.

5.4.3 U.S. EPA Data on Full-Scale Treatment of Aluminum Forming Wastewaters

U.S. EPA has documented case histories of full-scale treatment of aluminum forming wastewaters, which are included in Appendixes A, B, C, and D of this chapter for reference. The detailed theories and principles of the full-scale treatment processes can be found in the literature.[8–10]

Appendix A presents the actual data on a full-scale treatment of aluminum forming wastewater by emulsion breaking and oil–water separation. Either chemical emulsion breaking (CEB) or thermal emulsion breaking (TEB) can be used for breaking the emulsified oil in wastewater. Once the emulsified oil is freed as oil drops, it can be easily removed by an API oil–water separator (Figure 5.1), a parallel plate separator (Figure 5.2), an ultrafiltration unit, or a dissolved air flotation clarifier.[8–10] It is very encouraging to note that over 94% of TSS, COD, TOC, O&G (oil and grease), cadmium, chromium, copper, lead, nickel, zinc, acenaphthene, phenol, bis-(2-ethylhexyl)-phthalate, bi-n-butyl phthalate, anthracene, fluorine, phenanthrene, pyrene, endrin-aldehyde, and PCB can be successfully removed by these preliminary treatment processes.

Appendix B presents the U.S. EPA data on a full-scale treatment of aluminum forming wastewater by emulsion breaking and ultrafiltration. After emulsion breaking, various oil–water separation

TABLE 5.13
Removal of Pollutants by Lime Precipitation at Metal Processing Plant A[a]

Pollutant (mg/L)	Raw Wastewater	Treated Effluent	Percent Removal
pH (pH units)	2.8	7.1	—
TSS	24	11	54
Copper	180,000	2000	99
Zinc	110,000	8700	92

[a] Data are based on the average of three influent/effluent samples.

Source: U.S. EPA, *Development Document for Effluent Limitations Guidelines and Standards for the Aluminum Forming Point Source Category*, Vols. 1 & 2, U.S. EPA, Washington, DC, 1984; U.S. EPA, *Aluminum Forming Point Source Category*, 2008.

TABLE 5.14
Removal of Pollutants by a Combination of Lime Precipitation, Sedimentation, and Filtration at Plant B[a]

Pollutant (µg/L)	Raw Wastewater	Treated Effluent	Percent Removal
Chromium	5900	38	99
Copper	170	11	94
Nickel	3300	180	95
Zinc	2900	35	99
Iron	22900	400	98

[a] Treated effluent performance reported for the period 1974–1979.

Source: U.S. EPA, *Development Document for Effluent Limitations Guidelines and Standards for the Aluminum Forming Point Source Category*, Vols. 1 & 2, U.S. EPA, Washington, DC, 1984; U.S. EPA, *Aluminum Forming Point Source Category*, 2008.

processes can be used for separation of the freed oil together with other pollutants from the wastewater. This specific set of treatment data documents the efficiency of the emulsion breaking and ultrafiltration treatment system. Again, Appendix B shows that the preliminary treatment system of emulsion breaking and ultrafiltration can achieve over 90% of O&G, TSS, COD, TOC, benzene, 2,4,6-trichlorophenol, bis-(2-ethylhexyl) phthalate, diethyl phythalate, tetrachloroethylene, and PCB.

Appendix C documents the U.S. EPA data on full-scale treatment of aluminum forming wastewater by chemical precipitation and clarification. It is important to note that Appendixes A and B present the preliminary treatment performance by which main organics are removed. The preliminary treatment step will not remove heavy metals in a significant amount. Accordingly, the system presented in Appendix C is a secondary treatment system following the preliminary treatment step. Appendix C shows that in the secondary treatment system consisting of chemical precipitation, sedimentation clarification can further achieve over 90% removal of oil and grease, and total toxic organics (TTO), such as fluoranthene, methylene chloride, 2,4-dinitrophenol, n-nitrosodiphenylamine, chrysene, and pyrene. Although lead was removed by 69%, the removal of other major heavy metals (chromium, copper, mercury, nickel, and zinc) was insignificant.

Appendix D presents the U.S. EPA data on a full-scale treatment of aluminum forming wastewater by chromium reduction, chemical precipitation, and sedimentation clarification. Chromium reduction, as described in detail by Wang, Hung and Shammas,[9] is an important step prior to

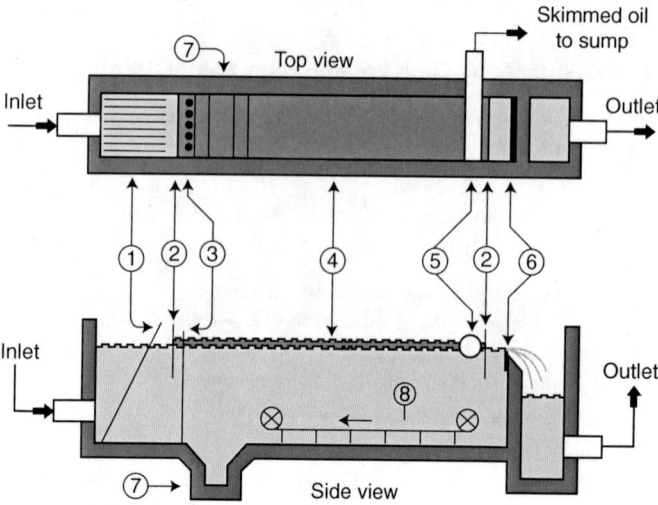

FIGURE 5.1 Typical gravimetric API oil–water separator. (*Source:* Wikipedia Encyclopedia, API Oil–Water Separator, http://en.wikipedia.org/wiki/API_oil-water_Separator.)

1 Trash trap (inclined rods)
2 Oil retention baffles
3 Flow distributors (vertical rods)
4 Oil layer
5 Slotted pipe skimmer
6 Adjustable overflow weir
7 Sludge sump
8 Chain and flight scraper

chemical precipitation and clarification. The treatment system documented in Appendix D consists of the following:

1. Chromium reduction using sulfuric acid
2. Neutralization and chemical precipitation using lime and/or sodium hydroxide
3. Sedimentation for clarification

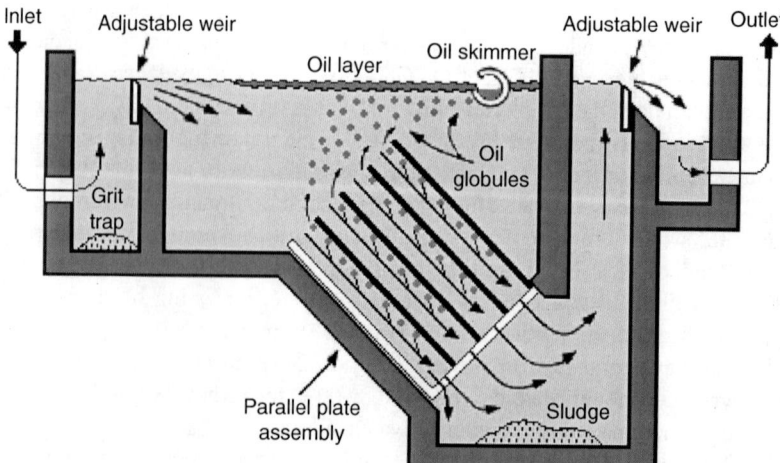

FIGURE 5.2 Typical parallel plate oil–water separator. (*Source:* Wikipedia Encyclopedia, API Oil–Water Separator, http://en.wikipedia.org/wiki/API_oil-water_Separator.)

Aluminum Forming Industry

TABLE 5.15
Removal of Pollutants by Sodium Hydroxide Precipitation

Pollutant (μg/L)	Raw Wastewater	Treated Effluent	Percent Removal
pH (pH units)	2.3	9	—
Chromium	74	BDL	93[a]
Copper	65	17	74
Iron	11,000	880	92
Lead	1300	120	91
Manganese	0.11	0.05	55
Nickel	61	31	49
Zinc	120	12	90
TSS	—	12,000	—

BDL, below detection limit.
[a] Approximate value.
Source: U.S. EPA, *Development Document for Effluent Limitations Guidelines and Standards for the Aluminum Forming Point Source Category*, Vols. 1 & 2, U.S. EPA, Washington, DC, 1984; U.S. EPA, *Aluminum Forming Point Source Category*, 2008.

It should be noted that dissolved air flotation (DAF) is a more effective process for clarification.[8–10] As shown in Appendix D, with an additional step of chromium reduction, the secondary treatment system effectively removed chromium (over 99%), copper (89%), cadmium (64%), lead (67%), and zinc (77%).

5.5 POLLUTANT REMOVABILITY

This section describes the treatment techniques currently used or available to remove or recover wastewater pollutants normally generated by aluminum forming facilities. In general, these pollutants are removed by oil removal (skimming, emulsion breaking, and flotation), chemical precipitation and sedimentation, or filtration.[6–13]

TABLE 5.16
Removal of Pollutants by Lime and Sodium Hydroxide Precipitations

Pollutant	Raw Wastewater	Treated Effluent	Percent Removal
pH (pH units)	9.4	8.3	—
Aluminum (mg/L)	35	0.35	99
Copper (μg/L)	670	BDL	99[a]
Iron (mg/L)	150	0.55	>99
Manganese (mg/L)	210	0.12	>99
Nickel (mg/L)	6100	BDL	>99[a]
Selenium (mg/L)	29,000	BDL	>99[a]
Titanium (mg/L)	130	BDL	>99[a]
Zinc (mg/L)	17,000	27	>99
TSS (mg/L)	3600	12	>99

BDL, below detection limit.
[a] Approximate value.
Source: U.S. EPA, *Development Document for Effluent Limitations Guidelines and Standards for the Aluminum Forming Point Source Category*, Vols. 1 & 2, U.S. EPA, Washington, DC, 1984; U.S. EPA, *Aluminum Forming Point Source Category*, 2008.

TABLE 5.17
Removal of Pollutants by Sulfide Precipitation at Three Plants

Pollutant	Raw Wastewater	Treated Effluent	Percent Removal
Plant 1			
pH (pH units)	5.9	8.5	—
Chromium, hexavalent (mg/L)	26,000	<14	>99
Chromium (mg/L)	32,000	<40	>99
Iron (mg/L)	0.52	0.10	81
Zinc (mg/L)	40,000	<70	>99
Plant 2			
pH (pH units)	7.7	7.4	—
Chromium, hexavalent (mg/L)	22	<20	>9
Chromium (mg/L)	2400	<100	>99
Iron (mg/L)	110	0.6	>99
Nickel (mg/L)	680	100	>85
Zinc (mg/L)	34,000	100	>99
Plant 3			
Chromium, hexavalent (mg/L)	11,000	BDL	>99[a]
Chromium (mg/L)	18,000	BDL	>99[a]
Copper (mg/L)	29	BDL	83[a]
Zinc (mg/L)	60	BDL	92[a]

BDL, below detection limit.
[a] Approximate value.
Source: U.S. EPA, references 1, 2, and 5.

Most of the pollutants may be effectively removed by precipitation of metal hydroxides or carbonates using a reaction with lime, sodium hydroxide, or sodium carbonate. For some, improved removals are provided by the use of sodium sulfide or ferrous sulfide to precipitate the pollutants as sulfide compounds with very low solubilities. After soluble metals are precipitated as insoluble flocs, one of the water–solid separators (such as dissolved air flotation, sedimentation, centrifugation, membrane filtration, and so on) can be used for flocs removal.[9–11] The effectiveness of pollutant removal by several different precipitation methods is summarized in Tables 5.15–5.17.

Table 5.18 presents the removability of pollutants by two types of skimming systems:

1. The API system
2. The TEB system

API stands for the American Petroleum Institute and TEB stands for Thermal Emulsion Breaking. Figures 5.1 and 5.2 show two typical types of oil–water separators, gravimetric and parallel plate.[7] A dissolved air flotation (DAF) clarifier is commonly used for polishing the effluent from an oil–water separator.[8–10,13]

5.6 TREATMENT TECHNOLOGY COSTS

The investment cost, operating, and maintenance costs, and energy costs for the application of control technologies to the wastewaters of the Aluminum Forming Industry have been analyzed.

Aluminum Forming Industry

TABLE 5.18
Removal of Pollutants by Two Types of Skimming Systems

Pollutant	Raw Wastewater	Treated Effluent	Percent Removal
API System[a]			
Oil and grease (mg/L)	230,000	15	>99
Chloroform (mg/L)	23	BDL	78[c]
Methylene chloride (µg/L)	13	12	8
Naphthalene (µg/L)	2300	BDL	>99[c]
N-nitrosodiphenylamine (µg/L)	59,000	180	>99
Bis(2-ethylhexyl) phthalate (µg/L)	11,000	27	>99
Butylbenzyl phthalate (µg/L)	BDL	BDL	NM
Di-n-octyl phthalate (µg/L)	19	BDL	74[c]
Anthracene-phenanthrene (µg/L)	16,000	14	>99
Toluene (µg/L)	20	12	40
TEB System[b]			
Oil and grease (mg/L)	2600	10	>99
Chloroform (µg/L)	BDL	BDL	NM
Methylene chloride (µg/L)	BDL	BDL	NM
Naphthalene (µg/L)	1800	BDL	>99[c]
Bis (2-ethylhexyl) phthalate (µg/L)	1600	18	99
Diethyl phthalate (µg/L)	17	BDL	71[c]
Anthracene-phenanthrene (µg/L)	140	BDL	96[c]

BDL, below detection limits; NM, not meaningful.
[a] API, American Petroleum Institute oil–water separator.
[b] TEB, thermal emulsion breaker.
[c] Approximate value.
Source: U.S. EPA, *Development Document for Effluent Limitations Guidelines and Standards for the Aluminum Forming Point Source Category*, Vols. 1 & 2, U.S. EPA, Washington, DC, 1984; U.S. EPA, *Aluminum Forming Point Source Category*, 2008.

These costs were developed to reflect the conventional use of technologies in this industry. Several unit operation/unit process configurations have been analyzed for the cost of application of technologies and to select the best practicable control technology (BPT) and best applicable technology (BAT) level of treatment. The detailed presentation of the applicable treatment technologies, cost methodology, and cost data are available in the literature.[1-4] Specifically the U.S. EPA Report W-83-13, "Cost Effectiveness Analysis of Effluent Guidelines and Standards for the Aluminum Forming Category" reports the results of a cost-effectiveness (CE) analysis of alternative water pollution control technologies for the aluminum forming category.[4] The primary cost of interest in the U.S. EPA report is the total annualized cost in complying with the regulations incurred by industry. The 1983 costs can be easily updated to the current cost,[12,13] or any future costs using the cost indexes announced periodically by the U.S. Army Corps of Engineers (U.S. ACE)[14] or *Engineering News Record* of McGraw-Hill Book Company, NY.

5.7 ALUMINUM FORMING INDUSTRIAL EFFLUENT LIMITATIONS

The effluent limitations of all aluminum forming operations representing (a) the degree of effluent reduction attainable by the application of the BPT currently available; and (b) the degree of effluent reduction attainable by the application of the BAT economically achievable, can be found from the

TABLE 5.19

U.S. EPA Aluminum Forming Industry Effluent Guidelines and Categorical Regulations Summary (April 6, 2001)

Industrial Category Category Description	40 CFR Reference	SIC Codes (Partial List)	Subparts	Promulgation Date	New Source Date	Regulated Parameters for Pretreatment
Aluminum Forming Processes by which aluminum or aluminum alloys are changed in size and shape. Processes include rolling, extrusion, forging, and drawing.	467	3353　3354　3355　3357　3463	A. Rolling with neat oils　B. Rolling with emulsions　C. Extrusion　D. Forging　E. Drawing with neat oils　F. Drawing with emulsions or soaps	10/24/83	11/22/82	Cr, CN, Zn, TTO

Source: U.S. EPA, *Development Document for Effluent Limitations Guidelines and Standards for the Aluminum Forming Point Source Category*, Vols. 1 & 2, U.S. EPA, Washington, DC, 1984; U.S. EPA, *Aluminum Forming Point Source Category*, 2008.

U.S. Code of Federal Register 40CFR–Chapter I, Part 467, p. 615, Aluminum Forming Point Source Category, available at http://www.access.gpo.gov.[5] Table 5.19 indicates that the regulated parameters for pretreatment of aluminum forming industrial wastewater are chromium, cyanide, zinc, and total toxic organics (TTO).

5.8 TECHNICAL TERMINOLOGIES OF ALUMINUM FORMING OPERATIONS AND POLLUTION CONTROL

1. Aluminum forming is a set of manufacturing operations in which aluminum and aluminum alloys are made into semifinished products by hot or cold working.
2. Ancillary operation is a manufacturing operation that has a large flow, discharges significant amounts of pollutants, and may not be present at every plant in a subcategory, but when present is an integral part of the aluminum forming process.
3. Contact cooling water is any wastewater that contacts the aluminum workpiece or the raw materials used in forming aluminum.
4. Continuous casting is the production of sheet, rod, or other long shapes by solidifying the metal while it is being poured through an open-ended mold using little or no contact cooling water. Continuous casting of rod and sheet generates spent lubricants, and rod casting also generates contact cooling water.
5. Degassing is the removal of dissolved hydrogen from the molten aluminum prior to casting. Chemicals are added and gases are bubbled through the molten aluminum. Sometimes a wet scrubber is used to remove excess chlorine gas.
6. Direct chill casting is the pouring of molten aluminum into a water-cooled mold. Contact cooling water is sprayed onto the aluminum as it is dropped into the mold, and the aluminum ingot falls into a water bath at the end of the casting process.
7. Drawing is the process of pulling metal through a die or succession of dies to reduce the metal's diameter or alter its shape. There are two aluminum forming subcategories based

Aluminum Forming Industry

on the drawing process. In the drawing with neat oils subcategory, the drawing process uses pure or neat oil as a lubricant. In the drawing with emulsions or soaps subcategory, the drawing process uses an emulsion or soap solution as a lubricant.
8. Emulsions are stable dispersions of two immiscible liquids. In the aluminum forming category this is usually an oil and water mixture.
9. Cleaning or etching is a chemical solution bath and a rinse or series of rinses designed to produce a desired surface finish on the workpiece. This term includes air pollution control scrubbers, which are sometimes used to control fumes from chemical solution baths. Conversion coating and anodizing when performed as an integral part of the aluminum forming operations are considered cleaning or etching operations. When conversion coating or anodizing are covered here they are not subject to regulation under the provisions of the U.S. 40 CFR Part 433, Metal Finishing.
10. Extrusion is the application of pressure to a billet of aluminum, forcing the aluminum to flow through a die orifice. The extrusion subcategory is based on the extrusion process.
11. Forging is the exertion of pressure on dies or rolls surrounding heated aluminum stock, forcing the stock to change shape and in the case where dies are used to take the shape of the die. The forging subcategory is based on the forging process.
12. Heat treatment is the application of heat of specified temperature and duration to change the physical properties of the metal.
13. Hot water seal is a heated water bath (heated to approximately 180°F) used to seal the surface coating on formed aluminum that has been anodized and coated. In establishing an effluent allowance for this operation, the hot water seal shall be classified as a cleaning or etching rinse.
14. In-process control technology is the conservation of chemicals and water throughout the production operations to reduce the amount of wastewater to be discharged.
15. Neat oil is pure oil with no or few impurities added. In aluminum forming its use is mostly as a lubricant.
16. Rolling is the reduction in thickness or diameter of a workpiece by passing it between lubricated steel rollers. There are two subcategories based on the rolling process. In the rolling with neat oils subcategory, pure or neat oils are used as lubricants for the rolling process. In the rolling with emulsions subcategory, emulsions are used as lubricants for the rolling process.
17. The term Total Toxic Organics (TTO) means the sum of the masses or concentrations of each of the following toxic organic compounds, which is found in the discharge at a concentration greater than 0.010 mg/L:
    i. p-Chloro-m-cresol
    ii. 2-Chlorophenol
    iii. 2,4-Dinitrotoluene
    iv. 1,2-Diphenylhydrazine
    v. Ethybenzene
    vi. Fluoranthene
    vii. Isophorone
    viii. Napthalene
    ix. N-Nitro sodi phenyl amine
    x. Phenol
    xi. Benzo (a) pyrene
    xii. Benzo (ghi) perylene
    xiii. Fluorene
    xiv. Phenanthrene
    xv. Dibenzo (a,h)
    xvi. Anthracene

xvii. Indeno (1,2,3-c,d) pyrene
xviii. Pyrene
xix. Tetrachloroethylene
xx. Toluene
xxi. Trichloroethylene
xxii. Endosulfan sulfate
xxiii. Bis(2-ethyl hexyl)phthalate
xxiv. Diethylphthalate
xxv. 3,4-Benzofluoranthene
xxvi. Benzo (k) fluoranthene
xxvii. Chrysene
xxviii. Acenaphthylene
xxix. Anthracene
xxx. Di-n-butyl phthalate
xxxi. Endrin
xxxii. Endrin aldehyde
xxxiii. PCB 1242, 1254, 1221
xxxiv. PCB 1232, 1248, 1260, 1016
xxxv. Acenaphthene

18. Stationary casting is the pouring of molten aluminum into molds and allowing the metal to air cool.
19. Wet scrubbers are air pollution control devices used to remove particulates and fumes from air by entraining the pollutants in a water spray.
20. BPT means the best practicable control technology currently available under the U.S. Federal Act Section 304(b)(1).
21. BAT means the best available technology economically achievable under the U.S. Federal Act Section 304(b)(2)(B).
22. BCT means the best conventional pollutant control technology under U.S. Federal Act Section 304(b)(4).
23. The production normalizing mass (kkg) for each core or ancillary operation is the mass (off-kkg or off-lb) processed through that operation.
24. The term off-kilogram (off-pound) shall mean the mass of aluminum or aluminum alloy removed from a forming or ancillary operation at the end of a process cycle for transfer to a different machine or process.

Aluminum Forming Industry

APPENDIX A
Full Scale Treatment of Aluminum Forming Wastewater by Emulsion Breaking and Oil-Water Separation

Removal Data
Sampling: Three 24-Hour or One 72-Hour Composite

Pollutant/Parameter	Concentration		Percent Removal
	Influent	Effluent	
Classical Pollutants (mg/L)			
Suspended solids	760	12	98
COD	80,000	830	99
TOC	39,000	260	99
Phenol	0.21	0.21	0
pH, pH units	NA	4.8	NM
Oil and grease	18,000	42	99
Toxic Pollutants (µg/L)			
Arsenic	BDL	BDL	NM
Cadmium	<200	5	>98
Chromium	<1000	20	>98
Copper	7000	BDL	99*
Cyanide	BDL	BDL	NM
Lead	<3000	30	99
Mercury	<70	BDL	NM
Nickel	<1000	40	>96
Zinc	<7000	200	>97
Acenaphthene	5700	6	>99
Benzene	BDL	BDL	NM
Chloroform	16	20	NM
Ethylbenzene	30	BDL	83
Methylene chloride	F400	330	>18
Phenol	90	ND	>99
Bis(2-ethylhexyl) phthalate	1200	44	94
Di-n-butyl phthalate	1300	49	94
Diethyl phthalate	820	65	92
Anthracene	700	ND	>99
Fluorene	330	ND	>99
Phenanthrene	1000	ND	>99
Pyrene	41	ND	>99
Tetrachloroethylene	20	14	30
Toluene	30	BDL	83*
4,4-DDE	BDL	BDL	NM
Endrin-aldehyde	14	ND	>99
alpha-BHC	BDL	ND	NM
beta-BHC	ND	BDL	NM
PCB-1242, 1254, 1221	76	BDL	97
PCB-1232, 1248, 1260, 1016	160	BDL	98

Blanks indicate data not available.
BDL, below detection limit.
ND, not detected.
NM, not meaningful.
* Approximate value.
Source: U.S. EPA. *Treatability Manual.* Vols. 1–3. U.S. EPA, Washington, DC, 1981.

APPENDIX B
Full Scale Treatment of Aluminum Forming Wastewater by Emulsion Breaking and Ultrafiltration

Removal Data

Sampling: Three 24-Hour or One 72-Hour Composite

Pollutant/Parameter	Concentration		Percent Removal
	Influent	Effluent	
Classical Pollutants (mg/L)			
Oil and grease	13	0.11	99
Suspended solids	2.6	0.019	99
COD	31	2.4	92
TOC	12	1.0	92
Phenol	0.022	0.016	27
pH, pH units	7.9	8.0	NM
Toxic Pollutants (µg/L)			
Arsenic	BDL	BDL	NM
Cadmium	ND	BDL	NM
Chromium	ND	<68	NM
Copper	ND	BDL	NM
Cyanide	BDL	BDL	NM
Lead	ND	BDL	NM
Mercury	ND	1	NM
Nickel	ND	<10	NM
Zinc	ND	BDL	NM
Acenaphthene	ND	3	NM
Benzene	40	ND	99*
2,4,6-Trichlorophenol	500	ND	99*
Chloroform	17	62	NM
Ethylbenzene	30	36	NM
Methylene chloride	67	320	NM
Naphthalene	ND	66	NM
Phenol	7900	9700	NM
Bis(2-ethylhexyl) phthalate	820	BDL	99*
Di-n-butyl phthalate	93	13	86
Diethyl phthalate	110	BDL	95*
Tetrachloroethylene	3000	200	93
Toluene	17	BDL	71*
4,4-DDE	7	BDL	64*
Alpha-endosulfan	BDL	BDL	NM
Endrin aldehyde	BDL	BDL	NM
Alpha-BHC	12	BDL	79*
Beta-BHC	5	BDL	50*
PCB-1242, 1254, 1221	110	BDL	98*
PCB-1232, 1248, 1260, 1016	360	BDL	99*

Blanks indicate data not available.
ND, not detected.
NM, not meaningful.
BDL, below detection limit.
* Approximate value.

Source: U.S. EPA. *Treatability Manual.* Vols. 1–3. U.S. EPA, Washington, DC, 1981.

APPENDIX C
Full Scale Treatment of Aluminum Forming Wastewater by Chemical Precipitation and Clarification

Removal Data
Sampling: Thee 24-Hour or One 72-Hour Composite

Pollutant/Parameter	Concentration		Percent Removal
	Influent	Effluent	
Classical Pollutants (mg/L)			
Oil and grease	86	15	99
Suspended soilds	450	710	NM
COD	260	280	NM
TOC	75	74	1
Phenol	0.003	0.002	33
pH, pH units	2.8	3.7	NM
Toxic Pollutants (µg/L)			
Chromium	900,000	790,000	12
Copper	2,200,000	2,200,000	0
Cyanide	BDL	BDL	NM
Lead	3200	1000	69
Mercury	<1	<1	NM
Nickel	2600	2400	8
Zinc	2,000,000	1,800,000	10
Fluoranthene	10	ND	>99
Methylene chloride	260	15	93
2,4-Dinitrophenol	37	ND	>99
N-nitrosodiphenylamine	67	ND	>99
Chrysene	10	ND	>99
Anthracene/phenanthrene	<26	BDL	NM
Pyrene	16	ND	>99

Blanks indicate data not available.
BDL, below detection limit.
ND, not detected.
NM, not meaningful.
Source: U.S. EPA. *Treatability Manual.* Vols. 1–3. U.S. EPA, Washington, DC, 1981.

APPENDIX D
Full Scale Treatment of Aluminum Forming Wastewater by Chromium Reduction, Chemical Precipitation, and Sedimentation Clarification

Removal Data

Sampling: Thee 24-Hour or One 72-Hour Composite

Pollutant/Parameter	Concentration		Percent Removal
	Influent	Effluent	
Classical Pollutants (mg/L)			
Oil and grease	5	< 95	NM
Suspended solids	<2	<5	NM
COD	20	30	NM
TOC	13	9.7	2.3
Phenol	0.003	0.009	NM
pH, pH units	2.6	9.8	NM
Toxic Pollutants (μg/L)			
Cadmium	2.8	BDL	64*
Chromium	100,000	90	>99
Copper	40	BDL	89*
Lead	30	BDL	67*
Mercury	3.4	<5	NM
Zinc	110	BDL	77*
Methylene chloride	30	60	NM
Bis(2-ethylhexyl) phthalate	ND	BDL	NM

Blanks indicate data not available.
BDL, below detection limit.
ND, not detected.
NM, not meaningful.
* Approximate value.
Source: U.S. EPA. *Treatability Manual.* Vols. 1–3. U.S. EPA, Washington, DC, 1981.

REFERENCES

1. U.S. EPA, *Development Document for Effluent Limitations Guidelines and Standards for the Aluminum Forming Point Source Category,* Vol. 1, final report W-84-14-A, U.S. EPA, Washington, DC, 1984.
2. U.S. EPA, *Development Document for Effluent Limitations Guidelines and Standards for the Aluminum Forming Point Source Category,* Vol. 2, final report W-84-14-B, U.S. EPA, Washington, DC, 1984.
3. U.S. Government Printing Office, Aluminum Forming Point Source Category, Code of Federal Regulations, Title 40, Vol. 27, Part 467, U.S. GPO, Washington, DC, July 1, 2003.
4. U.S. EPA, *Cost Effectiveness Analysis of Effluent Guidelines and Standards for the Aluminum Forming Category,* report W-83-13, U.S. EPA, Washington, DC, 1983.
5. U.S. EPA, *Aluminum Forming Point Source Category,* available at http://www.access.gpo.gov/nara/cfr/waisidx_03/40cfr467_03.html, 2008.
6. Higgins, T.E., *Pollution Prevention Handbook,* CRC Press, Boca Raton, Florida, 1995, pp. 386–388.
7. Wikipedia Encyclopedia, API Oil–Water Separator, available at http://en.wikipedia.org/wiki/API_oil-water_separator, March 9, 2008.
8. Wang, L.K., Hung Y.T., and Shammas N.K., Eds., *Physicochemical Treatment Processes,* Humana Press, Totowa, NJ, 2005.

9. Wang, L.K., Hung, Y.T., and Shammas, N.K., Eds., *Advanced Physicochemical Treatment Processes*, Humana Press, Totowa, NJ, 2006.
10. Wang, L.K., Hung, Y.T., and Shammas, N.K., Eds., *Advanced Physicochemical Treatment Technologies*, Humana Press, Totowa, NJ, 2007.
11. Wang, L.K., Shammas, N.K., and Hung, Y.T., Eds., *Biosolids Treatment Processes*, Humana Press, Totowa, NJ, 2007.
12. Wang, L.K., Shammas, N.K., and Hung, Y.T., Eds., *Biosolids Engineering and Management*, Humana Press, Totowa, NJ, 2008, pp. 396–398.
13. Wang, L.K., Hung, Y.T., Lo, H.H., and Yapijakis, C., Eds., *Handbook of Industrial and Hazardous Wastes Treatment*, Marcel Dekker, New York, 2004.
14. U.S. ACE, Yearly Average Cost Index for Utilities, in *Civil Works Construction Cost Index System Manual*, 110-2-1304, U.S. Army Corps of Engineers, Washington, DC, available at http://www.nww.usace.army.mil/cost, 2008.
15. U.S. EPA. *Treatability Manual*. Vols. 1, 2 and 3. Final reports EPA-600/2-82-001a, EPA-600/2-82-001b, and EPA-600/2-82-001c.s respectively. U.S. Environmental Protection Agency, Washington, DC, 1981.

6 Treatment of Nickel-Chromium Plating Wastes

*Lawrence K. Wang, Nazih K. Shammas,
Donald B. Aulenbach, and William A. Selke*

CONTENTS

6.1	Introduction	234
6.2	The Nickel-Chromium Plating Process	234
	6.2.1 Nickel Plating	234
	6.2.2 Chromium Plating	235
6.3	Sources of Pollution	235
	6.3.1 Environmental Impact of Nickel	236
	6.3.2 Environmental Impact of Chromium	236
6.4	Waste Minimization	237
	6.4.1 Assessment of Hazardous Waste	237
	6.4.2 Improved Procedures and Segregation of Wastes	237
	6.4.3 Material Substitution	238
	6.4.4 Extending Process Bath Life	238
	6.4.5 Dragout Reduction	238
	6.4.6 Reactive Rinses	239
6.5	Material Recovery and Recycling	239
	6.5.1 Dragout Recovery	239
	6.5.2 Evaporative Recovery	240
	6.5.3 Reverse Osmosis	240
	6.5.4 Ion Exchange	241
	6.5.5 Electrodialysis	241
	6.5.6 Electrolytic Recovery	242
	6.5.7 Deionized Water	242
6.6	Chemical Treatment	242
	6.6.1 Neutralization	242
	6.6.2 Hexavalent Chromium Reduction	243
	6.6.3 pH Adjustment and Hydroxide Precipitation	245
	6.6.4 Reduction and Flotation Combination	247
6.7	Conventional Reduction–Precipitation System	248
6.8	Modified Reduction–Flotation Wastewater Treatment System	249
6.9	Innovative Flotation–Filtration Wastewater Treatment Systems	251
	6.9.1 Flotation–Filtration System Using Conventional Chemicals	251
	6.9.2 Flotation–Filtration System Using Innovative Chemicals	251
	6.9.3 Flotation–Filtration Systems	252
6.10	Summary	255
References		257

6.1 INTRODUCTION

Applicable local, state, and federal environmental laws require that the waste generated by the nickel-chromium plating process be pretreated to provide a discharge acceptable to the public wastewater treatment system.

The specific purpose of this chapter is to describe the chemical and physical pretreatment methods required for nickel-chromium plating wastewater, to describe the upgrades needed by a municipal wastewater treatment system to manage this waste, and to relate the methods and upgrades to the operation of the total treatment system. Special emphasis is placed on presentation of the following:

1. The chemistry of nickel-chromium plating and waste generation
2. The type of pollutants and their sources
3. Waste minimization
4. Recovery and recycling
5. Conventional reduction–precipitation treatment systems
6. Modified reduction–flotation treatment systems
7. Innovative flotation–filtration treatment systems

6.2 THE NICKEL-CHROMIUM PLATING PROCESS

The nickel-chromium plating process includes the steps in which a ferrous base material is electroplated with nickel and chromium. The electroplating operations for plating the two metals are basically oxidation–reduction reactions. Typically, the part to be plated is the cathode, and the plating metal is the anode.

6.2.1 Nickel Plating

To plate nickel on iron parts, the iron parts form the cathodes, and the anode is a nickel bar. On the application of an electric current, the nickel bar anode oxidizes, dissolving in the electrolyte:

$$Ni \rightarrow Ni^{2+} + 2e^- \quad (6.1)$$

The resulting nickel ions are reduced at the cathode (the iron part) to form a nickel plate:

$$Ni^{2+} + 2e^- \rightarrow Ni \quad (6.2)$$

Nickel plating can also be accomplished by an electroless plating technique involving deposition of a metallic coating by a controlled chemical reduction that is catalyzed by the metal or alloy being deposited. A special feature of electroless plating is that no external electrical energy is required. The following are the basic ingredients in electroless plating solutions:

1. A source of metal, usually a salt
2. A reducer to reduce the metal to its base state
3. A chelating agent to hold the metal in solution so the metal will not plate out indiscriminately
4. Various buffers and other chemicals designed to maintain stability and increase bath life

Nickel electroless plating on a less noble metal is common.[1–7] For example, the source of nickel can be nickel sulfate. The reducer can be an organic substance, such as formaldehyde. A chelating agent (tartrate or equivalent) is generally required. The nickel salt is ionized in water:

$$NiSO_4 \rightarrow Ni^{2+} + SO_4^{2-} \quad (6.3)$$

Nickel-Chromium Wastes

There is then a redox reaction with the nickel and the formaldehyde:

$$Ni^{2+} + 2H_2CO + 4OH^- \longrightarrow Ni + 2HCO_2^- + 2H_2O + H_2 \tag{6.4}$$

The base metal nickel now begins to plate out on an appropriate surface, such as a less noble metal.

6.2.2 Chromium Plating

In chromium plating, the chromium is supplied to the plating baths as chromic acid. For example, plating baths can be prepared by adding hexavalent chromium in the form of either sodium dichromate ($Na_2Cr_2O_7 \cdot H_2O$) or chromium trioxide (CrO_3). When sodium dichromate is used it dissociates to produce the divalent dichromate ion ($Cr_2O_7^{2-}$). When chromium trioxide is used, it immediately dissolves in water to form chromic acid according to the following reaction[8-15]:

$$CrO_3 + H_2O \longrightarrow H_2CrO_4 \tag{6.5}$$

Chromic acid is considered a strong acid, although it never completely ionizes. Its ionization has been described as follows:

$$H_2CrO_4 \longrightarrow H^+ + HCrO_4^- \text{ (acid chromate ion)} \tag{6.6}$$
$$K_a = 0.83 \text{ at } 25°C$$

$$HCrO_4^- \longrightarrow H + CrO_4^{2-} \text{ (chromate ion)} \tag{6.7}$$
$$K_a = 3.2 \times 10^{-7} \text{ at } 25°C$$

Moreover, the dichromate ion ($Cr_2O_7^{2-}$) will exist in equilibrium with the acid chromate ion as follows:

$$Cr_2O_7^{2-} + H_2O \longrightarrow 2HCrO_4^- \tag{6.8}$$
$$K_a = 0.0302 \text{ at } 25°C$$

Theoretically, $HCrO_4^-$ is the predominant species between pH 1.5 and 4.0, $HCrO_4^-$ and CrO_4^{2-} exist in equal amounts at pH 6.5, and CrO_4^{2-} predominates at higher pH values. Chromium plating wastewater is generally somewhat acid, and the acid chromate ion $HCrO^-$ is predominant in this wastewater.

Chromating is one of the chemical conversion coating technologies. Chrome coatings are applied to previously deposited nickel for increased corrosion protection and to improve surface appearance. Chromate conversion coatings are formed by immersing the metal in an aqueous acidified chromate solution consisting substantially of chromic acid or water-soluble salts of chromic acid, together with various catalysts or activators.

6.3 SOURCES OF POLLUTION

A conceptual arrangement of the nickel-chromium plating process can be broken down into three general steps:

1. Surface preparation involving the conditioning of the base material for plating
2. Actual application of the plate by electroplating
3. The posttreatment steps

The major waste sources during normal nickel-chromium plating operations are alkaline cleaners, acid cleaners, plating baths, posttreatment baths, and auxiliary operation units.

The wastestreams generated by the plating process can be subdivided and classified into eight categories[1,5,6,15]:

1. Concentrated acid wastes
2. Concentrated phosphate cleaner wastes
3. Acid rinsewater
4. Alkaline rinsewater
5. Chromium rinsewater
6. Nickel rinsewater
7. Concentrated nickel wastes
8. Concentrated chromium wastes

In the above categories, there are seven major types of aqueous pollutants that must be pretreated and removed[5,15]:

1. Acidity
2. Alkalinity
3. Nickel
4. Chromium
5. Iron
6. Organics (COD, BOD)
7. Suspended solids

The environmental impact of the two most toxic pollutants, nickel and chromium, is briefly presented in the following.[1,16,17] Significant concentrations of these elements pass through conventional treatment plants.

6.3.1 Environmental Impact of Nickel

Nickel is toxic to aquatic organisms at levels typically observed in POTW (publicly owned treatment works) effluents:

1. 50% reproductive impairment of *Daphnia magna* at 0.095 mg/L
2. Morphological abnormalities in developing eggs of *Limnaea palustris* at 0.230 mg/L
3. 50% growth inhibition of aquatic bacteria at 0.020 mg/L

Because surface water is often used as a drinking water source, nickel passed through a POTW becomes a possible drinking water contaminant.

A U.S. Environmental Protection Agency (U.S. EPA) study of 165 sludges showed nickel concentrations ranging from 2 to 3520 mg/kg (dry basis).[18] Nickel toxicity may develop in plants from application of municipal wastewater biosolids on acid soils. Nickel reduces yields for a variety of crops including oats, mustard, turnips, and cabbage.

6.3.2 Environmental Impact of Chromium

Chromium can exist as either trivalent or hexavalent compounds in raw wastewater streams. The chromium that passes through the POTW is discharged to ambient surface water. Chromium is toxic to aquatic organisms at levels observed in POTW effluents[15]:

1. Trivalent chromium significantly impaired the reproduction of *Daphnia magna* at levels of 0.3 to 0.5 mg/L.
2. Hexavalent chromium retards growth of chinook salmon at 0.0002 mg/L. Hexavalent chromium is also corrosive and a potent human skin sensitizer.

Nickel-Chromium Wastes

Besides providing an environment for aquatic organisms, surface water is often used as a source of drinking water. The National Primary Drinking Water Standards are based on total chromium, the limit being 0.1 mg/L.[19]

A U.S. EPA study of 180 municipal wastewater sludges showed that municipal wastewater sludge contains 10 to 99,000 mg/kg (dry basis) of chromium. Most crops absorb relatively little chromium even when it is present in high levels in soils, but chromium in sludge has been shown to reduce crop yields in concentrations as low as 200 mg/kg.[18]

6.4 WASTE MINIMIZATION

All metal finishing facilities have one thing in common—the generation of metal-containing hazardous waste from the production processes. Reducing the volume of waste generated can save money and at the same time decreases future liabilities. Typical wastes generated are as follows:

1. Industrial wastewater and treatment residues
2. Spent plating baths
3. Spent process baths
4. Spent cleaners
5. Waste solvents and oil

This section identifies areas for reducing waste generation. It also suggests techniques available to metal finishers for waste reduction and is intended to help metal finishing shop owners decide whether waste reduction is a possibility.

Both state (Health and Safety Code) and federal (40 CFR, Part 262, Subpart D) regulations require that generators of hazardous waste file a biennial generator's report. Among other things, this report must include a description of the efforts undertaken and achievements accomplished during the reporting period to reduce the volume and toxicity of waste generated. The Uniform Hazardous Waste Manifest requires that large generators certify that they have a program in place to reduce the volume and toxicity of waste generated that is determined to be economically practicable. Small-quantity generators must certify that they have made a good faith effort to minimize waste generation and have selected the best affordable waste management method available.

As waste reduction methods reduce the amount of waste generated, and also the amount subject to regulation, these practices can help a shop comply with the requirements while also saving money. The shop's owner or manager must be committed to waste reduction and pass that commitment on to the employees, establish training for employees in waste reduction, hazardous material handling and emergency response, and establish incentive programs to encourage employees to design and use new waste reduction ideas. The following is a list of some common waste reduction methods for metal finishing electroplating shops.[20,21]

6.4.1 Assessment of Hazardous Waste

Waste assessments are used to list the sources, types, and amounts of hazardous waste generated to make it easier to pinpoint where wastes can be reduced.

Source reduction is usually the least expensive approach to minimizing waste. Many of these techniques involve housekeeping changes or minor inplant process modifications.

6.4.2 Improved Procedures and Segregation of Wastes

These may be summarized as follows:

1. Good housekeeping is the easiest and often the cheapest way to reduce waste. Keep work areas clean.

2. Improve inventory procedures to reduce the amount of off-specification materials generated.
3. Reduce quantities of raw materials to levels where materials will be used up just as new materials are arriving.
4. Designate protected raw material and hazardous waste storage areas with spill containment. Keep the areas clean and organized and give one person the responsibility for maintaining the areas.
5. Label containers as required and cover them to prevent contact with rainfall and avoid spills.
6. Use a "first-in, first-out" policy for raw materials to keep them from becoming too old to be used. Give one person responsibility for maintaining and distributing raw materials.
7. Use bench-scale testing for samples rather than process baths.
8. Designate one person to accept chemical samples and return unused samples to suppliers.
9. Limit bath mixing to trained personnel.
10. Segregate wastestreams for recycling and treatment, and keep nonhazardous material from becoming contaminated.
11. Prevent and contain spills and leaks by installing drip trays and splash guards around processing equipment.
12. Conduct periodic inspections of tanks, tank liners, and other equipment to avoid failures. Repair malfunctions when they are discovered. Use inspection logs to follow up on repairs.
13. Inspect plating racks for loose insulation that would cause increased dragout.
14. Use dry cleanup where possible to reduce the volume of wastewater.

6.4.3 Material Substitution

In summary:

1. Use process chemistries that are treatable or recyclable on site.
2. Use deionized water instead of tap water in process baths or rinsing operations to reduce chemical reactions with impurities in the tap water, which would increase sludge production.
3. Use nonchelated process chemistries rather than chelated chemistries to reduce sludge volume.
4. Replace cyanide process baths with noncyanide process baths to simplify the treatment required.
5. Use alkaline cleaners instead of solvents for degreasing operations; they can be treated on site and usually discharged to the sewer with permit authorization.

6.4.4 Extending Process Bath Life

This may be achieved with the following procedures:

1. Treatment of process baths can extend their useful life.
2. Bath replenishment extends the useful life of the bath.
3. Monitoring (using pH meters or conductivity meters) the process baths can determine the need for bath replenishment.

6.4.5 Dragout Reduction

Dragout reduction is achieved using the following steps:

1. Minimize bath concentrations to the lower end of their operating range.
2. Maximize bath operating temperatures to lower the solution's viscosity.
3. Use wetting agents (which reduce the surface tension of the solution) in process baths to decrease the amount of dragout.
4. Withdraw workpieces from tanks slowly to allow maximum drainage back into process tank.

Nickel-Chromium Wastes

5. Use air knives or spray rinses above process tanks to rinse excess solution off a workpiece and into the process bath.
6. Install drainage boards between process tanks and rinse tanks to direct dragout back into process tank.
7. Use dedicated dragout tanks after process baths to capture dragout.
8. Install rails above process tanks to hang workpiece racks for drainage prior to rinsing.
9. Use spray rinses as the initial rinse after the process tank and before the dip tank.
10. Use air agitation or workpiece agitation to improve rinse efficiency.
11. Install multiple rinse tanks (including counterflow rinse tanks) after process baths to improve rinse efficiency and reduce water consumption.

6.4.6 REACTIVE RINSES

The following steps should be applied:

1. Reuse the acid rinse effluent as influent for the alkaline rinse tank, thus allowing the fresh water feed to the alkaline rinse tank to be turned off (reactive rinsing). This can also be applied to process tank rinses.
2. Treat rinsewater effluent to recover process bath chemicals. This allows the reuse of the effluent for rinsing or neutralization prior to discharge.
3. Reuse the spent reagents from the process baths in the wastewater treatment process.
4. Recycle spent solvents on site or off site.
5. Use treatment technologies to recycle rinsewaters in a closed loop or open loop system.
6. Some recycling and most treatment processes require a permit. Be sure to contact the local Department of Health Services regional office to determine if there is a need for a permit to treat or recycle the wastes.
7. Pretreat process water to reduce the natural contaminants that contribute to the sludge volume.
8. Use treatment chemicals that reduce sludge generation (e.g., caustic soda instead of lime).
9. Use sludge dewatering equipment to reduce sludge volume.
10. Use treatment technologies (such as ion exchange, evaporation, and electrolytic metal recovery) that do not use standard precipitation/clarification methods that generate heavy metal sludges.

6.5 MATERIAL RECOVERY AND RECYCLING

Unlike the 1970s and 1980s when waste management costs were relatively inexpensive, today's metal finishers are facing increasingly higher disposal costs. This change is due in part to a decrease in the volume of available landfill space, which has resulted in escalating landfill fees and more stringent federal and state environmental regulations that mandate treatment prior to landfilling.

Metal finishers are seeing their profits shrink as waste management costs increase. To control waste disposal costs, metal finishers must focus on developing and implementing a facility-wide waste reduction program. In other words, as discussed in Section 6.4, metal finishers must consciously seek out ways to decrease the volume of waste that they generate.

One approach to waste reduction is to recover process materials for reuse. Materials used in metal finishing processes can be effectively recovered using available technologies such as dragout, evaporation, reverse osmosis, ion exchange, electrodialysis, and electrolytic recovery.[22–26]

6.5.1 DRAGOUT RECOVERY

Dragout recovery is a simple technology used by metal finishers to recover plating chemicals. It involves using drain boards, drip tanks, fog-spray tanks, or dragout tanks separately, or in combination, to capture plating chemicals dragged out of plating tanks from parts being plated.

Drain boards are widely used throughout the metal industry to capture plating solutions. Boards are suspended between process tanks and are constructed of plastic, plain or teflon-coated steel. Solutions drip on the boards and drain back into their respective processing tanks.[22,27]

In contrast, a drip tank recovers process chemicals by collecting dragout into a separate tank, from which it can be returned to the process as needed.

In a fog-spray tank, plating chemicals clinging to parts are recovered by washing them with a fine water-mist. The solution that collects in the fog-spray tank is returned to the process tank as needed. The added water helps to offset evaporative losses from the process tanks.

Dragout tanks are essentially rinse tanks. Dragout chemicals are captured in a water solution, which is returned to the process tank as needed.

The presence of airborne particles and other contaminants in recovered plating chemicals may necessitate treatment of the collected solution to remove the contaminants prior to solution reuse.

There are advantages and disadvantages to dragout recovery. Depending upon the solution, up to 60% of the materials carried out of a plating tank can be recovered for reuse; thus dragout can affect metal deposition and surface finish quality. Impurities can concentrate in the solutions causing a deteriorating effect on the plating process when returned to the plating bath.

6.5.2 Evaporative Recovery

A widely used metal salt recovery technique is evaporation. With evaporation, plating chemicals are concentrated by evaporating water from the solution. Evaporators may use heat or natural evaporation to remove water.[22,28] Additionally, evaporators may operate at atmospheric pressure or under vacuum.

Atmospheric evaporators are more commonly used. They are open systems that use process heat and warm air to evaporate water. These evaporators are relatively inexpensive, require low maintenance and are self-operating. Under the right conditions, they can evaporate water from virtually any plating bath or rinse. A packed-bed evaporator is an example of an atmospheric evaporator.

Vacuum evaporators are also used to recover plating chemicals. They are closed systems that use steam heat to evaporate water under a vacuum. This results in lower boiling temperature, with a reduction in thermal degradation of the solution. Like atmospheric evaporators, they require low maintenance and are self-operating. A climbing file evaporator is an example of a vacuum evaporator.

A typical evaporative recovery system consists of an evaporator, a feed pump, and a heat exchanger. Plating solution or rinsewater containing dilute plating chemicals is circulated through the evaporator. The water evaporates and concentrates the plating chemicals for reuse. In open evaporator systems, the water evaporates and mixes with air and is released to the atmosphere. It may be necessary to vent the contaminated airstream to a ventilation/scrubber treatment system prior to release. In enclosed evaporators the water is condensed from the air and can be reused in rinses, which further increases savings. Water reuse is preferred whenever possible.

As with all process equipment, the design size of an evaporator system is dependent upon volumetric flow, specifically the rinsewater flow rate required and the volume of process solution dragout. When operated properly, a commercial evaporator can attain a 99% material recovery rate.

There are drawbacks to using an evaporator to recover plating chemicals. For instance, impurities are concentrated along with recovered plating chemicals. These impurities can alter desired deposited metal characteristics, including surface finish quality. Vacuum evaporation can be used to avoid degradation of plating solutions containing additives that are sensitive to heat.

The evaporative recovery is a very energy-intensive process. Approximately 538 chu (970 Btu) are required to evaporate 1 lb of water at standard atmospheric pressure. Additional energy is required to raise the temperature of the solution to its boiling point.

6.5.3 Reverse Osmosis

Reverse osmosis (RO) recovers plating chemicals from plating rinsewater by removing water molecules with a semipermeable membrane. The membrane allows water molecules to pass through, but blocks metallic salts and additives.[29]

Like evaporators, RO works on most plating baths and rinse tanks. Most RO systems consist of a housing that contains a membrane and feed pump. There are four basic membrane designs: plate-and-frame, spiral-wound, tubular, and hollow-fiber. The most common types of membrane materials are cellulose acetate, polyether/amide, and polysulfones.[29]

Diluted or concentrated rinsewaters are circulated through the membrane at pressures greater than aqueous osmotic pressure. This action results in the separation of water from the plating chemicals. The recovered chemicals can be returned to the plating bath for reuse, and the permeate, which is similar to the condensate from an evaporator, can be used as make-up water. RO units work best on dilute solutions.[30]

The design and capacity of an RO unit is dependent upon the type of chemicals in the plating solution and the dragout solution rate. Certain chemicals require specific membranes. For instance, polyamide membranes work best on zinc chloride and nickel baths, and polyether/amide membranes are suggested for chromic acid and acid copper solutions. The flow rate across the membrane is very important. It should be set at a rate to obtain maximum product recovery. RO systems have a 95% recovery rate with some materials and with optimum membrane selection.[22]

There are advantages to using RO. Energy usage is much lower than for other recovery systems and plating chemicals can be recovered from temperature-sensitive solutions. However, RO also has limitations. The membrane is susceptible to fouling, which is often caused by the precipitation of suspended and dissolved solids that plug the membrane's pores. Also, as with evaporators, RO can concentrate impurities along with plating chemicals, which degrade plating quality.

6.5.4 Ion Exchange

Ion exchange is a molecular exchange process where metal ions in solution are removed by a chemical substitution reaction with an ion-exchange resin.[31] Ion exchange can be used with most plating baths. Metal cations exchange sites with sodium or hydrogen ions and anions (such as chromate) with hydroxyl ions. The exchange resin can generally be regenerated with an acid or alkaline solution and reused. When a cation exchange resin is regenerated, it produces a metal salt. For example, copper is removed from an ion exchange resin by passing sulfuric acid over the resin, producing copper sulfate. This salt can be added directly into the plating bath.[23,32]

The required size of an ion-exchange unit is dependent upon the composition and volume of plating dragout. Each ion-exchange resin has a maximum capacity for recovery of specific ions. The ion-exchange unit's size (volume of resin) is determined by the amount of metal to be removed from the recovered solutions.

Ion exchange has its drawbacks. Most commercially available resins are nonselective and, therefore, similarly charged ions can be exchanged by a given resin whether desired in the process or not. This means that certain contaminants cannot be removed by ion exchange and are returned to the plating tank with the metal salt.[22] The metal salt solution produced after regeneration is often a dilute solution that can only be put back into the process bath if evaporation is used to make room in the process tank. In addition, ion exchange is not a continuous process and system sizing must take into account resin regeneration time.

6.5.5 Electrodialysis

Electrodialysis units recover plating chemicals differently from the recovery units discussed thus far. In electrodialysis, electromotive forces selectively drive metal ions through an ion-selective membrane (in RO, pressure is the driving force; in ion exchange, the driving force is chemical attraction). The membranes are thin sheets of plastic material with either anionic or cationic characteristics.[33]

Electrodialysis units are constructed using a plate-and-frame technique similar to filter presses. Alternating sheets of anionic and cationic membranes are placed between two electrodes. The plating or rinse solution to be recovered (electrolyte) circulates past the system's electrodes. Hydrogen and oxygen evolve. Positive ions travel to the negative terminal and negative ions travel to the positive terminal. The electrolyte also provides overall electrical conductivity to the cell. In some units, the current is periodically reversed to reduce membrane fouling.

Electrodialysis is compatible with most plating baths, and the design size of a unit is dependent upon the rinsewater flow rate and concentration.[22]

Electrodialysis has advantages and disadvantages. For instance, the process requires very little energy and can recover highly concentrated solutions. On the other hand, similarly to other membrane processes, electrodialysis membranes are susceptible to fouling and must be regularly replaced.

6.5.6 Electrolytic Recovery

Electrolytic recovery (ER) is the oldest metal recovery technique. Metal ions are plated-out of solution electrochemically by reduction at the cathode.[34] There are essentially two types of cathodes used for this purpose: a conventional metal cathode and a high surface area cathode (HSAC). Both cathodes can effectively plate-out metals, such as gold, zinc, cadmium, copper, and nickel.[22]

Electrolytic recovery systems work best on concentrated solutions. For optimal plating efficiency, recovery tanks should be agitated ensuring that good mass transfer occurs at the electrodes. Another important factor to consider is the anode/cathode ratio. The cathode area (plating surface area) and mass transfer rate to the cathode greatly influence the efficiency of metal deposition.

Electrolytic recovery can be used with most plating baths. The amount of metal to be plated per square meter of cathode determines the electrolytic recovery unit's design capacity. Therefore, the volume and concentration of plating dragout greatly influences system design and size.[22,35]

There are advantages to the electrolytic recovery process. For instance, ER units can operate continuously, and the product is in a metallic form that is very suitable for reuse or resale. Electrolytic units are also mechanically reliable and self-operating. Very importantly, contaminants are not recovered and returned to the plating bath. Thus, electrolytically recovered metals are as pure as "virgin" plating raw material.

The major disadvantage to electrolytic recovery is high energy cost. Energy costs will vary, of course, with cathode efficiencies and local utility rates.[22]

6.5.7 Deionized Water

Using deionized water to prepare plating bath solutions is an effective way of preventing waste generation. Some groundwater and surface waters contain high concentrations of calcium, magnesium, chloride, and other soluble contaminants that may build up in process baths.[22] By using deionized water, buildup of these contaminants can be more easily controlled. Technologies such as RO and ion exchange can also be used to effectively remove soluble contaminants from incoming water.[36]

6.6 CHEMICAL TREATMENT

Treatment for the removal of chromium and nickel from electroplating wastewater involves neutralization, hexavalent chromium reduction, pH adjustment, hydroxide precipitation, and final solid–liquid separation.[15,37–48]

6.6.1 Neutralization

Excess acidity and alkalinity may be eliminated by simple neutralization by either a base or an acid. This is a simple stoichiometric chemical reaction of the following type[5,15,49]:

$$\text{Strong base + strong acid} \longrightarrow \text{salt + water} \tag{6.9}$$

Nickel-Chromium Wastes

Examples of this include the following:

1. Alkali

$$2\text{ NaOH} + \text{H}_2\text{SO}_4 \rightarrow \text{Na}_2\text{SO}_4 + 2\text{ HOH} \qquad (6.10)$$
$$\text{Base} \quad \text{Acid} \quad\quad \text{Salt} \quad\quad \text{Water}$$

2. Acid

$$\text{HNO}_3 + \text{NaOH} \rightarrow \text{NaNO}_3 + \text{HOH} \qquad (6.11)$$
$$\text{Acid} \quad\quad \text{Base} \quad\quad \text{Salt} \quad\quad \text{Water}$$

A slight excess of base may be titrated in the previous reactions to shift the pH to a slight basic condition. This is important for the precipitation of certain metal salts (such as nickel, iron, and trivalent chromium) as hydroxides.

6.6.2 Hexavalent Chromium Reduction

Chemical treatment of chromium wastewater is usually conducted in two steps. In the first step hexavalent chromium is reduced to trivalent chromium by the use of a chemical reducing agent. The trivalent chromium is precipitated during the second stage of treatment.[15]

Sulfur dioxide (SO_2), sodium bisulfite ($NaHSO_3$), and sodium metabisulfite ($Na_2S_2O_5$) are commonly used as reducing agents.[15,50] All these compounds react to produce sulfurous acid when added to water, according to the following reactions:

$$SO_2 + H_2O \rightarrow H_2SO_3 \qquad (6.12)$$
$$Na_2S_2O_5 + H_2O \rightarrow 2\ NaHSO_3 \qquad (6.13)$$
$$NaHSO_3 + H_2O \rightarrow H_2SO_3 + NaOH \qquad (6.14)$$

It is the sulfurous acid produced from these reactions that is responsible for the reduction of hexavalent chromium. The reaction is shown in the following equation:

$$2\ H_2CrO_4 + 3\ H_2SO_3 \rightarrow Cr_2(SO_4)_3 + 5\ H_2O \qquad (6.15)$$

The typical amber color of the hexavalent chromium solution will turn to a pale green once the chromium has been reduced to the trivalent state. Although this color change is a good indicator, redox control is usually employed.

The theoretical amount of sulfurous acid required to reduce a given amount of chromium can be calculated from the above equation. The actual amount of sulfurous acid required to treat a wastewater will be greater than this because other compounds and ions present in the wastewater may consume some of the acid. Primary among these is dissolved oxygen, which oxidizes sulfurous acid to sulfuric acid according to the following reaction:

$$H_2SO_3 + 0.5\ O_2 \rightarrow H_2SO_4 \qquad (6.16)$$

Each part of dissolved oxygen initially present in the wastewater produces 6.1 parts of sulfuric acid.

Undissociated sulfurous acid is responsible for the reduction of hexavalent chromium. Consequently, the reduction reaction is strongly pH-dependent because of the effect of pH on acid dissociation:

$$H_2SO_3 \rightarrow H^+ + HSO_3^- \qquad (6.17)$$
$$K_a = 1.72 \times 10^{-2} \text{ at } 25°C$$

$$HSO_3^- \rightarrow H^+ + SO_3^{2-} \quad (6.18)$$
$$K_a = 1.0 \times 10^{-7} \text{ at } 25°C$$

The dissociation as a function of pH and the effect of pH on reaction rate is shown in Figure 6.1 and Figure 6.2, respectively.[15] Obviously, the reaction proceeds much faster at low pH values, where the concentration of undissociated sulfurous acid is highest. As a result, chromium reduction processes are generally conducted at pH values of 2 to 3 to maximize reaction rates and minimize the volume of reaction vessels. Sulfuric acid is generally added to reduce the pH of the wastewater to the desired level and to maintain it at that level throughout treatment. If the pH is not maintained at the desired level but is allowed to increase during treatment, the reaction may not go to completion in the retention time available, and unreduced hexavalent chromium may exist in the effluent. The amount of acid required to depress the pH to the level selected for chrome reduction will depend on the alkalinity of the wastewater being treated. This acid requirement can be determined by titrating a sample of wastewater with sulfuric acid to the desired pH in the absence of a reducing agent.

In addition to the sulfuric acid required for pH adjustment, some amount of acid is consumed by the reduction reaction (Equation 8.15). If sulfur dioxide is used as the reducing agent, it will provide all the acid consumed by this reaction, and additional acid will not be required. However, if sodium bisulfite or sodium metabisulfite is used, additional acid must be supplied to satisfy the acid demand. This acid requirement is stoichiometric and can be calculated from Equations 6.19 to 6.22.

At pH 3.0 to 4.0:

$$3 \text{ NaHSO}_3 + 1.5 \text{ H}_2\text{SO}_4 + 2 \text{ H}_2\text{CrO}_4 \rightarrow \text{Cr}_2(\text{SO}_4)_3 + 1.5 \text{ Na}_2\text{SO}_4 + 5 \text{ H}_2\text{O} \quad (6.19)$$
$$1.5 \text{ Na}_2\text{S}_2\text{O}_5 + 1.5 \text{ H}_2\text{SO}_4 + 2 \text{ H}_2\text{CrO}_4 \rightarrow \text{Cr}_2(\text{SO}_4)_3 + 1.5 \text{ Na}_2\text{SO}_4 + 3.5 \text{ H}_2\text{O} \quad (6.20)$$

At pH 2.0:

$$3 \text{ NaHSO}_3 + 2 \text{ H}_2\text{SO}_4 + 2 \text{ H}_2\text{CrO}_4 \rightarrow \text{Cr}_2(\text{SO}_4)_3 + \text{Na}_2\text{SO}_4 + \text{NaHSO}_4 + 51 \text{ H}_2\text{O} \quad (6.21)$$
$$1.5 \text{ Na}_2\text{S}_2\text{O}_5 + 2 \text{ H}_2\text{SO}_4 + 2 \text{ H}_2\text{CrO}_4 \rightarrow \text{Cr}_2(\text{SO}_4)_3 + \text{Na}_2\text{SO}_4 + \text{NaHSO}_4 + 3.5 \text{ H}_2\text{O} \quad (6.22)$$

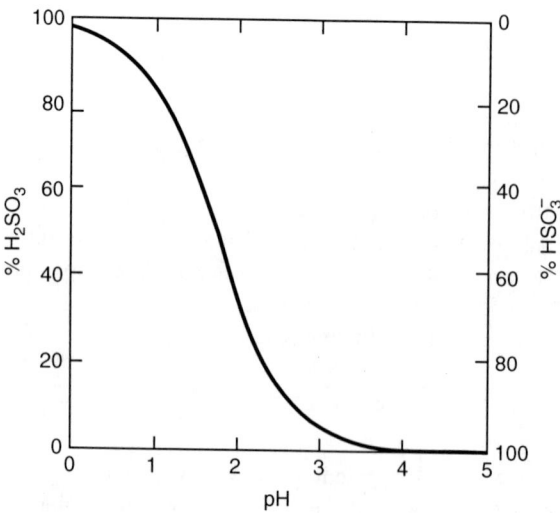

FIGURE 6.1 Relationship between H_2SO_3 and HSO_3^- at various pH values. (Taken from Krofta, M. and Wang, L.K., *Design of Innovative Flotation–Filtration Wastewater Treatment Systems for a Nickel-Chromium Plating Plant*, U.S. Department of Commerce, National Technical Information Service, Springfield, VA, Technical Report PB-88-200522/AS, January 1984.)

Nickel-Chromium Wastes

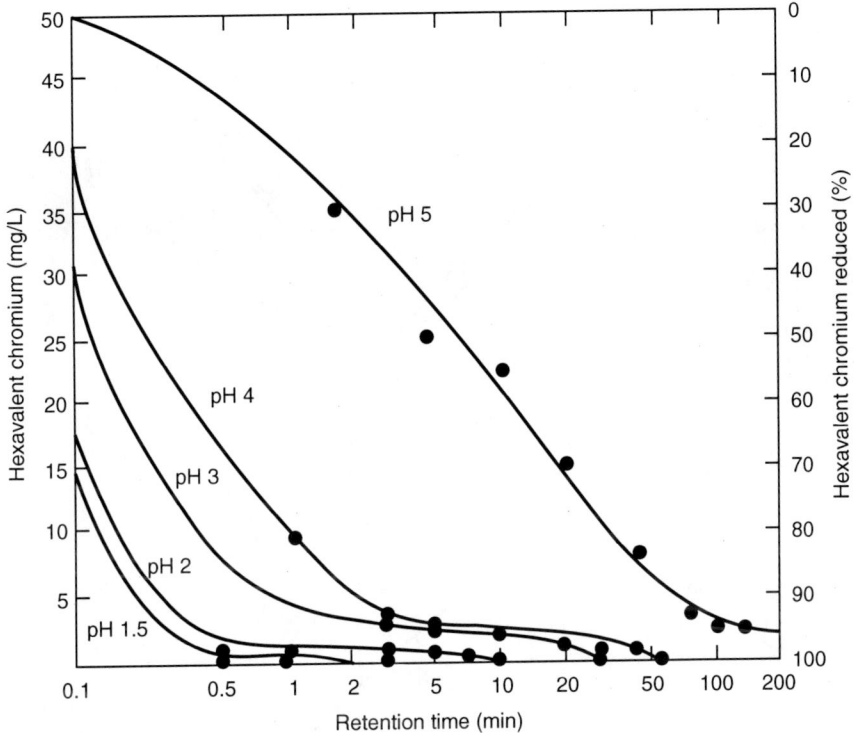

FIGURE 6.2 Rate of reduction of hexavalent chromium in the presence of excess SO_2 at various pH levels. (Taken from Krofta, M. and Wang, L.K., *Design of Innovative Flotation–Filtration Wastewater Treatment Systems for a Nickel-Chromium Plating Plant*, U.S. Department of Commerce, National Technical Information Service, Springfield, VA, Technical Report PB-88-200522/AS, January 1984.)

Similar equations can be developed for pH values between 2 and 3 as a function of the SO_4^{2-} and HSO_4^- distribution.

6.6.3 pH Adjustment and Hydroxide Precipitation

Wastewater pH is adjusted by addition of an acid or an alkali, depending on the purpose of the adjustment. The most common purposes of wastewater pH adjustment are the following:

1. Chemical precipitation of dissolved heavy metals, as illustrated by Figure 6.3
2. Pretreatment of metal-bearing wastewater before sulfide precipitation so that the formation of hazardous gaseous hydrogen sulfide does not occur
3. Neutralization of wastewater before discharge to either a stream or a sanitary sewer[37–48]

To accomplish hydroxide precipitation, an alkaline substance such as lime or sodium hydroxide is added to the wastewater to increase the pH to the optimum range of minimum solubility at which the metal precipitates as a hydroxide[51]:

$$M(II)^{2+} + Ca(OH)_2 \rightarrow M(II)(OH)_2 + Ca^{2+} \quad (6.23)$$
$$2\,M(III)^{3+} + 3\,Ca(OH)_2 \rightarrow 2\,M(III)(OH)_3 + 3\,Ca^{2+} \quad (6.24)$$

where M(II) = divalent metal and M(III) = trivalent metal.

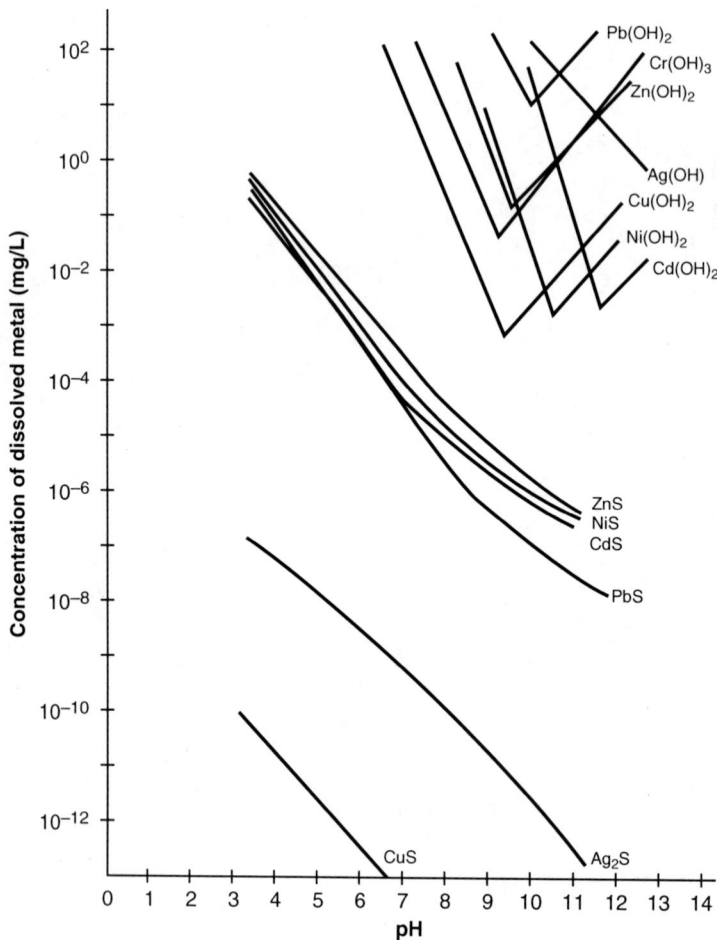

FIGURE 6.3 Solubility of metal hydroxides and sulfides. (Taken from Krofta, M. and Wang, L.K., *Design of Innovative Flotation–Filtration Wastewater Treatment Systems for a Nickel-Chromium Plating Plant*, U.S. Department of Commerce, National Technical Information Service, Springfield, VA, Technical Report PB-88-200522/AS, January 1984.)

The precipitated metal hydroxide can then be removed from the wastewater by clarification or other solid–water separation techniques.[52]

As a practical example, following the reduction of hexavalent chromium, sodium hydroxide, lime, or sodium hydroxide can be added to the wastewater to neutralize the pH and precipitate the trivalent chromium, nickel, iron, divalent, and other heavy metals. If lime is used, lime will react with heavy metals and with any residual sodium sulfate, sulfurous acid, or sodium bisulfite. The following reactions apply:

$$NiCl_2 + Ca(OH)_2 \rightarrow Ni(OH)_2 + CaCl_2 \tag{6.25}$$

$$NiSO_4 + Ca(OH)_2 \rightarrow Ni(OH)_2 + CaSO_4 \tag{6.26}$$

$$2\,Fe_2(SO_4)_3 + 6\,Ca(OH)_2 \rightarrow 4\,Fe(OH)_3 + 6\,CaSO_4 \tag{6.27}$$

$$Cr_2(SO_4)_3 + 3\,Ca(OH)_2 \rightarrow 2\,Cr(OH)_3 + 3\,CaSO_4 \tag{6.28}$$

$$H_2SO_4 + Ca(OH)_2 \rightarrow CaSO_4 + 2\,H_2O \tag{6.29}$$

$$2\,NaHSO_4 + Ca(OH)_2 \rightarrow CaSO_4 + Na_2SO_4 + 2\,H_2O \tag{6.30}$$

$$H_2SO_3 + Ca(OH)_2 \rightarrow CaSO_3 + 2\,H_2O \tag{6.31}$$

$$2\,NaHSO_3 + Ca(OH)_2 \rightarrow CaSO_3 + Na_2SO_3 + 2\,H_2O \tag{6.32}$$

Nickel-Chromium Wastes

Chromium hydroxide is an amphoteric compound and exhibits minimum solubility in the pH range of 7.5 to 10.0. Effluents from chromium reduction processes should be neutralized to the range of zero solubility (pH 8.5 to 9.0) to minimize the amount of soluble chromium remaining in solution.

It should be noted that if sodium hydroxide is used instead of lime, the chemical cost will be higher, less sludge will be produced, and effluent sulfate concentration will be higher.[15]

6.6.4 Reduction and Flotation Combination

Alternatively, hexavalent chromium can be reduced, precipitated, and floated by ferrous sulfide. By applying ferrous sulfide as a flotation aid to a plating waste with an initial hexavalent chromium concentration of 130 mg/L and total chromium concentration of 155 mg/L, an effluent quality of less than 0.05 mg/L of either chromium species can be achieved if a flotation–filtration wastewater treatment system is used.[15]

Ferrous sulfide acts as a reducing agent at pH 8 to 9 for reduction of hexavalent chromium and then precipitates the trivalent chromium as a hydroxide in one step without pH adjustment.[51,62] So, the hexavalent chromium in the nickel-chromium plating wastewater does not have to be isolated and pretreated by reduction to the trivalent form. The new process is applicable for removal of all heavy metals. All heavy metals other than chromium are removed as insoluble metal sulfides, M(II)S.

$$FeS + M(II)^{2+} \longrightarrow M(II)S + Fe^{2+} \quad (6.33)$$
$$6\,Fe^{2+} + Cr_2O_7^{2-} + 14\,H^+ \longrightarrow 2\,Cr^{3+} + 6\,Fe^{3+} + 7\,H_2O \quad (6.34)$$
$$Cr^{3+} + 3\,OH^- \longrightarrow Cr(OH)_3 \quad (6.35)$$
$$Fe^{3+} + 3\,OH^- \longrightarrow Fe(OH)_3 \quad (6.36)$$

M(II)S, $Cr(OH)_3$, and $Fe(OH)_3$ are all insoluble precipitates, which can be floated by dissolved air flotation (DAF).

This new method can eliminate the potential hazard of excess sulfide in the effluent and the formation of gaseous hydrogen sulfide. In operation, the FeS is added to wastewater to supply sufficient sulfide ions to precipitate metal sulfides that have lower solubilities than FeS. Typical reactions include the following[51,62]:

$$FeS + Ni^{2+} \longrightarrow NiS + Fe^{2+} \quad (6.37)$$
$$FeS + Zn^{2+} \longrightarrow ZnS + Fe^{2+} \quad (6.38)$$
$$FeS + Pb^{2+} \longrightarrow PbS + Fe^{2+} \quad (6.39)$$
$$FeS + Cd^{2+} \longrightarrow CdS + Fe^{2+} \quad (6.40)$$
$$FeS + Cu^{2+} \longrightarrow CuS + Fe^{2+} \quad (6.41)$$
$$FeS + 2\,Ag^+ \longrightarrow Ag_2S + Fe^{2+} \quad (6.42)$$

Ferrous sulfide can also react with metal hydroxide to form insoluble metal sulfide:

$$FeS + M(II)(OH)_2 \longrightarrow Fe(OH)_2 + M(II)S \quad (6.43)$$

Ferrous sulfide itself is also a relatively insoluble compound. Thus, the sulfide ion concentration is limited by its solubility, which amounts to only about 0.02 g/L, and the inherent problems associated with conventional sulfide precipitation are significantly minimized.

The newly developed flotation–filtration process involving the use of ferrous sulfide as a flotation aid offers a distinct advantage in the treatment of nickel-chromium plating wastewater that contains hexavalent chromium, nickel, iron, and other metals.

6.7 CONVENTIONAL REDUCTION–PRECIPITATION SYSTEM

A conventional system for treatment of nickel-chromium plating wastewater involves the use of the following unit processes[37–48]:

1. Neutralization
2. Chromium reduction
3. pH adjustment and hydroxide precipitation
4. Clarification (either sedimentation or DAF)
5. Sludge treatment (filter press and final disposal)

Figure 6.4 shows an example of an existing plating facility and its conventional reduction–precipitation wastewater treatment system in New Britain, TN.[15]

Initially the nickel-chromium plating process is designed to minimize the liquid loading to the waste treatment system. Counterflow rinsing, spray rinsing, and stagnant rinse recovery methods are employed in order to minimize the amount of wastes to be treated and allow as much treatment or retention time in the waste treatment system as is possible.

In the application of the previous chemical methods, a certain amount of steady-state continuity has been built into the system. To accomplish this, initial concentrated alkaline and acid rinse wastewaters are retained after dumping in the waste holding tank [T-91] (Figure 6.4) and acid chromium plating wastewater is stored in the waste holding tank [T-51]. Extremely concentrated chromium plating wastewater from rinse step No. 1 is sent to an evaporation tank [T-40] for

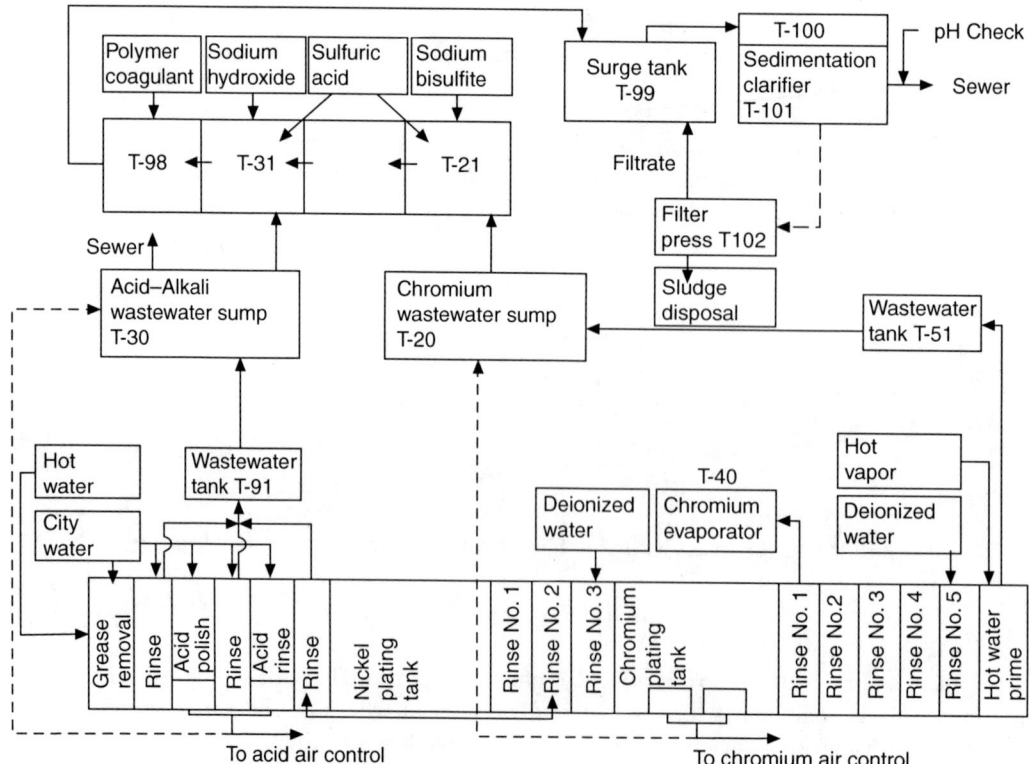

FIGURE 6.4 Conventional reduction–precipitation wastewater treatment system.

chromium recovery. In the case of the wastewater tank [T-51], the waste is slowly bled into the chromium wastewater sump [T-20] to minimize overloading of the total system. The alkaline and acid wastes in [T-.91] are neutralized and slowly bled directly to an acid–alkali wastewater sump [T-30]. It should be noted that the concentrated alkaline wastes are the result of alkaline cleaner replenishment and do not contain heavy metals.

Hexavalent chromium wastes resulting from rinsewater and the concentrated acid bleed accumulate in the chromium waste sump [T-20]. The chromium wastes are then pumped into the chromium treatment module [T-21] for reduction to the trivalent form. This pump is activated only if the oxidation–reduction potential (ORP) and pH are at the proper levels and if the level in the chromium wastewater sump [T-20] is sufficiently high.

Liquid flowing into the chromium treatment module [T-21] is monitored by a pH instrument that controls a feed pump to add the required amount of sulfuric acid from a storage tank. The sulfuric acid is needed to lower the pH to 2.0 to 2.5 for the desired reduction reaction to occur. An ORP instrument controls the injection rate of sodium metabisulfite solution from a metering pump to reduce hexavalent chromium (Cr^{6+}) to the trivalent state (Cr^{3+}).

The acid and alkali wastes are pumped from the acid–alkali wastewater sump [T-30] into the acid–alkali treatment module [T-31]. Metering pumps controlled by pH instruments feed either acid or caustic to the module as required to maintain an acceptable alkalinity for the formation of metal hydroxides prior to discharge to the precipitator consisting of a mixing tank [T-98], a surge tank [T-99], and a sedimentation clarifier [T-101]. The pH is adjusted to a value of 8.5 for optimum metal hydroxide formation and removal.

An ultrasonic transducer is installed on the pH probe mount in the acid–alkali treatment module [T-31]. This prevents fouling of the electrodes and provides a more closely controlled pH in the effluent discharged to the precipitator.

The first step in the precipitator is the addition of polyelectrolyte solution in the flash mix tank [T-98], surge tank [T-99], and then into the slow mix unit [T-100] containing a variable speed mixing paddle. The purpose of this unit is to coagulate and flocculate[53] the metal hydroxide precipitates.

From the slow mix unit [T-100], the waste flows into the lamellar portion of the sedimentation clarifier [T-101].[54,55] The lamella in the clarifier concentrates the metal hydroxide precipitates. Clarified effluent can be discharged to the sewer.

Concentrated metal hydroxide sludge is pumped from the clarifier to a polypropylene plate filter press [T-102]. The plate filter press[56] is of sufficient capacity without any buildup in the lamellar portion of the unit. This also prevents any overflow of precipitate to the sewer system. The metal hydroxides form a dense sludge cake suitable for disposal in an approved landfill. The liquid effluent from the plate filter is returned to the surge tank [T-99].

A sampling station is provided on the rear exterior wall of the facility for flow measurement and monitoring of the effluent stream.

6.8 MODIFIED REDUCTION–FLOTATION WASTEWATER TREATMENT SYSTEM

A modified reduction–flotation system (Figure 6.5) is very similar to the existing conventional reduction–precipitation system (Figure 6.4), except that a DAF clarifier [T-101F] is used for clarification[15,57] instead of using a conventional sedimentation clarifier (Tank T-101, Figure 6.4).

The flotation system consists of four major components: air supply, pressurizing pump, air dissolving tube, and flotation chamber.[57,58] According to Henry's Law, the solubility of gas (such as air) in aqueous solution increases with increasing pressure. The influent feedstream can be saturated at several times atmospheric pressure, 1.8 to 6 kg/cm^2 (25 to 85 psig), by a pressurizing pump. The pressurized feedstream is held at this high pressure for about 0.5 min in an air dissolving tube (i.e., a pressure vessel) designed to provide sufficient time for dissolution of air into the stream to be

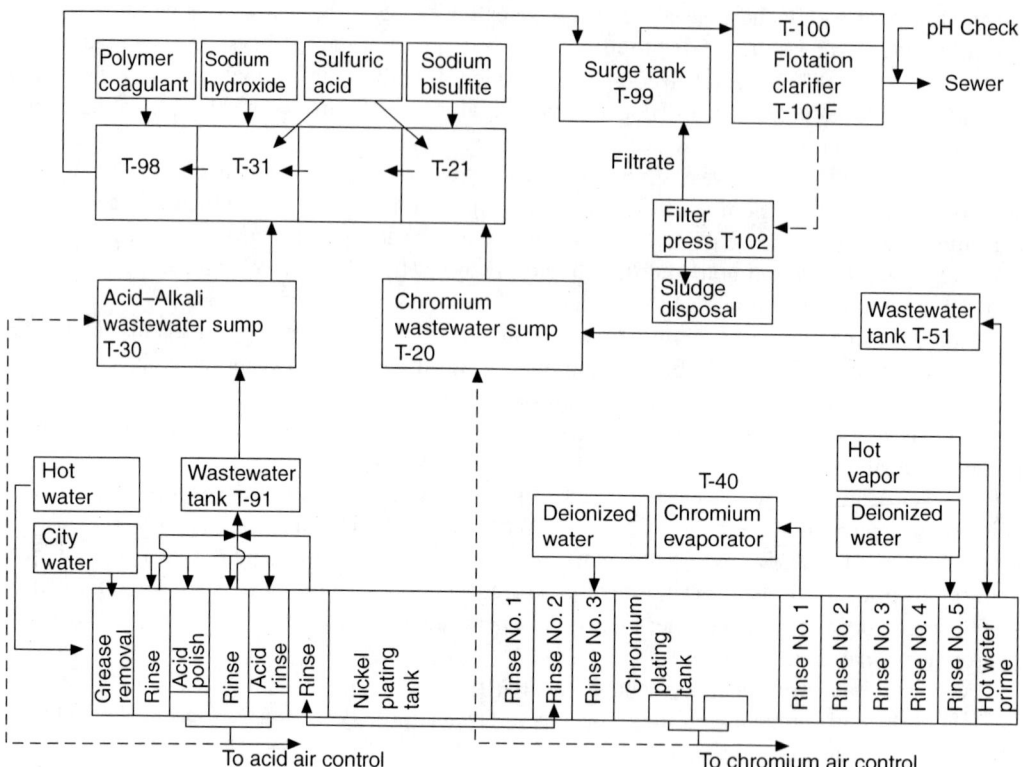

FIGURE 6.5 Modified reduction–precipitation wastewater treatment system.

treated. From the air dissolving tube, the stream is released back to atmospheric pressure in the flotation chamber.[15,57]

Most of the pressure drop occurs after a pressure reducing valve and in the transfer line between the air dissolving tube and the flotation chamber so that the turbulent effects of the depressurization can be minimized. The sudden reduction in pressure in the flotation chamber results in the release of microscopic air bubbles (average diameter 50 μm or smaller) which nucleate on suspended or colloidal particles in the process water in the flotation chamber. This results in agglomeration that, due to the entrained air, gives a net combined specific gravity less than that of water, causing the flotation phenomenon. The vertical rising rate of air bubbles ranges between 15 and 60 cm/min (0.5 to 2.0 ft/min). The floated material rises to the surface of the flotation chamber to form a floated layer. Specially designed flight scrapers or other skimming devices continuously remove the floated material. The surface sludge layer can in certain cases reach a thickness of many inches and can be relatively stable for a short period. The layer thickens with time, but undue delays in removal will cause a release of particulates back to the liquid. Clarified subnatant water (effluent) is drawn off from the flotation chamber and either recovered for reuse or discharged.

The retention time in the flotation chamber is usually about 3 to 5 min, depending on the characteristics of the process water and the performance of the flotation unit. The process effectiveness depends upon the attachment of air bubbles to the particles to be removed from the process water.[57] The attraction between the air bubbles and particles is primarily a result of the particle surface charges and bubble size distribution. The more uniform the distribution of water and microbubbles, the shallower the flotation unit can be.

A high-rate DAF unit with only 3 min of retention time can treat water and wastewater at an overflow rate of 2.4 L/sec/m² (3.5 gal/min/ft²) for a single unit and up to 7.2 L/sec/m² (10.5 gal/min/ft²)

Nickel-Chromium Wastes

for triple stacked units. The comparison between a flotation clarifier and a settler shows the following[59,60]:

1. The DAF unit floor space requirement is only 15% of the settler.
2. The DAF unit volume requirement is only 5% of the settler.
3. In DAF, higher biosolids densities are obtained than in sedimentation. Even in shallow flotation clarifiers a satisfactory biosolids density is attainable.
4. The degrees of clarification of both clarifiers are the same with the same flocculating chemical addition.
5. The operational cost of the DAF clarifier is slightly higher than that for the settler, but this is offset by the considerably lower cost of the installation's financing.
6. DAF clarifiers are mainly prefabricated in stainless steel for erection cost reduction, corrosion control, better construction flexibility, and possible future upgrades, contrary to *in situ* constructed heavy concrete sedimentation tanks.

It should be noted that the chemical reactions of the conventional reduction–precipitation system (Figure 6.4) and the modified reduction–flotation system are identical.

Comparatively, the modified reduction–flotation system will have lower annual total cost (amortized capital cost plus O&M cost) and will require less space, because the flotation unit is very shallow in depth and thus can be elevated. It is expected, however, that the treatment efficiency of the modified system will be higher due to the fact that the DAF clarifier can separate not only the suspended solids but also organics such as oil and grease, detergent, and so on.[57,58,61] Conventional sedimentation clarifiers can separate only insoluble suspended solids.

6.9 INNOVATIVE FLOTATION–FILTRATION WASTEWATER TREATMENT SYSTEMS

6.9.1 FLOTATION–FILTRATION SYSTEM USING CONVENTIONAL CHEMICALS

There are two innovative flotation–filtration wastewater treatment systems that are technically feasible for the treatment of the nickel-chromium plating wastewater.

The first system, shown in Figure 6.6, is identical to the conventional reduction–precipitation in chemistry (i.e., neutralization, chromium reduction, pH adjustment, metal hydroxide precipitation, and so on). However, a flotation–filtration clarifier (Tank T101SF, as shown in Figure 6.6) is used. The unit consists of rapid mixing, flocculation, high-rate DAF, and sand filtration.[15,57]

The treatment efficiency of this system (Figure 6.6) is much higher than that of the conventional reduction–precipitation wastewater treatment system (Figure 6.4).[15]

6.9.2 FLOTATION–FILTRATION SYSTEM USING INNOVATIVE CHEMICALS

Another innovative flotation–filtration wastewater treatment system adopts the innovative use of the chemical ferrous sulfide (FeS), which reduces the hexavalent chromium and allows separation of chromium hydroxide, nickel hydroxide, and ferric hydroxide in one single step at pH 8.5. Figure 6.7 illustrates the entire system. Again, a DAF–filtration clarifier plays the most important role in this wastewater treatment system.

It is seen from Figure 6.7 that this system is much simpler, more cost-effective, and easier to operate in comparison with all other process systems discussed earlier. The treatment efficiency of the new flotation–filtration system is expected to be higher than that of the conventional reduction–precipitation system. The new flotation–filtration system also requires much less land space.[15]

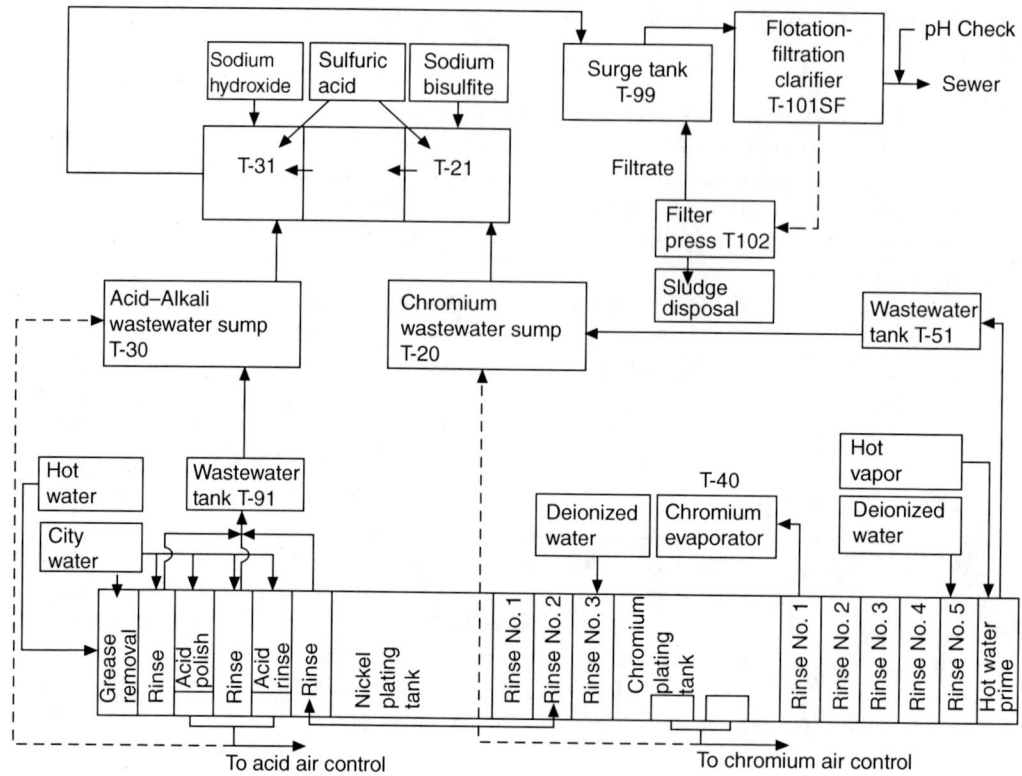

FIGURE 6.6 Innovative flotation–filtration wastewater treatment system using conventional chemicals.

6.9.3 FLOTATION–FILTRATION SYSTEMS

6.9.3.1 Combined Flotation–Filtration Unit

A combined flotation–filtration unit, shown in Figure 6.8, is an advanced water clarification system, using a combination of chemical flocculation, DAF, and rapid sand filtration in one unit. The average processing time from start to finish is less than 15 min.[15,57,58]

Its unique compact and efficient design is made possible by the use of the principle of zero velocity eliminating internal water turbulence (see below). The flocculated water thus stands still in the flotation tank for optimum clarification. The unit is complete with automatic backwash filter in which dirty backwash water is recycled back to the unit inlet for reprocessing. The average waste flow from the process is less than 1.0% of the incoming raw water.

The flotation unit maximum loading is 2.1 L/sec/m² (3.1 gal/min/ft²). The maximum filtration rate is 1.7 L/sec/m² (2.5 gal/min/ft²). Each filter compartment is backwashed at or more than 10.2 L/sec/m² (15 gal/min/ft²) during the backwash operation. The single-medium backwash filter consists of 28 mm (11 in.) high-grade silica sand. The effective size and uniformity coefficient for the sand are 0.35 mm and 1.55, respectively.

The following paragraphs briefly describe how the flotation–filtration unit shown in Figure 6.8, works.[15,57,58]

The influent raw water or wastewater enters the unit at the center near the bottom [1] and flows through a hydraulic rotary joint [2] and an inlet distributor [3] into the rapid mixing section of the slowly moving carriage. The entire moving carriage consists of a rapid mixer [3], flocculator [4], backwash pumps [5 & 6] and sludge discharge scoop [7]. To flocculate colloids and suspended solids, chemicals [8] are added at the inlet [1].

Nickel–Chromium Wastes

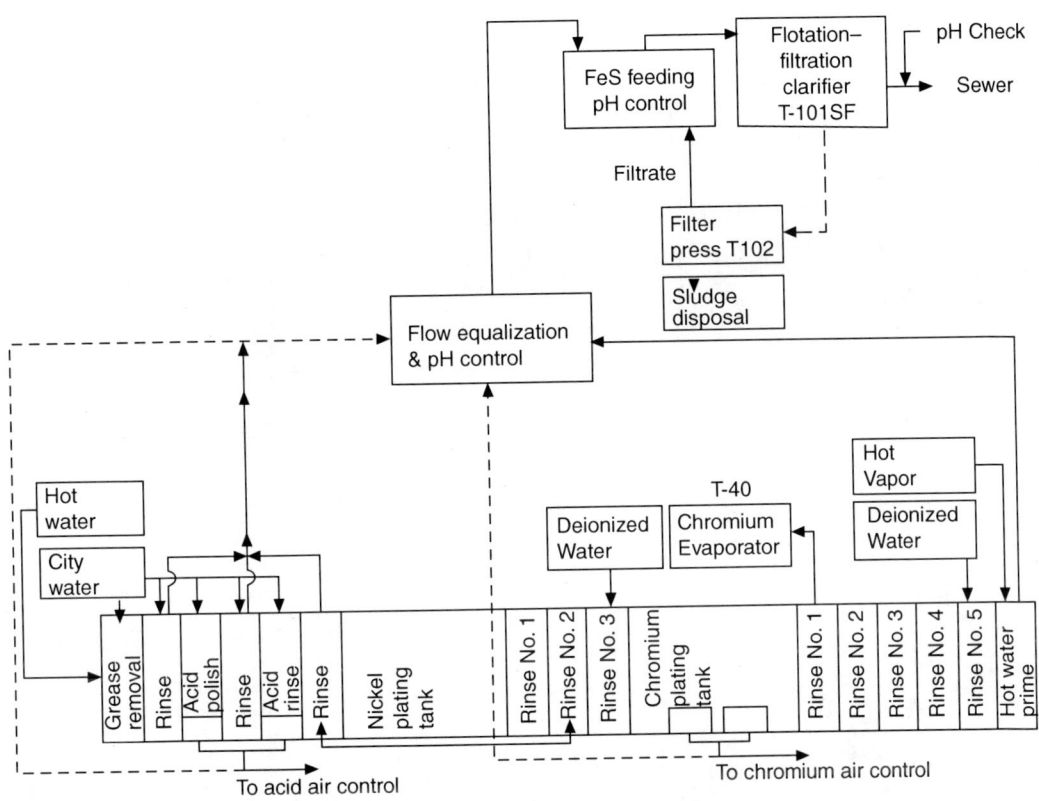

FIGURE 6.7 Innovative flotation–precipitation wastewater treatment system using an innovative chemical.

From the rapid mixing section [3] water enters the hydraulic flocculator [4], gradually building up the flocs by gentle mixing. The flocculated water moves from the flocculator into the flotation tank [9], clockwise, with the same velocity as the entire carriage including flocculator simultaneously moves counterclockwise. The outgoing flocculator effluent velocity is compensated by the opposite velocity of the moving carriage, resulting in a zero horizontal velocity of the flotation tank influent. The flocculated water thus stands still in the flotation tank for optimum clarification.

At the outlet of the flocculator [10], a small percentage of chemically pretreated raw water with microscopic air bubbles is added to the flotation tank [9] in order to float the insoluble flocs and suspended matter to the water surface. The floating sludge accumulated on the water surface is scooped off by a sludge discharge scoop [7] and discharged into the center sludge collector [14] where there is a sludge outlet [15] to an appropriate sludge treatment facility.

The small microscopic air bubbles are the product of raw water pressurized to 4 to 6 kg/cm^2 (55 to 85 psi) in the air dissolving tube (ADT) [32]. Water enters the ADT tangentially [33] at one end and is discharged at the opposite end. During its short passage, the water cycles inside the tube and passes repeatedly by an insert fed by compressed air. Very thorough mixing under pressure dissolves the air into the water.

The bottom of the unit is composed of multiple sections of sand filter [11] and clearwell [12]. The clarified flotation effluent passes through the sand filter downward and enters the clearwell through the circular hole underneath each sand filter section; the filter effluent then enters the center portion of the clearwell where there is an outlet for the effluent.[13]

For backwashing of the sandbeds, two pumps [5 & 6] are placed on the carriage. One pump [5] is at the center of the clearwell for pumping washing water during the backwash cycle through the individual clearwell compartments. The turbid backwash water is collected in a traveling hood [16],

254 Waste Treatment in the Metal Manufacturing, Forming, Coating, and Finishing Industries

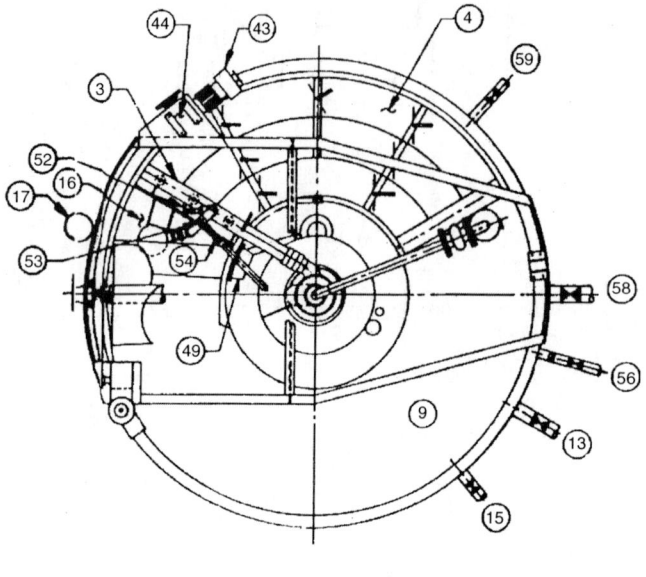

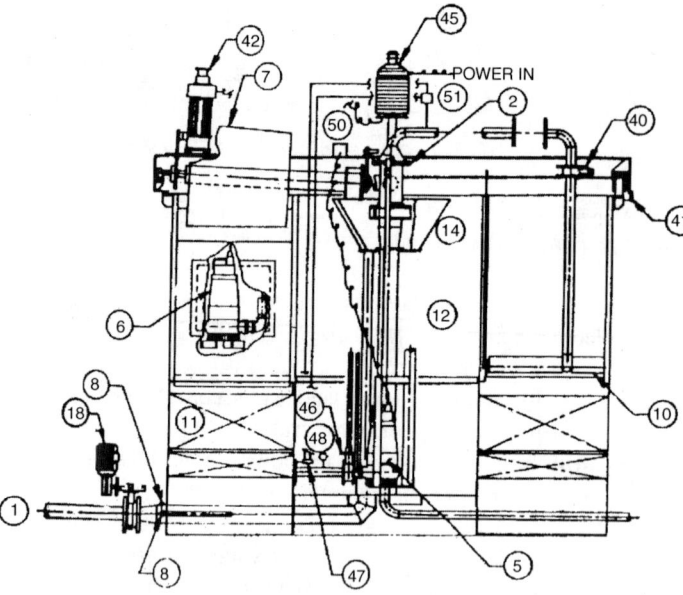

1. Raw water inlet
2. Hydraulic rotary joint
3. Rapid mixing section
4. Flocculator
5. Backwash washing pump
6. Backwash recycle pump
7. Sludge discharge scoop
8. Chemical lines
9. Flotation tank
10. Pressure release manifold
11. Sand filter (silica sand 11 in. depth)
12. Clear well
13. Clear well effluent outlet (gate valve)
14. Center sludge collector
15. Floated sludge outlet (gate valve)
16. Backwash hood area
17. Level control
18. Inlet motorized butterfly valve
19. Clear well electrodes
20. Control panel
21. Raw water inlet manual valve
22. Inlet flow meter
23. Inlet flow meter
24. Chemical pump discharge gate valve
25. Chemical pump
26. Pressure pump suction gate valve
27. Pressure pump discharge gate valve
28. Pressure pump
29. Pressure gate valve
30. Pressure gage 0–100 psi
31. Pressure gate valve
32. Safety valve set to 100 psi
33. Air dissolving tube
34. Air dissolving tube inlet orifice
35. Air compressor
36. Air filter
37. Air regulator
38. Air rotometer
39. Gate valve
40. Gate valve
41. Pressure reducing butterfly valve
42. Limit switch
43. Scoop drive motor
44. Carriage drive motor
45. Carriage drive wheel
46. Electrical rotary contact
47. Butterfly valve
48. Backwash cylinder
49. Pressure gage (0–100 psi)
50. Backwash shoe
51. Delay timer solenoid valve
52. Gate valve
53. Pressure gage (0–100 psi)
54. Gate valve
55. Clear well site tube
56. Clear well drain (gate valve)
57. Bleed off vent
58. Pressurized inlet (gate valve)
59. Pressurized suction pipe (gate valve)

FIGURE 6.8 Top and side views of the flotation–filtration unit. (Taken from Krofta, M. and Wang, L.K., Flotation Engineering, Technical Manual Lenox/1-06-2000/368, Lenox Institute of Water Technology, Lenox, MA, 2000.)

where the second backwash pump [6] collects the water and discharges it into the rapid mix inlet section [3] for reprocessing.

The flotation–filtration unit can be either manually operated or completely automated with a level control [17] that operates the inlet flow valve [18]. Filter backwashing can also be automated by a timer or head loss control [19].

6.9.3.2 Separate Flotation and Filtration Units

It is important to note that all flotation clarifiers[63,64] may be used for treatment of nickel-chromium plating wastes regardless of their shapes (rectangular or circular) or manufacturers. A filtration unit is an optional step for final polishing. The treatment efficiency of separate flotation and filtration units[65] will be similar to that of a combined flotation–filtration unit (Figure 6.8).

The authors of this chapter are introducing a modern technology involving the use of flotation and filtration for treating nickel-chromium plating wastes. The authors are not endorsing any manufacturer's products.

6.10 SUMMARY

Waste reduction methods reduce the amount of waste generated and also the amount subject to environmental regulations. Hence, these practices can help an electroplating shop comply with requirements and save money. The shop's owner or manager must be committed to waste reduction and pass that commitment on to the employees.

Technologies exist for capturing and reclaiming metal finishing waste, including rinsewaters. It is important to treat each recovery system with as much care as the plating baths. Regular maintenance and the use of trained operators will help ensure that the recovery system performs at its optimum design capacity. The recovery and reuse of the chemical solutions will add dollars into the metal finisher's pockets.

All four wastewater treatment systems introduced in this chapter are technically feasible for treating nickel-chromium plating wastewater in order to meet the maximum permissible concentrations shown in Table 6.1 for industrial wastewater discharge into a municipal sewerage system[15] or Table 6.2 for discharge to surface waters.[21]

TABLE 6.1
Maximum Permissible Concentrations of Industrial Wastewater Discharge into Municipal Systems

Parameter	Maximum Permissible Concentration (mg/L)
Arsenic	0.1
Boron	10.0
Barium	0.5
Cadmium	0.2
Copper	0.1
Cyanide	0.5
Lead	1.0
Mercury	0.5
Nickel	0.5
Silver	0.03
Chromium (total)	0.5

Continued

TABLE 6.1 (continued)

Parameter	Maximum Permissible Concentration (mg/L)
Vanadium	0.5
Zinc	0.5
Chloroform	1.0
BOD	1000
TSS	1000
COD	1500
Oil and grease (nonmineral)	300
Oil and grease (mineral)	100
Chlorinated hydrocarbons	0.02
Phenolic compounds	1.0
pH	5.5–9.5
Temperature	55.5°C

BOD, biochemical oxygen demand; COD, chemical oxygen demand; TSS, total suspended solids.

Source: Krofta, M. and Wang, L.K., *Design of Innovative Flotation–Filtration Wastewater Treatment Systems for a Nickel-Chromium Plating Plant*, U.S. Department of Commerce, National Technical Information Service, Springfield, VA, Technical Report PB-88-200522/AS, January 1984.

TABLE 6.2
Maximum Permissible Concentrations of Electroplating Wastewater Discharge to Surface Waters

Parameter	Maximum Permissible Value (mg/L)
pH	7–10
TSS	25
Oil and grease	10
Arsenic	0.1
Cadmium	0.1
Chromium (hexavalent)	0.1
Chromium (total)	0.5
Copper	0.5
Lead	0.2
Mercury	0.01
Nickel	0.5
Silver	0.5
Zinc	2
Total metals	10
Cyanides (free)	0.2
Fluorides	20
Trichloroethane	0.05
Trichloroethylene	0.05
Phosphorus	5

TSS, total suspended solids.
Source: See Table 6.1.

TABLE 6.3
Characteristics of Typical Wastewater Discharge of Conventional Reduction–Precipitation System[a]

Characteristic	Concentration (mg/L)	Amount of Pollutant Discharged (kg/d)
Chromium, C_r^{6+}	<0.006	<0.0003
Chromium, total	0.79	0.04
Iron	0.15	0.008
Nickel	0.21	0.011
TSS	<0.10	<0.51
Settleable solids	<0.1 mL/L	
pH	7.6–8.2	

TSS, total suspended solids.
[a] See Figure 6.4.

Table 6.3 shows the characteristics of a typical effluent discharge from a conventional reduction–precipitation system. The effluent quality meets industrial pretreatment requirements.

The modified reduction–flotation wastewater treatment system (Figure 6.5) will be very attractive if all or most of an existing wastewater treatment facilities are to be reused. The high-rate DAF clarifier is a very low-cost clarification unit.

The treatment efficiencies of the two innovative flotation–filtration wastewater treatment systems (Figures 6.6 and 6.7) are expected to be higher than those of the conventional reduction–precipitation system.

The innovative flotation–filtration wastewater treatment system (Figure 6.6) using conventional chemicals has the highest flexibility and best performance. When desirable, the innovative chemical FeS or equivalent can also be used.

Another innovative flotation–filtration wastewater treatment system using FeS (Figures 6.7 and 6.8) is highly recommended if a totally new system is to be designed and installed for treatment of nickel-chromium plating wastewater. This system is extremely compact, easy to operate, and cost-effective. Treatment efficiency is also excellent.

All flotation clarifiers[63] may be used for the treatment of nickel-chromium plating wastes regardless of their shapes (rectangular or circular) or manufacturers. A filtration unit is an optional step for final polishing. The treatment efficiency of separate flotation and filtration units[64] will be similar to that of a combined flotation–filtration unit (Figure 6.8).

REFERENCES

1. Elsevier Inc., *Metal Finishing Guidebook & Directory*, Elsevier, New York, 2005.
2. Dini, J.W., *Electrodeposition — The Materials Science of Coating and Substrates*, Noyes Publishing, Park Ridge, NJ, 1993.
3. American Society for Metals, *Surface Engineering*, ASM Metals Handbook, American Society for Metals, Vol. 5, Materials Park, OH, 1994.
4. Graves, B.A., *Hardware for Plating Hand-Tools*, Danaher Tool Group, Products Finishing Online, 2005. Available at http://www.pfonline.com/articles/049903.html.
5. Parthasaradhy, N.V., *Practical Electroplating Handbook*, Prentice-Hall, New York, 1989.
6. Durney, L.J., Ed., *The Electroplating Engineering Handbook*, American Society of Electroplated Plastics, 4th ed., Washington, DC, 1984.
7. Brenner, A., *Electrodeposition of Alloys*, Academic Press, New York, 1991.
8. Mandich, N.V., Important practical considerations in chromium plating—Part IV, *Met. Finish.*, 97, 9, 1999.

9. Mandich, N.V., Important practical considerations in chromium plating—Part V, *Met. Finish.*, 97, 10, 1999.
10. Mandich, N.V., Important practical considerations in chromium plating—Part VI, *Met. Finish.*, 98, 3, 2000.
11. Mandich, N.V. and Vyazovikina, N.V., Kinetics and mechanisms of the chromium anodic dissolution in the chromium plating solution in the transpassive range, *47th Meeting of International Society for Electrochemistry*, Veszprem, Hungary, September 1996.
12. Mandich, N.V., Chemistry and theory of chromium deposition—Part I: Chemistry, *Plating Surf. Finish.*, 84, 5, 1997.
13. Mandich, N.V., Chemistry and theory of chromium deposition—Part II: Theory of deposition, *Plating Surf. Finish.*, 84, 6, 1997.
14. Finishing.com, *Equipment Needed to Get Started in Chroming Business*, The home page of the finishing industry, 2005. Available at http://www.finishing.com/197/26.shtml.
15. Krofta, M. and Wang, L.K., *Design of Innovative Flotation–Filtration Wastewater Treatment Systems for a Nickel-Chromium Plating Plant*, U.S. Department of Commerce, National Technical Information Service, Springfield, VA, Technical Report PB-88-200522/AS, January 1984.
16. UNEP, *Environmental Aspects of the Metal Finishing Industry: A Technical Guide*, United Nations Environment Program, Paris, 1992.
17. U.S. EPA, Monitoring Industrial Wastewater, U.S. EPA, National Technical Information Service, Springfield, VA, 1973.
18. U.S. EPA, Process Design Manual for Sludge Treatment and Disposal, Report EPA 6251-174-006, U.S. EPA, National Technical Information Service, Springfield, VA, 1974.
19. SFDA, National Primary Drinking Water Regulations, Safe Drinking Water Act, Update March 2002.
20. CDHS, Hazardous Waste Generated by Metal Refinishing Facilities, Fact Sheet, California Department of Health Services, Toxic Substances Control Program, Alternative Technology Division, April 1990.
21. World Bank, Electroplating, in *Pollution Prevention Abatement Handbook*, World Bank Group, July 1998.
22. DCNR, Metal Recovery Technologies for the Metal Finishing Industry, Fact Sheet, Office of Waste Reduction Services, Environmental Services Division, State of Michigan, Departments of Commerce and Natural Resources (DCNR), Lansing, MI, November 1993.
23. Gouthro, R.P. and Vaz, L., Recovery and purification of nickel salts and chromic acid using the RECOFLO system, Eco-Tec, Technical Paper 145, presented at the Metal Finisher's Association of India, Mumbai, India, September 1999. Available at http://www.eco-tec.com/main/electroplate.htm.
24. Lowenheim, F.A., *Electroplating*, McGraw-Hill, New York, 1978.
25. Montgomery, D., *Light Metals Finishing Process Manual*, American Electroplaters & Surface Finishers Society, Orlando, FL, 1990.
26. Nordic Council of Ministers, *Possible Ways of Reducing Environmental Pollution from the Surface-Treatment Industry*, Oslo (1993).
27. Kushner, J.B. and Kushner, A.S., *Water and Waste Control for the Plating Shop*, Hanser Gardner, 3rd ed., (1994).
28. Wang, L.K., Shammas, N.K., Williford, C., Chen, W.-Y., and Sakellaropoulos, G.P., Evaporation processes, in *Advanced Physicochemical Treatment Processes*, Wang, L.K., Hung, Y.T., and Shammas, N.K., Eds., The Humana Press, Totowa, NJ, 2006, pp. 549–580.
29. Chian, E.S.K., Chen, J.P., Sheng, P.X., Ting, Y.P., and Wang, L.K., Reverse osmosis technology for desalination, in *Advanced Physicochemical Treatment Technologies*, Wang, L.K., Hung, Y.T., and Shammas, N.K., Eds., The Humana Press, Totowa, NJ, 2007, pp. 327–366.
30. Golomb, A., *Application of Reverse Osmosis to Electroplating Waste Treatment*, Ontario Research Foundation, Ontario, Canada, 1995.
31. Chen, J.P., Yang, L., Ng, W.J., Wang, L.K., and Thong, S.L., Ion exchange, in *Advanced Physicochemical Treatment Processes*, Wang, L.K., Hung, Y.T., and Shammas, N.K., Eds., The Humana Press, Totowa, NJ, 2006, pp. 261–292.
32. CAT, *Treatment of Electroplating Waste*, Center for Advanced Technology, Indore, India, 2005. Available at http://www.cat.ernet.in/technology/accel/indus/wsa/ctl/chem_effulent.html.
33. Marquínez, R., Pourcelly, G., Bauer, B., Ochoa, J.R., López, R., Viala, S., Mahiout, A., and Leinonen, H., Chromic acid recycling from rinsewater in galvanic plants by electro-electrodialysis, in *Waste Treatment, and Clean Technology*, The Global Symposium on Recycling, REWAS '04, Vol. II, Gaballah, I., Mishra, B., Solosabal, R., and Tanaka, M., Eds., Madrid, Spain, September 26–29, 2004.

34. Chen, J.P., Chang, S.Y., and Hung, Y.T., Electrolysis, in *Physicochemical Treatment Processes*, Wang, L.K., Hung, Y.T., and Shammas, N.K., Eds., The Humana Press, Totowa, NJ, 2005, pp. 359–378.
35. Van De Putte, B. Verhaege, M., Brughmans, S., and Vanrobaeys, D., Electrochemical recovery of nickel from electroplating effluents by an electro-active mixed bed resin cathode, in *Waste Treatment, and Clean Technology*, The Global Symposium on Recycling, REWAS '04, Vol. II, Gaballah, I., Mishra, B., Solosabal, R., and Tanaka, M., Eds., Madrid, Spain, September 26–29, 2004.
36. Wang, L.K., Hung, Y.T., and Shammas, N.K., Eds., *Advanced Physicochemical Treatment Processes*, The Humana Press, Totowa, NJ, 2006, 690 p.
37. Ayers, D.M., Davis, A.P., and Gietka, P.M., Removing Heavy Metals from Wastewater, Report, University of Maryland, Engineering Research Center, 1994.
38. NETCSC, Industrial Pretreatment and Hazardous Material Recognition for Small Communities, The National Environmental Training Center for Small Communities, West Virginia University, Morgan Town, WV, 1996.
39. U.S. EPA, Industrial Waste Treatment, U.S. EPA, Office of Water Programs, V1 and V2, 1995, 1996.
40. Wang, L.K., Hung, Y., Lo, H.H., and Yapijakis, C., Eds., *Handbook of Industrial and Hazardous Wastes Treatment*, 2nd ed., Marcel-Dekker, New York, 2004, 1345 p.
41. Guyer, H.H., *Industrial Processes and Waste Stream Management*, Wiley, New York, 1998.
42. Cheremisinoff, P.N., *Handbook of Water and Wastewater Treatment Technology*, Marcel-Dekker, 1994.
43. Cushnie, G.C., Jr, *Electroplating Wastewater Pollution Control Technology*, Noyes Publishing, Park Ridge, NJ, 1985.
44. Patterson, J.W., *Industrial Wastewater Treatment Technology*, 2nd ed., Butterworth, Boston, MA, 1985.
45. Johannes, R.D. et al., Electroplating/metal finishing wastewater treatment: Practical design guidelines, *Proc. 43rd Purdue Industrial Wastes Conference*, West Lafayette, IN, 1988.
46. Pattanayak, J., Mandich, N.V., Mondal, K., Wiltowski, T., and Lalvani, S.B., Removal of iron and nickel from solutions by applications of electric field, *Environ. Technol.*, 20, 317, 1999.
47. Pattanayak, J., Mandich, N.V., Mondal, K., Wiltowski, T., and Lalvani, S.B., Recovery of metallic impurities from plating solutions by electromigration, *Met. Finish.*, 98, 3, 2000.
48. Medina, B.Y., Torem, M.L., and De Mesquita, L.M.S., Removal of chromium III from liquid effluent streams by precipitate flotation, in *Waste Treatment, and Clean Technology*, The Global Symposium on Recycling, REWAS '04, Vol. II, Gaballah, I., Mishra, B., Solosabal, R., and Tanaka, M., Eds., Madrid, Spain, September 26–29, 2004.
49. Goel, R.K., Flora, J.R.V., and Chen, J.P., Flow equalization and neutralization, in *Physicochemical Treatment Processes*, Wang, L.K., Hung, Y.T., and Shammas, N.K., Eds., The Humana Press, Totowa, NJ, 2005.
50. Chamberlain, N.S. and Day, R.V., Technology of chrome reduction with sulfur dioxide, *Proc. Eleventh Purdue Industrial Waste Conference*, West Lafayette, IN, 129, 1956.
51. Wang, L.K., Vaccari, D.A., Li, Y., and Shammas, N.K., Chemical precipitation, in *Physicochemical Treatment Processes*, Wang, L.K., Hung, Y.T., and Shammas, N.K., Eds., The Humana Press, Totowa, NJ, 2005, pp. 141–198.
52. Wang, L.K., Hung, Y.T., and Shammas, N.K., Eds., *Physicochemical Treatment Processes*, The Humana Press, Totowa, NJ, 2005, 723 p.
53. Shammas, N.K., Coagulation and flocculation, in *Physicochemical Treatment Processes*, Wang, L.K., Hung, Y.T., and Shammas, N.K., Eds., The Humana Press, Totowa, NJ, 2005, pp. 103–140.
54. Shammas, N.K., Kumar, I.J., and Chang, S.Y., Sedimentation, in *Physicochemical Treatment Processes*, Wang, L.K., Hung, Y.T., and Shammas, N.K., Eds., The Humana Press, Totowa, NJ, 2005, pp. 379–430.
55. U.S. EPA, Process Design Manual for Suspended Solids Removal, Report EPA 625/1-75-003a, U.S. EPA, National Technical Information Service, Springfield, VA, 1975.
56. Wang, L.K., Shammas, N.K. and Hung, Y.T., Eds., *Biosolids Engineering and Management*, The Humana Press, Totowa, NJ, 2008, 788 p.
57. Wang, L.K., Fahey, E.M., and Wu, Z., Dissolved air flotation, in *Physicochemical Treatment Processes*, Wang, L.K., Hung, Y.T., and Shammas, N.K., Eds., The Humana Press, Totowa, NJ, 2005, pp. 431–500.
58. Krofta, M. and Wang, L.K., Flotation Engineering, Technical Manual Lenox/1-06-2000/368, Lenox Institute of Water Technology, Lenox, MA, 2000.
59. Shammas, N. and DeWitt, N., Flotation: a viable alternative to sedimentation in wastewater treatment plants, *Water Environment Federation 65th Annual Conf., Proc. Liquid Treatment Process Symposium*, New Orleans, LA, September 20–24, 1992, pp. 223–232.

60. Krofta, M. and Wang, L.K., Flotation Replaces Sedimentation in Water and Effluent Clarification, Report, Krofta Engineering Corp., Lenox, MA, 1990.
61. Uribe-Salas, A., Tinoco-Elvir, J.F., Pérez-Garibay, R., and Nava-Alonso, F., Experimental study on the flotation of Pb^{2+} and Ni^{2+} with sodium dodecylsulfate, in *Waste Treatment, and Clean Technology*, The Global Symposium on Recycling, REWAS '04, Vol. II, Gaballah, I., Mishra, B., Solosabal, R., and Tanaka, M., Eds., Madrid, Spain, September 26–29, 2004.
62. Wang, L.K., Kurylko, L., and Wang, M.H.S. Water and Wastewater Treatment, U.S. Patent 5,240,600, August 31, 1993.
63. Dissolved Air Flotation Equipment—Annual 2005–2006 Buyer's Guide, *Water & Waste Digest*, 45, 73–76, 2005.
64. Flotation Equipment—2005 Buyer's Guide, *Environ. Protect.*, 16, 79, 2005.
65. Olson, S., Dissolved air flotation—new applications, *J. NEWWA*, 133–151, 1994.

7 Waste Treatment and Management in the Coil Coating Industry

Lawrence K. Wang and Nazih K. Shammas

CONTENTS

7.1	General Description of Coil Coating Industry and Operations	262
7.2	Cleaning Operation of Coil Coating	263
	7.2.1 Mild Alkaline Cleaning	263
	7.2.2 Strong Alkaline Cleaning	264
	7.2.3 Acid Cleaning	264
	7.2.4 Special Cleaning	264
7.3	Conversion Coating Process of Coil Coating	264
	7.3.1 Chromate Conversion Coatings	265
	7.3.2 Phosphate Conversion Coatings	265
	7.3.3 Complex Oxide Conversion Coating	266
	7.3.4 No-Rinse Conversion Coatings	266
7.4	Painting Operation of Coil Coating	267
7.5	Subcategories of the Coil Coating Industry	267
	7.5.1 General Description of Subcategorization	267
	7.5.2 Coil Coating on Steel Subcategory	268
	7.5.3 Coil Coating on Zinc Coated Steel (Galvanized Steel) Subcategory	268
	7.5.4 Coil Coating on Aluminum Subcategory	269
7.6	Wastewater Characterization of the Coil Coating Industry	269
	7.6.1 Effluent Characteristics of Coil Coating on Steel Operation	269
	7.6.2 Effluent Characteristics of Coil Coating on Zinc Coated Steel (Galvanized Steel) Operation	269
	7.6.3 Effluent Characteristics of Coil Coating on Aluminum Operation	277
7.7	Plant-Specific Effluent Characterization Data	278
	7.7.1 Plant A: Coating Cold Rolled Steel and Galvanized Steel	279
	7.7.2 Plant B: Coating Both Cold Rolled Steel and Galvanized Steel	279
	7.7.3 Plant C: Coating Aluminum and Other Metals	279
7.8	Coil Coating Effluent Treatment Technologies	280
	7.8.1 Ion Exchange	281
	7.8.2 Electrochemical Chromium Regencration	282
	7.8.3 Oil Skimming	282
	7.8.4 Chromium Reduction and Chemical Precipitation	282
	7.8.5 Cyanide Destruction	283
	7.8.6 Oil–Water Separation, Biological Treatment, Powdered Activated Carbon Adsorption, and Clarification	283

7.8.7 Granular Bed Filtration and Granular Activated Carbon Filtration 284
7.8.8 Membrane Processes 286
7.8.9 Other Water–Solids Separation Technologies 286
7.8.10 Full-Scale Wastewater Treatment Case History: Steel Subcategory 289
7.8.11 Full-Scale Wastewater Treatment Case History: Galvanized Subcategory 291
7.8.12 Full-Scale Wastewater Treatment Case History: Aluminum Subcategory 292
7.9 Wastewater Treatment Levels versus Costs 293
    7.9.1 BPT Level Treatment 293
        7.9.1.1 Suggested BPT 293
        7.9.1.2 System Component of the Suggested BPT 293
        7.9.1.3 Unit Cost of the Suggested BPT 294
    7.9.2 BAT Level of Treatment 294
        7.9.2.1 Suggested BAT 294
        7.9.2.2 System Components of the Suggested BAT 294
        7.9.2.3 Unit Cost of the Suggested BAT 294
7.10 Multimedia Waste Management in the Coil Coating Industry 294
    7.10.1 Air Pollution Control 294
    7.10.2 Water Pollution Control 295
    7.10.3 Solid and Hazardous Wastes Management and Disposal 296
    7.10.4 Waste Minimization and Cleaner Production Alternatives
        for Roll and Coil Coating 296
7.11 Coil Coating Industry Liquid Effluent Limitations, Performance
Standards, and Pretreatment Standards 297
    7.11.1 Effluent Limitations, Performance Standards, and Pretreatment
        Standards of the Steel Basis Material Subcategory 297
    7.11.2 Effluent Limitations and Performance Standards of the Galvanized
        Basis Material Subcategory 298
    7.11.3 Effluent Limitations and Performance Standards of the Aluminum
        Basis Material Subcategory 300
    7.11.4 Effluent Limitations, Performance Standards, and Pretreatment
        Standards of Canmaking Subcategory 302
7.12 Technical Terminologies of Coil Coating Operations and Pollution Control 304
References 306

7.1 GENERAL DESCRIPTION OF COIL COATING INDUSTRY AND OPERATIONS

The U.S. coil coating industry consists of about 80 plants processing approximately 1.5 billion square meters of painted coil each year. Facilities vary in size and corporate structure, ranging from independent shops to captive operations. Independent shops obtain untreated coil, conversion coating chemicals, and paints, and produce a wide variety of coated coil. Typically, the annual production at these coil coating plants is low compared to that from the captive coating operations. The captive coil coating operation is usually an integral part of a large corporation engaged in many other kinds of metal production and finishing.

The coil coating sequence, regardless of basis material or conversion coating process used, consists of three operational steps:

1. Cleaning
2. Conversion coating
3. Painting

There are three types of cleaning operations used in coil coating, and they can be used alone or in combination. These cleaning operations are as follows:

1. Mild alkaline cleaning
2. Strong alkaline cleaning
3. Acid cleaning

There are four basic types of conversion coating operations, and the use of one precludes the use of the others on the same coil:

1. Chromating
2. Phosphating
3. Use of complex oxides
4. No-rinse conversion coating

Some of these conversion coating operations are designed for use on specific basis materials. The painting operation is performed by roll coating and is independent of the basis material and conversion coating. Some specialized coatings are supplied without conversion-coating the basis material. For example, Zincrometal is a specialized coating consisting of two coats of special paints that do not require conversion coating. In this process, coils are cleaned and dried, and then receive two coats of the special paints.

The selection of basis material, conversion coating, and paint formulation is an art based upon experience. The variables that are typically involved in the selection are appearance, color, gloss, corrosion resistance, abrasion resistance, process line capability, availability of raw materials, customer preference, and cost. Some basis materials inherently work better with certain conversion coatings, and some conversion coatings work better with certain paint formulations. On the whole, however, the choice of which combination to use on a basis material is limited only by plant and customer preferences.[1-4]

The following subsections describe the coil coating processes in more detail.

7.2 CLEANING OPERATION OF COIL COATING

Coil coating requires that the basis material be clean. A thoroughly clean coil ensures efficient conversion coating and a resulting uniform surface for painting. The soils, oils, and oxide coatings found on a typical coil originate from rolling mill operations and storage conditions prior to coil coating. Such substances can stop the conversion coating reaction, cause a coating void on part of the basis material, and result in the production of a nonuniform coating. Cleaning operations must chemically and physically remove these interfering substances without degrading the surface of the basis material. Excessive cleaning can roughen a basically smooth surface to a point where a paint film will not provide optimum protective properties.

7.2.1 Mild Alkaline Cleaning

Aluminum and galvanized steel are prone to develop an oxide coating that acts as a barrier to chemical conversion coatings. However, these oxide films are easier to remove than rust and, therefore, require a less vigorous cleaning process. A mild alkaline cleaner is usually applied with power spray equipment to remove the oxide coating and other interfering substances. The cleaning solutions normally used consist of combinations of sodium carbonates, phosphates, silicates, and hydroxides. These compounds give the solution its alkaline character and emulsify the removed soils. Soap and detergents may be added to the solution to lower the surface and interfacial tension.

7.2.2 STRONG ALKALINE CLEANING

A good cleaning solution also rinses easily. Solutions may be made stronger with the addition of more sodium hydroxide.

A spray rinse follows either the mild alkaline cleaning step or strong alkaline cleaning step. Spray rinsing is conducive to the fast line speeds that make coil coating an economical coating procedure. The spray rinse physically removes alkaline cleaning residues and soil by both the physical impingement of the water and the diluting action of the water. The rinsewater is usually maintained at approximately 66°C (150°F) to keep the coil warm for the subsequent conversion coating reactions and to help the rinsing action. The rinsing action prevents contamination of the conversion coating bath with cleaning residues that are dragged out on the strip and that could be subsequently deposited in the conversion coating solutions. The rinsing step also keeps the surface of the metal wet and active, which permits faster conversion coating film formation.

7.2.3 ACID CLEANING

Steel, unless adequately protected with a film of oil subsequent to rolling mill operations, has a tendency to form surface rust rather quickly. This rust on the surface of the metal prevents proper conversion coating. A traditional method of removing rust is an acid applied by power spray equipment. The spraying action cleans both by physical impingement and the etching action of the acid. The power spray action is followed by a brush scrub, which further removes soil loosened by the acid. The brush scrub is followed by a strong alkaline spray wash, which removes all traces of the acid and neutralizes the surface.[1-5]

7.2.4 SPECIAL CLEANING

The no-rinse conversion coating and the Zincrometal processes require a coil that is clean, warm, and dry. For these processes a squeegee roll and forced air drying are used to assure a clean, dry coil following alkaline cleaning and rinsing.

7.3 CONVERSION COATING PROCESS OF COIL COATING

The basic objective of the conversion coating process is to provide a corrosion-resistant film that is integrally bonded chemically and physically to the base metal and that provides a smooth and chemically inert surface for subsequent application of a variety of paint films. The conversion coating processes effectively render the surface of the basis material electrically neutral and immune to galvanic corrosion. Conversion coating on basis material coils does not involve the use of applied electric current to coat the basis material. The coating mechanisms are chemical reactions that occur between solution and basis material.[1-4]

Four types of conversion coatings are normally used in coil coating:

1. Chromate conversion coatings
2. Phosphate conversion coatings
3. Complex oxides conversion coatings
4. No-rinse conversion coatings

Chromate conversion coatings, phosphate conversion coatings, and complex oxide conversion coatings are applied in basically the same manner. No-rinse conversion coatings are roll applied and use quite different chemical solutions than phosphating, chromating, or complex oxides solutions. However, the dried film is used as basis for paint application similar to phosphating, chromating, and complex oxide conversion coating films.

7.3.1 CHROMATE CONVERSION COATINGS

Chromate conversion coatings can be applied to both aluminum and galvanized surfaces but are generally applied only to aluminum surfaces. These coatings produce an amorphous layer of chromium chromate complexes and aluminum ions. The coatings offer unusually good corrosion-inhibiting properties but are not as abrasion resistant as phosphate coatings. Scratched or abraded films retain a great deal of protective value because the hexavalent chromium content of the film is slowly leached by moisture, providing a self-healing effect. Under limited applications, these coatings can serve as the finished surface without being painted. If further finishing is required, it is necessary to select an organic finishing system that has good adhesive properties. Chromate conversion coatings are extremely smooth, electrically neutral, and quite resistant to chemical attack.

Chromate conversion coatings for aluminum are carried out in acidic solutions. These solutions usually contain one chromium salt, such as sodium chromate or chromic acid and a strong oxidizing agent such as hydrofluoric acid or nitric acid. The final film usually contains both products and reactants and water of hydration. Chromate films are formed by the chemical reaction of hexavalent chromium with a metal surface in the presence of "accelerators" such as cyanides, acetates, formates, sulfates, chlorides, fluorides, nitrates, phosphates, and sulfamates.

Chromate conversion coating requires that the basis material be alkaline-cleaned and spray-rinsed with warm water. The cleaning and rinsing assures a clean, warm, wet surface on which the conversion coating process takes place. Once the film is formed, it is rinsed with water followed by a chromic acid sealing rinse. This latter rinse seals the free pore area of the coating by forming a chromium chromate gel. Also, the sealing rinse more thoroughly removes precipitated deposits that may have been formed by hard water in previous operations. The coil is then subjected to a forced air drying step to assure a uniformly dry surface for the following painting operation.

7.3.2 PHOSPHATE CONVERSION COATINGS

Phosphate conversion coatings provide a highly crystalline, electrically neutral bond between a base metal and paint film. The most widespread use of phosphate coatings is to prolong the useful life of paint finishes. Phosphate coatings are primarily used on steel and galvanized surfaces but can also be applied to aluminum. Basically, there are three types of phosphate coatings:

1. Iron phosphate coating
2. Zinc phosphate coating
3. Manganese phosphate coating

Manganese phosphate coatings are not used in coil coating operations because they are relatively slow in forming and, as such, are not amenable to the high production speeds of coil coaters.

The remaining two phosphate coatings are applied by spraying or immersing the coil, with the major difference between them being the weight and thickness of the dried coating. Iron phosphate coatings are the thinnest and lightest and generally the cheapest. Iron phosphate solutions are applied chiefly as a base for paint films. Spray application of iron phosphating solutions is most commonly used. The coating weights range from 0.22 to 0.86 g/m^2.

Zinc phosphate coatings are quite versatile and can be used as a base for paint or oil, as an aid to cold forming, to increase wear resistance, and to provide rust-proofing. Zinc phosphate coatings can be applied by spray or immersion, with applied coating weights ranging from 1.08 to 10.8 g/m^2 for spray coating and from 1.61 to 43.1 g/m^2 for immersion coating.

Phosphate coatings are formed in the metal surface, incorporating metal ions dissolved from the surface. This creates a coating that is integrally bonded to the base metal. In this respect, phosphate coatings differ from electrodeposited coatings, which are superimposed on the metal. Most metal phosphates are insoluble in water but soluble in mineral acids. Phosphating solutions consist of metal phosphates dissolved in carefully balanced solutions of phosphoric acid. As long as the acid concentration of the bath remains above a critical point, the metal ions remain in solution. Accelerators speed up film formation and prevent the polarization effect of hydrogen on the surface of the metal. The accelerators commonly used include nitrites, nitrates, chlorates, and peroxides. Cobalt and nickel nitrite accelerators are the most widely used and develop a coarse crystalline structure. The peroxides are relatively unstable and difficult to control, whereas chlorate accelerators generate a fine sludge that may cause dusty or powdery deposits.

After phosphating, the coil is passed through a recirculating hot water spray rinse. The rinsing action removes excess acid and unreacted products, thereby stopping the conversion coating reaction. Insufficient rinsing could cause blistering under the subsequent paint film from the galvanic action of the residual acid and metal salts.

The basis material is then passed through an acid sealing rinse comprising up to 0.1% by volume of phosphoric acid, chromic acid, and various metallic conditioning agents, notably zinc. This solution seals the free pore area of the coating by forming a chromium chromate gel. Also, this acidic sealing rinse more thoroughly removes precipitated deposits formed by hard water in the previous rinses. Modified chromic acid rinses have been used extensively in the industry. These rinses are prepared by reducing chromic acid with an organic reductant to form a mixture of trivalent chromium and hexavalent chromium in the form of a complex chromium chromate.

7.3.3 Complex Oxide Conversion Coating

Complex oxide conversion coatings can be applied to aluminum and galvanized surfaces but are generally applied to only galvanized surfaces. The nature of the film and the chemical and physical actions of its formation are a function and a reinforcement of the naturally occurring protective oxide coating that is found on galvanized surfaces. The physical properties of the complex oxide conversion coating film are comparable to those of chromate conversion coating films and phosphate conversion coating films.

Complex oxide film is formed in a basic solution, whereas the films described earlier are formed in an acidic solution. Complex oxide conversion coating reactions do not contain either hexavalent or trivalent chromium ions. However, the sealing rinse contains much greater quantities of hexavalent and trivalent chromium ions than do the sealing rinses associated with phosphate conversion coatings and chromate conversion coatings.

7.3.4 No-Rinse Conversion Coatings

Recent developments in chromate conversion coating solutions have resulted in a solution that can be applied to cold rolled steel, galvanized steel, or aluminum without the need for any rinsing after the coating has formed on the basis material. The basis material must first be alkaline cleaned, thoroughly rinsed, and forced-air dried prior to conversion coating. The conversion coating solution is applied with a roll mechanism used in roll coating paint. Once the solution is roll coated onto the basis material, the coil is forced-air dried at approximately 66°C. The no-rinse solutions are formulated in such a way that once a film is formed and dried, there are no residual or detrimental products left on the coating that could interfere with normal coil coating paint formulations.

Although no-rinse conversion coatings currently represent a small proportion of the conversion coating techniques that are used, they offer several advantages, including fewer process steps in a physically smaller process line, higher line speeds, application of a very uniform thickness by roll coating rather than spray or dip coating, and reduction of waste treatment requirements because of the reduced use of chromium compounds. Disadvantages include roll coating mechanism wear possibly reducing quality, the closer coordination of the entire line that is needed, difficulty in adaptation, and the hazardous organic acids content of the no-rinse conversion coating chemicals.

7.4 PAINTING OPERATION OF COIL COATING

Roll coating of paint is the final process in a coil coating line. Roll coating is an economical method to paint large areas of metal with a variety of finishes and to produce a uniform and high-quality coating. The reverse roll procedure for coils is used by the coil coating industry, and allows both sides of the coil to be painted simultaneously.

The paint formulations used in the coil coating industry have high pigmentation levels (providing hiding power), adhesion, and flexibility. Most coatings of this type are thermosetting and are based on vinyl, acrylic, and epoxy functional aromatic polyethers, and some reactive monomer or other resin with reactive functions, such as melamine formaldehyde resins. Also, a variety of copolymers of butadiene with styrene or maleic anhydride are used in coating formulations. These coatings are cured by oxidation mechanisms during baking, similar to those that harden drying oils.

After paint application, all coils are cured in an oven. Curing temperatures depend upon basis material, conversion coating, paint formulation, and line speed. Typical temperatures range from ~93°C to a maximum of ~454°C. Upon leaving the oven, the coils are quenched with water to induce rapid cooling prior to rewinding.

The quench is necessary for all basis materials, conversion coatings, and paint formulations. A coil that is rewound when too warm will develop internal and external stresses, causing a possible degradation of the appearance of the paint film and of the forming properties of the coil. The volume of water used in the quench often has the largest flow rate of all of the coil-coating processes. However, the water is often circulated to a cooling tower for heat dissipation and reuse.

The finished coils are used in a variety of industries. The building products industry utilizes prefinished coils to fabricate exterior siding, window and door frames, storm windows, storm gutters, and various other trim and accessory building products. The food and beverage industries utilize various types of coils and finishes to safely and economically package and ship a wide variety of food and beverage products. Until recently, the automotive and appliance industries have made limited use of prefinished coils. These industries have relied on postassembly finishing of their products. Recently, the automotive industry has begun using a cold rolled steel coil coated on one side with a finish called Zincrometal. This coating is applied to the under surfaces of exterior automobile sheet metal to protect them from corrosion. The appliance industry uses prefinished coils in constructing certain models of refrigerator exteriors to provide a finished product that minimizes the costly and labor-intensive painting operation after forming.

Coil coating operations are located throughout the country, usually in well established industrial centers. Compared to some other industries, coil coating operations are not physically large. Coil coating operations use large quantities of water and are often a significant contributor to municipal waste treatment systems or surface waters. In addition, the curing ovens from coil coating operations are a source of air pollution in the form of reactive hydrocarbons.

7.5 SUBCATEGORIES OF THE COIL COATING INDUSTRY

7.5.1 GENERAL DESCRIPTION OF SUBCATEGORIZATION

The primary purpose of subcategorization is to establish groupings within the coil coating industry such that each group has a uniform set of effluent limitations. Although subcategorization is based on wastewater characteristics, a review of the other subcategorization factors reveals that the basis material used and the processes performed on these basis materials are the principal factors affecting the wastewater characteristics of plants in the coil coating industry. The coil coating industry is therefore divided into the following three subcategories:

1. Coil coating on steel
2. Coil coating on zinc-coated steel (galvanized)
3. Coil coating on aluminum or aluminized steel

Of all coil coating plants in the U.S., about 36% of the plants pretreat their industrial effluents and directly discharge their pretreated effluents to the receiving waters, and the remaining 54% of the plants pretreat and discharge their effluents to the municipal wastewater treatment plants for further treatment. The following subsections describe the above subcategories.[1–3]

7.5.2 Coil Coating on Steel Subcategory

In the U.S., 59 facilities in the coil coating industry were surveyed for process type and pollutant levels. Of these, 38 plants are in the coil coating on steel subcategory. Ten facilities coat steel alone and the remaining 28 coat a combination of steel coils and coils from the other subcategories. The production rate is approximately 85,000 m^2/hr. Operations used at these facilities include acid cleaning, strong alkaline cleaning, phosphating, no-rinse conversion coating, roll coating, and Zincrometal coating. Water usage rates for the general operations at steel coating facilities are listed in Table 7.1.

7.5.3 Coil Coating on Zinc Coated Steel (Galvanized Steel) Subcategory

Within the 59 plants surveyed, 17 coil coat on galvanized steel with a production of ~60 × 10^3 m^2/hr. Only two facilities produce coated galvanized steel alone. Operations used at the galvanized coating facilities include mild alkaline cleaning, phosphating, chromating, complex oxide treatment, no-rinse conversion coating, roll coating, and Zincrometal coating. Table 7.1 also presents water usage data for the general operations at galvanized coating facilities.[1,2]

TABLE 7.1
Summary of Water Usage Rates for the Coil Coating Industry by Subcategory

Operation	Number of Plants Sampled	Water Use (L/m^2) Range	Mean
Steel			
Cleaning	11	0.04–7.3	1.9
Conversion coating	8	0.04–0.76	0.43
Quenching	4	2.0–5.7	4.0
All operations	13	0.37–13	4.5
Galvanized			
Cleaning	10	0.17–8.8	1.9
Conversion coating	10	0.03–0.98	0.49
Quenching	5	0.44–5.1	2.7
All operations	12	0.65–8.4	3.6
Aluminum			
Cleaning	12	0.21–2.0	0.97
Conversion coating	12	0.18–1.8	0.56
Quenching	9	1.2–3.5	2.3
All operations	15	0.26–5.8	2.5

Source: U.S. EPA, *Development Document for Effluent Limitations Guidelines and Standards for the Coil Coating Point Source Category*, (Canmaking Subcategory), Final report 440/1-83/071, Washington, DC, November 1983; U.S. EPA, *Coil Coating Forming Point Source Category*, available at http://www.access.gpo.gov/nara/cfr/waisidx_03/40cfr467_03.html, 2008.

7.5.4 COIL COATING ON ALUMINUM SUBCATEGORY

Thirty-nine of the facilities in the U.S. coil coat on aluminum with a production rate of 90×10^3 m²/hr. Nineteen facilities coat only aluminum coils. The aluminum coating facilities use mild alkaline cleaning, phosphating, chromating, complex oxide treatment, no-rinse conversion coating, and roll coating. Water usage rates for the general processes in this subcategory are listed in Table 7.1.

Water is used in virtually all coil coating operations. It provides the mechanism for removing undesirable compounds from the basis material, is the medium for the chemical reactions that occur on the basis material, and cools the basis material following baking. Water is the medium that permits the high degree of automation associated with coil coating and the high quality of the finished product. The nature of coil coating operations, the large amount of basis material processed, and the quantity and type of chemicals used produces a large volume of wastewater that requires treatment before discharge.

Wastewater generation occurs for each basis material (steel, galvanized and aluminum) and for each functional operation (cleaning, conversion coating, and painting). The wastewater generated by the three functional operations may be handled in one of the following ways:

1. It may flow directly to a municipal wastewater treatment system or surface water.
2. It may flow directly to an onsite waste treatment system and then to a municipal wastewater treatment system or surface water.
3. It may be reused directly or following intermediate treatment.
4. It may undergo a combination of the above processes.

Coil coating operations that produce wastewater are characterized by the pollutant constituents associated with respective basis materials. The constituents in the raw wastewaters include ions of the basis material, oil and grease found on the basis material, components of the cleaning and conversion coating solutions, and the paints and solvents used in roll coating the basis materials. The following tables present wastewater characterization data for each subcategory. The data presented are the results of verification analysis of the industry. Prior to verification sampling, a screening program was conducted to identify the presence or absence of the 129 priority pollutants. Those pollutants detected in screening at a concentration greater than 10 µg/L were further studied in the verification analysis. The minimum detection limit in the verification analysis for pesticides was 5 µg/L and for all other toxic pollutants, 10 µg/L. Any value below its detection limit is presented in the following tables as below detection limit (BDL).

Tables 7.2 through 7.5 present raw wastewater characterization data for each general process in each subcategory and for the wastewater in each subcategory when combined into a single representative stream as a whole. Table 7.6 presents raw wastewater flow data for each subcategory.

7.6 WASTEWATER CHARACTERIZATION OF THE COIL COATING INDUSTRY

7.6.1 EFFLUENT CHARACTERISTICS OF COIL COATING ON STEEL OPERATION

Wastewaters from the coil coating on steel subcategory generally have higher levels of phosphorus than that from the other subcategories because of the use of concentrated phosphate alkaline cleaners. Oil and grease in this subcategory are also found in larger concentrations than the other basis materials' wastewater because of the increased raw material protection needed to inhibit rust. This can often cause an increase in the number of hydrocarbons found in the wastewater. Suspended solids may be at higher levels because of the adhering dirt in the oil.[1-3]

7.6.2 EFFLUENT CHARACTERISTICS OF COIL COATING ON ZINC COATED STEEL (GALVANIZED STEEL) OPERATION

Coil coating on galvanized steel generally produces significant suspended solids concentrations in wastewater. Another pollutant problem is the high concentration of dissolved zinc and iron in the

TABLE 7.2
Toxic and Classical Pollutants in Raw Wastewater of the Steel Subcategory, Verification Data

Pollutant	Number of Samples	Number of Detections	Range of Samples	Mean[a] of Samples	Number of Samples	Number of Detections	Range of Samples	Mean[a] of Samples
		Cleaning Operations				Conversion Operations		
Toxic Organic Pollutants (μg/L)								
1,1-Trichloroethane	6	5	ND–BDL	BDL	8	3	ND–40	BDL
1,1-Dichloroethane	5	0	—	—	7	1	ND–77	11
1,1-Dichloroethylene	2	0	—	—	2	0	—	—
1,2-trans-Dichloroethylene	2	0	—	—	2	0	—	—
2,4-Dimethylphenol	3	0	—	—	3	0	—	—
Fluoranthene	9	1	ND–68	BDL	7	1	ND–BDL	BDL
Isophorone	9	1	ND–18	BDL	7	0	—	—
Naphthalene	9	2	ND–20	BDL	7	4	ND–BDL	BDL
Phenol	3	0	—	—	3	0	—	—
Bis(2-ethylhexyl)phthalate	9	7	ND–150	34	7	5	ND–110	20
Butyl benzyl phthalate	9	1	ND–360	40	7	0	—	—
Di-n-butyl phthalate	9	5	ND–30	BDL	7	3	ND–14	BDL
Di-n-octyl phthalate	9	3	ND–BDL	BDL	7	1	ND–760	110
Diethyl phthalate	9	6	ND–210	46	7	6	ND–180	100
Dimethyl phthalate	9	0	—	—	7	0	—	—
1,2-Benzanthracene	9	2	ND–30	BDL	7	0	—	—
Benzo(a)pyrene	9	0	—	—	7	0	—	—
3,4-Benzo fluoranthene	9	0	—	—	7	0	—	—
Benzo(k)fluoranthene	9	0	—	—	7	0	—	—
Chrysene	9	2	ND–30	BDL	7	0	—	—
Acenaphthylene	9	1	ND–BDL	BDL	7	1	ND–BDL	BDL
Anthracene	9	7	ND–280	51	7	3	ND–BDL	BDL
1,1,2-Benzoperylene	9	0	—	—	7	0	—	—

Pollutant	n	n_d	Range	Mean	n	n_d	Range	Mean
Fluorene	9	1	ND-BDL	ND	7	2	ND-BDL	BDL
Phenanthrene	9	7	ND-280	51	7	3	ND-BDL	BDL
1,2,5,6-Dibenzanthracene	9	0	—	—	7	0	—	—
Ideno(1,2,3-cd)pyrene	9	0	—	—	7	0	—	—
Pyrene	9	0	—	—	7	0	—	—
Toluene	3	0	—	—	3	0	—	—
Trichloroethylene	6	4	ND-22	BDL	8	3	ND-89	13
Toxic Metals and Inorganics (µg/L)								
Cadmium	9	2	ND-BDL	BDL	8	3	ND-73	10
Chromium, total	9	8	ND-620	210	8	8	280-920,000	320,000
Chromium, hexavalent	9	0	—	—	8	7	ND-410,000	110,000
Copper	9	9	21-180	70	8	6	ND-160	41
Cyanide, total	8	5	ND-120	28	7	1	ND-92	12
Cyanide, amn. to chlorine	8	3	ND-99	17	7	1	ND-12	BDL
Lead	9	4	ND-1100	240	8	3	ND-3600	530
Nickel	9	5	ND-210	38	8	4	ND-19,000	4,000
Zinc	9	9	220-42,000	10,000	8	8	530-140,000	54,000
Classical Pollutants (mg/L)								
Aluminum	9	7	ND-0.85	0.35	8	5	ND-11	1.9
Fluorides	9	9	0.18-3.4	1.3	8	8	1.1-74	31
Iron	9	9	0.93-80	25	8	8	3.3-77	19
Manganese	9	9	0.26-1.7	0.8	8	8	0.11-1.5	0.61
Oil and grease	9	9	9.8-1600	520	8	6	ND-18	6.5
Phenols, total	9	5	ND-0.27	0.18	7	4	ND-0.23	0.038
Phosphorus	7	7	11-78	46	7	6	9.7-70	41
TDS	4	4	1100-17,000	9300	3	3	3300-3500	3400
TSS	9	9	52-440	220	8	8	27-250	130

BDL, below detection limit; ND, not detected.

[a] BDL was calculated in the mean concentration as equal to zero.

Source: U.S. EPA, *Development Document for Effluent Limitations Guidelines and Standards for the Coil Coating Point Source Category*, (Canmaking Subcategory), Final report 440/1-83/071, Washington, DC, November 1983; U.S. EPA, *Coil Coating Forming Point Source Category*, available at http://www.access.gpo.gov/nara/cfr/waisidx_03/40cfr467_03.html, 2008.

TABLE 7.3
Toxic and Classical Pollutants in Raw Wastewater of the Galvanized Subcategory, Verification Data

Pollutant	Number of Samples	Number of Detections	Range of Samples	Mean[a] of Samples	Number of Samples	Number of Detections	Range of Samples	Mean[a] of Samples
	Cleaning Operations				Conversion Operations			
Toxic Organic Pollutants (μg/L)								
1,1,1-Trichloroethane	10	4	ND–BDL	BDL	10	4	ND–140	21
1,1-Dichloroethane	1	0	—	—	1	0	—	—
1,1-Dichloroethylene	10	0	—	—	10	1	ND–BDL	BDL
1,2-*trans*-dichloroethylene	10	0	—	—	10	2	ND–15	BDL
2,4-Dimethyl phenol	2	0	—	—	2	0	—	—
Fluoranthene	10	3	ND–BDL	BDL	10	1	ND–23	BDL
Isophorone	10	1	ND–47	BDL	10	1	ND–520	52
Naphthalene	10	2	ND–38	BDL	10	5	ND–15	BDL
Phenol	4	0	—	—	4	0	—	—
Bis(2-ethylhexyl) phthalate	10	9	ND–340	110	10	9	ND–1200	220
Butyl benzyl phthalate	10	1	ND–130	13	10	3	ND–BDL	BDL
Di-n-butyl phthalate	10	7	ND–170	30	10	3	ND–20	BDL
Di-n-octyl phthalate	10	1	ND–BDL	BDL	10	0	—	—
Diethyl phthalate	10	8	ND–420	110	10	9	ND–300	77
Dimethyl phthalate	10	0	—	—	10	0	—	—
1,2-Benzanthracene	10	4	ND–27	13	10	1	ND–BDL	BDL
Benzo(a)pyrene	10	0	—	—	10	0	—	—
3,4-Benzo fluoranthene	10	0	—	—	10	0	—	—
Benzo(k)fluoranthene	10	0	—	—	10	0	—	—
Chrysene	10	4	ND–27	BDL	10	1	ND–BDL	BDL
Acenaphthylene	10	0	—	—	10	1	ND–BDL	BDL
Anthracene	10	3	ND–250	27	10	3	ND–290	29
1,1,2-Benzoperylene	10	0	—	—	10	0	—	—

Pollutant							
Fluorene	10	4	—	—	10	1	—
(cont.)							BDL
Phenanthrene	10	3	ND–85	13	10	3	ND–290	29
1,2,5,6-Dibenzanthracene	10	0	ND–47	BDL	10	0	—	—
Ideno(1,2,3-cd)pyrene	10	0	—	—	10	0	—	—
Pyrene	10	3	ND–BDL	BDL	10	1	ND–11	BDL
Toluene	4	0	—	—	4	0	—	—
Trichloroethylene	10	2	ND–BDL	BDL	10	2	ND–110	14
Toxic Metals and Inorganics (µg/L)								
Cadmium	10	8	ND–120	36	10	5	ND–110	21
Chromium, total	10	9	ND–610	280	10	10	3400–780,000	290,000
Chromium, hexavalent	9	1	ND–260	29	10	10	50–310,000	140,000
Copper	10	9	ND–57	27	10	8	ND–140	25
Cyanide, total	10	4	ND–43	BDL	10	5	ND–470	150
Cyanide, amn. to chlorine	10	3	ND–21	BDL	10	4	ND–330	48
Lead	10	9	ND–2600	1,400	10	10	BDL–1300	560
Nickel	10	1	ND–150	15	10	6	ND–31,000	4600
Zinc	10	10	690–120,000	63,000	10	10	33,000–710,000	220,000
Classical Pollutants (mg/L)								
Aluminum	10	9	ND–4.9	2.2	10	9	ND–11	3.2
Fluorides	10	10	0.16–16	2.5	10	10	1.5–71	16
Iron	10	10	0.19–17	4.8	10	10	0.84–21	6.6
Manganese	10	9	ND–0.73	0.17	10	10	0.035–1.3	0.25
Oil and grease	10	10	10–970	270	10	10	1.3–110	19
Phenols, total	9	7	ND–0.079	0.029	10	7	ND–0.067	0.015
Phosphorus	9	9	9.2–56	33	7	7	3.8–66	33
TDS	1	1	2001	—	1	1	2,500	—
TSS	10	10	15–630	250	10	10	68–450	250

BDL, below detection limit; ND, not detected.

[a] BDL was calculated in the mean concentration as equal to zero.

Source: U.S. EPA, *Development Document for Effluent Limitations Guidelines and Standards for the Coil Coating Point Source Category,* (Canmaking Subcategory), Final report 440/1-83/071, Washington, DC, November 1983; U.S. EPA, *Coil Coating Forming Point Source Category,* available at http://www.access.gpo.gov/nara/cfr/waisidx_03/40cfr467_03.html, 2008.

TABLE 7.4
Toxic and Classical Pollutants in Raw Wastewater of the Aluminum Subcategory, Verification Data

Pollutant	Cleaning Operations				Conversion Operations			
	Number of Samples	Number of Detections	Range of Samples	Mean[a] of Samples	Number of Samples	Number of Detections	Range of Samples	Mean[a] of Samples
Toxic Organics (μg/L)								
Fluoranthene	12	0			12	0		
Isophorone	12	0			12	0		
Naphthalene	9	3	ND–BDL	BDL	12	3	ND–BDL	BDL
Phenol	2	0			2	0		
Bis(2-ethylhexyl)phthalate	12	10	ND–450	no	12	9	ND–300	37
Butyl benzyl phthalate	12	0			12	0		
Di-n-butyl phthalate	12	2	ND–12	BDL	12	2	ND–BDL	BDL
Di-n-octyl phthalate	12	0			12	1	ND–BDL	BDL
Diethyl phthalate	12	7	ND–450	99	12	9	ND–200	57
Dimethyl phthalate	12	2	ND–BDL	BDL	12	1	ND–110	BDL
1,2-Benzanthracene	12	0			12	0		
Benzo(a)pyrene	12	3	ND–BDL	BDL	12	2	ND–BDL	BDL
3,4-Benzo fluoranthene	12	0			12	0		
Benzo(k)fluoranthene	12	0			12	0		
Chrysene	12	0			12	0		
Acenaphthylene	12	0			12	0		
Anthracene	12	2	ND–BDL	BDL	12	4	ND–BDL	BDL
1,1,2-Benzoperylene	12	0			12	0		
Fluorene	12	1	ND–BDL	BDL	12	0		
Phenanthrene	12	2	ND–BDL	BDL	12	4	ND–BDL	BDL

Waste Treatment and Management in the Coil Coating Industry

Pollutant								
1,2,5,6-Dibenz anthracene	12	0			12	0		
Ideno(1,2,3-cd)pyrene	12	0			12	0		
Pyrene	12	0			12	0		
Toluene	2	0			2	0		
Toxic Metals and Inorganics (µg/L)								
Cadmium	12	3	ND–21	BDL	12	3	ND–19	BDL
Chromium, total	12	9	ND–6000	980	12	12	15,000–960,000	270,000
Chromium, hexavalent	11	1	ND–6600	600	12	12	11,000–330,000	120,000
Copper	12	9	ND–210	63	12	10	ND–980	160
Cyanide, total	12	9	ND–260	30	12	9	ND–7500	2400
Cyanide, amn. to chlorine	12	8	ND–240	25	9	6	ND–7000	1400
Lead	12	5	ND–220	58	12	2	ND–400	48
Nickel	12	0			12	4	ND–260	40
Zinc	12	10	ND–14,000	1300	12	12	16–43,000	8800
Classical pollutants, (mg/L)								
Aluminum	12	12	8.6–940	400	12	12	11–410	160
Fluorides	12	9	ND–9.5	1.5	12	12	17–510	210
Iron	12	12	0.077–0.69	0.34	12	12	0.8–87	21
Manganese	12	9	ND–15	3.8	12	12	0.049–12	1.4
Oil and grease	12	9	ND–2800	400	12	9	ND–60	7.1
Phenols, total	12	11	ND–0.16	0.043	12	8	ND–0.14	0.02
phosphorus	9	6	ND–100	42	2	2	13–16	14
TSS	12	12	6.0–970	180	12	12	4.2–1200	160

BDL, below detection limit; ND, not detected.

[a] BDL was calculated as equal to zero in the mean concentration.

Source: U.S. EPA, *Development Document for Effluent Limitations Guidelines and Standards for the Coil Coating Point Source Category*, (Canmaking Subcategory). Final report 440/1-83/071, Washington, DC, November 1983; U.S. EPA, *Coil Coating Forming Point Source Category*, available at http://www.access.gpo.gov/nara/cfr/waisidx_03/40cfr467_03.html, 2008.

TABLE 7.5
Toxic and Classical Pollutants in Quenching Raw Wastewater of All Subcategories, Verification Data

Pollutant	Number of Samples	Number of Detections	Range of Samples	Mean[a] of Samples
Toxic Organic Pollutants (µg/L)				
1,1,1-Trichloroethane	9	4	ND–3100	400
1,1-Dichloroethane	3	0	—	—
1,1-Dichloroethylene	6	1	ND–36	BDL
1,2-*trans*-Dichloroethylene	6	1	ND–43	BDL
2,4-Dimethylphenol	3	0	—	—
Fluoranthene	18	1	ND–BDL	BDL
Isophorone	18	0	—	—
Naphthalene	18	3	ND–BDL	BDL
Phenol	7	0	—	—
Bis(2-ethylhexyl)phthalate	18	14	ND–880	72
Butyl benzyl phthalate	18	2	ND–15	BDL
Di-n-butyl phthalate	18	6	ND–20	BDL
Di-n-octyl phthalate	18	1	ND–BDL	BDL
Diethyl phthalate	18	15	ND–330	64
Dimethyl phthalate	18	2	ND–BDL	BDL
1,2-Benzanthracene	18	0	—	—
Benzo(a)pyrene	18	1	ND–BDL	BDL
3,4-Benzo fluoranthene	18	1	ND–BDL	BDL
Benzo(k)fluoranthene	18	1	ND–BDL	BDL
Chrysene	18	0	—	—
Acenaphthylene	18	0	—	—
Anthracene	18	2	ND–BDL	BDL
1,1,2-Benzoperylene	18	1	ND–BDL	BDL
Fluorene	18	0	—	—
Phenanthrene	18	2	ND–BDL	BDL
1,2,5,6-Dibenzanthracene	18	0	—	—
Ideno(1,2,3-cd)pyrene	18	0	—	—
Pyrene	18	0	—	—
Toluene	7	0	—	—
Trichloroethylene	9	5	ND–3100	410
Toxic Metals and Inorganics (µg/L)				
Cadmium	20	3	ND–270	15
Chromium, total	20	15	ND–440	43
Chromium, hexavalent	20	0	—	—
Copper	20	7	ND–17	BDL
Cyanide, total	20	17	ND–200	33
Cyanide, amn. to chlor.	20	11	ND–80	14
Lead	20	2	ND–64	BDL
Nickel	20	1	ND–190	BDL
Zinc	20	20	14–5000	610

Continued

TABLE 7.5 (continued)

Pollutant	Number of Samples	Number of Detections	Range of Samples	Mean[a] of Samples
Classical Pollutants (mg/L)				
Aluminum	20	8	ND–1.4	0.38
Fluorides	20	20	0.15–11	1.6
Iron	20	20	0.018–1.6	0.37
Manganese	20	15	ND–0.78	0.14
Oil and grease	20	15	ND–26	5.3
Phenols, total	20	15	ND–0.04	0.012
Phosphorus	18	11	ND–15	1.2
TDS	3	3	99–1100	440
TSS	20	18	ND–24	6.2

BDL, below detection limit; ND, not detected.

[a] BDL was calculated as equal to zero in the mean concentration.

Source: U.S. EPA, *Development Document for Effluent Limitations Guidelines and Standards for the Coil Coating Point Source Category*, (Canmaking Subcategory), Final report 440/1-83/071, Washington, DC, November 1983; U.S. EPA, *Coil Coating Forming Point Source Category*, available at http://www.access.gpo.gov/nara/cfr/waisidx_03/40cfr467_03.html, 2008.

wastewater as a result of the dissolved metals from the cleaning operation. Significant concentrations of hexavalent chromium are generally expected in all three subcategory wastewaters.

7.6.3 Effluent Characteristics of Coil Coating on Aluminum Operation

Wastewaters from the coil coating on aluminum subcategory contain higher levels of cyanide and fluorides than the other subcategories as a result of chromating solutions containing cyanide ions

TABLE 7.6
Wastewater Flows (m³/day) for the Coil Coating Industry

Operation	Number of Samples	Flow Range	Flow Mean
Steel			
Cleaning	9	7.7–650	170
Conversion coating	8	1.4–75	38
Galvanized			
Cleaning	10	15–330	110
Conversion coating	10	1.8–75	36
Aluminum			
Cleaning	12	11–160	83
Conversion coating	12	15–60	39
Total industry			
Quenching	20	36–1100	320

Source: U.S. EPA, *Development Document for Effluent Limitations Guidelines and Standards for the Coil Coating Point Source Category*, (Canmaking Subcategory), Final report 440/1-83/071, Washington, DC, November 1983; U.S. EPA, *Coil Coating Forming Point Source Category*, available at http://www.access.gpo.gov/nara/cfr/waisidx_03/40cfr467_03.html, 2008.

and hydrofluoric acid. Aluminum wastewater is also more acidic and contains more dissolved aluminum. This is due to the acidic nature of the chromating solutions that dissolve more aluminum than the phosphating solutions.

Painting wastewater generally consists of quench water. Wastewater from this operation is generally less toxic than wastewater from the other general operations; normally, only the following pollutants are expected to exceed 10 µg/L: oil and grease, fluorides, TSS, iron, zinc, bis(2-ethylhexyl) phthalate, and diethyl phthalate.

7.7 PLANT-SPECIFIC EFFLUENT CHARACTERIZATION DATA

A limited amount of plant-specific data for the coil coating industry is available. Data available in the reference documents on the effluent streams for the plants discussed in the following subsections are summarized in Table 7.7. These data are verification data. All three subcategories are represented by the facilities.[1]

TABLE 7.7
Plant-Specific Effluent Concentrations, Verification Data

Pollutant	Steel Subcategory Plant A	Galvanized Subcategory Plant B	Aluminum Subcategory Plant C
Toxic Organic Pollutants (µg/L)			
1,1,1-Trichloroethane	BDL	BDL	—
1,1-Dichloroethane	ND	—	—
1,1-Dichloroethylene	—	ND	—
1,2-*trans*-Dichloroethylene	—	ND	—
2,4-Dimethylphenol	ND	—	—
Fluoranthene	BDL	BDL	ND
Isophorone	ND	BDL	ND
Naphthalene	BDL	BDL	BDL
Phenol	—	—	—
Bis(2-ethylhexyl)phthalate	BDL	42	15
Butyl benzyl phthalate	—	ND	ND
Di-n-butyl phthalate	BDL	BDL	ND
Di-n-octyl phthalate	ND	ND	ND
Diethyl phthalate	BDL	330	140
Dimethyl phthalate	BDL	ND	BDL
1,2-Benzanthracene	BDL	BDL	ND
Benzo(a)pyrene	—	ND	ND
3,4-Benzo fluoranthene	ND	ND	ND
Benzo(k)fluoranthene	ND	ND	ND
Chrysene	BDL	BDL	ND
Acenaphthylene	ND	ND	ND
Anthracene	12	BDL	ND
1,1,2-Benzoperylene	—	ND	ND
Fluorene	BDL	BDL	ND
Phenanthrene	12	BDL	ND
1,2,5,6-Dibenzanthracene	—	ND	ND
Ideno(1,2,3-cd)pyrene	—	ND	ND
Pyrene	BDL	BDL	ND
Toluene	—	—	—
Trichloroethylene	BDL	ND	—

Continued

TABLE 7.7 (continued)

Pollutant	Steel Subcategory Plant A	Galvanized Subcategory Plant B	Aluminum Subcategory Plant C
Toxic Metals and Inorganics (µg/L)			
Cadmium	ND	ND	BDL
Chromium, total	1700	1300	BDL
Chromium, hexavalent	600	ND	ND
Copper	120	BDL	ND
Cyanide, total	BDL	ND	14
Cyanide, amn. to chlor.	ND	ND	ND
Lead	11	ND	ND
Nickel	10	15	ND
Zinc	290	2900	390
Classical Pollutants (mg/L)			
Aluminum	37	3900	5900
Fluorides	—	—	2300
Iron	700	250	130
Manganese	90	BDL	BDL
Oil and grease	3000	10,000	5900
Phenols, total	BDL	33	24
Phosphorus	4600	1700	BDL
TSS	460,000	27,000	8600

BDL, below detection limit; ND, not detected.

Source: U.S. EPA, *Development Document for Effluent Limitations Guidelines and Standards for the Coil Coating Point Source Category*, (Canmaking Subcategory), Final report 440/1-83/071, Washington, DC, November 1983; U.S. EPA, *Coil Coating Forming Point Source Category*, available at http://www.access.gpo.gov/nara/cfr/waisidx_03/40cfr467_03.html, 2008.

7.7.1 PLANT A: COATING COLD ROLLED STEEL AND GALVANIZED STEEL

This site coats cold rolled steel and galvanized steel. The data presented are the analyses of the effluent from the cold rolled steel operations. Approximately 11 million m^2 of steel material are cleaned, coated and painted annually in the U.S. The plant uses water at a rate of 1.2 L/m^2 of product and produces 1630 m^2/hr of coated steel coil.

7.7.2 PLANT B: COATING BOTH COLD ROLLED STEEL AND GALVANIZED STEEL

This facility coats both cold rolled steel and galvanized steel. The data presented are the analyses of the effluent from the galvanized steel operations. Approximately 22 million m^2 of galvanized steel are cleaned and coated and 45 million m^2 painted annually in the U.S. Water is used at a rate of 0.63 L/m^2 of product and the production rate of painted galvanized steel is 2700 m^2/hr.

7.7.3 PLANT C: COATING ALUMINUM AND OTHER METALS

No production information is available for this facility. The data presented are the analyses of the effluent from the aluminum operations. Treatment consists of lagooning and sedimentation.

Tables 7.8 and 7.9 present the major pollutants and combined wastewater characteristics, respectively, of coil coating wastewater streams.

TABLE 7.8
Major Pollutants in Coil Coating Wastewater That Must Be Monitored and Removed

Industrial Category	40 CFR Reference	SIC Codes (Partial List)	Subparts	Promulgation Date	New Source Date	Regulated Parameters for Pretreatment
Coil Coating	465	3411	A—Steel basis material	12/01/82	1/12/81	Cr, Cn, Zn, TTO, oil and grease, Mn, F, P, Cu
Processes by which long thin strips of metal (coils) are cleaned and painted with an organic paint, includes canmaking		3412	B—Galvanized basis material			
		3479	C—Aluminum basis material			
		3497	D—Canmaking	11/17/83	2/10/83	

Source: U.S. EPA, *Development Document for Effluent Limitations Guidelines and Standards for the Coil Coating Point Source Category*, (Canmaking Subcategory), Final report 440/1-83/071, Washington, DC, November 1983; U.S. EPA, *Coil Coating Forming Point Source Category*, available at http://www.access.gpo.gov/nara/cfr/waisidx_03/40cfr467_03.html, 2008.

7.8 COIL COATING EFFLUENT TREATMENT TECHNOLOGIES

This section describes the treatment technologies currently in use to recover or remove wastewater pollutants normally found at coil coating facilities. The treatment processes can be divided into six categories: recovery techniques, oil removal, dissolved inorganics removal, cyanide destruction, trace organics removal, and solids removal.[5–14] Adoption of specific treatment processes will depend on the following:

1. The wastewater characteristics of a specific wastewater stream to be treated
2. The effluent discharge limitations imposed by the Federal and local governments

TABLE 7.9
Heavy Metal Concentration Ranges of Combined Coil Coating Wastewater

Pollutant	Minimum Concentration (mg/L)	Maximum Concentration (mg/L)
Cadmium	<0.1	3.83
Chromium	<0.1	116
Copper	<0.1	108
Lead	<0.1	29.2
Nickel	<0.1	27.5
Zinc	<0.1	337
Iron	<0.1	263
Manganese	<0.1	5.98
TSS	4.6	4390

Source: U.S. EPA, *Development Document for Effluent Limitations Guidelines and Standards for the Coil Coating Point Source Category*, (Canmaking Subcategory), Final report 440/1-83/071, Washington, DC, November 1983; U.S. EPA, *Coil Coating Forming Point Source Category*, available at http://www.access.gpo.gov/nara/cfr/waisidx_03/40cfr467_03.html, 2008.

7.8.1 Ion Exchange

Recovery of process chemicals in coil coating plants is applicable to chromating baths and sealing rinses. Recovery techniques currently in use include ion exchange and electrochemical chromium regeneration.[8,9]

Other possible recovery processes that are not currently in use include evaporation and insoluble starch xanthate. Ion exchange columns are used at four facilities within the coil coating industry. The wastewater stream is filtered to remove solids and then flows through a column of ion exchange resin, which retains copper, iron, and trivalent chromium. The stream then passes through an anion exchanger, which retains hexavalent chromium. Several columns may be necessary to achieve the desired levels. By regenerating the exchange resin, the life expectancy of the column is extended. In some regeneration procedures, hexavalent chromium is removed by conversion to sodium dichromate with sodium hydroxide. The sodium dichromate is then passed through a cation exchanger, which converts it to chromic acid for reuse. The cation exchanger can be regenerated with sulfuric acid.[9] Figure 7.1 illustrates how an ion exchange process can be applied to coil coating effluent treatment and Table 7.10 introduces some anticipated ion exchange capabilities for removal of heavy metals from coil coating wastewater streams.

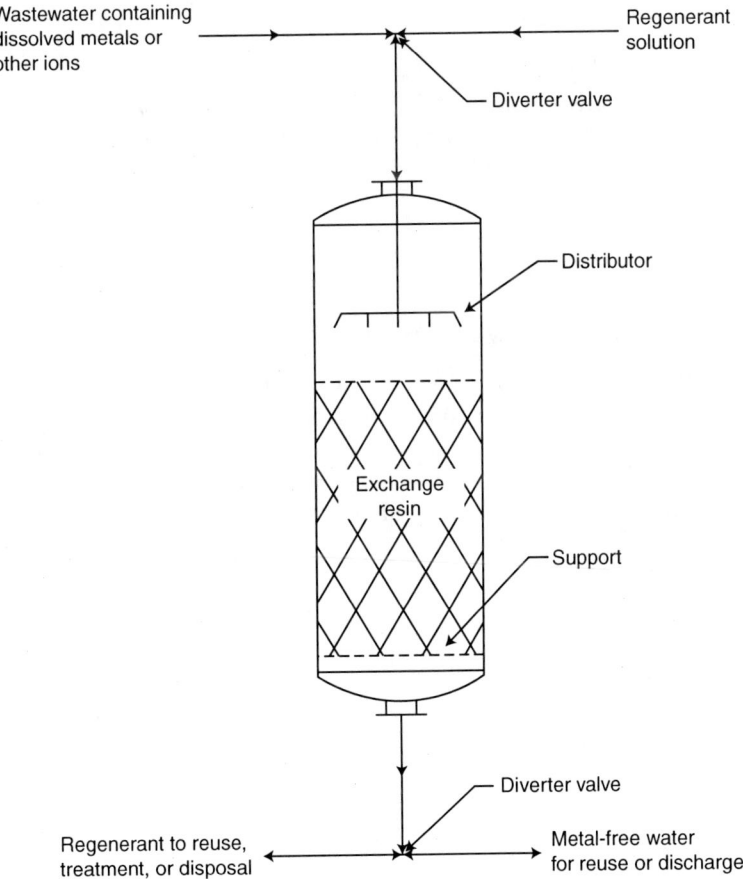

FIGURE 7.1 Ion exchange process. [From U.S. EPA, *Development Document for Effluent Limitations Guidelines and Standards for the Coil Coating Point Source Category*, (Canmaking Subcategory), Final report 440/1-83/071, Washington, DC, November 1983; U.S. EPA, *Coil Coating Forming Point Source Category*, available at http://www.access.gpo.gov/nara/cfr/waisidx_03/40cfr467_03.html, 2008.]

TABLE 7.10
Removal of Heavy Metals and Sulfate from Coil Coating Wastewater by Ion Exchange Process

Parameter (mg/L)	Plant A		Plant B	
	Prior to Purification	After Purification	Prior to Purification	After Purification
Aluminum	5.6	0.20	—	—
Cadmium	5.7	0.00	—	—
Chromium^{3+}	3.1	0.01	—	—
Chromium^{6+}	7.1	0.01	—	—
Copper	4.5	0.09	43.0	0.10
Cyanide	9.8	0.04	3.40	0.09
Gold	—	—	2.30	0.10
Iron	7.4	0.01	—	—
Lead	—	—	1.70	0.01
Manganese	4.4	0.00	—	—
Nickel	6.2	0.00	1.60	0.01
Silver	1.5	0.00	9.10	0.01
SO$_4$	—	—	210.00	2.00
Tin	1.7	0.00	1.10	0.10
Zinc	14.8	0.40	—	—

Source: U.S. EPA, *Development Document for Effluent Limitations Guidelines and Standards for the Coil Coating Point Source Category*, (Canmaking Subcategory), Final report 440/1-83/071, Washington, DC, November 1983; U.S. EPA, *Coil Coating Forming Point Source Category*, available at http://www.access.gpo.gov/nara/cfr/waisidx_03/40cfr467_03.html, 2008.

7.8.2 ELECTROCHEMICAL CHROMIUM REGENERATION

Electrochemical chromium regeneration oxidizes trivalent chromium to hexavalent chromium by electrooxidation. This system can be used with the wastewater or the dragout sludge from a settling basin. One coil coating operation presently uses this technique for chromic acid regeneration. This system offers relatively low energy consumption, operation at normal bath temperatures, elimination of metallic sludges, and regeneration of chromic acid.[8–10]

7.8.3 OIL SKIMMING

Oils occurring in wastewaters from the coil coating industry generally come from cutting fluids, lubricants, and preservative coatings used in metal fabrication operations. Oil skimming is the only current method used in this industry to remove oil. Oil flotation, as shown in Figure 7.2, has been suggested for this industry to achieve low oil concentrations or to remove emulsified oils, but is not in current practice.[8,9] Table 7.11 presents the treatment results of the emulsion breaking process.

Oil skimming as a pretreatment method is effective in removing naturally floating waste material. It can also improve the performance of subsequent downstream treatments. Many coil coating plants employ this treatment process.

7.8.4 CHROMIUM REDUCTION AND CHEMICAL PRECIPITATION

The dissolved inorganic pollutants for the coil coating category are hexavalent chromium, chromium (total), copper, lead, nickel, zinc, cadmium, iron, and phosphorus. Removal of these inorganics is often

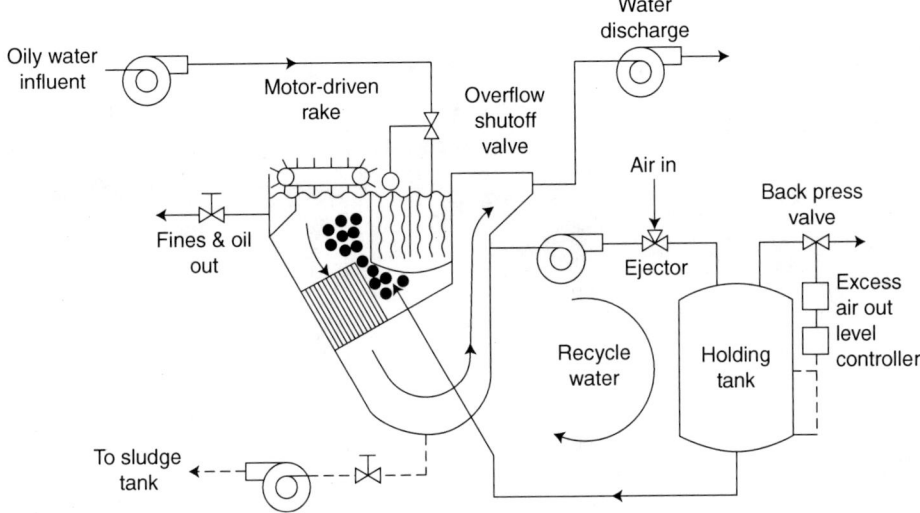

FIGURE 7.2 Dissolved air flotation process. [From U.S. EPA, *Development Document for Effluent Limitations Guidelines and Standards for the Coil Coating Point Source Category*, (Canmaking Subcategory), Final report 440/1-83/071, Washington, DC, November 1983; U.S. EPA, *Coil Coating Forming Point Source Category*, available at http://www.access.gpo.gov/nara/cfr/waisidx_03/40cfr467_03.html, 2008.]

a major step toward detoxifying wastewater. Chromium reduction, which can be carried out chemically or electrochemically, is frequently a preliminary step, as shown in Figure 7.3. The next major step in the classic treatment system is chemical precipitation, which is often accomplished by the addition of lime, sodium sulfide, sodium hydroxide, sodium carbonate, or ammonia. These additives result in the precipitation of metal hydroxides. The chemical reactions of the chromium reduction and chemical precipitation system for coil coating wastewater can be found in the literature.[8,9]

7.8.5 Cyanide Destruction

Cyanide destruction in coil coating facilities is necessary to reduce the cyanide concentration in wastewater from the plating and cleaning baths. Cyanide is generally destroyed by oxidation. Alkaline chlorination is the standard technique used in the coil coating industry, but oxidation by ozone, hydrogen peroxide, or by electrochemical means has been suggested. These alternative techniques, however, have not been demonstrated at this time. The reader is referred to the literature[9] for details of the cyanide destruction and removal system.

7.8.6 Oil–Water Separation, Biological Treatment, Powdered Activated Carbon Adsorption, and Clarification

Plant sampling data show that organic compounds tend to be removed in standard biological wastewater treatment process equipment. Oil separation not only removes oil but also removes organics that are more soluble in the oil than in water. Combined powdered activated carbon (PAC) adsorption and clarification also removes organic solids by adsorption on inorganic solids. PAC adsorption to remove organics has been demonstrated in the electroplating industry, but is not commonly used in the coil coating industry. Wang, Hung, and Shammas[9] have introduced detailed processes for PAC and oil–water separation. Table 7.12 indicates that many toxic organic substances are removed simultaneously when oil is removed during coil coating wastewater treatment.

Clarification by either sedimentation or dissolved air flotation is the most common solid–water separation technique used for the removal of precipitates. In this process application, clarification

TABLE 7.11
Removal of Total Suspended Solids and Oil and Grease from Coil Coating Wastewater by Emulsion Breaking and Clarification Process

Parameter	Concentration (mg/L)		Reference
	Influent	Effluent	
Oil and grease	6060	98	Sampling data[a]
TSS	2612	46	
Oil and grease	13,000	277	Sampling data[b]
	18,400	—	
	21,300	189	
TSS	540	121	
	680	59	
	1060	140	
Oil and grease	2300	52	Sampling data[c]
	12,500	27	
	13,800	18	
TSS	1650	187	
	2200	153	
	3470	63	
Oil and grease	7200	80	

[a] Oil and grease and total suspended solids were taken as grab samples before and after batch emulsion breaking treatment, which used alum and polymer on emulsified rolling oil wastewater.

[b] Oil and grease (grab) and total suspended solids (grab) samples were taken on three consecutive days from emulsified rolling oil wastewater. A commercial demulsifier was used in this batch treatment.

[c] Oil and grease (grab) and total suspended solids (composite) samples were taken on three consecutive days from emulsified rolling oil wastewater. A commercial demulsifier (polymer) was used in this batch treatment.

[d] This result is from a full-scale batch chemical treatment system for emulsified oils from a steel rolling mill.

Source: U.S. EPA, *Development Document for Effluent Limitations Guidelines and Standards for the Coil Coating Point Source Category*, (Canmaking Subcategory), Final report 440/1-83/071, Washington, DC, November 1983; U.S. EPA, *Coil Coating Forming Point Source Category*, available at http://www.access.gpo.gov/nara/cfr/waisidx_03/40cfr467_03.html, 2008.

(sedimentation or dissolved air flotation) is preceded by chemical precipitation, which converts dissolved pollutants to a solid form, and by coagulation, which enhances separation by coagulating suspended precipitates into larger flocs. The major advantage of clarification is the simplicity of the process. Clarification is used in 55 coil coating plants in various forms, including ponds, lagoons, slant tube clarifiers, flotation clarifiers, and Lamella clarifiers.

7.8.7 Granular Bed Filtration and Granular Activated Carbon Filtration

Granular bed filters are used in ten coil coating plants to remove residual solids from the clarifier effluent, and are considered to be tertiary or advanced wastewater treatment. Chemicals may be added upstream to enhance the solids removal. Pressure filtration is also used in this industry to reduce the solids concentration in clarifier effluent and to remove excess water from the clarifier sludge. Figure 7.4 shows a granular bed filter and Table 7.13 presents the heavy metal removal data of a lime clarification and filtration system.

Granular activated carbon (GAC) and peat adsorption are two tertiary wastewater filtration processes using GAC and peat, respectively, as the media for removing not only insoluble suspended solids, but also dissolved organic solids.[8] Tables 7.14 and 7.15 report the adsorption efficiencies for

Waste Treatment and Management in the Coil Coating Industry

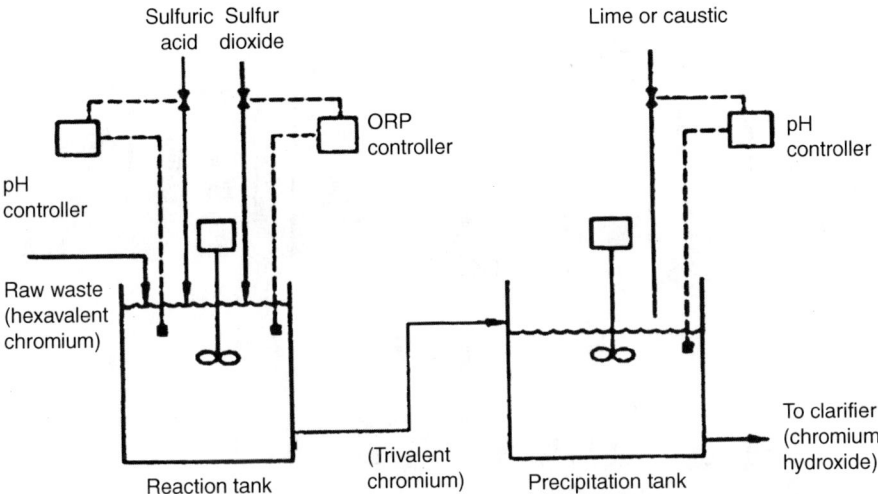

FIGURE 7.3 Chromium reduction and precipitation system. [From U.S. EPA, *Development Document for Effluent Limitations Guidelines and Standards for the Coil Coating Point Source Category*, (Canmaking Subcategory), Final report 440/1-83/071, Washington, DC, November 1983; U.S. EPA, *Coil Coating Forming Point Source Category*, available at http://www.access.gpo.gov/nara/cfr/waisidx_03/40cfr467_03.html, 2008.]

TABLE 7.12
Removal of Total Toxic Organics and Oil and Grease from Coil Coating Wastewater by Oil–Water Separation

	Pollutant Parameter	Influent Concentration (mg/L)	Effluent Concentration (mg/L)
001	Acenaphthene	5.7	ND
038	Ethylbenzene	0.089	0.01
055	Naphthalene	0.75	0.23
062	N-nitrosodiphenylamine	1.5	0.091
065	Phenol	0.18	0.04
066	Bis(2-ethylhexyl)phthalate	1.25	0.01
068	Di-n-butyl phthalate	1.27	0.019
078/081	Anthracene/phenanthrene	2.0	0.1
080	Fluorene	0.76	0.035
084	Pyrene	0.075	0.01
085	Tetrachloroethylene	4.2	0.1
086	Toluene	0.16	0.02
087	Trichloroethylene	4.8	0.01
097	Endosulfan sulfate	0.012	ND
098	Endrin	0.066	0.005
107	PCB 1254[a]	1.1	0.005
110	PCB 1248[b]	1.8	0.005

[a] PCB 1242, PCB 1254, PCB 1221, PCB 1232 reported together.
[b] PCB 1248, PCB 1260, PCB 1016 reported together.

Source: U.S. EPA, *Development Document for Effluent Limitations Guidelines and Standards for the Coil Coating Point Source Category*, (Canmaking Subcategory), Final report 440/1-83/071, Washington, DC, November 1983; U.S. EPA, *Coil Coating Forming Point Source Category*, available at http://www.access.gpo.gov/nara/cfr/waisidx_03/40cfr467_03.html, 2008.

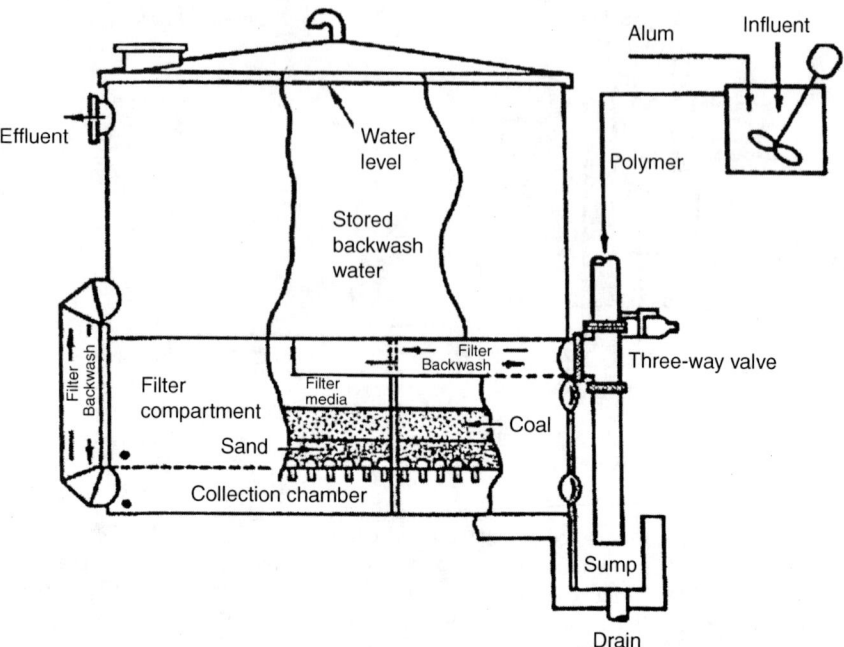

FIGURE 7.4 Granular bed filtration. [From U.S. EPA, *Development Document for Effluent Limitations Guidelines and Standards for the Coil Coating Point Source Category*, (Canmaking Subcategory), Final report 440/1-83/071, Washington, DC, November 1983; U.S. EPA, *Coil Coating Forming Point Source Category*, available at http://www.access.gpo.gov/nara/cfr/waisidx_03/40cfr467_03.html, 2008.]

removing mercury by GAC and heavy metals by peat, respectively. Figure 7.5 is a schematic of the granular activated carbon filtration process.

7.8.8 Membrane Processes

Membrane processes include microfiltration (MF), ultrafiltration (UF), nanofiltration (NF), reverse osmosis (RO), electrodialysis (ED), electrodialysis reversal (EDR), and so on.[9–10,13] MF, UF, NF, and RO are also called membrane filtration processes, of which only UF has been widely used by the coil coating industry for wastewater treatment. Membrane filtration is a physical unit process used to segregate dissolved or suspended solids from a liquid stream on the basis of molecular size. Figures 7.6 and 7.7 illustrate how a UF process works. The ultrafilter membrane forms a molecular screen that separates molecular particles based on their differences in size, shape, and chemical structure. A hydrostatic pressure, ranging from 34 to 690 kPa (5 to 100 psi), is applied to the upstream side of a membrane unit, which acts as a filter passing small particles while blocking (rejecting) larger emulsified and suspended matter. The pores of UF membranes are much smaller than the retained particles, thereby preventing the particles from clogging the membrane. In contrast to ordinary filtration, the concentrated retained particles are continuously washed off the membrane filter rather than being held by the filter. Tables 7.16 and 7.17 show some treatability data of UF in treating the coil coating wastewater streams.

7.8.9 Other Water–Solids Separation Technologies

Other sludge dewatering technologies used include vacuum filtration, centrifugation, and sludge bed drying. No pollutant removability data are currently available for this industry.

TABLE 7.13
Removal of Heavy Metals from Coil Coating Wastewater by Lime Precipitation, Clarification, and Filtration

Parameters	No. of Plants	Range (mg/L)	Mean ± s.d.	Mean + 2 s.d.
For 1979–Treated Wastewater				
Chromium	47	0.015–0.13	0.045 ± 0.029	0.10
Copper	12	0.01–0.03	0.019 ± 0.006	0.03
Nickel	47	0.08–0.64	0.22 ± 0.13	0.48
Zinc	47	0.08–0.53	0.17 ± 0.09	0.35
Iron				
For 1978–Treated Wastewater				
Chromium	47	0.01–0.07	0.06 ± 0.10	0.26
Copper	28	0.005–0.055	0.016 ± 0.010	0.04
Nickel	47	0.10–0.92	0.20 ± 0.14	0.48
Zinc	47	0.08–2.35	0.23 ± 0.34	0.91
Iron	21	0.26–1.1	0.49 ± 0.18	0.85
Raw Waste				
Chromium	5	32.0–72.0		
Copper	5	0.08–0.45		
Nickel	5	1.65–20.0		
Zinc	5	33.2–32.0		
Iron	5	10.0–95.0		

s.d., standard deviation.

Source: U.S. EPA, *Development Document for Effluent Limitations Guidelines and Standards for the Coil Coating Point Source Category*, (Canmaking Subcategory), Final report 440/1-83/071, Washington, DC, November 1983; U.S. EPA, *Coil Coating Forming Point Source Category*, available at http://www.access.gpo.gov/nara/cfr/waisidx_03/40cfr467_03.html, 2008.

TABLE 7.14
Removal of Mercury from Coil Coating Wastewater by Granular Activated Carbon Filtration

Plant	Mercury Levels (mg/L)	
	In	Out
A	28.0	0.9
B	0.36	0.015
C	0.008	0.0005

Source: U.S. EPA, *Development Document for Effluent Limitations Guidelines and Standards for the Coil Coating Point Source Category*, (Canmaking Subcategory), Final report 440/1-83/071, Washington, DC, November 1983; U.S. EPA, *Coil Coating Forming Point Source Category*, available at http://www.access.gpo.gov/nara/cfr/waisidx_03/40cfr467_03.html, 2008.

TABLE 7.15
Removal of Heavy Metals from Coil Coating Wastewater by Peat Adsorption

Pollutant (mg/L)	In	Out
Chromium^{6+}	35,000	0.04
Copper	250	0.24
Cyanide	36.0	0.7
Lead	20.0	0.025
Mercury	1.0	0.02
Nickel	2.5	0.07
Silver	1.0	0.05
Antimony	2.5	0.9
Zinc	1.5	0.25

Source: U.S. EPA, *Development Document for Effluent Limitations Guidelines and Standards for the Coil Coating Point Source Category*, (Canmaking Subcategory), Final report 440/1-83/071, Washington, DC, November 1983; U.S. EPA, *Coil Coating Forming Point Source Category*, available at http://www.access.gpo.gov/nara/cfr/waisidx_03/40cfr467_03.html, 2008.

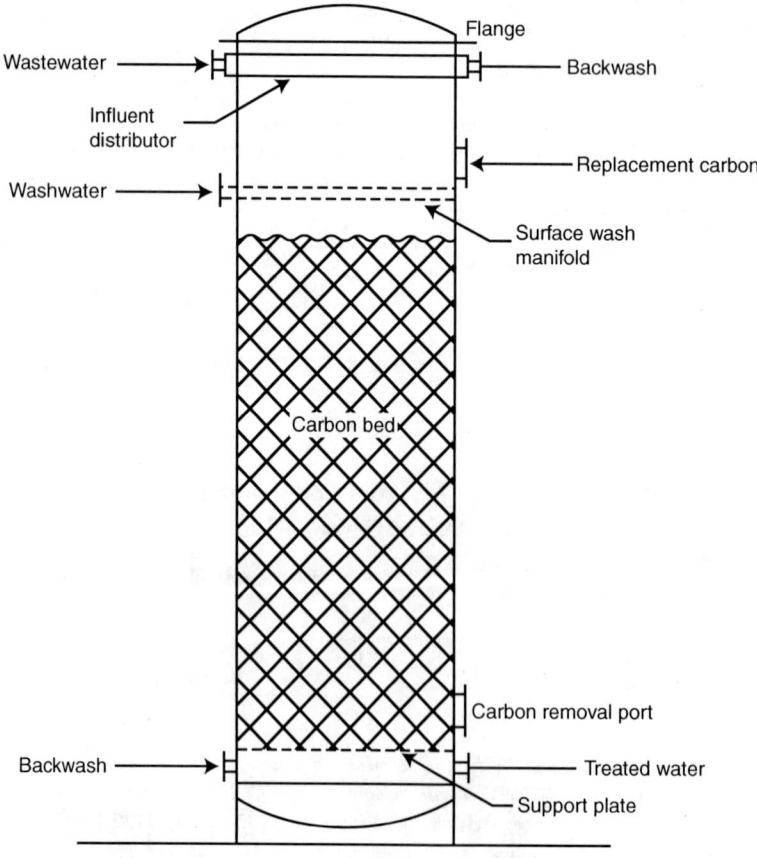

FIGURE 7.5 Schematic of granular activated carbon filtration process. [From U.S. EPA, *Development Document for Effluent Limitations Guidelines and Standards for the Coil Coating Point Source Category*, (Canmaking Subcategory), Final report 440/1-83/071, Washington, DC, November 1983; U.S. EPA, *Coil Coating Forming Point Source Category*, available at http://www.access.gpo.gov/nara/cfr/waisidx_03/40cfr467_03.html, 2008.]

Waste Treatment and Management in the Coil Coating Industry

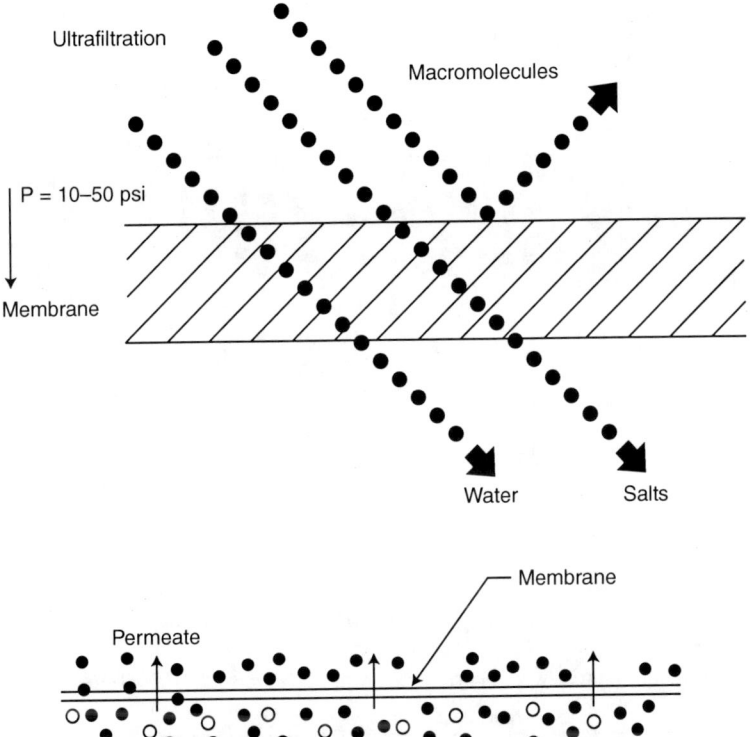

FIGURE 7.6 Membrane ultrafiltration. [From U.S. EPA, *Development Document for Effluent Limitations Guidelines and Standards for the Coil Coating Point Source Category*, (Canmaking Subcategory), Final report 440/1-83/071, Washington, DC, November 1983; U.S. EPA, *Coil Coating Forming Point Source Category*, available at http://www.access.gpo.gov/nara/cfr/waisidx_03/40cfr467_03.html, 2008.]

7.8.10 Full-Scale Wastewater Treatment Case History: Steel Subcategory

Hamilton Standard of the U.S. EPA has reported several coil coating plants' wastewater treatment case histories.[8,9] A full-scale wastewater treatment plant system has performed well for treatment of the wastewater generated from coil coating steel subcategory operations. The process principles and operational data of the full-scale treatment of a steel subcategory wastewater are summarized herein for the convenience of readers:

1. The process flow scheme consists of chromium reduction, lime precipitation, and clarification.
2. The sources of theories and principles for chromium reduction using an acid, chemical precipitation using a base, and clarification are detailed in Refs. 8 to 10.
3. The flow rate of the wastewater treatment facility is 174,000 m^3/d.
4. The acid used for chromium reduction is sulfuric acid.
5. The base used for neutralization and chemical precipitation is lime (note: sodium hydroxide can also be used for neutralization and chemical precipitation).

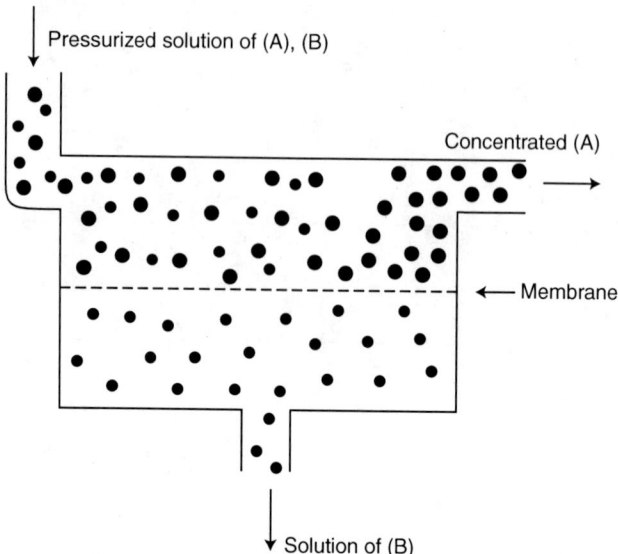

FIGURE 7.7 Schematic of membrane ultrafiltration process. [From U.S. EPA, *Development Document for Effluent Limitations Guidelines and Standards for the Coil Coating Point Source Category*, (Canmaking Subcategory), Final report 440/1-83/071, Washington, DC, November 1983; U.S. EPA, *Coil Coating Forming Point Source Category*, available at http://www.access.gpo.gov/nara/cfr/waisidx_03/40cfr467_03.html, 2008.]

TABLE 7.16
Removal of COD, TSS, TS, and Oil and Grease from Coil Coating Wastewater by Ultrafiltration

Parameter	Feed (mg/L)	Permeate (mg/L)
Oil (freon extractable)	1230	4
COD	8920	148
TSS	1380	13
Total solids	2900	296

Source: U.S. EPA, *Development Document for Effluent Limitations Guidelines and Standards for the Coil Coating Point Source Category*, (Canmaking Subcategory), Final report 440/1-83/071, Washington, DC, November 1983; U.S. EPA, *Coil Coating Forming Point Source Category*, available at http://www.access.gpo.gov/nara/cfr/waisidx_03/40cfr467_03.html, 2008.

TABLE 7.17
Removal of Heavy Metals from Coil Coating Wastewater by Membrane Filtration

Specific Metal, mg/L	Manufacturers' Guarantee	Plant 19066		Plant 31022		Predicted Performance
		In	Out	In	Out	
Aluminum	0.5	—	—	—	—	—
Chromium^{6+}	0.02	0.46	0.01	5.25	<0.005	—
Chromium (total)	0.03	4.13	0.018	98.4	0.057	0.05
Copper	0.1	18.8	0.043	8.00	0.222	0.20

Continued

TABLE 7.17 (continued)

Specific Metal, mg/L	Manufacturers' Guarantee	Plant 19066		Plant 31022		Predicted Performance
Iron	0.1	288	0.3	21.1	0.263	0.30
Lead	0.05	0.652	0.01	0.288	0.01	0.05
Cyanide	0.02	<0.005	<0.005	<0.005	<0.005	0.02
Nickel	0.1	9.56	0.017	194	0.352	0.40
Zinc	0.1	2.09	0.046	5.00	0.051	0.10
TSS	—	632	0.1	13.0	8.0	1.0

Source: U.S. EPA, *Development Document for Effluent Limitations Guidelines and Standards for the Coil Coating Point Source Category*, (Canmaking Subcategory), Final report 440/1-83/071, Washington, DC, November 1983; U.S. EPA, *Coil Coating Forming Point Source Category*, available at http://www.access.gpo.gov/nara/cfr/waisidx_03/40cfr467_03.html, 2008.

---

6. The type of clarification used comprises eight sedimentation tanks (note: dissolved air flotation can also be used for clarification).
7. The sedimentation hydraulic detention time is 3.9 hr.
8. The sedimentation hydraulic loading rate is 733 L/hr/m².
9. The operation mode is continuous, 24 hr/d.
10. The pollutant removal data are as follows:

	Initial Concentration	Reduction (%)
TSS	34 mg/L	82
Iron	2 mg/L	50
Tin	0.02 mg/L	55
Oil and grease	20 mg/L	10
Cobalt	0.5 mg/L	60
Cadmium	8 µg/L	99
Lead	200 µg/L	>99
1,1,1-Trichloroethane	2400 µg/L	88
Trichloroethylene	2700 µg/L	93
1,1-Dichloroethylene	530 µg/L	87
1,2-*trans*-Dichloroethylene	16 µg/L	38
Ethylbenzene	2 µg/L	>99
Isophorone	170 µg/L	35
Tetrachloroethylene	4 µg/L	50
Toluene	29 µg/L	83

7.8.11 Full-Scale Wastewater Treatment Case History: Galvanized Subcategory

A full-scale wastewater treatment plant system has performed well for treatment of the wastewater generated from coil coating galvanized subcategory operations. The process principles and operational data of the full-scale treatment of a galvanized subcategory wastewater are summarized as follows:

1. The process flow diagram consists of chromium reduction, chemical precipitation, and clarification.
2. The sources of theories and principles for chromium reduction using an acid, chemical precipitation using a base, and clarification can be found in Refs. 8 to 10.

3. The flow rate of the wastewater treatment facility is 174,000 m^3/d.
4. The acid used for chromium reduction is sulfuric acid.
5. The base used for neutralization and chemical precipitation is lime (note: sodium hydroxide can also be used for neutralization and chemical precipitation).
6. The type of clarification used comprises eight sedimentation tanks (note: dissolved air flotation can also be used for clarification).
7. The sedimentation hydraulic detention time is 3.9 hr.
8. The sedimentation hydraulic loading rate is 733 L/hr/m^2.
9. The operation mode is continuous, 24 hr/d.
10. The pollutant removal data are as follows:

	Initial Concentration	Reduction (%)
TSS	170 mg/L	88
Iron	44 mg/L	96
Aluminum	1.8 mg/L	62
Oil and grease	54 mg/L	61
Manganese	0.38 mg/L	76
Copper	14 µg/L	>99
Chromium	1300 µg/L	92
Lead	260 µg/L	>99
Zinc	2000 µg/L	95
1,1,1-Trichloroethane	3100 µg/L	19
Trichloroethylene	3800 µg/L	21
1,2-*trans*-Dichloroethylene	34 µg/L	44

7.8.12 FULL-SCALE WASTEWATER TREATMENT CASE HISTORY: ALUMINUM SUBCATEGORY

A full-scale wastewater treatment plant system has performed well for treatment of the wastewater generated from coil coating aluminum subcategory operations. The process principles and operational data of the full-scale treatment of the aluminum subcategory wastewater are summarized as follows:

1. The process flow diagram consists of chromium reduction, chemical precipitation, and clarification.
2. The sources of theories and principles for chromium reduction using an acid, chemical precipitation using a base, and clarification are detailed in Refs. 8 to 10.
3. The flow rate of the wastewater treatment facility is 3930 L/d.
4. The acid used for chromium reduction is sulfuric acid.
5. The base used for neutralization and chemical precipitation is sodium hydroxide (note: lime can also be used for neutralization and chemical precipitation).
6. The type of clarification used consists of tube plate settlers (note: dissolved air flotation can also be used for clarification).
7. The operation mode is continuous, 24 hr/d.
8. The pollutant removal data are as follows:

	Initial Concentration	Reduction (%)
TSS	530 mg/L	93
Iron	7.3 mg/L	99
Phosphorus	46 mg/L	97
Oil and grease	1400 mg/L	98

Continued

	Initial Concentration	Reduction (%)
Phenol, total	0.07 mg/L	71
Aluminum	530 mg/L	99
Manganese	1.1 mg/L	99
Cadmium	5.5 µg/L	>99
Chromium	330,000 µg/L	99
Copper	220 µg/L	95
Nickel	95 µg/L	>99
Zinc	19,000 µg/L	>99
Lead	115 µg/L	>99
Bis(2-ethylhexyl)phthalate	140 µg/L	96
Diethyl phthalate	240 µg/L	99
Hexavalent chromium	140,000 µg/L	>99

7.9 WASTEWATER TREATMENT LEVELS VERSUS COSTS

The investment cost, operating and maintenance costs, and energy costs for the application of control technologies to the wastewater of the coil coating industry have been analyzed. These costs were developed to reflect the practical application of technologies in this industry. A detailed presentation of the cost methodology and cost data is available in the literature.[1-19]

Application of the wastewater treatment technologies can fall into one of the following legal categories:

1. The best practicable control technology (BPT) currently available under the U.S. Federal Act Section 304(b)(1)
2. The best available technology (BAT) economically achievable under the U.S. Federal Act Section 304(b)(2)(B)
3. The best conventional pollutant control technology (BCT), under U.S. Federal Act Section 304(b)(4)

The available industry-specific cost information is characterized as follows. Unit operation/unit process configurations have been analyzed for the cost of application to the wastewater of this industry. Recommended unit process configurations for BPT and BAT levels of treatment and their costs are summarized briefly in the following sections.

7.9.1 BPT Level Treatment

7.9.1.1 Suggested BPT

The BPT treatment train for the steel, galvanized, and aluminum subcategories of wastewater consists of chemical oxidation of cyanide and chemical reduction of chromium for cyanide- and chromium-bearing wastestreams; oil skimming, chemical precipitation with lime, and sedimentation of combined wastestreams; and a vacuum filter to dewater sludge. For the purpose of cost estimates, cyanide oxidation was assumed to be a required treatment process only for the aluminum subcategory, because of the presence of cyanide in the chromating bath applied to aluminum. Chromium reduction was included in the system costs for all subcategories to treat chromium wastes from the chromic acid sealer and conversion coating rinses, where appropriate.

7.9.1.2 System Component of the Suggested BPT

Cyanide oxidation consists of a reaction with sodium hypochlorite under alkaline conditions in either a batch or continuous system. A complete system includes reactors, sensors, controls, mixers, and chemical feed equipment. Chromium reduction consists of reaction with sulfur dioxide under acid conditions for continuous systems and reaction with sodium bisulfite under acid conditions for batch systems. A complete system consists of reaction tanks, mixers and controls, and chemical feed equipment. Oil is separated from process wastewater by gravity in a baffled rectangular concrete tank and removed by a skimming device. Chemical precipitation and sedimentation may be by either a continuous or batch treatment system. A continuous system includes chemical storage and feeding equipment, a mix tank for reagent feed addition, a flocculator, and settling tank with associated equipment. A batch treatment system consists of dual tanks and chemical storage and feeding equipment.

7.9.1.3 Unit Cost of the Suggested BPT

Total annual unit costs consisting of annual cost of capital, depreciation, operation and maintenance cost, and energy cost for medium, low, and high flow rates are summarized in Table 7.18.

7.9.2 BAT LEVEL OF TREATMENT

7.9.2.1 Suggested BAT

The BAT level of treatment consists of all components of BPT except segregation and recirculation of quench wastewater. The combined wastewater after sedimentation is treated in multimedia filters and then discharged.

7.9.2.2 System Components of the Suggested BAT

Quench waste recirculation requires installation of a cooling tower to lower the temperature of the quench wastewater stream. The multimedia filter system for the final polishing of effluent includes a backwash mechanism, pumps, control media, and the filter structure.

7.9.2.3 Unit Cost of the Suggested BAT

Total annual unit costs for the complete BAT system, which includes components described in the BPT system for the three different flow rates, are summarized in Table 7.19.

7.10 MULTIMEDIA WASTE MANAGEMENT IN THE COIL COATING INDUSTRY

7.10.1 AIR POLLUTION CONTROL

In the U.S., the Clean Air Act regulates the emission of volatile organic compounds (VOCs) (40 CFR Part 60) and hazardous air pollutants (HAPs) (40 CFR Part 61 and 40 CFR Part 63).

TABLE 7.18
Total Annual Unit Cost for BPT Level of Treatment in 2008 U.S. Dollars

Coil Coating Subcategory	Flow[a] (L/min)	Cost ($/m³)	Flow[a] (L/min)	Cost ($/m³)	Flow[a] (L/min)	Cost ($/m³)
Steel	4	21	342	0.94	657	0.47
Galvanized	26	5.6	131	1.4	526	0.70
Aluminum	13	12	158	1.9	394	1.2

[a] For flows less than 342 L/min treatment is by batch system.

Source: U.S. EPA, *Development Document for Effluent Limitations Guidelines and Standards for the Coil Coating Point Source Category*, (Canmaking Subcategory), Final report 440/1-83/071, Washington, DC, November 1983; U.S. EPA, *Coil Coating Forming Point Source Category*, available at http://www.access.gpo.gov/nara/cfr/waisidx_03/40cfr467_03.html, 2008.

TABLE 7.19
Total Annual Unit Cost for BAT Level of Treatment in 2008 U.S. Dollars

Coil Coating Subcategory	Flow (L/min)	Cost ($/m³)	Flow (L/min)	Cost ($/m³)	Flow (L/min)	Cost ($/m³)
Steel	2	NA	137	2.6	263	1.6
Galvanized	23	28	46	7.0	184	2.3
Aluminum	5	47	55	7.0	138	3.7

Note: All costs are for batch treatment systems.

Source: U.S. EPA, *Development Document for Effluent Limitations Guidelines and Standards for the Coil Coating Point Source Category*, (Canmaking Subcategory), Final report 440/1-83/071, Washington, DC, November 1983; U.S. EPA, *Coil Coating Forming Point Source Category*, available at http://www.access.gpo.gov/nara/cfr/waisidx_03/40cfr467_03.html, 2008.

Depending on the solvent content of the coating material used with roll and coil methods, solvents can evaporate and produce sufficient VOC and HAP emissions to subject an operator to major source requirements and Title V permitting requirements.[15] The Act also provides specific standards of performance to control emissions from coil coating operations (40 CFR Part 60 TT). Controlling VOC emissions from roll and coil coating areas can be accomplished in several ways. First, a coating material with a lower VOC content can be used. Second, air pollution control equipment can be attached to the ventilation system to capture VOCs prior to their release into the atmosphere. Roll and coil coating systems apply coating materials, which may include solvents classified as volatile organic compounds and/or hazardous air pollutants. The solvents evaporate and may accumulate above limits allowed by Clean Air Act Title V permits. Ventilation and exhaust systems must operate properly to ensure the vapors are removed from the coating area. Air pollution control equipment should be attached to exhaust systems to recover or destroy volatile organic compounds instead of releasing them to the air. All air pollution control technologies are discussed in two recent books published by Wang, Pereira, and Hung.[6,7]

7.10.2 Water Pollution Control

As part of the Clean Water Act, Effluent Guidelines and Standards for Coil Coating (40 CFR Part 465) have been established that limit concentrations of heavy metals, toxic organics, and conventional pollutants in wastewater streams in the U.S.[15] The organic solvents often contained in liquid coatings used with roll and coil coating application methods may be classified as toxic organics. These materials can enter the wastewater when cleaning coatings from containers or equipment. Actual limits for effluent constituents are dependent on the size of the operation and the amount of wastewater generated from the facility. If the facility discharges directly to receiving waters, these limits will be established through the facility's National Pollutant Discharge Elimination System (NPDES) permit (40 CFR Part 122). Facilities that are indirect dischargers releasing to a POTW must meet limits in the POTW's discharge agreement. Roll and coil coating systems utilize liquid coating materials and solvent and water rinses, which can contaminate water streams. Contamination may occur when cleaning equipment or from accidental spills or leaks from equipment. Contaminated water streams may contain pollutants or heavy metals in concentrations that exceed the limits established by facility NPDES or POTW discharge agreement permits. In such cases, effluent may not be directly released to water systems or to publicly owned treatment works without pretreatment.

Wastewater streams with concentrations exceeding permit limits will require pretreatment prior to discharge to receiving waters or to publicly owned treatment works. Pretreatment may include separation of liquid wastes to remove big suspended solids, oils, solvents, and so on, as discussed in Section 7.8.

7.10.3 Solid and Hazardous Wastes Management and Disposal

Under the Resource Conservation and Recovery Act (RCRA), organic finishing facilities are required to manage listed and characteristic hazardous wastes (40 CFR Part 261) in the U.S.[15] Liquid coatings used with roll and coil coating application methods may contain constituents listed or characterized as hazardous wastes. Materials contaminated with the coatings, such as roller surface covers, conveyor components, and rags or other materials used for cleaning, may require treatment as hazardous waste depending on their formulation. Hazardous waste management (40 CFR Part 262) includes obtaining permits for the facility in order to generate wastes, meeting accumulation limits for waste storage areas, and manifesting waste containers for offsite disposal.

Roll and coil coating systems utilize liquid coating materials with organic solvents, which must be stored, manifested, and disposed of according to 40 CFR Part 262 if classified as hazardous waste under 40 CFR Part 261.

Responsibilities will vary according to the amount of hazardous waste generated; facilities generating at least 100 kg of hazardous waste per month must comply with the Federal and the State hazardous waste generator requirements. Each state or region is primarily responsible for the regulation of nonhazardous solid wastes (those not governed by the hazardous waste provisions of RCRA). State environmental agencies should be contacted for specific guidance.

7.10.4 Waste Minimization and Cleaner Production Alternatives for Roll and Coil Coating

"Cleaner production" means the conceptual and procedural approach to production that demands that all phases of the life-cycle of a product or of a process should be addressed with the objectives of prevention of pollution and minimization of short- and long-term risks to humans and the environment.

The following pollution prevention and cleaner production alternatives are recommended by the Paints and Coatings Resource Center of U.S. EPA[15]:

1. Liquid coating materials with low organic solvent content should be used to minimize the amount of volatile organic compounds that will be volatized and to reduce the volume of solid and liquid hazardous waste created.
2. The use of roll and coil coating systems leads to pollution prevention over traditional spray application systems due to their higher transfer efficiency (>95%) and lower volatilization of organic solvents.
3. Paint jobs should be scheduled to minimize changing colors in roll and coil coating equipment. Paint with light colors first, then darker ones; the lighter coating does not need to be completely removed from the equipment, but can blend into the darker coating. As most roll and coil lines apply only one color, this is typically not an issue.
4. Roll and coil coating equipment should be cleaned regularly to prevent coating materials from drying on rollers and feed lines. Water should be used in cleaning steps to reduce the amount of organic solvents used and the amount of hazardous waste generated. Initial cleaning should be performed with used solvents, saving fresh solvents for final cleaning stages.
5. Nonhazardous coating solids and water should be segregated from hazardous solvents and thinners, and containers labeled to prevent mixing. Separation of the materials reduces the amount of hazardous waste that is produced. Coating material solids can be dried and treated as a solid waste allowing for disposal in a landfill.
6. Roll and coil coating equipment should be maintained to sustain proper operation. Valves, gages, and rollers should be checked to ensure they are in proper working order.
7. Roll and coil coating areas should be kept clean so that problems with equipment can be found and fixed quickly, and accidents prevented.

8. Employees should be trained on safe handling of materials and wastes and encouraged to continuously improve. Training familiarizes workers with their responsibilities, which reduces spills and accidents.

7.11 COIL COATING INDUSTRY LIQUID EFFLUENT LIMITATIONS, PERFORMANCE STANDARDS, AND PRETREATMENT STANDARDS

Table 7.8 indicates that the major regulated parameters for pretreatment of coil coating industrial wastewater are chromium, cyanide, zinc, total toxic organics (TTO), oil and grease, manganese, fluoride, phosphorus, and copper. The effluent limitations of all coil coating operations represent the following:

1. The degree of effluent reduction attainable by the application of the BPT currently available
2. The degree of effluent reduction attainable by the application of the BAT economically achievable, which can be found from the U.S. Code of Federal Register 40CFR–Chapter I, Part 465, Coil Coating Point Source Category[4]

Also documented in this chapter for reference by the readers are the latest U.S. performance standards and pretreatment standards of four subcategories:

1. Steel basis material
2. Galvanized basis material
3. Aluminum basis material
4. Canmaking

7.11.1 EFFLUENT LIMITATIONS, PERFORMANCE STANDARDS, AND PRETREATMENT STANDARDS OF THE STEEL BASIS MATERIAL SUBCATEGORY

Table 7.20 shows the effluent limitations of the steel basis material subcategory that represents the degree of effluent reduction attainable by the application of the BPT currently available. Table 7.21

TABLE 7.20
The Effluent Limitations of the Steel Basis Material Subcategory That Represent the Degree of Effluent Reduction Attainable by the Application of the BPT Currently Available

Pollutant or Pollutant Property	BPT Effluent Limitations [mg/m² (lb/10⁶ ft²) of Area Processed]			
	Maximum for Any 1 Day		Maximum for Monthly Average	
Chromium	1.16	(0.24)	0.47	(0.096)
Cyanide	0.80	(0.17)	0.33	(0.068)
Zinc	3.66	(0.75)	1.54	(0.32)
Iron	3.39	(0.70)	1.74	(0.36)
Oil and grease	55.1	(11.3)	33.1	(6.77)
TSS	113.0	(23.1)	55.1	(11.3)
pH	7.5–10.0[a]	7.5–10.0[a]	7.5–10.0[a]	7.5–10.0[a]

[a] Within this range at all times.
Source: U.S. EPA. *Code of Federal Regulations (CFR). Title 40 Protection of Environment. Part 465. Coil Coating Point Source Category.* Washington, DC, 2008.

TABLE 7.21
The Effluent Limitations of the Steel Basis Material Subcategory That Represent the Degree of Effluent Reduction Attainable by the Application of the BAT Economically Achievable

Pollutant or Pollutant Property	BPT Effluent Limitations [mg/m² (lb/10⁶ ft²) of Area Processed]			
	Maximum for Any 1 Day		Maximum for Monthly Average	
Chromium	0.50	(0.10)	0.20	(0.041)
Cyanide	0.34	(0.07)	0.14	(0.029)
Zinc	1.56	(0.32)	0.66	(0.14)
Iron	1.45	(0.30)	0.74	(0.15)

Source: U.S. EPA. *Code of Federal Regulations (CFR). Title 40 Protection of Environment. Part 465. Coil Coating Point Source Category.* Washington, DC, 2008.

shows the effluent limitations of the steel basis material subcategory that represents the degree of effluent reduction attainable by the application of the BAT economically achievable. Table 7.22 lists the new source performance standards of the steel basis material subcategory that establish the quantity or quality of pollutants or pollutant properties. Table 7.23 lists the pretreatment standards for existing sources of the steel basis material subcategory effluents. Table 7.24 lists the pretreatment standards for new sources of the steel basis material subcategory effluents.

7.11.2 Effluent Limitations and Performance Standards of the Galvanized Basis Material Subcategory

Table 7.25 shows the effluent limitations of the galvanized basis material subcategory that represents the degree of effluent reduction attainable by the application of the BPT currently available. Table 7.26 shows the effluent limitations of the galvanized basis material subcategory that represents the degree of effluent reduction attainable by the application of the BAT economically achievable.

TABLE 7.22
The New Source Performance Standards (NSPS) of the Steel Basis Material Subcategory That Establish the Quantity or Quality of Pollutants or Pollutant Properties

Pollutant or Pollutant Property	NSPS [mg/m² (lb/10⁶ ft²) of Area Processed]			
	Maximum for Any 1 Day		Maximum for Monthly Average	
Chromium	0.12	(0.024)	0.047	(0.01)
Cyanide	0.063	(0.013)	0.025	(0.005)
Zinc	0.33	(0.066)	0.14	(0.027)
Iron	0.39	(0.086)	0.20	(0.041)
Oil and grease	3.16	(0.65)	3.16	(0.65)
TSS	4.74	(0.97)	3.79	(0.78)
pH	7.5–10.0ᵃ	7.5–10.0ᵃ	7.5–10.0ᵃ	7.5–10.0ᵃ

ᵃ Within this range at all times.

Source: U.S. EPA. *Code of Federal Regulations (CFR). Title 40 Protection of Environment. Part 465. Coil Coating Point Source Category.* Washington, DC, 2008.

TABLE 7.23
The Pretreatment Standards (PSES) for Existing Sources of the Steel Basis Material Subcategory Effluents

	PSES [mg/m² (lb/10⁶ ft²) of Area Processed]			
Pollutant or Pollutant Property	Maximum for Any 1 Day		Maximum for Monthly Average	
Chromium	0.50	(0.10)	0.20	(0.041)
Cyanide	0.34	(0.07)	0.14	(0.029)
Zinc	1.56	(0.32)	0.66	(0.14)

Source: U.S. EPA. *Code of Federal Regulations (CFR)*. Title 40 Protection of Environment. Part 465. Coil Coating Point Source Category. Washington, DC, 2008.

TABLE 7.24
The Pretreatment Standards (PSNS) for New Sources of the Steel Basis Material Subcategory Effluents

	PSNS [mg/m² (lb/10⁶ ft²) of Area Processed]			
Pollutant or Pollutant Property	Maximum for Any 1 Day		Maximum for Monthly Average	
Chromium	0.12	(0.024)	0.047	(0.01)
Cyanide	0.063	(0.013)	0.025	(0.005)
Zinc	0.33	(0.066)	0.14	(0.027)

Source: U.S. EPA. *Code of Federal Regulations (CFR)*. Title 40 Protection of Environment. Part 465. Coil Coating Point Source Category. Washington, DC, 2008.

TABLE 7.25
The Effluent Limitations of the Galvanized Basis Material Subcategory That Represent the Degree of Effluent Reduction Attainable by the Application of the BPT Currently Available

	BPT Effluent Limitations [mg/m² (lb/10⁶ ft²) of Area Processed]			
Pollutant or Pollutant Property	Maximum for Any 1 Day		Maximum for Monthly Average	
Chromium	1.10	(0.23)	0.45	(0.091)
Copper	4.96	(1.02)	2.61	(0.54)
Cyanide	0.76	(0.16)	0.32	(0.064)
Zinc	3.47	(0.71)	1.46	(0.30)
Iron	3.21	(0.66)	1.65	(0.34)
Oil and grease	52.2	(10.7)	31.3	(6.42)
TSS	107.0	(21.9)	52.2	(10.7)
pH	7.5–10.0[a]	7.5–10.0[a]	7.5–10.0[a]	7.5–10.0[a]

[a] Within this range at all times.

Source: U.S. EPA. *Code of Federal Regulations (CFR)*. Title 40 Protection of Environment. Part 465. Coil Coating Point Source Category. Washington, DC, 2008.

TABLE 7.26
The Effluent Limitations of the Galvanized Basis Material Subcategory That Represent the Degree of Effluent Reduction Attainable by the Application of the BAT Economically Achievable

Pollutant or Pollutant Property	BAT Effluent Limitations [mg/m² (lb/10⁶ ft²) of Area Processed]			
	Maximum for Any 1 Day		Maximum for Monthly Average	
Chromium	0.37	(0.077)	0.16	(0.031)
Copper	1.71	(0.35)	0.90	(0.19)
Cyanide	0.26	(0.053)	0.11	(0.022)
Zinc	1.20	(0.25)	0.51	(0.11)
Iron	1.10	(0.23)	0.57	(0.12)

Source: U.S. EPA. *Code of Federal Regulations (CFR). Title 40 Protection of Environment. Part 465. Coil Coating Point Source Category.* Washington, DC, 2008.

Table 7.27 lists the new source performance standards of the galvanized basis material subcategory that establish the quantity or quality of pollutants or pollutant properties. Table 7.28 lists the pretreatment standards for existing sources of the galvanized basis material subcategory effluents. Table 7.29 lists the pretreatment standards for new sources of the galvanized basis material subcategory effluents.

7.11.3 EFFLUENT LIMITATIONS AND PERFORMANCE STANDARDS OF THE ALUMINUM BASIS MATERIAL SUBCATEGORY

Table 7.30 shows the effluent limitations of the aluminum basis material subcategory that represents the degree of effluent reduction attainable by the application of the BPT currently available. Table 7.31 shows the effluent limitations of the aluminum basis material subcategory that represents

TABLE 7.27
The New Source Performance Standards (NSPS) of the Galvanized Basis Material Subcategory That Establish the Quantity or Quality of Pollutants or Pollutant Properties

Pollutant or Pollutant Property	NSPS [mg/m² (lb/10⁶ ft²) of Area Processed]			
	Maximum for Any 1 Day		Maximum for Monthly Average	
Chromium	0.13	(0.027)	0.052	(0.011)
Copper	0.44	(0.090)	0.21	(0.043)
Cyanide	0.07	(0.015)	0.028	(0.006)
Zinc	0.35	(0.08)	0.15	(0.030)
Iron	0.43	(0.09)	0.22	(0.045)
Oil and grease	3.43	(0.71)	3.43	(0.702)
TSS	5.15	(1.06)	4.12	(0.84)
pH	7.5–10.0[a]	7.5–10.0[a]	7.5–10.0[a]	7.5–10.0[a]

[a] Within this range at all times.

Source: U.S. EPA. *Code of Federal Regulations (CFR). Title 40 Protection of Environment. Part 465. Coil Coating Point Source Category.* Washington, DC, 2008.

TABLE 7.28
The Pretreatment Standards for Existing Sources (PSES) of the Galvanized Basis Material Subcategory Effluents

Pollutant or Pollutant Property	PSES [mg/m² (lb/10⁶ ft²) of Area Processed]			
	Maximum for Any 1 Day		Maximum for Monthly Average	
Chromium	0.37	(0.077)	0.16	(0.031)
Copper	1.71	(0.35)	0.90	(0.19)
Cyanide	0.26	(0.053)	0.11	(0.022)
Zinc	1.20	(0.25)	0.51	(0.11)

Source: U.S. EPA. *Code of Federal Regulations (CFR). Title 40 Protection of Environment. Part 465. Coil Coating Point Source Category.* Washington, DC, 2008.

TABLE 7.29
The Pretreatment Standards for New Sources (PSNS) of the Galvanized Basis Material Subcategory Effluents

Pollutant or Pollutant Property	PSNS [mg/m² (lb/10⁶ ft²) of Area Processed]			
	Maximum for Any 1 Day		Maximum for Monthly Average	
Chromium	0.13	(0.027)	0.052	(0.011)
Copper	0.44	(0.090)	0.21	(0.043)
Cyanide	0.07	(0.015)	0.028	(0.006)
Zinc	0.35	(0.072)	0.15	(0.030)

Source: U.S. EPA. *Code of Federal Regulations (CFR). Title 40 Protection of Environment. Part 465. Coil Coating Point Source Category.* Washington, DC, 2008.

TABLE 7.30
The Effluent Limitations of the Aluminum Basis Material Subcategory That Represent the Degree of Effluent Reduction Attainable by the Application of the BPT Currently Available

Pollutant or Pollutant Property	BPT Effluent Limitations [mg/m² (lb/10⁶ ft²) of Area Processed]			
	Maximum for Any 1 Day		Maximum for Monthly Average	
Chromium	1.42	(0.29)	0.58	(0.12)
Cyanide	0.98	(0.20)	0.41	(0.083)
Zinc	4.48	(0.92)	1.89	(0.39)
Aluminum	15.3	(3.14)	6.26	(1.28)
Oil and grease	67.3	(13.8)	40.4	(8.27)
TSS	138.0	(28.3)	67.3	(13.8)
pH	7.5–10.0[a]	7.5–10.0[a]	7.5–10.0[a]	7.5–10.0[a]

[a] Within this range at all times.

Source: U.S. EPA. *Code of Federal Regulations (CFR). Title 40 Protection of Environment. Part 465. Coil Coating Point Source Category.* Washington, DC, 2008.

TABLE 7.31
The Effluent Limitations of the Aluminum Basis Material Subcategory That Represent the Degree of Effluent Reduction Attainable by the Application of the BAT Economically Achievable

Pollutant or Pollutant Property	BAT Effluent Limitations [mg/m² (lb/10⁶ ft²) of Area Processed]			
	Maximum for Any 1 Day		Maximum for Monthly Average	
Chromium	0.42	(0.085)	0.17	(0.034)
Cyanide	0.29	(0.059)	0.12	(0.024)
Zinc	1.32	(0.27)	0.56	(0.12)
Aluminum	4.49	(0.92)	1.84	(0.38)

Source: U.S. EPA. *Code of Federal Regulations (CFR). Title 40 Protection of Environment. Part 465. Coil Coating Point Source Category.* Washington, DC, 2008.

the degree of effluent reduction attainable by the application of the BAT economically achievable. Table 7.32 lists the new source performance standards of the aluminum basis material subcategory that establish the quantity or quality of pollutants or pollutant properties. Table 7.33 lists the pretreatment standards for existing sources of the aluminum basis material subcategory effluents. Table 7.34 lists the pretreatment standards for new sources of the aluminum basis material subcategory effluents.

7.11.4 EFFLUENT LIMITATIONS, PERFORMANCE STANDARDS, AND PRETREATMENT STANDARDS OF CANMAKING SUBCATEGORY

Table 7.35 shows the effluent limitations of the canmaking subcategory that represents the degree of effluent reduction attainable by the application of the best practicable control technology (BPT) currently available. Table 7.36 shows the effluent limitations of the canmaking subcategory

TABLE 7.32
The New Source Performance Standards (NSPS) of the Aluminum Basis Material Subcategory That Establish the Quantity or Quality of Pollutants or Pollutant Properties

Pollutant or Pollutant Property	NSPS [mg/m² (lb/10⁶ ft²) of Area Processed]			
	Maximum for Any 1 Day		Maximum for Monthly Average	
Chromium	0.18	(0.037)	0.072	(0.015)
Cyanide	0.095	(0.020)	0.038	(0.008)
Zinc	0.49	(0.10)	0.20	(0.041)
Aluminum	1.44	(0.30)	0.59	(0.121)
Oil and grease	4.75	(0.98)	4.75	(0.98)
TSS	7.13	(1.46)	5.70	(1.17)
pH	7.5–10.0[a]	7.5–10.0[a]	7.5–10.0[a]	7.5–10.0[a]

[a] Within this range at all times.

Source: U.S. EPA. *Code of Federal Regulations (CFR). Title 40 Protection of Environment. Part 465. Coil Coating Point Source Category.* Washington, DC, 2008.

TABLE 7.33
The Pretreatment Standards for Existing Sources (PSES) of the Aluminum Basis Material Subcategory Effluents

Pollutant or Pollutant Property	PSES [mg/m² (lb/10⁶ ft²) of Area Processed]			
	Maximum for Any 1 Day		Maximum for Monthly Average	
Chromium	0.42	(0.085)	0.17	(0.034)
Cyanide	0.29	(0.059)	0.12	(0.024)
Zinc	1.32	(0.27)	0.56	(0.12)

Source: U.S. EPA. *Code of Federal Regulations (CFR). Title 40 Protection of Environment. Part 465. Coil Coating Point Source Category.* Washington, DC, 2008.

TABLE 7.34
The Pretreatment Standards for New Sources (PSNS) of the Aluminum Basis Material Subcategory Effluents

Pollutant or Pollutant Property	PSNS [mg/m² (lb/10⁶ ft²) of Area Processed]			
	Maximum for Any 1 Day		Maximum for Monthly Average	
Chromium	0.18	(0.037)	0.072	(0.015)
Cyanide	0.095	(0.02)	0.038	(0.008)
Zinc	0.49	(0.10)	0.20	(0.041)

Source: U.S. EPA. *Code of Federal Regulations (CFR). Title 40 Protection of Environment. Part 465. Coil Coating Point Source Category.* Washington, DC, 2008.

TABLE 7.35
The Effluent Limitations of the Canmaking Subcategory That Represent the Degree of Effluent Reduction Attainable by the Application of the BPT Currently Available

Pollutant or Pollutant Property	BPT Effluent Limitations [g (lb)/10⁶ Cans Manufactured]	
	Maximum for Any 1 Day	Maximum for Monthly Average
Chromium	94.60 (0.209)	38.70 (0.085)
Zinc	313.90 (0.692)	131.15 (0.289)
Aluminum	1382.45 (3.048)	688.00 (1.517)
Fluoride	12,792.50 (28.203)	5676.00 (12.514)
Phosphorus	3590.50 (7.916)	1468.45 (3.237)
Oil and grease	4300.00 (9.480)	2580.00 (5.688)
TSS	8815.00 (19.434)	4192.50 (9.243)
pH	7.0–10.0[a]	7.0–10.0[a]

[a] Within this range at all times.

Source: U.S. EPA. *Code of Federal Regulations (CFR). Title 40 Protection of Environment. Part 465. Coil Coating Point Source Category.* Washington, DC, 2008.

TABLE 7.36
The Effluent Limitations of the Canmaking Subcategory That Represent the Degree of Effluent Reduction Attainable by the Application of the BAT Economically Achievable

Pollutant or Pollutant Property	BAT Effluent Limitations [g (lb)/10^6 Cans Manufactured]	
	Maximum for Any 1 Day	Maximum for Monthly Average
Chromium	36.92 (0.081)	15.10 (0.033)
Zinc	122.49 (0.270)	51.18 (0.113)
Aluminum	539.48 (1.189)	268.48 (0.592)
Fluoride	4992.05 (11.001)	2214.96 (4.883)
Phosphorus	1401.13 (3.089)	573.04 (1.263)

Source: U.S. EPA. *Code of Federal Regulations (CFR). Title 40 Protection of Environment. Part 465. Coil Coating Point Source Category.* Washington, DC, 2008.

that represents the degree of effluent reduction attainable by the application of the best available technology (BAT) economically achievable. Table 7.37 lists the new source performance standards of the canmaking subcategory that establish the quantity or quality of pollutants or pollutant properties. Table 7.38 lists the pretreatment standards for existing sources of the canmaking subcategory effluents. Table 7.39 lists the pretreatment standards for new sources of the canmaking subcategory effluents.

7.12 TECHNICAL TERMINOLOGIES OF COIL COATING OPERATIONS AND POLLUTION CONTROL

1. "Cleaner Production" means the conceptual and procedural approach to production that demands that all phases of the life-cycle of a product or of a process should be addressed with the objectives of prevention of pollution and minimization of short- and long-term risks to humans and the environment.

TABLE 7.37
The New Source Performance Standards (NSPS) of the Canmaking Subcategory That Establish the Quantity or Quality of Pollutants or Pollutant Properties

Pollutant or Pollutant Property	NSPS [g (lb)/10^6 Cans Manufactured]	
	Maximum for Any 1 Day	Maximum for Monthly Average
Chromium	27.98 (0.062)	11.45 (0.025)
Zinc	92.86 (0.205)	38.80 (0.086)
Aluminum	408.95 (0.902)	203.52 (0.449)
Fluoride	3784.20 (8.343)	1679.04 (3.702)
Phosphorus	1062.12 (2.342)	434.39 (0.958)
Oil and grease	1272.00 (2.804)	763.20 (1.683)
TSS	2607.60 (5.749)	1240.20 (2.734)
pH	7.0–10.0[a]	7.0–10.0[a]

[a] Within this range at all times.

Source: U.S. EPA. *Code of Federal Regulations (CFR). Title 40 Protection of Environment. Part 465. Coil Coating Point Source Category.* Washington, DC, 2008.

TABLE 7.38
The Pretreatment Standards for Existing Sources (PSES) of the Canmaking Subcategory Effluents

Pollutant or Pollutant Property	PSES [g (lb)/10^6 Cans Manufactured]	
	Maximum for Any 1 Day	Maximum for Monthly Average
Chromium	36.92 (0.081)	15.10 (0.033)
Copper	159.41 (0.351)	83.90 (0.185)
Zinc	122.49 (0.270)	51.18 (0.113)
Fluoride	4992.05 (11.001)	2214.96 (4.883)
Phosphorus	1401.13 (3.089)	573.04 (1.263)
Manganese	57.05 (0.126)	24.33 (0.053)
TTO	26.85 (0.059)	12.59 (0.028)
Oil and grease (for alternative monitoring)	1678.00 (3.699)	1006.80 (2.220)

Source: U.S. EPA. *Code of Federal Regulations (CFR). Title 40 Protection of Environment. Part 465. Coil Coating Point Source Category.* Washington, DC, 2008.

2. "Coil" means a strip of basis material rolled into a roll for handling.
3. "Coil coating" means the process of converting basis material strip into coated stock. Usually cleaning, conversion coating, and painting are performed on the basis material. Coil coating processes perform any two or more of the three operations.
4. "Basis material" means the coiled strip that is processed.
5. "Area processed" means the area actually exposed to process solutions. Usually this includes both sides of the metal strip.
6. "Steel basis material" means cold rolled steel, hot rolled steel, and chrome, nickel and tin coated steels that are processed in coil coating.
7. "Galvanized basis material" means zinc coated steel, galvanized brass, and other copper base strip that is processed in coil coating.
8. "Aluminum basis material" means aluminum, aluminum alloys, and aluminum coated steels that are processed in coil coating.

TABLE 7.39
The Pretreatment Standards for New Sources (PSNS) of the Canmaking Subcategory Effluents

Pollutant or Pollutant Property	PSNS [g (lb)/10^6 Cans Manufactured]	
	Maximum for Any 1 Day	Maximum for Monthly Average
Chromium	27.98 (0.0617)	11.45 (0.025)
Copper	120.84 (0.267)	63.60 (0.140)
Zinc	92.86 (0.205)	38.80 (0.086)
Fluoride	3784.20 (8.345)	1679.04 (3.702)
Phosphorus	1062.12 (2.342)	434.39 (0.958)
Manganese	43.25 (0.095)	18.44 (0.041)
TTO	20.35 (0.045)	9.54 (0.0210)
Oil and grease (for alternative monitoring)	1272.00 (2.804)	763.20 (1.683)

Source: U.S. EPA. *Code of Federal Regulations (CFR). Title 40 Protection of Environment. Part 465. Coil Coating Point Source Category.* Washington, DC, 2008.

9. The term "can" means a container formed from sheet metal and consisting of a body and two ends or a body and a top.
10. The term "canmaking" means the manufacturing process or processes used to manufacture a can from a basic metal.
11. The term "total toxic organics (TTO)" means the sum of the mass of each of the following toxic organic compounds, which are found at a concentration greater than 0.010 mg/L:

   a) 1,1,1-Trichloroethane
   b) 1,1-Dichloroethane
   c) 1,1,2,2-Tetrachloroethane
   d) Bis(2-chloroethyl)ether
   e) Chloroform
   f) 1,1-Dichloroethylene
   g) Methylene chloride (dichloromethane)
   h) Pentachlorophenol
   i) Bis(2-ethylhexyl)phthalate
   j) Butyl benzyl-phthalate
   k) Di-n-butyl phthalate
   l) Phenanthrene
   m) Tetrachloroethylene
   n) Toluene

REFERENCES

1. U.S. EPA, *Development Document for Effluent Limitations Guidelines and Standards for the Coil Coating Point Source Category*, (Canmaking Subcategory), final report 440/1-83/071, U.S. EPA, Washington, DC, November 1983.
2. U.S. EPA, *EPA Self-Audit and Inspection Guide Organic Finishing of Metals, Paints and Coatings Research Center*, U.S. EPA, Washington, DC, available at www.paintcenter.org/ctc/Rollcoi2.cfm, 2008.
3. U.S. Government Printing Office, *Coil Coating Point Source Category, Code of Federal Regulations, Title 40, Part 465*, U.S. GPO, Washington, DC, July 1, 2003.
4. U.S. EPA, *Coil Coating Forming Point Source Category*, available at http://www.access.gpo.gov/nara/cfr/waisidx_03/40cfr467_03.html, 2008.
5. Higgins, T.E., *Pollution Prevention Handbook*, CRC Press, Boca Raton, FL, 1995.
6. Wang, L.K., Pereira, N.C., and Hung, Y.T., Eds., *Air Pollution Control Engineering*, Humana Press, Totowa, NJ, 2004.
7. Wang, L.K., Pereira, N.C., and Hung, Y.T., Eds., *Advanced Air and Noise Pollution Control*, Humana Press, Totowa, NJ, 2005.
8. Wang, L.K., Hung, Y.T., and Shammas, N.K., Eds., *Physicochemical Treatment Processes*, Humana Press, Totowa, NJ, 2005.
9. Wang, L.K., Hung, Y.T., and Shammas, N.K., Eds., *Advanced Physicochemical Treatment Processes*, Humana Press, Totowa, NJ, 2006.
10. Wang, L.K., Hung, Y.T., and Shammas, N.K., Eds., *Advanced Physicochemical Treatment Technologies*, Humana Press, Totowa, NJ, 2007.
11. Wang, L.K., Shammas, N.K., and Hung, Y.T., Eds., *Biosolids Treatment Processes*, Humana Press, Totowa, NJ, 2007.
12. Wang, L.K., Shammas, N.K., and Hung, Y.T., Eds., *Biosolids Engineering and Management*, Humana Press, Totowa, NJ, 2008, pp. 396–398.
13. Wang, L.K., Hung, Y.T., Lo, H.H., and Yapijakis C., Eds., *Handbook of Industrial and Hazardous Wastes Treatment*, Marcel Dekker, Inc., New York, NY, 2004.
14. U.S. ACE, Yearly Average Cost Index for Utilities, in *Civil Works Construction Cost Index System Manual*, 110-2-1304, U.S. Army Corps of Engineers, Washington, DC, available at http://www.nww.usace.army.mil/cost, 2008.
15. U.S. EPA. *Code of Federal Regulations (CFR). Title 40 Protection of Environment. Part 465. Coil Coating Point Source Category*. www.epa.gov. U.S. Environmental Protection Agency, Washington, DC, 2008.

8 Waste Treatment in the Porcelain Enameling Industry

Lawrence K. Wang and Nazih K. Shammas

CONTENTS

8.1 Industry Description 308
    8.1.1 Historical Cultural Development 308
    8.1.2 Industrial Technology Development 308
8.2 Porcelain Enameling Process Steps 309
    8.2.1 Enamel Slip Preparation 309
    8.2.2 Base Material Surface Preparation 309
    8.2.3 Enamel Application and Firing 309
8.3 Subcategory Descriptions of the Porcelain Enameling Industry 310
    8.3.1 Porcelain Enameling on Steel Subcategory 310
    8.3.2 Porcelain Enameling on Cast Iron Subcategory 311
    8.3.3 Porcelain Enameling on Aluminum Subcategory 311
    8.3.4 Porcelain Enameling on Copper Subcategory 312
    8.3.5 Porcelain Enameling on Continuous Strip Subdivision 312
8.4 Wastewater Characterization of the Porcelain Enameling Industry ... 312
    8.4.1 Wastewater from the Porcelain Enameling on Steel Subcategory ... 313
    8.4.2 Wastewater from the Porcelain Enameling on Cast Iron Subcategory ... 314
    8.4.3 Wastewater from the Porcelain Enameling on Aluminum Subcategory ... 314
    8.4.4 Wastewater from the Porcelain Enameling on Copper Subcategory ... 314
8.5 Specific Descriptions of Porcelain Enameling Industrial Plants 318
    8.5.1 Industrial Plants of Porcelain Enameling on Steel Subcategory ... 323
        8.5.1.1 Plant 40053 323
        8.5.1.2 Plant 41062 323
        8.5.1.3 Plant 36030 323
    8.5.2 Industrial Plants of the Porcelain Enameling on Aluminum Subcategory ... 324
        8.5.2.1 Plant 11045 324
        8.5.2.2 Plant 47051 324
        8.5.2.3 Plant 33077 324
    8.5.3 Industrial Plants of the Porcelain Enameling on Cast Iron Subcategory ... 326
        8.5.3.1 Plant 15712 326
        8.5.3.2 Plant 40053 326
    8.5.4 Industrial Plants of the Porcelain Enameling on Copper Subcategory ... 326
8.6 Pollutant Removability of Porcelain Enameling Industry Wastewater ... 329
    8.6.1 Equalization and Neutralization 329
    8.6.2 Clarification by Sedimentation (Settling) or Flotation 330
    8.6.3 Chemical Addition, Precipitation, Coagulation, and Flocculation ... 331

		8.6.4	Granular Bed Filtration, Granular Activated Carbon Filtration, and Membrane Filtration	331

        8.6.5 Sludge Concentration and Dewatering 331
8.7 Pollution Prevention and Cleaner Production in the Porcelain Enameling Industry 331
8.8 Costs for Treatment of Porcelain Enameling Industrial Wastewaters 332
        8.8.1 BPT Level Treatment. 332
                8.8.1.1 Suggested BPT for Treating Porcelain Enameling Industrial Wastes 332
                8.8.1.2 Suggested BPT System Components for Treating Porcelain Enameling Industrial Wastes 332
                8.8.1.3 BPT Unit Cost for Treating Porcelain Enameling Industrial Wastes 332
        8.8.2 BAT Level Treatment 333
                8.8.2.1 Suggested BAT for Treating Porcelain Enameling Industrial Wastes 333
                8.8.2.2 Suggested BAT System Components for Treating Porcelain Enameling Industrial Wastes 333
                8.8.2.3 BAT Unit Cost for Treating Porcelain Enameling Industrial Wastes 333
8.9 Porcelain Enameling Point Source Discharge Effluent Limitations, Performance Standards, and Pretreatment Standards 333
        8.9.1 U.S. Environmental Regulations for the Steel Basis Material Subcategory 333
        8.9.2 U.S. Environmental Regulations for the Cast Iron Basis Material Subcategory 334
        8.9.3 U.S. Environmental Regulations for the Aluminum Basis Material Subcategory 335
        8.9.4 U.S. Environmental Regulations for the Copper Basis Material Subcategory 337
8.10 Technical Terminologies Used in the Porcelain Enameling Industry 342
References 344

8.1 INDUSTRY DESCRIPTION

8.1.1 HISTORICAL CULTURAL DEVELOPMENT

Enameling is an old and widelyadopted technology.[1-9] The ancient Egyptians applied enamels to pottery and stone objects. The ancient Greeks, Celts, Russians, and Chinese also used enameling processes on metal objects.[9]

Enameling was also used to decorate glass vessels during the Roman period, and there is evidence of this as early as the late Republican and early Imperial periods in the Levantine, Egypt, Britain, and the Black Sea.[1] Enamel powder could be produced in two ways, either through the powdering of colored glass, or the mixing of colorless glass with colorants such as a metallic oxide.[2] Designs were either painted freehand or over the top of outline incisions, and the technique probably originated in metalworking. Once painted, enameled glass vessels needed to be fired at a temperature high enough to melt the applied powder, but low enough that the fabric of the vessel itself was not melted. Production is thought to have come to a peak in the Claudian period and persisted for some 300 years, although archaeological evidence for this technique is limited to some 40 vessels or vessel fragments.[1]

8.1.2 INDUSTRIAL TECHNOLOGY DEVELOPMENT

Porcelain enameling began in the U.S. in the late 1800s. Following the Depression, the manufacture of porcelain enameled refrigerators, stoves, and other household items expanded manyfold.[3-7] The demand for porcelain enamel products and finishes reached a peak in the 1960s and 1970s. The majority of the porcelain enameling plants in the U.S. are located east of the Mississippi River.[3]

Porcelain Enameling Industry

The porcelain enameling industry consists of at least 116 plants enameling approximately 150 million square meters (150 km^2) of steel, iron, aluminum, and copper each year (each coat of multiple coats is considered in this total). Porcelain enameling is the application of glass-like coatings to the metals mentioned above. The purpose of the coating is to improve resistance to chemicals, abrasion, and water, and to improve thermal stability, electrical resistance, and appearance. The coating applied to the metal, called a "slip," is composed of one of many combinations of frits (glassy raw materials), clays, coloring oxides, water, and special additives such as suspending agents. These vitreous inorganic coatings are applied to the metal by a variety of methods such as spraying, drying, and flow coating, and are bonded to the metal at temperatures over 500°C (over 1000°F).

Several processes are used in the porcelain enameling industry regardless of the metal being coated. These processes, discussed below, include preparation of the enamel slip, surface preparation of the base material, and enamel application and firing to fuse the coating to the metal.[3-6]

8.2 PORCELAIN ENAMELING PROCESS STEPS

8.2.1 ENAMEL SLIP PREPARATION

The preparation of the enamel slip includes ball milling the frit and raw materials to the appropriate consistency. Frit is the glassy raw material that makes up the backbone of porcelain enameling. Most frit is manufactured outside the operation but some plants do include captive operations. Other raw materials, such as clay and gums, are mixed into the frit by the ball mill that then releases this mixture to the coating operation.[3-6]

8.2.2 BASE MATERIAL SURFACE PREPARATION

In order for the porcelain enamel to form a good bond with the workpiece, the base metal to be coated must be properly prepared. Depending on the type of metal being finished, one or more preparation processes are performed. Solvent cleaning removes oil, greases, and fingerprints from the metal by exposing it to nonflammable solvents such as trichloroethylene or 1,1,2-trichloro-ethane at their boiling points. This process may also be combined with water to provide a two-phase cleaning system for solvent-soluble and water-soluble contaminants.

Alkaline cleaning removes oils and soils from the workpieces by the detergent nature of the solution. Soaking, spraying, and electrolytic alkaline cleaning are the most common methods used, with the electrolytic process providing the cleanest surface. If aluminum is the metal being coated, a stronger alkaline solution is often used as a mild etch that removes the surface oxides.

Acid treatment is used to remove rust, scale, and oxides from the base and may be carried out in the form of acid cleaning, pickling, or etching. Each option involves a slightly stronger acid solution. Generally, sulfuric acid is used for this treatment, although other acids may be applied.

Nickel deposition is a common step when enameling steel in order to improve the bonding of the enamel to the metal. Nickel is normally deposited after the part has been acid treated and rinsed. Neutralization normally follows acid pickling and nickel deposition to remove the last traces of acid left on the metal. Chromate cleaning and grit blasting may also be used to prepare the base metal prior to the coating process. When used, grit blasting is normally the sole preparation step because it cleans the metal and roughens the surface, providing a good base for bonding.[3-6]

8.2.3 ENAMEL APPLICATION AND FIRING

Once the workpiece has undergone proper base metal preparation and the enamel slip has been prepared, the next step is the actual application of the porcelain enamel. Included among the application methods are air spraying, electrostatic spraying, dip coating, flow coating, powder coating, and silk screening. After each coating is applied, the part is fired in a furnace to achieve a fusion between the enamel coating and the base metal or substrate.

Air spraying is the most widely used method for enamel application. In this process, the enamel is atomized and propelled by air onto the base metal to form an enamel coating. Overspraying is a common problem with this technique, because the atomized particles may not adhere to the part. Spray booths to collect this oversprayed enamel are necessary. A modification of this technique is the electrostatic spray coating method where the atomized particles are charged at 70,000 to 100,000 V and directed toward the grounded part. This charge increases the adhering efficiency but does not eliminate the need for spray booth collectors. Other advantages such as edging and the coating of both sides at once are also applicable.

Dip coating consists simply of dipping the workpiece in an enamel bath and allowing it to drain. Flow coating floods the piece with enamel and then recycles the unused, recovered enamel. Powder coating is the dusting of a red hot cast iron workpiece with porcelain enamel in the form of a dry powder. The glass powder melts as it strikes the hot surface. Silk screening is used to apply a decorative pattern on a porcelain enameled piece.[3-6]

Porcelain enameling plants are located primarily in the states of Wisconsin, Illinois, Indiana, Michigan, Ohio, Pennsylvania, Kentucky, and Tennessee. Of the facilities, 76% discharge to publicly owned treatment works (POTWs), 22% to streams or rivers, and 2% to both. Approximately 10% of the plants recycle, with an average recycle of 9.6 m^3/hr, which represents 46% of the average process water usage rate of 20.8 m^3/hr. The total porcelain enamel applied each year by all plants is estimated at 150×10^6 m^2.

Of the 130 porcelain enameling industrial plants studied, 30 plants treat their wastewaters for direct discharge into receiving waters, and 100 plants pretreat their wastewaters for discharge into POTWs.

8.3 SUBCATEGORY DESCRIPTIONS OF THE PORCELAIN ENAMELING INDUSTRY

The porcelain enameling industry consists of four subcategories:

1. Porcelain enameling on steel
2. Porcelain enameling on iron
3. Porcelain enameling on aluminum
4. Porcelain enameling on copper

This subcategorization was chosen on the basis of the base metals used. Other possible subcategories (dependent on wastewater characterization, manufacturing processes, products, water use, and so on) were considered, but all were found to be directly related to the base metal used. In addition to the four subcategories selected, steel and aluminum base metals may be further divided into two segments, sheet and strip, to account for the significant water-saving potential of continuous operations relative to individual sheet processing. However, because only two porcelain enameling facilities treat strip, no separate division is necessary at this time.

In general, only 10% of the porcelain enameling facilities enamel more than one type of base metal. Over 70% of the plants enamel solely on steel, 10% on aluminum, and 8% on iron. Less than 1% of the plants enamel copper, strip steel, or strip aluminum separately.[3-6]

8.3.1 PORCELAIN ENAMELING ON STEEL SUBCATEGORY

Steel is by far the most widely used base metal for porcelain enameling, with the average yearly production of a plant being 1.34×10^6 m^2 (14.4×10^6 ft^2). This figure represents the area of enamel applied. For multiple coats, the area for each coat is considered. Among the products that use porcelain enameled steel are the following: cooking and heating equipment such as ranges, home laundry equipment (washers and dryers), refrigerators, freezers, dishwashers, water heaters, process vessels, architectural panels, plumbing fixtures, and various appliance parts.

Several processes are used when enameling on steel. The parts to be coated are first alkaline cleaned and rinsed to remove soils. An acid treatment step and rinse follow in that sulfuric acid, ferric sulfate in conjunction with sulfuric acid, or muriatic acid are used for oxide removal. A nickel deposition step and rinse ensues, followed by a neutralization operation, which removes any remaining traces of acid.

Following surface preparation and drying, the part is ready for enamel application. Steel parts are either sprayed, dipped, or flow coated. The enamel slip can be applied in a single coating operation (referred to as direct-on), or a ground coat and a cover coat may be applied separately. For the direct-on process, corners and edges are usually reinforced (precoated) to ensure coverage. For either case, each coat is fired at a temperature of approximately 820°C (1500°F). The total thickness of sheet steel enamels involving a ground coat and cover coat is in the range of 0.13 to 0.20 mm (5 to 9 mils).

When the direct-on process is utilized, surface preparation requirements are more critical to ensure effective enamel adhesion. The acid etch is often deeper and the nickel deposition is always thicker. Typically, the nickel coating is 0.01 to 0.02 g/m² for direct-on coating as compared to 0.002 to 0.007 g/m² for two-coat applications. A few porcelain enamelers prefer to omit the nickel deposition step. Although the nickel enhances enamel bonding, product quality requirements may not require nickel deposition. The omission of the nickel step necessitates the utilization of a heavy acid etch to ensure a clean, properly conditioned surface for enamel bonding.[3-6]

8.3.2 Porcelain Enameling on Cast Iron Subcategory

Cast iron is porcelain enameled primarily for plumbing fixtures for the sanitary products industry. It is also used for cookware and for various appliance parts such as grates for gas ranges. The average yearly production of a plant is 1.56×10^6 m² (16.8×10^6 ft²). This figure represents the areas of enamel applied. For multiple coats, the area for each coat is considered.

The porcelain enameling of cast iron is a process in which water is not generally used for metal preparation but is sometimes used for coating application. The casting to be coated is blasted with sand or a combination of grit and sand to produce a smooth, velvety surface. The parts are then brushed off and any rough edges are removed by grinding.

The ground coat is then applied by spraying, dipping, or flow coating. If only one coat is required, a heavy ground coat is applied. If there is to be a ground coat and a top coat, a thin layer of enamel is used for the ground coat. The ground coat is then fired. The firing period is longer than for sheet steel because of the greater mass of the enameled body, and firing temperature is reduced to avoid excessive baking. When the cast is removed from the furnace and still red hot, the top coat is applied by powder coating. The enamel in powder form is dusted on the hot part and fused to the surface. The total thickness of dry process coatings is approximately 0.50 mm (20 mils).

8.3.3 Porcelain Enameling on Aluminum Subcategory

Porcelain enameling on aluminum finds use in the cookware and housewares industry. It is also used for panels and signs. The average yearly production for a plant in this subcategory is 0.25×10^6 m² (2.7×10^6 ft²). This figure represents the area of enamel applied. For multiple coats, the area for each coat is considered.[3-6]

Although all aluminum parts can be coated in a similar fashion, the surface preparation can vary from company to company. The choice of surface preparation methodology is based upon the alloy type of the base metal and the cleanliness requirements involved. Pure aluminum requires only a cleaning step. A heat-treatable alloy may require a pickling step in addition to cleaning. Porcelain enameling on a high-magnesium alloy could necessitate a chromate cleaning process. This chromate coating retards the oxidation of the magnesium in this high-strength alloy.

Nearly all aluminum parts are first treated in an alkaline solution. In some cases, this is only a cleaner for removing grease and soils; sometimes it is a mild etchant to remove a layer of metal and its oxides. Frequently, this is all the surface preparation that is necessary. Any further preparation steps are to remove residual oxides (e.g., chemical deoxidizing with nitric acid) or to impart a thin protective layer on the metal (alkaline chromate treatment). The users of such processes were limited in the plants studied.

Aluminum does not require a ground coat. Enamel is generally applied by spraying, with firing accomplished by heating to 450 to 550°C (850 to 1040°F) for 2 to 10 min.

8.3.4 PORCELAIN ENAMELING ON COPPER SUBCATEGORY

Porcelain enameling on copper represents a very small part of the porcelain enameling category. It is not practiced by many firms and the ones involved do it on a small scale. Enameled copper is used mostly for ornamental purposes, such as jewelry, decorative ware, and metal sculpture. The average yearly production of a plant in this subcategory is 1.4×10^4 m² (1.5×10^4 ft²).

As it is essential to remove all the oil and grease on the copper before coating, the part is first alkaline cleaned, degreased, or annealed. After cleaning, the part is then typically pickled for oxide removal.

Enamel application involves two processes: a ground coat or backing coat and a cover coat to prevent the copper base from being taken into solution with the enamel and causing discoloration. This ground coat is applied by either spraying or dipping. The cover coat can be applied by powder coating or with silk screening to achieve patterns.

8.3.5 PORCELAIN ENAMELING ON CONTINUOUS STRIP SUBDIVISION

In addition to the above subcategories, porcelain enameling on continuous strip is a subdivision within this industry. However, because there are only two plants in the U.S. producing this product, a separate subcategory is not necessary. These plants start with coils of steel, aluminum, or aluminized steel, porcelain enamel them and either recoil them for sale to metal fabricators or shear them into pieces for use as architectural panels or chalkboards. The estimated production was 2.0×10^6 m² (22×10^6 ft²). This figure represents the area of enamel applied. For multiple coats, the area for each coat is considered.

The surface preparation operations for strip are dependent upon whether the basis material is steel or aluminum. The surface preparation steps for steel strip are minimal in comparison to porcelain enameling on steel sheets because precleaned strip steel is used. Steel strip is nickel immersion plated prior to the enameling step. Surface preparation for aluminum involves only cleaning. The enamel for either basis material is applied by means of spray guns that are aimed at the surface of the moving strip. Two coats are normally applied, the strip being fired after each coat.

8.4 WASTEWATER CHARACTERIZATION OF THE PORCELAIN ENAMELING INDUSTRY

This section presents water uses and discharges, and waste constituents emanating from the porcelain enameling category. Published literature, data collection portfolio (DCP) responses, and screening and verification sampling data have been used to obtain the relevant information. The screening analysis of the porcelain enameling category consisted of a sampling program for all 129 priority pollutants. Those pollutants that were detected in the screening program were further studied in the verification analysis. Only those pollutants included in the verification program are presented in the following tables. The minimum detection limit for toxic pollutants is 10 µg/L and any value below 10 µg/L is presented in the following tables as below detection limit (BDL). Table 8.1 presents wastewater flow data on a subcategory and stream basis for the porcelain enameling industry.[3–10]

TABLE 8.1
Wastewater Flows from the Porcelain Enameling (PE) Industry

Stream	Number of Samples	Flow Range (m³/d)	Flow Median (m³/d)	Flow Mean (m³/d)
PE on steel				
Alkaline cleaning	21	1.64–122	30.3	47.2
Acid etch	21	0.556–56.2	19.6	23.7
Nickel flash	12	19.1–31.2	24.8	25.2
Neutralization	8	0.999–19.8	15.1	14.1
Coating	21	0.783–505	4.03	107
Total raw waste	21	11.3–711	175	197
PE on iron				
Coating	7	0.636–7.21	1.23	2.88
PE on aluminum				
Alkaline cleaning	8	19.2–217	169	131
Coating	8	4.84–546	297	285
Total raw waste	8	68.2–223	197	160
PE on copper				
Acid etch	3	6.14–7.27	7.27	6.89
Coating	4	0.008–1.27	0.636	0.638
Total raw waste	4	1.27–7.90	7.02	5.81

Source: U.S. EPA, *Development Document for Effluent Limitations Guidelines and Standards for the Porcelain Enameling Point Source Category*, Washington, DC, 1982; U.S. EPA, *Porcelain Enameling Point Source Category*, 2008.

8.4.1 WASTEWATER FROM THE PORCELAIN ENAMELING ON STEEL SUBCATEGORY

Wastewater from porcelain enameling on steel is generated by base metal surface preparation, enamel application, ball milling, and related operations. The constituents in the wastewater include the base material being coated (iron), as well as the components of the surface treatment solutions and enamels being applied.

Water rinses are used in surface preparation operations such as acid pickling, alkaline cleaning, and nickel deposition to remove any process solution film left from the previous bath. A water rinse may also follow the neutralization step. Another common water use is in the ball milling process, which uses water as the vehicle for the enamel ingredients, as a cooling medium, and for cleaning the equipment. Coating application processes normally use wet spray booths to capture oversprayed enamel particles. Water wash spray booths use a water curtain into which the enamel particles are blown and captured.

The major sources of waste generated by this subcategory are the process solutions used in basis material preparation, the base metal being coated, and the enamel being prepared. Alkaline cleaning solution varies with the type of soil being removed. Wastewaters from this operation contain constituents of the cleaning solution as well as oil and grease. These wastewaters also contain iron but in lesser concentrations than those from the acid pickling process. Alkaline cleaning wastes enter the wastestream in three ways: during the rinse step, from the cleaning bath overflow, and in the batch dump of the spent alkaline bath.

Acid treatment is typically sulfuric acid with lesser amounts of hydrochloric, phosphoric, and nitric acids being used. Acid solutions develop a high metallic content due to the dissolution of the steel itself during the pickling operation. As a result, the baths are frequently dumped, introducing large amounts of iron into the wastestream. Also present in significant concentrations are phosphorus and manganese. The stream has a low pH.

Nickel deposition can place large amounts of nickel and iron into the wastestream by batch dumping and dragout. The neutralization step eases the pH burden and adds little additional loading of any pollutant.

The introduction of enamel into the wastestream results in an increase in the concentration of metals, but these metals (antimony, titanium, zirconium, tin, cobalt, and manganese) are in solid form whereas the metals generated by surface preparation are normally in dissolved form. These solid metals increase the suspended solids concentration of the stream. Other metals that may be found in the enamel preparation and application wastestream in significant amounts include aluminum, copper, iron, lead, nickel, and zinc. Table 8.2 presents pollutant sampling data for the processes used in the porcelain enameling on steel industry.

8.4.2 Wastewater from the Porcelain Enameling on Cast Iron Subcategory

There are two different types of cast iron porcelain enameling:

1. Dry process enameling cast iron, which uses no water and does not produce wastewater
2. Wet process enameling cast iron, which uses water for ball milling and enamel application

These processes are very similar to the ones described for the steel subcategory. Surface preparation involves sand or grit blasting and uses water only in an air scrubber operation. Ball milling uses water as a vehicle for the enamel slip ingredients, as cooling water, and for equipment cleanup. Coating application uses water as a trap for the excess enamel particles during the spray step. Wastewater constituents in significant concentrations in the streams emanating from this subcategory include suspended solids, aluminum, iron, copper, lead, manganese, nickel, titanium, zinc, and cobalt. All of these metals are the result of the enamel carryover via spray booth blowdown or ball mill washdown.

Table 8.3 presents wastewater characterization data for the streams in this cast iron subcategory.

8.4.3 Wastewater from the Porcelain Enameling on Aluminum Subcategory

Wastewaters from this subcategory come from surface preparation, enamel application, ball milling, and related operations. Constituents of this wastewater include aluminum and components of the surface preparation solutions and the enamels being applied.[3-6]

Water is used in this subcategory as solution makeup and for rinsing in the surface preparation process, as the vehicle for the coating in the application process (normally done by spray coating), and for cooling and cleanup in the ball milling operation.

The surface preparation process contributes pollutants to the wastewater by the continuous overflow of the cleaning bath (if a continuous process), by the batch dumping of spent solutions, and by the rinsing steps directly following the process. Generally, significant quantities of dirt and grease are removed during this cleaning process. Also entering the wastestream is a considerable amount of aluminum. When an alkaline cleaning process is used, the wastewater contains significant concentrations of suspended solids, phosphorus, and aluminum. Acids used to deoxidize the surface normally remove a larger amount of aluminum than alkaline treatments and, therefore, increase the dissolved aluminum concentration. The enamel preparation and application steps contribute significant amounts of suspended solids and metals, particularly cadmium, lead, titanium, zinc, aluminum, barium, iron, selenium, and antimony due to use of these metals in the enamel itself. There are also high levels of fluorides and phosphorus.

Table 8.4 presents classical and toxic pollutant concentrations for the porcelain enameling on aluminum subcategory.

8.4.4 Wastewater from the Porcelain Enameling on Copper Subcategory

Wastewater from this subcategory is generated as in the previous subcategories, by surface preparation, enamel application, ball milling, and related operations. Wastewater constituents generally consist of copper and the components used to form the enamel.

TABLE 8.2
Wastewater Characterization of the Porcelain Enameling on Steel Subcategory

Pollutant	Number of Samples	Alkaline Cleaning			Number of Samples	Acid Etch		
		Number of Detections	Range of Detections	Average of Detections		Number of Detections	Range of Detections	Average of Detections
Classical Pollutants (mg/L)								
TSS	20	20	6–650	140	19	19	1.9–310	32
Total phosphorus	20	19	0.29–92	15	9	9	0.56–12	7.0
Total phenols	20	19	0.006–0.69	0.08	19	15	0.005–0.95	0.38
Oil and grease	11	11	3–63	17	10	10	1–17	4.5
pH (pH units)	38	38	2–11.7	7.9	26	26	2–7.5	2.9
Fluorides	21	21	0.23–1.8	0.80	21	21	0.14–1.1	0.58
Aluminum	21	16	0.08–3.2	0.43	21	15	0.05–3.2	0.4
Iron	18	18	0.028–1500	90	21	21	13–10,000	2100
Manganese	19	14	0.005–4.5	0.50	21	20	0.06–53	8.8
Titanium	21	0	—	—	21	1	0.05	—
Cobalt	21	2	0.001–0.11	0.05	21	18	0.017–0.38	0.11
Toxic Pollutants (μg/L)								
Metals and inorganics								
Antimony	21	0	—	—	21	0	—	—
Arsenic	21	0	—	—	21	0	—	—
Beryllium	21	0	—	—	21	0	—	—
Cadmium	21	3	BDL–84	36	21	1	14	—
Chromium	21	9	BDL–260	47	21	21	11–3100	590
Copper	21	19	BDL–220	63	21	21	BDL–380	75
Cyanide	8	0	—	—	7	0	—	—
Lead	21	0	—	—	21	5	50–130	85
Nickel	19	7	14–25,000	3600	21	17	87–25,000	5000
Selenium	21	19	BDL–210	110	21	1	210	—
Zinc	20	18	13–810	90	21	21	17–250	110

Continued

TABLE 8.2 (continued)

Pollutant	Nickel Flash				Neutralization			
	Number of Samples	Number of Detections	Range of Detections	Average of Detections	Number of Samples	Number of Detections	Range of Detections	Average of Detections
Classical Pollutants (mg/L)								
TSS	10	10	2–310	56	6	6	8.0–57	30
Total phosphorus	6	6	1.1–8.3	4.5	7	6	0.04–7.5	1.7
Total phenols	10	8	0.008–0.095	0.04	6	6	0.004–0.50	0.10
Oil and grease	7	7	1–18	5.1	6	6	1–3.8	2.7
pH (pH units)	20	20	2.0–6.2	3.0	14	14	6–9.7	8.4
Fluorides	12	12	0.27–0.82	0.55	7	7	0.32–1.1	0.69
Aluminum	12	6	0.04–0.33	0.19	7	2	0.04–0.34	0.19
Iron	12	12	57–1500	640	7	7	1.8–44	13
Manganese	12	12	0.27–7.6	3.2	7	7	0.016–0.25	0.09
Titanium	12	0	—	—	7	0	—	—
Cobalt	12	12	0.01–0.46	0.18	7	0	—	—
Toxic Pollutants (µg/L)								
Metals and inorganics								
Antimony	12	0	—	—	7	0	—	—
Arsenic	12	0	—	—	7	0	—	—
Beryllium	12	0	—	—	7	0	—	—
Cadmium	12	1	12	—	7	0	—	—
Chromium	12	12	19–260	88	7	3	12–32	25
Copper	12	11	8–79	33	7	3	10–14	11
Cyanide	7	0	—	—	6	0	—	—
Lead	12	0	—	—	7	0	—	—
Nickel	12	12	2900–280,000	76,000	7	7	75–9500	1700
Selenium	12	1	210	—	7	0	—	—
Zinc	12	12	36–1300	200	7	7	BDL–25	13
Silver	3	2	8–54	31	—	—	—	—

	Coating		Total Raw Waste	
Classical Pollutants (mg/L)				
TSS	21	360–320,000	18	66–16,000
Total phosphorus	19	0.49–9.8	9	0.82–13
Total phenols	16	0.005–0.066	17	0.006–0.29
Oil and grease	15	1–98	16	2.3–38
pH (pH units)	30	5.8–12.5	36	2–12.5
Fluorides	21	1.9–120	20	0.61–30
Aluminum	21	5.2–1500	20	0.45–210
Iron	21	0.45–620	17	1.5–670
Manganese	21	0.45–400	18	0.08–61
Titanium	21	4.3–1600	20	0.17–1200
Cobalt	21	0.31–350	20	0.17–9.2
Toxic Pollutants (µg/L)				
Metals and inorganics				
Antimony	21	920–1.0 × 10^6	9	52–22,000
Arsenic	21	52–3500	9	BDL–2500
Beryllium	21	14–120	8	BDL–BDL
Cadmium	21	BDL–54,000	16	BDL–5300
Chromium	21	BDL–37,000	20	22–840
Copper	21	160–55,000	20	34–2200
Cyanide	12	55	0	
Lead	21	47–11,000	19	BDL–830
Nickel	21	390–360,200	18	82–32,000
Selenium	21	120–17,000	15	BDL–13,000
Zinc	21	1100–1.3 × 10^6	19	78–45,000
Organics				
Toluene	3	—	0	—
1,2-Dichlorobenzene	1	—	0	—
Chloroform	3	BDL–BDL	0	—
Dichlorobromomethane	3	BDL–BDL	0	—
1,1,2,2-Tetrachloroethane	2	BDL–BDL	0	—
Tetrachloroethylene	3	BDL–BDL	0	—

	Coating		Total Raw Waste	
			18	2800
		2.7	9	5.3
		0.019	18	0.05
		23	10	13
		8.5	36	6.2
		37	20	8.3
		180	20	26
		56	17	270
		58	18	9.1
		270	20	88
		38	20	2.9
		10,000	20	4600
		840	20	550
		50	20	BDL
		6600	20	440
		2100	20	160
		5400	20	600
			7	
		3100	20	320
		33000	18	1700
		2900	20	1700
		13,0000	19	12,000
			2	
			0	
		BDL	0	
		BDL	0	
		BDL	0	
		BDL	0	

BDL, below detection limit.

Source: U.S. EPA, *Development Document for Effluent Limitations Guidelines and Standards for the Porcelain Enameling Point Source Category*, Washington, DC, 1982; U.S. EPA, *Porcelain Enameling Point Source Category*, available at http://www.access.gpo.gov/nara/cfr/waisidx_03/40cfr467_03.html, 2008.

TABLE 8.3
Wastewater Characterization of the Porcelain Enameling on Cast Iron Subcategory

	Coating			
Pollutant	Number of Samples	Number of Detections	Range of Detections	Average of Detections
Classical Parameters (mg/L)				
TSS	7	7	6600–81,000	27,000
Total phosphorus	7	6	0.49–2.1	1.1
Total phenols	6	6	0.008–0.038	0.02
Oil and grease	3	3	1.0–9.5	4.7
pH (pH units)	14	14	7.9–11.4	9.5
Fluorides	7	7	2.0–120	41
Aluminum	7	7	0.38–1200	340
Iron	6	5	18–150	56
Manganese	7	7	0.003–65	15
Titanium	7	4	0.02–100	44
Cobalt	7	7	0.044–95	24
Toxic Pollutants (µg/L)				
Metals and inorganics				
Antimony	7	1	6000	
Arsenic	7	3	1900–2800	2400
Beryllium	7	4	BDL–120	49
Cadmium	7	4	14–9600	2700
Chromium	7	7	BDL–1100	430
Copper	7	7	BDL–8800	2600
Cyanide	3	1	BDL	—
Lead	7	7	490–880,000	170,000
Nickel	7	4	250–67,000	33,000
Selenium	7	7	430–160,000	29,000
Zinc	7	6	680–650,000	130,000

BDL, below detection limit.
Source: U.S. EPA, *Development Document for Effluent Limitations Guidelines and Standards for the Porcelain Enameling Point Source Category*, Washington, DC, 1982; U.S. EPA, *Porcelain Enameling Point Source Category*, 2008.

Water is used to rinse the workpieces after various operations, as a constituent of the enamel slip, in spray booths, and in cleaning, cooling, and air scrubbing. Pollutants such as dirt and grease enter the wastestream from the surface preparation and rinsing steps. Acid pickling adds dissolved copper to the wastestreams. Enamel preparation and application may add high concentrations of aluminum, titanium, manganese, nickel, zinc, and cobalt, as well as fluorides, antimony, copper, lead, and iron for the porcelain enameling on copper subcategory on a stream basis, as shown in Table 8.5.

8.5 SPECIFIC DESCRIPTIONS OF PORCELAIN ENAMELING INDUSTRIAL PLANTS

Only a limited amount of information is available on specific plants within this industry. This section describes the treatment practice and wastewater composition at nine plants: three that enamel on steel, three on aluminum, two on cast iron, and one on copper. The major treatment operation used is a settling technique. Treatment operations are not necessarily listed in this narrative in the same order that they are used at the plants. Wastewater composition data were obtained from verification sampling.[3–6]

TABLE 8.4
Wastewater Characterization of the Porcelain Enameling on Aluminum Subcategory

Pollutant	Number of Samples	Number of Detections	Range of Detections	Average of Detections
		Alkaline Cleaning		
Classical Pollutants (mg/L)				
TSS	8	8	1.0–180	40
Total phosphorus	8	8	0.41–24	8.5
Total phenols	8	7	0.005–0.02	0.008
Oil and grease	8	4	3–11	6.8
pH (pH units)	16	16	6.3–11	8.7
Fluorides	8	8	0.72–0.98	0.88
Aluminum	8	7	0.68–26	6.6
Barium	8	0	—	—
Iron	8	8	0.01–0.33	0.1
Manganese	8	3	0.02–0.18	0.11
Titanium	8	0	—	—
Cobalt	8	0	—	—
Toxic Pollutants (µg/L)				
Metals and inorganics				
Antimony	8	0	—	—
Arsenic	8	0	—	—
Beryllium	8	0	—	—
Cadmium	8	1	BDL	—
Chromium	8	2	BDL–18	12
Copper	8	2	21–56	38
Cyanide	8	2	15–180	95
Lead	8	2	40–4300	2200
Nickel	8	0	—	—
Selenium	8	0	—	—
Zinc	8	7	19–540	210
Organics				
Bis(2-ethylhexyl)phthalate	8	0	—	—
Di-n-octyl phthalate	8	0	—	—
Toluene	3	0	—	—
		Coating		
Classical Pollutants (mg/L)				
TSS	8	8	55–650	330
Total phosphorus	8	8	0.38–65	9.8
Total phenols	8	5	0.005–0.02	0.01
Oil and grease	8	3	2.3–4.7	3.4
pH (pH units)	16	16	7.0–10	8.9
Fluorides	8	8	0.92–1.9	1.2
Aluminum	8	8	0.25–2.1	0.62
Barium	8	8	0.11–1.4	0.59
Iron	8	8	0.11–0.94	0.33
Manganese	8	2	0.003–0.01	0.007
Titanium	8	8	3.1–30	10
Cobalt	8	1	0.03	—
Toxic Pollutants (µg/L)				
Metals and inorganics				
Antimony	8	2	210–360	280
Arsenic	8	0	—	—

Continued

TABLE 8.4 (continued)

Pollutant	Number of Samples	Number of Detections	Range of Detections	Average of Detections
Beryllium	8	0	—	—
Cadmium	8	7	290–54,000	11,000
Chromium	8	8	BDL–39	24
Copper	8	6	BDL–180	57
Cyanide	8	1	BDL	—
Lead	8	8	3500–38,000	15,000
Nickel	8	0	—	—
Selenium	8	4	530–7100	2200
Zinc	8	8	150–2000	740
Organics				
Bis(2-ethylhexyl)phthalate	8	0	—	—
Di-n-octyl phthalate	8	0	—	—
Toluene	3	0	—	—

← Total Raw Waste →

Pollutant	Number of Samples	Number of Detections	Range of Detections	Average of Detections
Classical Pollutants (mg/L)				
TSS	8	8	12–190	110
Total phosphorus	8	8	0.88–24	9.3
Total phenols	8	8	0.001–0.015	0.007
Oil and grease	8	5	1.7–11	5.8
pH (pH units)	16	16	6.3–10.4	8.7
Fluorides	8	8	0.74–0.98	0.89
Aluminum	8	8	0.08–10	3.8
Barium	8	8	0.01–0.24	0.10
Iron	8	8	0.02–0.71	0.24
Manganese	8	5	0.002–0.13	0.04
Titanium	8	8	0.09–6.1	2.6
Cobalt	8	1	0.005	—
Toxic Pollutants (µg/L)				
Metals and inorganics				
Antimony	8	2	150–260	210
Arsenic	8	0	—	—
Beryllium	8	0		
Cadmium	8	7	BDL–5200	2200
Chromium	8	8	BDL–13	BDL
Copper	8	6	BDL–130	48
Cyanide	8	2	BDL–140	73
Lead	8	8	150–12,000	3900
Nickel	8	0	—	—
Selenium	8	4	110–630	400
Zinc	8	8	120–530	300
Organics				
Bis(2-ethylhexyl)phthalate	8	0	—	—
Di-n-octyl phthalate	8	0	—	—
Toluene	3	0	—	—

BDL, below detection limit.

Source: U.S. EPA, *Development Document for Effluent Limitations Guidelines and Standards for the Porcelain Enameling Point Source Category*, Washington, DC, 1982; U.S. EPA, *Porcelain Enameling Point Source Category*, 2008.

Porcelain Enameling Industry

TABLE 8.5
Wastewater Characterization of the Porcelain Enameling on Copper Subcategory

Pollutant	Number of Samples	Number of Detections	Range of Detections	Average of Detections
		Acid Etch		
Classical Pollutants (mg/L)				
TSS	2	2	14–24	19
Total phosphorus	2	1	0.52	—
Total phenols	2	1	0.006	—
Oil and grease	1	1	200	—
pH (pH units)	5	5	1.8–6.6	5.7
Fluorides	2	2	0.11–0.12	0.12
Aluminum	3	3	0.0002–0.17	0.073
Iron	3	3	0.15–51	27
Manganese	3	3	0.01–0.26	0.09
Titanium	3	0	—	—
Cobalt	3	0	—	—
Toxic Pollutants (µg/L)				
Metals and inorganics				
Antimony	3	0	—	—
Arsenic	3	1	BDL	—
Beryllium	3	0	—	—
Cadmium	3	1	22	—
Chromium	3	3	BDL–60	26
Copper	3	3	9700–820,000	280,000
Cyanide	2	0	—	—
Lead	3	1	770	—
Nickel	3	1	120	—
Selenium	3	1	BDL	—
Zinc	3	3	49–2400	890
Organics				
Carbon tetrachloride	1	0	—	—
1,1,1-Trichloroethane	1	1	BDL	—
1,1,2-Trichloroethane	1	0	—	—
1,1-Dichloroethylene	1	0	—	—
Methylene chloride	1	0	—	—
Methyl chloride	1	0	—	—
Trichloroethylene	1	1	BDL	—
Toluene	2	0	—	—
Chloroform	3	2	BDL–BDL	BDL
Dichlorobromomethane	2	2	BDL–BDL	BDL
1,1,2,2-Tetrachloroethane	2	2	BDL–BDL	BDL
Tetrachloroethylene	4	0	—	—
		Coating		
Classical Pollutants (mg/L)				
TSS	3	3	14,000–94,000	46,000
Total phosphorus	2	2	1–71	36
Total phenols	3	0	—	—
Oil and grease	3	3	2.0–98	37
pH (pH units)	7	7	7.6–10	8.8

Continued

TABLE 8.5 (continued)

Pollutant	Number of Samples	Number of Detections	Range of Detections	Average of Detections
Fluorides	3	3	46–66	56
Aluminum	4	4	96–200	140
Iron	4	4	15–29	22
Manganese	4	4	6.2–120	68
Titanium	4	4	3.6–560	220
Cobalt	4	4	20–64	46
Toxic Pollutants (µg/L)				
Metals and inorganics				
Antimony	4	4	1600–3500	2300
Arsenic	4	2	420–3800	2100
Beryllium	4	3	BDL–59	34
Cadmium	4	4	97–2800	830
Chromium	4	4	200–3000	1000
Copper	4	4	4700–10,000	6900
Cyanide	3	1	55	—
Lead	4	4	2300–440,000	110,000
Nickel	4	4	20,000–49,000	37,000
Selenium	4	4	200–810	570
Zinc	4	4	1100–200,000	84,000
Organics				
1,2-Dichlorobenzene	1	0	—	—
Toluene	3	0	—	—
Carbon tetrachloride	1	1	BDL	—
1,1,1-Trichloroethane	1	1	BDL	—
1,1,2-Trichloroethane	1	0	—	—
1,1-Dichloroethylene	1	0	—	—
Methylene chloride	1	0	—	—
Methyl chloride	1	0	—	—
Trichloroethylene	1	1	BDL	—
Chloroform	4	4	BDL–BDL	BDL
Dichlorobromomethane	3	2	BDL–BDL	BDL
1,1,2,2-Tetrachloroethane	2	2	BDL–BDL	BDL
Tetrachloroethylene	4	2	BDL–BDL	BDL
	←———	Total Raw Waste	———→	
Classical Pollutants (mg/L)				
TSS	3	3	1100–94,000	33,000
Total phosphorus	1	1	0.08	—
Total phenols	3	1	0.006	—
Oil and grease	2	2	2–190	96
pH (pH units)	7	7	1.8–10	7.7
Fluorides	3	3	3.8–56	22
Aluminum	4	4	0.12–200	55
Iron	4	4	1.4–48	27
Manganese	4	4	0.27–120	33
Titanium	4	4	0.004–560	150
Cobalt	4	4	0.024–64	18

Continued

TABLE 8.5 (continued)

Pollutant	Number of Samples	Number of Detections	Range of Detections	Average of Detections
Toxic Pollutants (µg/L)				
Metals and inorganics				
Antimony	4	4	BDL–2400	690
Arsenic	4	2	BDL–420	210
Beryllium	4	3	BDL–35	13
Cadmium	4	4	BDL–220	69
Chromium	4	4	23–630	190
Copper	4	4	7100–810,000	210,000
Cyanide	t3	1	BDL	—
Lead	4	4	190–4800	1700
Nickel	4	4	140–49,000	14,000
Selenium	4	4	BDL–810	230
Zinc	4	4	2400–200,000	53,000
Organics				
Toluene	3	0	—	—
Chloroform	4	4	BDL–BDL	BDL
Dichlorobromomethane	3	3	BDL–BDL	BDL
1,1,1-Trichloroethane	1	1	BDL	—
1,1,2-Trichloroethane	1	0	—	—
1,1-Dichloroethylene	1	0	—	—
Methylene chloride	1	0	—	—
Methyl chloride	1	0	—	—
Trichloroethylene	1	1	BDL	—
1,1,2,2-Tetrachloroethane	2	2	BDL–BDL	BDL
Tetrachloroethylene	4	2	BDL–BDL	BDL
Carbon tetrachloride	1	1	BDL	—

BDL, below detection limit.
Source: U.S. EPA, *Development Document for Effluent Limitations Guidelines and Standards for the Porcelain Enameling Point Source Category*, Washington, DC, 1982; U.S. EPA, *Porcelain Enameling Point Source Category*, 2008.

8.5.1 Industrial Plants of Porcelain Enameling on Steel Subcategory

8.5.1.1 Plant 40053

This facility is involved with porcelain enameling on both steel and cast iron. Data presented in Tables 8.6 and 8.7 are for the coating on steel subcategory only.

8.5.1.2 Plant 41062

This plant produces 130 m^2/hr of enameled steel and operates 3500 hr/yr. It uses 0.0036 m^3 water/m^2 of product to coat the steel. Average process water flow is 0.144 m^3/hr for coating operations and 0.734 m^3/hr for metal preparation. The primary treatment in-place for process wastewater is clarification and settling. Other water treatment practices employed are pH adjustment with lime or acid, sludge applied to landfill, polyelectrolyte coagulation, and inorganic coagulation.

8.5.1.3 Plant 36030

This facility produces 110 m^2/hr of enameled steel and operates 4000 hr/yr. It uses 0.0042 m^3 water/m^2 of product in coating operations. Average process water flow is 1.69 m^3/hr for coating operations and 0.466 m^3/hr for metal preparation. The primary in-place treatment is chemical coagulation and clarification. Clarification can be either settling or dissolved air flotation.

TABLE 8.6
Water Use (m³ Water/m² Product) in the Porcelain Enameling on Steel Subcategory

Process[a]	Plant Identification					
	33617	40063	47033	40053	36030	41062
Alkaline cleaning	0.00094	0.0032	0.10	0.0056	0.010	0.029
Acid etch	0.00014	0.0026	0.038	0.012	0.0051	0.0071
Nickel flash	0.00033	0.0027	0.021	c	c	e
Neutralization	0.00011	0.0016	0.0056	c	0.00075	c
Ball milling	0.00004	0.017	0.0010	0.0013	0.0031	0.0053
Coating	0.00066	0.011	b	d	0.0013	d

[a] Because of differences in area prepared and coated, these data cannot be added directly for each process to obtain overall subcategory water usage.
[b] Uses dip coating and spray coating with a dry booth.
[c] No rinsing involved.
[d] Dry spray booths.
[e] Nickel flash not used at this plant.

Source: U.S. EPA, *Development Document for Effluent Limitations Guidelines and Standards for the Porcelain Enameling Point Source Category*, Washington, DC, 1982; U.S. EPA, *Porcelain Enameling Point Source Category*, 2008.

Table 8.6 gives the water use for each process in the production of porcelain enameled steel for the above plants. Pollutant concentrations for treated effluents are presented in Table 8.7.[3–6]

8.5.2 Industrial Plants of the Porcelain Enameling on Aluminum Subcategory

8.5.2.1 Plant 11045

This facility produces 210 m²/hr of enameled aluminum and uses 0.015 m³ water/m² of product for coating operations. The average process flow rate is 1.33 m³/hr for metal preparation operations and 0.716 m³/hr for coating operations. The primary in-place treatment for process wastewater is chemical coagulation and clarification (i.e., settling).

8.5.2.2 Plant 47051

This plant produces 290 m²/hr of enameled aluminum for 6400 hr/yr. It uses 0.018 m³ water/m² product for coating and ball milling purposes. The average process flow rate is 12.5 m³/hr for metal preparation and 1.59 m³/hr for coating and ball milling. In-place treatment consists primarily of chemical coagulation, clarification (settling), and final pH adjustment.

8.5.2.3 Plant 33077

This facility produces 360 m²/hr of porcelain enameled aluminum for 4000 hr/yr, and uses 0.038 m³ of process water/m² of product coated. The mixed wastewater stream is treated by equalization (settling), pH adjustment (lime or acid), polyelectrolyte coagulation, clarification, and contractor removal of the resulting sludge prior to discharge to a surface stream. Process water flow for this production consists of 8.12 m³/hr and 4.37 m³/hr for surface preparation and coating operations, respectively.[3–6]

Table 8.8 gives the water use for each process in the production of porcelain enameled aluminum for the above plants. Pollutant concentrations for treated effluents are presented in Tables 8.9 and 8.10.[10]

TABLE 8.7
Effluent Concentrations of Pollutants Found in Steel Subcategory Plants

Pollutant (mg/L)	Plant Identification		
	40053[a]	41062[b]	36030[b]
Aluminum	190	2100	130,000
Antimony	—	—	9700
Arsenic	—	ND	—
Cadmium	ND	75	550
Chromium	12	10	630
Cobalt	22	ND	32,000
Copper	52	13	3500
Fluoride	930	2300	58,000
Iron	250,000	240	630,000
Lead	ND	ND	3500
Manganese	910	BDL	51,000
Nickel	2700	14	29,000
Phenols, total	24	36	—
Phosphorus	11,000	770	3600
Selenium	ND	ND	590
Titanium	ND	160	660,000
Zinc	140	230	180,000
Oil and grease	—	1700	—
TSS	51,000	11,000	57,000,000[c]
pH (pH units)	2.1–3.2	8.1–9.1	—

Dashes indicate data not available.
BDL, below detection limit; ND, not detected.
[a] In-place treatment not available.
[b] In-place treatment consists of clarification/settling.
[c] As reported in reference; currently under review.
Source: U.S. EPA, *Development Document for Effluent Limitations Guidelines and Standards for the Porcelain Enameling Point Source Category*, Washington, DC, 1982; U.S. EPA, *Porcelain Enameling Point Source Category*, 2008.

TABLE 8.8
Water Use (m^3 Water/m^2 Product) in the Porcelain Enameling on Aluminum Subcategory

Process[a]	Plant Identification		
	11045	33077	47051
Surface preparation	0.029	0.140	0.042
Ball milling	0.041	0.014	0.0029
Coating	0.019	0.014	—[b]

[a] Because of differences in area prepared and coated, these data cannot be added for each process to obtain overall subcategory water usage.
[b] This plant employs dry spray booths.
Source: U.S. EPA, *Development Document for Effluent Limitations Guidelines and Standards for the Porcelain Enameling Point Source Category*, Washington, DC, 1982; U.S. EPA, *Porcelain Enameling Point Source Category*, 2008.

TABLE 8.9
Effluent Concentrations of Pollutants Found in Aluminum Subcategory Plants

	Plant Identification		
Pollutant (µg/L)	11045[a]	33077[a]	47051[a]
Aluminum	3600	76	6800
Antimony	—	ND	—
Arsenic	—	ND	—
Barium	240	200	300
Cadmium	1100	350	BDL
Chromium, total	BDL	BDL	57
Chromium-hexavalent	—	ND	ND
Cobalt	—	—	BDL
Copper	84	ND	52
Fluoride	930	1800	390
Iron	460	24	360
Lead	5300	210	97
Manganese	28	ND	80
Nickel	—	—	57
Phenols, total	BDL	BDL	BDL
Phosphorus	1900	1900	—
Selenium	—	28	—
Titanium	3900	130	ND
Zinc	210	390	290
Oil and grease	3300	ND	72,000
TSS	140,000	13,000	310
pH (pH units)	7.3–9.3	9–9.3	7.1–10.2

Dashes indicate data not available.
BDL, below detection limit; ND, not detected.
[a] In-place treatment consists of clarification/settling.
Source: U.S. EPA, *Development Document for Effluent Limitations Guidelines and Standards for the Porcelain Enameling Point Source Category*, Washington, DC, 1982; U.S. EPA, *Porcelain Enameling Point Source Category*, 2008.

8.5.3 Industrial Plants of the Porcelain Enameling on Cast Iron Subcategory

8.5.3.1 Plant 15712

This facility produces 9.1 m²/yr of porcelain enameled cast iron. The primary in-place treatment for process wastewater is chemical coagulation, clarification (settling), and skimming.

8.5.3.2 Plant 40053

This facility is involved with porcelain enameling on both steel and cast iron. Data presented in Tables 8.11 and 8.12 are for the coating on cast iron subcategory only. Table 8.11 gives the water use for each process in the production of porcelain enameled cast iron for the above plants. Pollutant concentrations in the treated effluent are presented in Table 8.12.

8.5.4 Industrial Plants of the Porcelain Enameling on Copper Subcategory

Plant 36030 enamels both copper and steel. It uses 0.042 m³ water/m² product in all coating operations. Process wastewater flow is 0.466 m³/hr for metal preparation and 1.69 m³/hr for coating

Porcelain Enameling Industry

TABLE 8.10
Full Scale Treatment of Porcelain Enameling on Aluminum Subcategory Wastewater by Equalization and Chemical Precipitation Using Lime and Polymer

Removal Data Sampling: [16-hr Composite, Flow Proportion (1 hr)]

	Concentration		Percent Removal	Detection Limit
	Influent	Effluent		
Classical Pollutants (mg/L)				
pH, minimum	8.9	9.4	—	—
pH, maximum	10.5	10.0	—	—
Fluorides	1.8	2.0	NM	0.1
Phosphorus	12	0.89	92	0.003
TSS	53	ND	>99	5.0
Iron	2.0	0.038	98	0.005
Titanium	1.2	ND	>99	—
Manganese	0.017	ND	>99	0.005
Phenols, total	0.006	ND	>99	0.005
Aluminum	1.2	ND	>99	0.04
Barium	0.23	0.20	13	—
Toxic Pollutants (µg/L)				
Cadmium	2900	57	98	2.0
Chromium, total	11	ND	>99	3.0
Copper	4.0	ND	>99	1.0
Lead	1200	ND	>99	30
Zinc	220	540	NM	1.0
Cyanide, total	160	ND	>99	5.0
Selenium	300	ND	>99	—

Dashes indicate data not available.
ND, not detected; NM, not meaningful.

Plant: 33077
Wastewater flow rate: 965 m^3/d
Chemical dosage(s): lime: 47,200 kg/yr; polymer: 320 kg/yr
Unit configuration: continuous operation (16 hr/day)
Source: U.S. EPA, references 5 and 7.

TABLE 8.11
Water Use (m^3 Water/m^2 Product) in the Porcelain Enameling on Cast Iron Subcategory

	Plant Identification	
Process	15712	40053
Surface preparation	a	a
Ball milling	0.00001	0.0013
Coating application	0.00028	b

a Surface preparation consists of dry operations.
b This plant uses dry spray booths.
Source: U.S. EPA, references 5 and 7.

TABLE 8.12
Effluent Concentrations of Pollutants Found in Cast Iron Subcategory Plants

	Plant Identification	
Pollutant (µg/L)	15712[a]	40053[b]
Aluminum	190,000	190,000
Cadmium	—	3600
Chromium, total	19	740
Cobalt	6200	50,000
Copper	BDL	6000
Fluoride	2200	89,000
Iron	13,000	80,000
Lead	110,000	5600
Manganese	BDL	35,000
Nickel	—	44,000
Phenols, total	20	20
Phosphorus	1200	980
Selenium	63,000	590
Titanium	—	58,000
Zinc	470	250,000
TSS	16,000,000	21,000,000
pH (pH units)	8.8–10.7	8.8–9.0

Analytic methods: V.7.3.16, Data set 2.
Dashes indicate data not available.
BDL, below detection limit.
[a] In-place treatment consists of clarification/settling.
[b] In-place treatment not available.
Source: U.S. EPA, *Development Document for Effluent Limitations Guidelines and Standards for the Porcelain Enameling Point Source Category*, Washington, DC, 1982; U.S. EPA, *Porcelain Enameling Point Source Category*, 2008.

and ball milling. The production rate for porcelain enameling on copper is 10 m²/hr for 4000 hr/yr. The primary in-place treatment is clarification and settling.

Table 8.13 gives the water use for each process in the production of porcelain enameled copper for two plants. Pollutant concentrations in the treated effluent are given in Table 8.14.

TABLE 8.13
Water Use (m³ Water/m² Product) in the Porcelain Enameling on Copper Subcategory

	Plant Identification	
	36030	06031
Acid etch	0.057	0.087
Ball milling	0.0037	[a]
Coating	0.0015	0.00017

[a] Ball milling operations at this facility generated no wastewater.
Source: U.S. EPA, *Development Document for Effluent Limitations Guidelines and Standards for the Porcelain Enameling Point Source Category*, Washington, DC, 1982; U.S. EPA, *Porcelain Enameling Point Source Category*, 2008.

TABLE 8.14
Effluent Concentrations of Pollutants Found in Copper Subcategory Plants

Pollutant (µg/L)	Plant Identification
	36030[a]
Aluminum	130,000
Antimony	6400
Cadmium	550
Chromium, total	1100
Cobalt	32,000
Copper	3500
Fluoride	58,000
Iron	630,000
Lead	2100
Manganese	51,000
Nickel	29,000
Phosphorus	3600
Selenium	590
Titanium	660,000
Zinc	180,000
TSS	57,000,000

[a] In-place treatment consists of clarification/settling.

Source: U.S. EPA, *Development Document for Effluent Limitations Guidelines and Standards for the Porcelain Enameling Point Source Category*, Washington, DC, 1982; U.S. EPA, *Porcelain Enameling Point Source Category*, 2008.

8.6 POLLUTANT REMOVABILITY OF PORCELAIN ENAMELING INDUSTRY WASTEWATER

Treatment technologies used in the porcelain enameling industry are generally chosen to remove the major wastewater components, such as suspended solids and toxic metals. Table 8.15 presents a summary of the treatment and disposal techniques used by this industry. Usually more than one treatment methods will be used at each facility.[3–6,10]

Some type of clarification technology is used in a large portion of the plants. Clarification can be either sedimentation or dissolved air flotation.[10–12] pH adjustment by chemical addition is another common treatment that is used to neutralize the alkaline or acid wastes. Coagulants are sometimes used to aid settling or flotation. Once the clarification (either settling or flotation) nears completion, filtration techniques are used to concentrate the sludge, which is then landfilled or contractor hauled.[13–15] Oils may be treated in a similar manner. Tables 8.8, 8.10, 8.12, 8.14, and 8.15 in the plant-specific section give treated effluent concentrations. Table 8.15 is a summary of the common wastewater treatment technologies used in porcelain enameling industry. Brief descriptions of the common treatment practices and the water reuse and recycle techniques follow.

8.6.1 Equalization and Neutralization

Raw wastewaters are commonly collected in equalization basins to even out the flow and the pollutant contaminant load. This permits uniform and controlled operation of subsequent treatment facilities. Wastes in this industry generally require pH adjustment, which can be performed in mixed equalization basins or in separate neutralization reactor basins following equalization.[10]

TABLE 8.15
Treatment Technologies in Current Use in the Porcelain Enameling Industry

Treatment Method	Number of Plants Using the Method by Subcategory				Total Plants
	Steel	Iron	Aluminum	Copper	
Skimming	2	—	—	—	2
Settling tank	33	7	5	1	46
Clarifier	17	—	2	—	19
Sedimentation lagoon	10	—	—	—	10
Tube/plate settler	3	—	—	—	3
Equalization	24	2	2	—	28
pH adjustment—lime	15	1	2	—	18
pH adjustment—caustic	6	2	—	—	8
pH adjustment—acid	6	—	1	—	7
pH adjustment—carbonate	2	—	1	—	3
pH adjustment—final	5	—	1	—	6
Coagulant—polyelectrolyte	10	1	1	—	12
Coagulant—inorganic	3	1	—	—	4
Chrome reduction	2	—	1	—	3
Emulsion breaking	1	—	—	—	1
Chlorination	1	—	—	—	1
Ultrafiltration	2	—	—	—	2
Pressure filtration	5	1	—	—	6
Vacuum filtration	5	—	—	—	5
Filtration	3	—	—	—	3
Aeration	2	—	—	—	2
Trickling filter	1	—	—	—	1
Centrifugation sludge	1	—	—	—	1
Material recovery	1	2	—	—	3
Air pollution control	1	—	—	—	1
Process reuse—oil	1	—	—	—	1
Contract removal—oil	7	—	—	—	7
Contract removal—sludge	5	—	1	—	6
Landfill—oil	2	—	—	—	2
Landfill—sludge	20	2	—	—	22
Sludge drying bed	3	—	—	—	3
Sludge thickening	—	—	1	—	1

Dashes indicate no plants using this method.
Source: U.S. EPA, references 5 and 7.

8.6.2 Clarification by Sedimentation (Settling) or Flotation

Sedimentation and dissolved air flotation are the most common clarification processes for removal of precipitates. Either sedimentation or flotation is often preceded by chemical coagulation or precipitation, which converts dissolved pollutants to a suspended form, and by flocculation, which enhances clarification by flocculating suspended solids into larger, more easily separating particles. Simple sedimentation normally requires a long retention time to adequately reduce the solids content. The detention time of dissolved air flotation, however, is much shorter. When chemicals are used, retention times are reduced and clarification removal efficiency of either sedimentation or flotation is increased. A properly operated clarification system is capable of efficient removal of suspended solids, metal hydroxides, and other wastewater impurities.[10–12]

Porcelain Enameling Industry

8.6.3 CHEMICAL ADDITION, PRECIPITATION, COAGULATION, AND FLOCCULATION

Chemical precipitation is used in porcelain enameling to precipitate dissolved metals and phosphates. Chemical precipitation can be utilized to permit removal of metal ions such as iron, lead, tin, copper, zinc, cadmium, aluminum, mercury, manganese, cobalt, antimony, arsenic, beryllium, molybdenum, and trivalent chromium. Removal efficiency can approach 100% for the reduction of heavy metal ions. Porcelain enameling plants commonly use lime, caustic, and carbonate for chemical precipitation and pH adjustment. Coagulants used in the industry include alum, ferric chloride, ferric sulfate, and polymers.[10–12]

8.6.4 GRANULAR BED FILTRATION, GRANULAR ACTIVATED CARBON FILTRATION, AND MEMBRANE FILTRATION

Granular bed filters are used in porcelain enameling wastewater treatment to remove residual solids from clarifier effluent (sedimentation effluent or flotation effluent). Filtration polishes the effluent and reduces suspended solids and insoluble precipitated metals to very low levels. Fine sand and coal are media commonly utilized in granular bed filtration. The filter is backwashed after becoming loaded with solids and the backwash is returned to the treatment plant influent for removal of solids in the clarification step.[10–12]

When granular activated carbons (GAC) are used as the filter media, the GAC filter can also remove dissolved organics (such as TTO, total toxic organics; or VOC, volatile organic compounds).[10,11]

Recently, membrane filtration has become popular for treating industrial effluent. Membrane filtration includes microfiltration (MF), ultrafiltration (UF), nanofiltration (NF), and reverse osmosis (RO).[11,12]

8.6.5 SLUDGE CONCENTRATION AND DEWATERING

Sludges from clarifiers can be thickened in gravity thickeners or mechanically thickened by centrifuges. Thickened sludges can be further dewatered on one of a number of dewatering operations including vacuum filters, pressure filters, and belt filter presses. Dewatered sludges are disposed generally to landfills that must be properly constructed to conform to provisions of the Resource Conservation and Recovery Act (RCRA) and regulations governing disposal of hazardous wastes.[13–15]

8.7 POLLUTION PREVENTION AND CLEANER PRODUCTION IN THE PORCELAIN ENAMELING INDUSTRY

Many facilities in this industry use in-plant technology to reduce or eliminate the waste load, requiring end-of-pipe treatment and thereby improve the quality of the effluent discharge and reduce treatment costs. In-plant technology involves water reuse, process material conservation, reclamation of waste enamel, process modifications, material substitutions, improved rinse techniques, and good housekeeping practices.[3–6,15]

Water reuse is practiced at several plants in this industry. Water that may be reused for such purposes as rinse water, makeup water, and cleanup water includes air conditioning water, acid treatment rinsewater, and noncontact cooling water. Reuse of acid rinsewater in alkaline rinses has been demonstrated at many electroplating plants.

Process material conservation is practiced by the recovery, reuse, or purification of the materials used in the processes. In the nickel deposition process the nickel solution is filtered to reduce its iron content, giving a longer life to the solution. Because the bath is dumped less often, the pollutant load is reduced.

The use of dry spray booths can also reduce the wastewater volume from the plant as well as increasing excess enamel recovery and reuse. Overspray is captured on filter screens and then swept up and reused in the enamel slip. Several plants use this and other, similar processes to recover the enamel raw material.

Process modifications, material substitutions, improved rinsing techniques, and good housekeeping procedures may also significantly reduce the amount of wastewater released.

8.8 COSTS FOR TREATMENT OF PORCELAIN ENAMELING INDUSTRIAL WASTEWATERS

The investment cost, operating and maintenance costs, and energy costs for the application of control technologies to the wastewater of the porcelain enameling industry have been analyzed. These costs were developed to reflect the conventional use of technologies in this industry. The detailed presentation of the cost methodology and cost information is characterized as follows.[3–6,16]

Unit operation and unit process configurations have been analyzed for the cost of application to the wastewater of this industry. Recommended unit process configurations for BPT (best practicable control technology) and BAT (best available technology) level of treatment and their costs are summarized briefly in the following sections.

8.8.1 BPT Level Treatment

8.8.1.1 Suggested BPT for Treating Porcelain Enameling Industrial Wastes

The BPT treatment for the steel, aluminum, and copper subcategories of this industry consists of settling the coating waste separately, chemical reduction of hexavalent chromium in the metal preparation stream, equalization of all other enameling wastewaters by combining the wastewater streams, and using chemical precipitation and sedimentation to remove metals and solids. For the purpose of the BPT system cost estimates, chromium reduction was included for the aluminum subcategory, because aluminum is the only subcategory that has chromium in the wastewater. The treatment for the cast iron subcategory consists of presettling of coating wastewaters, and chemical precipitation and sedimentation (or dissolved air flotation or membrane filtration) to remove metals and solids.[3–6,10–12]

8.8.1.2 Suggested BPT System Components for Treating Porcelain Enameling Industrial Wastes

A reinforced concrete sump and associated pumping equipment is required for sedimentation of coating wastewaters. Chromium reduction for the aluminum subcategory wastewater is achieved either by batch treatment or continuous treatment. The continuous treatment system consists of a single reaction tank, sulfur dioxide and sulfuric acid storage and feeding equipment, and mixers and controls. The batch treatment system consists of dual reaction tanks, chemical feed equipment, and mixers and controls. Chemical precipitation and sedimentation is either by batch or continuous treatment. The continuous treatment system includes lime storage and feed equipment, a flocculator, and a clarification basin with sludge rakes and pumps. The batch treatment system includes only reaction settling tanks and sludge pumps. The sludge from settling tanks is dewatered with vacuum filters and hauled away for offsite disposal.[10–12]

8.8.1.3 BPT Unit Cost for Treating Porcelain Enameling Industrial Wastes

Total annual unit costs consisting of annual cost of capital, depreciation, operation, and maintenance cost, and energy cost for average, low, and high flow rates are summarized in Table 8.16. The total capital cost for the treatment system includes the cost of the components discussed above and subsidiary costs including engineering, line segregation, administration, and interest expenses during construction.

TABLE 8.16
Total Annual Unit Cost for BPT Level of Wastewater Treatment[a] in 2008 U.S. Dollars[b]

Subcategory	Flow (L/min)	Cost ($/m³)	Flow (L/min)	Cost ($/m³)	Flow (L/min)	Cost ($/m³)
Steel	63	7.72	315	3.74	946	2.81
Aluminum	64	6.08	202	3.04	317	2.34
Copper	3	82.6	6	48.7	10	36.0
Cast iron	0.3	511	4	72.1	22	21.1

[a] For flows less than 315 L/min treatment is by batch system.
[b] Cost was updated to 2008 using U.S. ACE Cost Index for Utilities.[16]
Source: U.S. EPA, references 5 and 7.

8.8.2 BAT Level Treatment

8.8.2.1 Suggested BAT for Treating Porcelain Enameling Industrial Wastes

The BAT level of treatment consists of all components of BPT and the addition of a multimedia filter to treat the effluent from the sedimentation process.

8.8.2.2 Suggested BAT System Components for Treating Porcelain Enameling Industrial Wastes

The filtration system consists of a granular bed multimedia filter unit, a GAC filter unit, and/or a membrane filtration unit.[10–12]

8.8.2.3 BAT Unit Cost for Treating Porcelain Enameling Industrial Wastes

The total annual unit cost for the complete BAT system, which includes components described in the BPT system for those different flow rates, is summarized in Table 8.17.

8.9 PORCELAIN ENAMELING POINT SOURCE DISCHARGE EFFLUENT LIMITATIONS, PERFORMANCE STANDARDS, AND PRETREATMENT STANDARDS

8.9.1 U.S. Environmental Regulations for the Steel Basis Material Subcategory

Table 8.18 documents the current (May 2008) effluent limitations of the steel basis material subcategory that represent the degree of effluent reduction attainable by the application of the BPT currently available.[3–7]

TABLE 8.17
Total Annual Unit Cost for BAT Level of Wastewater Treatment[a] in 2008 U.S. Dollars[b]

Subcategory	Flow (L/min)	Cost ($/m³)	Flow (L/min)	Cost ($/m³)	Flow (L/min)	Cost ($/m³)
Steel	32	14.0	189	5.15	631	3.51
Aluminum	49	8.89	153	4.21	272	3.04
Copper	3	122	5	63.9	8	45.6
Cast iron	0.3	165	4	82.1	22	23.2

[a] For flows less than 189 L/min treatment is by batch system.
[b] Costs was updated to 2008 using U.S. ACE Cost Index for Utilities.[16]
Source: U.S. EPA, references 5 and 7.

TABLE 8.18
Effluent Limitations of the Steel Basis Material Subcategory That Represent the Degree of Effluent Reduction Attainable by the Application of the BPT Currently Available

Pollutant	Maximum for Any 1 Day		Maximum for Monthly Average	
	Metal Preparation	Coating Operation	Metal Preparation	Coating Operation
	(mg/m^2 of Area Processed or Coated)			
Chromium	16.82	3.41	6.81	1.38
Lead	6.01	1.21	5.21	1.06
Nickel	56.46	11.43	40.05	8.11
Zinc	53.26	10.78	22.43	4.54
Aluminum	182.20	36.87	74.47	15.07
Iron	112.12	22.69	56.06	11.34
Oil and grease	800.84	162.10	480.51	97.23
TSS	1642.00	332.20	800.90	162.00
pH	7.5–10.0[a]	7.5–10.0[a]	7.5–10.0[a]	7.5–10.0[a]
	($lb/10^6 ft^2$ of Area Processed or Coated)			
Chromium	3.45	0.07	1.40	0.29
Lead	1.23	0.25	1.07	0.22
Nickel	11.57	2.34	8.20	1.66
Zinc	10.91	2.21	4.60	0.93
Aluminum	37.32	7.55	15.26	3.09
Iron	22.96	4.65	11.48	2.32
Oil and grease	164.03	33.19	98.42	19.92
TSS	337.00	68.10	164.00	33.20
pH	7.5–10.0[a]	7.5–10.0[a]	7.5–10.0[a]	7.5–10.0[a]

[a] Within this range at all times.

Source: U.S. EPA, *Development Document for Effluent Limitations Guidelines and Standards for the Porcelain Enameling Point Source Category*, Washington, DC, 1982; U.S. EPA, *Porcelain Enameling Point Source Category*, available at http://www.access.gpo.gov/nara/cfr/waisidx_03/40cfr466_03.html, 2008.

Table 8.19 documents the current (May 2008) effluent limitations of the steel basis material subcategory that represent the degree of effluent reduction attainable by the application of the BAT economically achievable.

Table 8.20 presents the new source performance standards (NSPS) of the steel basis material subcategory. Any new source must achieve the NSPS.

Any existing source of the steel basis material subcategory that introduces pollutants into a POTW must achieve the pretreatment standards listed in Table 8.21B. In cases where a POTW finds it necessary to impose mass effluent pretreatment standards, the equivalent mass pretreatment standards are provided in Table 8.21B.[3–7]

Any new source of the steel basis material subcategory that introduces pollutants into a POTW must achieve the pretreatment standards listed in Table 8.22.

8.9.2 U.S. Environmental Regulations for the Cast Iron Basis Material Subcategory

There shall be no discharge of process wastewater pollutants from any metal preparation operations in the cast iron basis material subcategory. The discharge of process wastewater pollutants

TABLE 8.19
Effluent Limitations of the Steel Basis Material Subcategory That Represent the Degree of Effluent Reduction Attainable by the Application of the BAT Economically Achievable

Pollutant	Maximum for Any 1 Day		Maximum for Monthly Average	
	Metal Preparation	Coating Operation	Metal Preparation	Coating Operation
	(mg/m² of Area Processed or Coated)			
Chromium	16.82	0.53	6.81	0.22
Lead	6.01	0.19	5.21	0.16
Nickel	56.50	1.78	40.05	1.26
Zinc	53.30	1.68	22.43	0.71
Aluminum	182.00	5.74	74.48	2.35
Iron	112.12	3.53	56.06	1.77
	(lb/10⁶ ft² of Area Processed or Coated)			
Chromium	3.45	0.11	1.4	0.05
Lead	1.23	0.04	1.07	0.03
Nickel	11.57	0.37	8.20	0.26
Zinc	10.91	0.35	4.60	0.15
Aluminum	37.32	1.18	15.26	0.48
Iron	22.96	0.72	11.48	0.36

Source: U.S. EPA, *Development Document for Effluent Limitations Guidelines and Standards for the Porcelain Enameling Point Source Category*, Washington, DC, 1982; U.S. EPA, *Porcelain Enameling Point Source Category*, available at http://www.access.gpo.gov/nara/cfr/waisidx_03/40cfr466_03.html, 2008.

from all porcelain enameling coating operations shall not exceed the values set forth in Tables 8.23 through 8.27.

Table 8.23 documents the current (May 2008) effluent limitations of the cast iron basis material subcategory that represent the degree of effluent reduction attainable by the application of the BPT currently available.

Table 8.24 documents the current (May 2008) effluent limitations of the cast iron basis material subcategory that represent the degree of effluent reduction attainable by the application of the best available technology (BAT) economically achievable.

Table 8.25 presents the NSPS of the cast iron basis material subcategory. Any new source must achieve the NSPS.

Any existing source of the cast iron basis material subcategory that introduces pollutants into a POTW must achieve the pretreatment standards listed in Table 8.26a. In cases where a POTW finds it necessary to impose mass effluent pretreatment standards, the equivalent mass pretreatment standards are provided in Table 8.26b.[7]

Any new source of the cast iron basis material subcategory that introduces pollutants into a POTW must achieve the pretreatment standards listed in Table 8.27.

8.9.3 U.S. Environmental Regulations for the Aluminum Basis Material Subcategory

Table 8.28 documents the current (May 2008) effluent limitations of the aluminum basis material subcategory that represent the degree of effluent reduction attainable by the application of the BPT currently available.

Table 8.29 documents the current (May 2008) effluent limitations of the aluminum basis material subcategory that represent the degree of effluent reduction attainable by the application of the BAT economically achievable.

TABLE 8.20
New Source Performance Standards (NSPS) of the Steel Basis Material Subcategory

	Maximum for Any 1 Day		Maximum for Monthly Average	
Pollutant	Metal Preparation	Coating Operation	Metal Preparation	Coating Operation
	(mg/m² of Area Processed or Coated)			
Chromium	3.37	0.47	1.5	0.19
Lead	1.0	0.13	0.9	0.11
Nickel	12.0	1.51	6.3	0.79
Zinc	10.2	1.29	4.2	0.53
Aluminum	30.3	3.82	12.4	1.56
Iron	28.0	3.53	14.0	1.77
Oil and grease	100.0	12.60	100.0	12.60
TSS	150.0	18.91	120.0	15.12
pH	7.5–10.0[a]	7.5–10.0[a]	7.5–10.0[a]	7.5–10.0[a]
	(lb/10⁶ ft² of Area Processed or Coated)			
Chromium	0.76	0.10	0.31	0.04
Lead	0.21	0.03	0.19	0.03
Nickel	2.46	0.31	1.29	0.16
Zinc	2.09	0.27	0.86	0.11
Aluminum	6.21	0.78	2.54	0.32
Iron	5.74	0.72	2.87	0.36
Oil and grease	20.48	2.58	20.48	2.58
TSS	30.72	3.87	24.58	3.10
pH	7.5–10.0[a]	7.5–10.0[a]	7.5–10.0[a]	7.5–10.0[a]

Note: Any new source must achieve the NSPS.
[a] Within this range at all times.
Source: U.S. EPA, *Development Document for Effluent Limitations Guidelines and Standards for the Porcelain Enameling Point Source Category*, Washington, DC, 1982; U.S. EPA, *Porcelain Enameling Point Source Category*, available at http://www.access.gpo.gov/nara/cfr/waisidx_03/40cfr466_03.html, 2008.

Table 8.30 presents the NSPS of the aluminum basis material subcategory. Any new source must achieve the NSPS.

Any existing source of the aluminum basis material subcategory that introduces pollutants into a publicly owned treatment works (POTW) must achieve the pretreatment standards listed in Table 8.31a. In cases where POTW find it necessary to impose mass effluent pretreatment standards, the equivalent mass pretreatment standards are provided in Table 8.31B.[7]

Any new source of the aluminum basis material subcategory that introduces pollutants into a POTW must achieve the pretreatment standards listed in Table 8.32.

TABLE 8.21A
Effluent Pretreatment Standards of an Existing Source of the Steel Basis Material Subcategory That Introduces Pollutants into a POTW

Pollutant	Maximum for Any 1 Day (mg/L)	Maximum for Monthly Average (mg/L)
Chromium	0.42	0.17
Lead	0.15	0.13
Nickel	1.41	1.00
Zinc	1.33	0.56

TABLE 8.21B
Mass Effluent Pretreatment Standards of an Existing Source of the Steel Basis Material Subcategory That Introduces Pollutants into a POTW

	Maximum for Any 1 Day		Maximum for Monthly Average	
Pollutant	Metal Preparation	Coating Operation	Metal Preparation	Coating Operation
	(mg/m^2 of Area Processed or Coated)			
Chromium	16.82	0.53	6.81	0.22
Lead	6.01	0.19	5.21	0.16
Nickel	56.5	1.78	40.1	1.26
Zinc	53.3	1.68	22.5	0.71
	($lb/10^6 ft^2$ of Area Processed or Coated)			
Chromium	3.45	0.11	1.4	0.05
Lead	1.23	0.04	1.07	0.03
Nickel	11.6	0.37	8.20	0.26
Zinc	10.9	0.35	4.6	0.15

Source: U.S. EPA, *Development Document for Effluent Limitations Guidelines and Standards for the Porcelain Enameling Point Source Category*, Washington, DC, 1982; U.S. EPA, *Porcelain Enameling Point Source Category*, available at http://www.access.gpo.gov/nara/cfr/waisidx_03/40cfr466_03.html, 2008.

8.9.4 U.S. ENVIRONMENTAL REGULATIONS FOR THE COPPER BASIS MATERIAL SUBCATEGORY

Table 8.33 presents the NSPS of the copper basis material subcategory. Any new source must achieve the NSPS.

Any new source of the copper basis material subcategory that introduces pollutants into a POTW must achieve the pretreatment standards listed in Table 8.34.[7]

TABLE 8.22
Effluent Pretreatment Standards of a New Source of the Steel Basis Material Subcategory That Introduces Pollutants into a POTW

	Maximum for Any 1 Day		Maximum for Monthly Average	
Pollutant	Metal Preparation	Coating Operation	Metal Preparation	Coating Operation
	(mg/m^2 of Area Processed or Coated)			
Chromium	3.7	0.47	1.5	0.19
Lead	1.0	0.13	0.9	0.11
Nickel	12.0	1.51	6.3	0.79
Zinc	10.2	1.29	4.2	0.53
	($lb/10^6 ft^2$ of Area Processed or Coated)			
Chromium	0.76	0.10	0.31	0.04
Lead	0.2	0.03	0.19	0.002
Nickel	2.46	0.31	1.29	0.16
Zinc	2.09	0.27	0.86	0.11

Source: U.S. EPA, *Development Document for Effluent Limitations Guidelines and Standards for the Porcelain Enameling Point Source Category*, Washington, DC, 1982; U.S. EPA, *Porcelain Enameling Point Source Category*, available at http://www.access.gpo.gov/nara/cfr/waisidx_03/40cfr466_03.html, 2008.

TABLE 8.23
Effluent Limitations of the Cast Iron Basis Material Subcategory That Represent the Degree of Effluent Reduction Attainable by the Application of the BPT Currently Available

Pollutant	Maximum for Any 1 Day [mg/m² (lb/10⁶ ft²) of Area Coated]		Maximum for Monthly Average [mg/m² (lb/10⁶ ft²) of Area Coated]	
Chromium	0.29	(0.06)	0.12	(0.024)
Lead	0.11	(0.02)	0.09	(0.02)
Nickel	0.98	(0.02)	0.7	(0.15)
Zinc	0.93	(0.19)	0.39	(0.08)
Aluminum	3.16	(0.65)	1.29	(0.27)
Iron	0.86	(0.18)	0.44	(0.09)
Oil and grease	13.86	(2.84)	8.32	(1.71)
TSS	28.42	(5.82)	13.86	(2.84)
pH	7.5–10.0[a]	7.5–10.0[a]	7.5–10.0[a]	7.5–10.0[a]

[a] Within this range at all times.
Source: U.S. EPA, references 5 and 7.

TABLE 8.24
Effluent Limitations of the Cast Iron Basis Material Subcategory That Represent the Degree of Effluent Reduction Attainable by the Application of the BAT Economically Achievable

Pollutant	Maximum for Any 1 Day [mg/m² (lb/10⁶ ft²) of Area Coated]		Maximum for Monthly Average [mg/m² (lb/10⁶ ft²) of Area Coated]	
Chromium	0.53	(0.11)	0.22	(0.05)
Lead	0.19	(0.04)	0.16	(0.03)
Nickel	1.78	(0.37)	1.26	(0.26)
Zinc	1.68	(0.35)	0.71	(0.15)
Aluminum	5.74	(1.18)	2.35	(0.48)
Iron	1.55	(0.32)	0.79	(0.16)

Source: U.S. EPA; *Porcelain Enameling Point Source Category, Code of Federal Regulations*, Title 40, Volume 27, Part 466, Washington, DC, July 1, 2003. [47 FR 53184, Nov. 24, 1982, as amended at 50 FR 36543, Sept. 6, 1985].

TABLE 8.25
New Source Performance Standards (NSPS) of the Cast Iron Basis Material Subcategory. Any New Source Must Achieve the NSPS

Pollutant	Maximum for Any 1 Day [mg/m² (lb/10⁶ ft²) of Area Coated]		Maximum for Monthly Average [mg/m² (lb/10⁶ ft²) of Area Coated]	
Chromium	0.47	(0.10)	0.19	(0.04)
Lead	0.13	(0.03)	0.11	(0.02)
Nickel	0.69	(0.14)	0.47	(0.10)
Zinc	1.29	(0.27)	0.53	(0.11)
Aluminum	3.82	(0.78)	1.56	(0.32)
Iron	1.55	(0.32)	0.79	(0.16)
Oil and grease	12.60	(2.58)	12.60	(2.58)

Continued

TABLE 8.25 (continued)

Pollutant	Maximum for Any 1 Day [mg/m² (lb/10⁶ ft²) of Area Coated]		Maximum for Monthly Average [mg/m² (lb/10⁶ ft²) of Area Coated]	
TSS	18.91	(3.87)	15.12	(3.10)
pH	7.5–10.0[a]	7.5–10.0[a]	7.5–10.0[a]	7.5–10.0[a]

[a] Within this range at all times.

Source: U.S. EPA; *Porcelain Enameling Point Source Category, Code of Federal Regulations*, Title 40, Volume 27, Part 466, Washington, DC, July 1, 2003. [47 FR 53184, Nov. 24, 1982, as amended at 50 FR 36544, Sept. 6, 1985].

TABLE 8.26A
Effluent Pretreatment Standards of an Existing Source of the Cast Iron Basis Material Subcategory That Introduces Pollutants into a POTW

Pollutant	Maximum for Any 1 Day (mg/L)	Maximum for Monthly Average (mg/L)
Chromium	0.42	0.17
Lead	0.15	0.13
Nickel	1.41	1.00
Zinc	1.33	0.56

Source: U.S. EPA, references 5 and 7.

TABLE 8.26B
Mass Effluent Pretreatment Standards of an Existing Source of the Cast Iron Basis Material Subcategory That Introduces Pollutants into a POTW

Pollutant	Maximum for Any 1 Day [mg/m² (lb/10⁶ ft²) of Area Coated]		Maximum for Monthly Average [mg/m² (lb/10⁶ ft²) of Area Coated]	
Chromium	0.53	(0.11)	0.22	(0.05)
Lead	0.19	(0.04)	0.16	(0.03)
Nickel	1.78	(0.37)	1.26	(0.26)
Zinc	1.68	(0.35)	0.71	(0.15)

Source: U.S. EPA; *Porcelain Enameling Point Source Category, Code of Federal Regulations*, Title 40, Volume 27, Part 466, Washington, DC, July 1, 2003. [47 FR 53184, Nov. 24, 1982, as amended at 50 FR 36544, Sept. 6, 1985].

TABLE 8.27
Effluent Pretreatment Standards of a New Source of the Cast Iron Basis Material Subcategory That Introduces Pollutants into a POTW

Pollutant	Maximum for Any 1 Day [mg/m² (lb/10⁶ ft²) of Area Coated]		Maximum for Monthly Average [mg/m² (lb/10⁶ ft²) of Area Coated]	
Chromium	0.47	(0.10)	0.19	(0.04)
Lead	0.13	(0.03)	0.11	(0.02)
Nickel	0.69	(0.14)	0.47	(0.10)
Zinc	1.29	(0.27)	0.53	(0.11)

Source: U.S. EPA; *Porcelain Enameling Point Source Category, Code of Federal Regulations*, Title 40, Volume 27, Part 466, Washington, DC, July 1, 2003. [47 FR 53184, Nov. 24, 1982, as amended at 50 FR 36544, Sept. 6, 1985].

TABLE 8.28
Effluent Limitations of the Aluminum Basis Material Subcategory That Represent the Degree of Effluent Reduction Attainable by the Application of the BPT Currently Available

Pollutant	Maximum for Any 1 Day		Maximum for Monthly Average	
	Metal Preparation	Coating Operation	Metal Preparation	Coating Operation
	(mg/m² of Area Processed or Coated)			
Chromium	16.34	6.32	6.63	2.56
Lead	5.84	2.26	5.06	1.96
Nickel	54.85	21.21	38.90	15.04
Zinc	51.73	20.01	21.79	8.43
Aluminum	176.98	68.44	72.35	27.98
Iron	47.85	18.50	24.51	9.48
Oil and grease	777.92	300.84	466.76	108.50
TSS	1594.74	616.68	777.92	300.82
pH	7.5–10.0[a]	7.5–10.0[a]	7.5–10.0[a]	7.5–10.0[a]
	(lb/10⁶ ft² of Area Processed or Coated)			
Chromium	3.35	1.30	1.37	0.53
Lead	1.20	0.47	1.04	0.40
Nickel	11.24	4.35	7.97	3.08
Zinc	10.6	4.10	4.46	1.73
Aluminum	36.25	14.02	14.82	5.73
Iron	9.80	3.79	5.02	1.94
Oil and grease	159.33	61.61	95.60	36.97
TSS	326.62	126.33	159.33	61.61
pH	7.5–10.0[a]	7.5–10.0[a]	7.5–10.0[a]	7.5–10.0[a]

[a] Within this range at all times.

TABLE 8.29
Effluent Limitations of the Aluminum Basis Material Subcategory That Represent the Degree of Effluent Reduction Attainable by the Application of the BAT Economically Achievable

Pollutant	Maximum for Any 1 Day		Maximum for Monthly Average	
	Metal Preparation	Coating Operation	Metal Preparation	Coating Operation
	(mg/m² of Area Processed or Coated)			
Chromium	16.34	0.53	6.62	0.22
Lead	5.84	0.19	5.06	0.16
Nickel	54.85	1.78	38.90	1.26
Zinc	51.74	1.68	21.79	1.71
Aluminum	176.98	5.74	72.35	2.35
Iron	47.85	1.55	24.51	0.80
	(lb/10⁶ ft² of Area Processed or Coated)			
Chromium	3.35	0.11	1.36	0.05
Lead	1.20	0.04	1.04	0.03
Nickel	11.24	0.37	7.97	0.26
Zinc	10.60	0.35	4.46	0.35

Continued

TABLE 8.29 (continued)

	Maximum for Any 1 Day		Maximum for Monthly Average	
Pollutant	Metal Preparation	Coating Operation	Metal Preparation	Coating Operation
Aluminum	36.25	1.18	14.82	0.48
Iron	9.80	0.32	5.02	0.16

Source: U.S. EPA; *Porcelain Enameling Point Source Category, Code of Federal Regulations*, Title 40, Volume 27, Part 466, Washington, DC, July 1, 2003. [47 FR 53184, Nov. 24, 1982, as amended at 50 FR 36544, Sept. 6, 1985].

TABLE 8.30
New Source Performance Standards (NSPS) of the Aluminum Basis Material Subcategory

	Maximum for Any 1 Day		Maximum for Monthly Average	
Pollutant	Metal Preparation	Coating Operation	Metal Preparation	Coating Operation
	(mg/m^2 of Area Processed or Coated)			
Chromium	3.60	0.47	1.46	0.19
Lead	0.97	0.13	0.88	0.11
Nickel	5.35	0.69	3.60	0.47
Zinc	9.92	1.29	4.09	0.53
Aluminum	29.46	3.82	12.06	1.56
Iron	11.96	1.55	6.13	0.79
Oil and grease	97.24	12.60	97.24	12.60
TSS	145.86	18.91	116.69	15.12
pH	7.5–10.0[a]	7.5–10.0[a]	7.5–10.0[a]	7.5–10.0[a]
	($lb/10^6 ft^2$ of Area Processed or Coated)			
Chromium	0.74	0.10	0.30	0.04
Lead	0.20	0.03	0.18	0.20
Nickel	1.10	0.14	0.74	0.10
Zinc	2.03	0.27	0.84	0.11
Aluminum	6.03	0.78	2.47	0.32
Iron	2.45	0.32	1.26	0.16
Oil and grease	19.92	2.58	19.92	2.58
TSS	29.88	3.87	23.90	3.10
pH	7.5–10.0[a]	7.5–10.0[a]	7.5–10.0[a]	7.5–10.0[a]

Note: Any new source must achieve the NSPS.
[a] Within this range at all times.

TABLE 8.31A
Effluent Pretreatment Standards of an Existing Source of the Aluminum Basis Material Subcategory That Introduces Pollutants into a POTW

Pollutant	Maximum for Any 1 Day (mg/L)	Maximum for Monthly Average (mg/L)
Chromium	0.42	0.17
Lead	0.15	0.13
Nickel	1.41	1.00
Zinc	1.33	0.56

TABLE 8.31B
Mass Effluent Pretreatment Standards of an Existing Source of the Aluminum Basis Material Subcategory That Introduces Pollutants into a POTW

Pollutant	Maximum for Any 1 Day		Maximum for Monthly Average	
	Metal Preparation	Coating Operation	Metal Preparation	Coating Operation
	(mg/m² of Area Processed or Coated)			
Chromium	16.34	0.53	6.62	0.22
Lead	5.84	0.19	5.06	0.16
Nickel	54.85	1.78	38.9	1.26
Zinc	51.74	1.68	21.79	1.71
	(lb/10⁶ ft² of Area Processed or Coated)			
Chromium	3.35	0.11	1.36	0.05
Lead	1.20	0.04	1.04	0.03
Nickel	11.24	0.37	7.97	0.25
Zinc	10.6	0.35	4.46	0.35

Source: U.S. EPA; *Porcelain Enameling Point Source Category, Code of Federal Regulations*, Title 40, Volume 27, Part 466, Washington, DC, July 1, 2003. [47 FR 53184, Nov. 24, 1982, as amended at 50 FR 36544, Sept. 6, 1985].

TABLE 8.32
Effluent Pretreatment Standards of a New Source of the Aluminum Basis Material Subcategory That Introduces Pollutants into a POTW

Pollutant	Maximum for Any 1 Day		Maximum for Monthly Average	
	Metal Preparation	Coating Operation	Metal Preparation	Coating Operation
	(mg/m² of Area Processed or Coated)			
Chromium	3.60	0.47	1.46	0.19
Lead	0.97	0.13	0.88	0.11
Nickel	5.35	0.69	3.60	0.47
Zinc	9.92	1.29	4.09	0.53
	(lb/10⁶ ft² of Area Processed or Coated)			
Chromium	0.74	0.10	0.30	0.04
Lead	0.20	0.03	0.18	0.02
Nickel	1.10	0.14	0.74	0.10
Zinc	2.03	0.27	0.84	0.11

Source: U.S. EPA; *Porcelain Enameling Point Source Category, Code of Federal Regulations*, Title 40, Volume 27, Part 466, Washington, DC, July 1, 2003. [47 FR 53184, Nov. 24, 1982, as amended at 50 FR 36545, Sept. 6, 1985].

8.10 TECHNICAL TERMINOLOGIES USED IN THE PORCELAIN ENAMELING INDUSTRY

1. *Porcelain enameling.* This is the entire process of applying a fused vitreous enamel coating to a metal basis material. Usually this includes metal preparation and coating operations.[3–7]
2. *Basis material.* This is the metal part or base onto which porcelain enamel is applied.
3. *Area processed.* This is the total basis material area exposed to processing solutions.

TABLE 8.33
New Source Performance Standards (NSPS) of the Copper Basis Material Subcategory

	Maximum for Any 1 Day		Maximum for Monthly Average	
Pollutant	Metal Preparation	Coating Operation	Metal Preparation	Coating Operation
	(mg/m² of Area Processed or Coated)			
Chromium	6.23	0.46	2.52	0.19
Lead	1.69	0.13	1.52	0.11
Nickel	9.25	0.69	6.23	0.47
Zinc	17.16	1.29	7.07	0.53
Aluminum	50.97	3.82	20.86	1.56
Iron	20.69	1.55	10.60	0.79
Oil and grease	168.23	12.60	168.23	12.60
TSS	252.35	18.91	201.88	15.12
pH	7.5–10.0[a]	7.5–10.0[a]	7.5–10.0[a]	7.5–10.0[a]
	(lb/10⁶ ft² of Area Processed or Coated)			
Chromium	1.28	0.10	0.52	0.04
Lead	0.35	0.03	0.31	0.03
Nickel	1.90	0.14	1.28	0.10
Zinc	3.52	0.27	1.45	0.11
Aluminum	10.44	0.78	4.27	0.32
Iron	4.24	0.32	2.17	0.16
Oil and grease	34.46	2.58	34.46	2.58
TSS	51.69	3.87	41.35	3.10
pH	7.5–10.0[a]	7.5–10.0[a]	7.5–10.0[a]	7.5–10.0[a]

Note: Any new source must achieve the NSPS

[a] Within this range at all times.

Source: U.S. EPA; *Porcelain Enameling Point Source Category, Code of Federal Regulations*, Title 40, Volume 27, Part 466, Washington, DC, July 1, 2003. [47 FR 53184, Nov. 24, 1982, as amended at 50 FR 36545, Sept. 6, 1985].

TABLE 8.34
Effluent Pretreatment Standards of a New Source of the Copper Basis Material Subcategory That Introduces Pollutants into a POTW

	Maximum for Any 1 Day		Maximum for Monthly Average	
Pollutant	Metal Preparation	Coating Operation	Metal Preparation	Coating Operation
	(mg/m² of Area Processed or Coated)			
Chromium	6.23	0.46	2.52	0.19
Lead	1.69	0.13	1.52	0.11
Nickel	9.25	0.69	6.23	0.47
Zinc	17.16	1.29	7.07	0.53
	(lb/10⁶ ft² of Area Processed or Coated)			
Chromium	1.28	0.10	0.52	0.04
Lead	0.35	0.03	0.31	0.02
Nickel	1.90	0.14	1.28	0.10
Zinc	3.52	0.27	1.45	0.11

Source: U.S. EPA; *Porcelain Enameling Point Source Category, Code of Federal Regulations*, Title 40, Volume 27, Part 466, Washington, DC, July 1, 2003. [47 FR 53184, Nov. 24, 1982, as amended at 50 FR 36545, Sept. 6, 1985].

4. *Area coated.* This is the area of basis material covered by each coating of enamel.
5. *Coating operations.* This includes all of the operations associated with preparation and application of the vitreous coating. Usually this incorporates ball milling, slip transport, application of slip to the workpieces, cleaning and recovery of faulty parts, and firing (fusing) of the enamel coat.
6. *Metal preparation.* This comprises any and all of the metal processing steps preparatory to applying the enamel slip. Usually this includes cleaning, pickling, and applying a nickel flash or chemical coating.
7. *Control authority.* This is defined as the POTW if it has an approved pretreatment program; in the absence of such a program, this is the NPDES State if it has an approved pretreatment program or U.S. EPA if the State does not have an approved program.
8. *Precious metal.* This means gold, silver, or platinum group metals, and the principal alloys of those metals.

REFERENCES

1. Rutti, B., Early Enameled Glass, in *Roman Glass: Two Centuries of Art and Invention*, Newby M. and Painter K., Eds., Society of Antiquaries of London, London, 1991.
2. Gudenrath, W., Enameled Glass Vessels, 1425 BCE to 1800: the Decorating Process, *Journal of Glass Studies*, 48, 374, 2006.
3. U.S. EPA, *Draft Development Document for Effluent Limitations Guidelines and Standards for the Porcelain Enameling Point Source Category*, EPA-440179072a, U.S. EPA, Washington, DC, 1979.
4. U.S. EPA, *Proposed Development Document for Effluent Limitations Guidelines and Standards for the Porcelain Enameling Point Source Category*, EPA-440181072b, U.S. EPA, Washington, DC, 1981.
5. U.S. EPA, *Development Document for Effluent Limitations Guidelines and Standards for the Porcelain Enameling Point Source Category*, final report EPA-440182072, U.S. EPA, Washington, DC, 1982.
6. U.S. Government Printing Office, *Porcelain Enameling Point Source Category*, Code of Federal Regulations, Title 40, Volume 27, Part 466, U.S. GPO, Washington, DC, July 1, 2003.
7. U.S. EPA, *Porcelain Enameling Point Source Category*, available at http://www.access.gpo.gov/nara/cfr/waisidx_03/40cfr466_03.html, 2008.
8. Higgins, T. E., *Pollution Prevention Handbook*, CRC Press, Boca Raton, FL, 1995.
9. Wikipedia Encyclopedia, *Vitreoud Enamel*, available at http://en.wikipedia.org/wiki/Vitreous_enamel, March, 2008.
10. Wang, L.K., Hung, Y.T., and Shammas, N.K., Eds., *Physicochemical Treatment Processes*, Humana Press, Totowa, NJ, 2005.
11. Wang, L.K., Hung, Y.T., and Shammas, N.K., Eds., *Advanced Physicochemical Treatment Processes*, Humana Press, Totowa, NJ, 2006.
12. Wang L.K., Hung Y.T. and Shammas N.K., Eds., *Advanced Physicochemical Treatment Technologies*, Humana Press, Totowa, NJ, 710 pages, 2007.
13. Wang, L.K., Shammas, N.K., and Hung, Y.T., Eds., *Biosolids Treatment Processes*, Humana Press, Totowa, NJ, 2007.
14. Wang, L.K., Shammas, N.K., and Hung, Y.T., Eds., *Biosolids Engineering and Management*, Humana Press, Totowa, NJ, 2008, pp. 396–398.
15. Wang, L.K., Hung, Y.T., Lo, H.H., and Yapijakis C., Eds., *Handbook of Industrial and Hazardous Wastes Treatment*, Marcel Dekker, Inc., New York, NY, 2004.
16. U.S. ACE, Yearly Average Cost Index for Utilities, in *Civil Works Construction Cost Index System Manual*, 110-2-1304, U.S. Army Corps of Engineers, Washington, DC, available at http://www.nww.usace.army.miL/cost, 2008.

9 Treatment of Metal Finishing Industry Wastes

Nazih K. Shammas and Lawrence K. Wang

CONTENTS

9.1	Industry Description	346
	9.1.1 General Description	346
	9.1.2 Subcategory Descriptions	350
9.2	Wastewater Characterization	350
	9.2.1 "Common Metals" Subcategory	352
	9.2.2 "Precious Metals" Subcategory	352
	9.2.3 "Complexed Metals" Subcategory	352
	9.2.4 "Cyanide" Subcategory	352
	9.2.5 "Hexavalent Chromium" Subcategory	356
	9.2.6 "Oils" Subcategory	356
	9.2.7 "Solvent" Subcategory	356
9.3	Source Reduction	359
	9.3.1 Chemical Substitution	359
	9.3.2 Waste Segregation	360
	9.3.3 Process Modifications to Reduce Drag-Out Loss	361
	9.3.4 Waste Reduction Costs and Benefits	365
9.4	Pollutant Removabilty	366
	9.4.1 Common Metals	366
	9.4.2 Precious Metals	369
	9.4.3 Complexed Metal Wastes	370
	9.4.4 Hexavalent Chromium	370
	9.4.5 Cyanide	370
	9.4.6 Oils	370
	9.4.7 Solvents	371
9.5	Treatment Technologies	371
	9.5.1 Neutralization	371
	9.5.2 Cyanide-Containing Wastes	371
	9.5.3 Chromium-Containing Wastes	374
	9.5.4 Arsenic- and Selenium-Containing Wastes	374
	9.5.5 Other Metals Wastes	376
9.6	Costs	377
	9.6.1 Typical Treatment Options	377
	9.6.2 Costs	377
References		379

9.1 INDUSTRY DESCRIPTION

The metal finishing industry is one of many industries subject to regulation under the Resource Conservation and Recovery Act (RCRA)[1,2] and the Hazardous and Solid Waste Amendments (HSWA).[3] The metal finishing industry has also been subject to extensive regulation under the Clean Water Act (CWA).[4] Compliance with these regulations requires highly coordinated regulatory, scientific, and engineering analyses to minimize costs.[5]

9.1.1 General Description

The metal finishing industry comprises 44 unit operations involving the machining, fabrication, and finishing of metal products (Standard Industrial Classification (SIC) groups 34 through 39). There are approximately 160,000 manufacturing facilities in the U.S. that are classified as being part of the metal finishing industry.[6] These facilities are engaged in the manufacturing of a variety of products constructed primarily by using metals. The operations performed usually begin with a raw stock in the form of rods, bars, sheets, castings, forgings, and so on, and can progress to sophisticated surface-finishing operations. The facilities vary in size from small job shops employing fewer than ten people to large plants employing thousands of production workers. Wide variations also exist in the age of the facilities and the number and type of operations performed within facilities. Because of the differences in size and processes, production facilities are custom-tailored to the specific needs of each plant. The possible variations in unit operations within the metal finishing industry are extensive. Some complex products could require the use of nearly all 44 possible unit operations, but a simple product might require only a single operation. Each of the 44 individual unit operations is listed with a brief description in the following[7]:

1. *Electroplating* is the production of a thin coating of one metal upon another by electrodeposition.
2. *Electroless plating* is a chemical reduction process that depends upon the catalytic reduction of a metallic ion in an aqueous solution containing a reducing agent and the subsequent deposition of metal without the use of external electric energy.
3. *Anodizing* is an electrolytic oxidation process that converts the surface of the metal to an insoluble oxide.
4. *Chemical conversion coatings* are applied to previously deposited metal or basis material for increased corrosion protection, lubricity, preparation of the surface for additional coatings, or formulation of a special surface appearance. This operation includes chromating, phosphating, metal coloring, and passivating.
5. *Etching and chemical milling* are used to produce specific design configurations and tolerances on parts by controlled dissolution with chemical reagents or etchants.
6. *Cleaning* involves the removal of oil, grease, and dirt from the surface of the basis material using water with or without a detergent or other dispersing material.
7. *Machining* is the general process of removing stock from a workpiece by forcing a cutting tool through the workpiece, removing a chip of basis material. Machining operations such as turning, milling, drilling, boring, tapping, planing, broaching, sawing and cutoff, shaving, threading, reaming, shaping, slotting, hobbing, filing, and chamfering are included in this definition.
8. *Grinding* is the process of removing stock from a workpiece by the use of a tool consisting of abrasive grains held by a rigid or semirigid binder. The processes included in this unit operation are sanding (or cleaning to remove rough edges or excess material), surface finishing, and separating (as in cutoff or slicing operations).
9. *Polishing* is an abrading operation used to remove or smooth out surface defects (scratches, pits, tool marks, and so on) that adversely affect the appearance or function of a part. The operation usually referred to as buffing is included in the polishing operation.

10. *Barrel finishing* or tumbling is a controlled method of processing parts to remove burrs, scale, flash, and oxides, as well as to improve surface finish.
11. *Burnishing* is the process of finish sizing or smooth finishing a workpiece (previously machined or ground) by displacement, rather than removal, of minute surface irregularities. It is accomplished with a smooth point or line-contact and fixed or rotating tools.
12. *Impact deformation* is the process of applying an impact force to a workpiece such that the workpiece is permanently deformed or shaped. Impact deformation operations include shot peening, forging, high-energy forming, heading, and stamping.
13. *Pressure deformation* is the process of applying force (at a slower rate than an impact force) to permanently deform or shape a workpiece. Pressure deformation includes operations such as roiling, drawing, bending, embossing, coining, swaging, sizing, extruding, squeezing, spinning, seaming, staking, piercing, necking, reducing, forming, crimping, coiling, twisting, winding, flaring, or weaving.
14. *Shearing* is the process of severing or cutting a workpiece by forcing a sharp edge or opposed sharp edges into the workpiece, stressing the material to the point of shear failure and separation.
15. *Heat treating* is the modification of the physical properties of a workpiece through the application of controlled heating and cooling cycles. Such operations as tempering, carburizing, cyaniding, nitriding, annealing, normalizing, austenizing, quenching, austempering, siliconizing, martempering, and malleabilizing are included in this definition.
16. *Thermal cutting* is the process of cutting, slotting, or piercing a workpiece using an oxyacetylene oxygen lance or electric arc cutting tool.
17. *Welding* is the process of joining two or more pieces of material by applying heat, pressure, or both, with or without filler material, to produce a localized union through fusion or recrystallization across the interface. Included in this process are gas welding, resistance welding, arc welding, cold welding, electron beam welding, and laser beam welding.
18. *Brazing* is the process of joining metals by flowing a thin, capillary thickness layer of nonferrous filler metal into the space between them. Bonding results from the intimate contact produced by the dissolution of a small amount of base metal in the molten filler metal, without fusion of the base metal. The term brazing is used where the temperature exceeds 425°C (800°F).
19. *Soldering* is the process of joining metals by flowing a thin, capillary thickness layer of nonferrous filler metal into the space between them. Bonding results from the intimate contact produced by the dissolution of a small amount of base metal in the molten filler metal, without fusion of the base metal. The term soldering is used where the temperature range falls below 425°C (800°F).
20. *Flame spraying* is the process of applying a metallic coating to a workpiece using finely powdered fragments of wire and suitable fluxes, which are projected together through a cone of flame onto the workpiece.
21. *Sand blasting* is the process of removing stock, including surface films, from a workpiece by the use of abrasive grains pneumatically impinged against the workpiece. The abrasive grains used include sand, metal shot, slag, silica, pumice, or natural materials such as walnut shells.
22. *Abrasive jet machining* is a mechanical process for cutting hard, brittle materials. It is similar to sand blasting but uses much finer abrasives carried at high velocities (150 to 910 m/s [500 to 3000 ft/sec]) by a liquid or gas stream. Uses include frosting glass, removing metal oxides, deburring, and drilling and cutting thin sections of metal.
23. *Electrical discharge machining* is a process that can remove metal with good dimensional control from any metal. It cannot be used for machining glass, ceramics, or other non-conducting materials. Electrical discharge machining is also known as spark machining or

electronic erosion. The operation was developed primarily for machining carbides, hard nonferrous alloys, and other hard-to-machine materials.

24. *Electrochemical machining* is a process based on the same principles used in electroplating, except the workpiece is the anode and the tool is the cathode. Electrolyte is pumped between the electrodes and a potential applied, resulting in rapid removal of metal.
25. *Electron beam machining* is a thermoelectric process in which heat is generated by high-velocity electrons impinging the workpiece, converting the beam into thermal energy. At the point where the energy of the electrons is focused, the beam has sufficient thermal energy to vaporize the material locally. The process is generally carried out in a vacuum. The process results in X-ray emission, so the work area needs to be shielded to absorb radiation. At present the process is used for drilling holes as small as 0.05 mm (0.002 in.) in any known material, cutting slots, shaping small parts, and machining sapphire jewel bearings.
26. *Laser beam machining* is the process of using a highly focused, monochromatic collimated beam of light to remove material at the point of impingement on a workpiece. Laser beam machining is a thermoelectric process, and material removal is largely accomplished by evaporation, although some material is removed in the liquid state at high velocity. Because the metal removal rate is very small, this process is used for such jobs as drilling microscopic holes in carbides or diamond wire drawing dies, and for removing metal in the balancing of high-speed rotating machinery.
27. *Plasma arc machining* is the process of material removal or shaping of a workpiece by a high-velocity jet of high-temperature ionized gas. A gas (nitrogen, argon, or hydrogen) is passed through an electric arc, causing it to become ionized and raising its temperature in excess of 16,000°C (30,000°F). The relatively narrow plasma jet melts and displaces the workpiece material in its path.
28. *Ultrasonic machining* is a mechanical process designed to remove material by the use of abrasive grains, which are carried in a liquid between the tool and the work, and which bombard the work surface at high velocity. This action gradually chips away minute particles of material in a pattern controlled by the tool shape and contour. Operations that can be performed include drilling, tapping, coining, and the making of openings in all types of dies.
29. *Sintering* is the process of forming a mechanical part from a powdered metal by fusing the particles together under pressure and heat. The temperature is maintained below the melting point of the basis metal.
30. *Laminating* is the process of adhesive bonding of layers of metal, plastic, or wood to form a part.
31. *Hot dip coating* is the process of coating a metallic workpiece with another metal by immersion in a molten bath to provide a protective film. Galvanizing (hot dip zinc) is the most common hot dip coating.
32. *Sputtering* is the process of covering a metallic or nonmetallic workpiece with thin films of metal. The surface to be coated is bombarded with positive ions in a gas discharge tube, which is evacuated to a low pressure.
33. *Vapor plating* is the process of decomposing a metal or compound on a heated surface by reduction or decomposition of a volatile compound at a temperature below the melting point of either the deposit or the basis material.
34. *Thermal infusion* is the process of applying a fused zinc, cadmium, or other metal coating to a ferrous workpiece by imbuing the surface of the workpiece with metal powder or dust in the presence of heat.
35. *Salt bath descaling* is the process of removing surface oxides or scale from a workpiece by immersion of the workpiece in a molten salt bath or a hot salt solution. The work is immersed in the molten salt [temperatures range from 400 to 540°C (750 to 1000°F)],

quenched with water, and then dipped in acid. Oxidizing, reducing, and electrolytic baths are available, and the particular type needed depends on the oxide to be removed.

36. *Solvent degreasing* is a process for removing oils and grease from the surfaces of a workpiece by the use of organic solvents, such as aliphatic petroleum, aromatics, oxygenated hydrocarbons, halogenated hydrocarbons, and combinations of these classes of solvents. However, ultrasonic vibration is sometimes used with liquid solvent to decrease the required immersion time for complex shapes. Solvent cleaning is often used as a precleaning operation, for example, prior to the alkaline cleaning that precedes plating, as a final cleaning of precision parts, or as a surface preparation for some painting operations.
37. *Paint stripping* is the process of removing an organic coating from a workpiece. The stripping of such coatings is usually performed with caustic, acid, solvent, or molten salt.
38. *Painting* is the process of applying an organic coating to a workpiece. This process includes the application of coatings such as paint, varnish, lacquer, shellac, and plastics by methods such as spraying, dipping, brushing, roll coating, lithographing, and wiping. Other processes included in this unit operation are printing, silk screening, and stenciling.
39. *Electrostatic painting* is the application of electrostatically charged paint particles to an oppositely charged workpiece followed by thermal fusing of the paint particles to form a cohesive paint film. Both water-borne and solvent-borne coatings can be sprayed electrostatically.
40. *Electropainting* is the process of coating a workpiece by either making it anodic or cathodic in a bath that is generally an aqueous emulsion of the coating material. The electrodeposition bath contains stabilized resin, dispersed pigment, surfactants, and sometimes organic solvents in water.
41. *Vacuum metalizing* is the process of coating a workpiece with metal by flash-heating metal vapor in a high- vacuum chamber containing the workpiece. The vapor condenses on all exposed surfaces.
42. *Assembly* is the fitting together of previously manufactured parts or components into a complete machine, unit of a machine, or structure.
43. *Calibration* is the application of thermal, electrical, or mechanical energy to set or establish reference points for a component or complete assembly.
44. *Testing* is the application of thermal, electrical, or mechanical energy to determine the suitability or functionality of a component or complete assembly.

Table 9.1 presents an industry summary for the metal finishing industry, including the total number of subcategories, number of subcategories studied, and the type and number of dischargers.

TABLE 9.1
Metal Finishing Industry Summary

Item	Number
Total subcategories	51
Subcategories studied	28
Discharges in industry	98,418
Direct	20,632
Indirect	77,586
Zero discharge	200

Source: From U.S. EPA, Treatability Manual, Vol. II, Industrial Descriptions, Report EPA-600/2-82-001b, U.S. EPA, Washington, DC, September 1981.

9.1.2 SUBCATEGORY DESCRIPTIONS

The primary purpose of subcategorization is to establish groupings within the metal-finishing industry such that each subcategory has a uniform set of quantifiable effluent limitations. Several bases were considered in establishing subcategories within the metal finishing industry. These included the following:

1. Raw waste characteristics
2. Manufacturing processes
3. Raw materials
4. Product type or production volume
5. Size and age of facility
6. Number of employees
7. Water usage
8. Individual plant characteristics

After these subcategorization bases were evaluated, raw waste characterization was selected as the basis for subcategorization. The raw waste characterization is divided into two components, inorganic and organic wastes. These components are further subdivided into the specific types of wastes that occur within the components. Inorganics include common metals, precious metals, complexed metals, hexavalent chromium, and cyanide. Organics include oils and solvents.

Table 9.2 lists the unit operations associated with each of the seven industry subcategories (raw waste characteristics). Common metals are found in the raw waste of all 44 unit operations. Precious metals are found in only seven unit operations; cornplexed metals are found in three unit operations; hexavalent chromium is found in seven unit operations; and cyanide is found in eight unit operations. Within the organics, oils are found in 22 unit operations and solvents are found in nine unit operations. A unit operation will often be found in more than one subcategory.

9.2 WASTEWATER CHARACTERIZATION

In this section, the uses of water in the metal finishing industry are presented, and the waste constituents are identified and quantified.

Water is used for rinsing workpieces, washing away spills, air scrubbing, process fluid replenishment, cooling and lubrication, washing of equipment and workpieces, quenching, spray booths, and assembly and testing. Unit operations with significant water usage include electroplating, electroless plating, anodizing, conversion coating, etching, cleaning, machining, grinding, tumbling, heat treating, welding, sand blasting, salt bath descaling, paint stripping, painting, electrostatic painting, electroplating, and testing. Unit operations with zero discharge include electron beam machining, laser beam machining, plasma arc machining, ultrasonic machining, sintering, sputtering, vapor plating, thermal infusion, vacuum metalizing, and calibration.[7]

Table 9.3 displays the ranges of flows in the metal finishing industry. Approximately 81% of the plants have flows of between 1.9 and 57 m^3/h (67 to 2000 ft^3/h). For those plants with common metals wastestreams, the average contribution of these streams to the total wastewater flow within a particular plant is 62.4% (range, 0.007 to 100%). All of the plants have a wastestream requiring common metals treatment.

Of the plants, 4.8% have production processes that generate precious metals wastewater. The average precious metals wastewater flow is 21.5% of total plant flow.

The average contribution of the complexed metal streams to total plant flow is 22.2%. The percentage was computed from data for plants whose complexed metal streams could be segregated from the total stream.

Of the plants, 42.5% have segregated hexavalent chromium wastestreams. The average flow contribution of these wastestreams to the total wastewater stream is 28.7%. At those plants with cyanide

TABLE 9.2
Subcharacterization of Unit Operations

Industry Subcategory (Raw Waste Characteristics)	Unit Operations	
Common Metals		
All 44 unit operations		
Precious Metals		
Electroplating	Etching	Burnishing
Electroless plating	Cleaning	
Conversion coating	Polishing	
Complexed Metals		
Electroless plating		
Etching		
Cleaning		
Hexavalent Chromium		
Electroplating	Etching	Electrostatic painting
Anodizing	Cleaning	
Conversion coating	Tumbling	
Cyanide		
Electroplating	Cleaning	Heat treating
Electroless plating	Tumbling	Electrochemical machining
Conversion coating	Burnishing	
Oils		
Cleaning	Pressure deformation	Solvent degreasing
Machining	Shearing	Paint stripping
Grinding	Heat treating	Painting
Polishing	Other abrasive jet machining	Assembly
Tumbling	Electrostatic painting	Calibration
Burnishing	Electrical discharge machining	Testing
Impact deformation	Electrochemical machining	
Solvents		
Cleaning	Solvent degreasing	Electrostatic painting
Heat treating	Paint stripping	Electropainting
Electrochemical machining	Painting	Assembly

Source: From U.S. EPA, Treatability Manual, Vol. II, Industrial Descriptions, Report EPA-600/2-82-001b, U.S. EPA, Washington, DC, September 1981.

wastes, the average contribution of the cyanide-bearing stream to the total wastewater generated is 28.8% (range, 0.1 to 100%). Of the plants, 31.2% have segregated cyanide-bearing wastes.

Segregated oily wastewater is defined as oil waste collected from machine sumps and process tanks. The water is segregated from other wastewaters until it has been treated by an oily waste removal system. Of the plants, 12.4% are known to segregate their oily wastes. The average contribution of these wastes to the total plant wastewater flow is 6.6% (range, ca. 0.0 to 55.4%).

In order to characterize the wastestreams in each subcategory, raw waste data were collected. Discrete samples of raw wastes were taken for each subcategory and analyses were performed on the samples. The results of these analyses are presented for each subcategory in Tables 9.4 to 9.9. In each table, data are presented on the number of detections of a pollutant, the number of samples analyzed, the median concentration, the range in concentrations, and the mean concentration of

TABLE 9.3
Wastewater Flow Characterization of the Metal Finishing Industry

Flow of Plants (m³/h)	Percentage of Plants Represented by This Flow
<0.38	2.8
0.38–1.9	5.0
1.9–3.8	13
3.8–9.5	17
9.5–19	20.7
19–28	10.7
28–38	10.7
38–57	9.1
57–95	5.0
95–190	3.8
190–380	0.7
>380	1.5

Source: From U.S. EPA, Treatability Manual, Vol. II, Industrial Descriptions, Report EPA-600/2-82-001b, U.S. EPA, Washington, DC, September 1981.

those samples detected. The minimum detection limit for the toxic pollutants in the sampling program was 1 µg/L and any value below this is listed in the six tables as BDL, indicating "below detection limit."

9.2.1 "Common Metals" Subcategory

Pollutant parameters found in the "common metals" subcategory of raw wastestream from sampled plants are shown in Table 9.4. The major constituents shown are parameters that originate in process solutions (such as from plating or galvanizing) and enter wastewaters by drag-out to rinses. These metals appear in wastestreams in widely varying concentrations.

9.2.2 "Precious Metals" Subcategory

Table 9.5 shows the concentrations of pollutant parameters found in the "precious metals" subcategory of raw wastestreams. The major constituents are silver and gold, which are much more commonly used in metal finishing industry operations than palladium and rhodium. Because of their high cost, precious metals are of special interest to metal finishers.

9.2.3 "Complexed Metals" Subcategory

The concentrations of metals found in the "complexed metals" subcategory of raw wastestreams are presented in Table 9.6. Complexed metals may occur in a number of unit operations, but come primarily from electroless and immersion plating. The most commonly used metals in these operations are copper, nickel, and tin. Wastewaters containing complexing agents must be segregated and treated independently of other wastes in order to prevent further complexing of free metals in the other streams.

9.2.4 "Cyanide" Subcategory

Cyanide has been used extensively in the surface-finishing industry for many years; however, it is a hazardous substance that must be handled with caution. The use of cyanide in plating and stripping solutions stems from its ability to weakly complex many metals typically used in plating. Metal

TABLE 9.4
Concentrations of Pollutants Found in the "Common Metals" Subcategory of Raw Wastewater

Pollutant	Number of Samples	Number of Detections	Range of Detections	Median of Detections	Mean of Detections
Toxic Pollutants (concentrations shown in µg/L)					
Metals and inorganics					
Antimony	106	22	1–430	6	34
Arsenic	105	31	2–64	10	16
Beryllium	27	23	1–44	5	9
Cadmium	108	60	BDL–19,000	8	1,000
Chromium	105	89	3–35,000	180	16,000
Copper	108	105	3–500,000	180	16,000
Lead	108	73	3–42,000	120	1,400
Mercury	99	32	BDL–400	10	18
Nickel	108	88	4–420,000	200	24,000
Selenium	26	21	1–60	5	9
Thallium	26	21	1–62	3	10
Zinc	108	107	9–330,000	290	19,000
Phthalates					
Bis(2-ethylhexyl)phthalate	93	91	BDL–1,900	6	57
Butyl benzyl phthalate	65	38	BDL–10	BDL	1
Di-n-butyl phthalate	89	79	BDL–10	BDL	BDL
Di-n-octyl phthalate	65	25	BDL–10	BDL	BDL
Diethyl phthalate	83	66	BDL–240	5	31
Dimethyl phthalate	65	7	BDL–10	BDL	2
Nitrogen compounds					
3,3-dichlorobenzidene	4	1	BDL		
N-nitroso-di-n-propylamine	4	1	570		
Phenols					
2-Nitrophenol	4	1	24		
Phenol	23	15	BDL–1,000	45	240
Aromatics					
Benzene	6	4	BDL–16	7	8
Ethylbenzene	37	9	BDL–1,200	250	340
Toluene	39	17	2–690	77	140
Polycyclic aromatic hydrocarbons					
Fluoranthene	4	1	74		
Isophorone	4	4	13–310	180	170
Napthalene	89	61	BDL–2,000	1	83
Anthracene	82	56	BDL–30	1	2
Fluorene	2	2	BDL–160		80
Phenanthrene	71	55	BDL–30	1	2
Pyrene	4	1	190		
Halogenated aliphatics					
Carbon tetrachloride	57	37	BDL–1	BDL	BDL
1,2-Dichloroethane	4	1	3		
1,1,1-Trichloroethane	57	43	BDL–550	BDL	18
1,1,2-Trichloroethane	57	21	BDL–3	BDL	BDL
Chloroform	65	48	BDL–140	BDL	5
1,1-Dichloroethylene	58	4	BDL–110	BDL	20

Continued

TABLE 9.4 (continued)

Pollutant	Number of Samples	Number of Detections	Range of Detections	Median of Detections	Mean of Detections
Halogenated aliphatics					
1,2-*Trans*-dichloroethylene	5	3	1–5	2	3
1,2-Dichloropropylene	4	1	2		
Methylene chloride	80	27	BDL–570	BDL	53
Methyl chloride	74	3	BDL–60	3	21
Methyl bromide	4	1	2		
Dichlorobromomethane	5	2	3–8		
Chlorodibromomethane	4	1	8		
Tetrachloroethylene	59	23	BDL–66	BDL	6
Trichloroethylene	77	49	BDL–480	BDL	22
Pesticides and metabolites					
Dieldrin	4	1	BDL		
α-Endosulfan	4	1	9		
Endrin aldehyde	4	1	BDL		
α-BHC	4	1	BDL		
β-BHC	4	1	4		
δ-BHC	4	1	BDL		
Classical Pollutants (concentrations shown in mg/L)					
TSS	107	104	0.56–11,000	63	520
Aluminum	8	6	0.03–200	0.29	62
Barium	4	3	0.027–0.071	0.03	0.043
Calcium	3	3	25–76	52	51
Cobalt	4	4	0.009–0.023	0.02	0.017
Fluorides	7	3	0.021–36	1.1	5.3
Iron	85	76	0.035–490	1.9	28
Magnesium	88	87	5.6–31	14	16
Manganese	4	4	0.059–0.5	0.085	0.22
Molybdenum	7	7	0.031–0.3	0.27	0.2
Phosphorous	4	3	0.007–77	3	7.9
Sodium	4	3	17–310	140	160
Tin	4	4	0.002–15	0.86	3.7
Titanium	5	2	0.006–0.08	0.03	0.039
Vanadium	7	3	0.01–0.22	0.036	0.087
Yttrium	4	3	0.002–0.02	0.018	0.013

BDL, below detection limit; TSS, total suspended solids; BHC, a chemical that is the sum of isomers of 1,2,3,4,5,6,-hexachlorocyclohexane, such as lindane $C_6H_6Cl_6$.

Source: From U.S. EPA, Development Document for Effluent Limitations Guidelines and Standards for the Metal Finishing Point Source Category, Report EPA-440/ 1-80/091, U.S. EPA, Washington, DC, 1980.

---

deposits produced from cyanide plating solutions are finer grained than those plated from an acidic solution. In addition, cyanide-based plating solutions tend to be more tolerant of impurities than other solutions, offering preferred finishes over a wide range of conditions. In particular, cyanide is used in the following applications:

1. Cyanide-based strippers are used to selectively remove plated deposits from the base metal without attacking the substrate.
2. Cyanide-based electrolytic alkaline descalers are used to remove heavy scale from steel.
3. Cyanide-based dips are often used before plating or after stripping processes to remove metallic smuts on the surface of parts.

TABLE 9.5
Concentrations of Pollutants Found in the "Precious Metals" Subcategory of Raw Wastewater

Pollutant	Number of Samples	Number of Detections	Range of Detections	Median of Detections	Mean of Detections
Classical Pollutants (concentrations shown in mg/L)					
Silver	15	12	0.033–600	0.38	86
Gold	15	9	0.56–43	0.86	15
Palladium	13	3	0.09–0.12	0.09	0.10
Rhodium	12	1	0.22		

Source: U.S. EPA, Development Document for Effluent Limitations Guidelines and Standards for the Metal Finishing Point Source Category, Report EPA-440/1-80/091, U.S. EPA, Washington, DC, 1980.

Cyanide-based metal finishing solutions usually operate at basic pH levels to avoid decomposition of the complexed cyanide and the formation of highly toxic hydrogen cyanide gas.

The cyanide concentrations found in the "cyanide" subcategory of raw wastestreams are shown in Table 9.7. The levels of cyanide range from 0.045 to 500 mg/L. Streams with high cyanide concentrations normally originate in electroplating and heat-treating processes. Cyanide-bearing wastestreams should be segregated and treated before being combined with other raw wastestreams.

TABLE 9.6
Concentrations of Pollutants Found in the "Complexed Metals" Subcategory of Raw Wastewater

Pollutant	Number of Samples	Number of Detections	Range of Detections	Median of Detections	Mean of Detections
Toxic Pollutants (concentrations shown in µg/L)					
Cadmium	31	9	1–3,600	67	850
Copper	31	28	10–63,000	6,700	11,000
Lead	31	10	2–3,600	420	1,200
Nickel	31	25	26–290,000	3,200	28,000
Zinc	31	31	23–18,000	210	3,000
Classical Pollutants (concentrations shown in mg/L)					
Aluminum	1	1	0.1		
Calcium	1	1	17		
Iron	31	31	0.038–99	0.74	9.9
Magnesium	1	1	2		
Manganese	1	1	0.1		
Phosphorus	31	31	0.023–100	8.2	23
Sodium	1	1	110		
Tin	31	10	0.013–6	0.68	1.6

Source: U.S. EPA, Development Document for Effluent Limitations Guidelines and Standards for the Metal Finishing Point Source Category, Report EPA-440/1-80/091, U.S. EPA, Washington, DC, 1980.

TABLE 9.7
Concentrations of Pollutants Found in the "Cyanide" Subcategory of Raw Wastewater

Pollutant	Number of Samples	Number of Detections	Range of Detections	Median of Detections	Mean of Detections
Toxic Pollutants (concentrations shown in µg/L)					
Cyanide	20	20	45–500,000	45,000	110,000
Cyanide, amenable to chlorination	19	18	5–460,000	4,500	86,000

Source: U.S. EPA, Development Document for Effluent Limitations Guidelines and Standards for the Metal Finishing Point Source Category, Report EPA-440/ 1-80/091, U.S. EPA, Washington, DC, 1980.

9.2.5 "Hexavalent Chromium" Subcategory

Concentrations of hexavalent chromium from metal finishing raw wastes are shown in Table 9.8. Hexavalent chromium enters wastewater as a result of many unit operations and can be very concentrated. Because of its high toxicity, it requires separate treatment so that it can be efficiently removed from the wastewater.

9.2.6 "Oils" Subcategory

Pollutant parameters and their concentrations found in the "oily waste" subcategory streams are shown in Table 9.9. The oily waste subcategory for the metal finishing industry is characterized by both concentrated and dilute oily wastestreams that consist of a mixture of free oils, emulsified oils, greases, and other assorted organics. The appropriate treatment for oily wastestreams is dependent on the concentration levels of the wastes, but oily wastes normally receive specific treatment for oil removal prior to solids removal waste treatment.

The majority of the pollutants listed in Table 9.9 are priority organics that are used either as solvents or as oil additives to extend the useful life of the oils. Organic priority pollutants, such as solvents, should be segregated and disposed of or reclaimed separately. However, when they are present in wastewater streams, they are most often at the highest concentration in the oily wastestream, because organic pollutants generally have a higher solubility in hydrocarbons than in water. Oily wastes will normally receive treatment for oil removal before being directed to waste treatment for solids removal.

9.2.7 "Solvent" Subcategory

The "solvent" subcategory of raw wastes includes solvents generated in the metal finishing industry by the dumping of spent solvents from degreasing equipment (including sumps, water traps, and stills).

TABLE 9.8
Concentrations of Pollutants Found in the "Hexavalent Chromium" Subcategory of Raw Wastewater

Pollutant	Number of Samples	Number of Detections	Range of Detections	Median of Detections	Mean of Detections
Toxic Pollutants (concentrations shown in µg/L)					
Chromium, hexavalent	49	41	5–13,000,000	20,000	420,000

Source: U.S. EPA, Development Document for Effluent Limitations Guidelines and Standards for the Metal Finishing Point Source Category, Report EPA-440/ 1-80/091, U.S. EPA, Washington, DC, 1980.

TABLE 9.9
Concentrations of Pollutants Found in the "Oils" Subcategory of Raw Wastewater

Toxic Pollutants (Concentrations Shown in µg/L)	Number of Samples	Number of Detections	Range of Detection	Median of Detections	Mean of Detections
Phthalates					
Bis(2-ethylhexyl) phthalate	37	20	2–9,300	73	820
Butyl benzyl phthalate	37	9	1–10,000	130	1,600
Di-n-butyl phthalate	37	19	1–3100	16	270
Di-n-octyl phthalate	37	3	4–120	—	62
Diethyl phthalate	37	9	1–1,900	40	420
Dimethyl phthalate	37	34	1–1,200	1	400
Ethers					
Bis(chloromethyl)ether	37	1	9	—	—
Bis(2-chloroethyl)ether	37	2	4–10	—	7
Bis(2-chloroisopropyl)ether	37	1	4	—	—
Bis(2-chloroethoxy)methane	37	1	3	—	—
Nitrogen compounds					
1,2-Diphenylhydrazine	37	2	5–12	—	8
Phenols					
2,4,6-Trichlorophenol	37	3	10–1,000	10	610
Perachlorometacresol	37	8	4–800,000	2,300	100,000
2-Chlorophenol	37	2	76–620	—	350
2,4-Dichlorophenol	37	2	10–68	—	39
2,4-Dimethylphenol	37	6	1–31,000	10	5,200
2-Nitrophenol	37	3	10–320	35	120
4-Nitrophenol	37	1	10	—	—
2,4-Dinitrophenol	37	3	10–10,000	13	3,300
N-Nitrosodiphenylamine	37	5	4–900	750	490
Pentachlorophenol	37	3	10–50,000	5,200	18,000
Phenol	27	3	3–6,600	440	1,700
4,6-Dinitro-o-cresol	37	2	10–5,700	—	2,800
Aromatics					
Benzene	37	18	1–110	8	12
Chlorobenzene	37	2	11–610	—	310
Nitrobenzene	37	2	1–10	—	5
Toluene	37	25	1–37,000	33	1,800
Ethylbenzene	37	16	1–5,500	12	380
Polynuclear aromatic hydrocarbons					
Acenaphthane	37	2	57–5,700	—	2,900
2-Chloronaphthalene	37	1	130	—	—
Fluoranthene	37	8	1–55,000	110	8,300
Naphthalene	37	10	1–260	100	36
Benzo (a) pyrene	37	1	10	—	—
Chrysene	37	3	1–73	2	25
Acenaphthalene	37	3	77–1,000	140	410
Anthracene	43	7	3–2,000	34	360
Fluorine	37	7	1–760	75	180
Phenanthrene	37	8	2–2,000	28	400
Pyrene	37	5	31–150	75	79

Continued

TABLE 9.9 (continued)

Toxic Pollutants (Concentrations Shown in μg/L)	Number of Samples	Number of Detections	Range of Detection	Median of Detections	Mean of Detections
Halogenated hydrocarbons					
Carbon tetrachloride	37	5	1–10,000	97	2,600
1,2-Dichloroethane	37	6	9–2,100	1,400	1,100
1,1,1-Trichloroethane	37	18	1–1,300,000	260	75,000
1,1-Dichloroethane	37	11	2–1,100	600	460
1,1,2-Trichloroethane	37	4	6–1,300	10	330
1,1,2,2-Tetrachloroethane	37	2	6–570	—	290
Chloroform	37	19	2–690	10	58
1,1-Dichloroethylene	37	12	2–10,000	200	1,500
1,2-Trans-dichloroethylene	43	9	8–1,700	88	510
Methylene chloride	37	29	5–7,600	92	600
Methyl chloride	37	4	1–4,700	9	1,200
Bromoform	37	1	10	—	—
Dichlorobromomethane	37	2	1–10	—	5
Trichlorofluoromethane	37	2	260–290	—	280
Chlorodibromomethane	37	3	1–10	2	4
Tetrachloroethylene	37	18	1–110,000	10	8,900
Trichloroethylene	37	11	1–130,000	110	23,000
Pesticides and metabolites					
Aldrin	37	2	4–11	—	7
Dialdrene	37	1	3	—	—
Chlordane	37	2	1–13	—	7
4,4-DDT (DichloroDiphenyl Trichloroethane)	37	2	2–10	—	6
4,4-DDE (Dichlorodiphenyl DichloroEthylene)	37	4	BDL–53	2	14
4,4-DDD (DichloroDiphenyl Dichloroethane)	37	3	1–10	4	5
α-Endosulfan	37	2	8–28	—	18
β-Endosulfan	37	2	BDL–6	—	3
Endosulfan sulfate	37	4	1–16	11	10
Endrin	37	2	7–10	—	8
Endrin aldehyde	37	2	10–14	—	12
Heptachlor	37	1	BDL	—	—
Heptachlor epoxide	37	1	BDL	—	—
α-BHC (lindane)	37	3	4–18	13	12
β-BHC (lindane)	37	3	1–9	7	6
δ-BHC (lindane)	37	2	4–11	—	7
Polychlorinated biphenyls					
Aroclor 1254	37	2	76–1,100	—	590
Aroclor 1248	37	2	160–1,800	—	580
Classical Pollutants (concentrations shown in mg/L)					
Ammonia	37	10	0.46–270	7.9	46
Biochemical oxygen demand (BOD)	37	21	10–17,000	1,400	3,200
Chemical oxygen demand (COD)	37	16	310–1,500,000	12,000	120,000
Oil and grease	37	37	65–800,000	6,100	41,000
Phenols, total	37	34	0.002–49	0.24	2.5

Continued

TABLE 9.9 (continued)

Toxic Pollutants (Concentrations Shown in µg/L)	Number of Samples	Number of Detections	Range of Detection	Median of Detections	Mean of Detections
Total dissolved solids, TDS	37	9	250–4,900	1,600	2,000
Total organic carbon, TOC	37	37	3–560,000	1,600	28,000
Total suspended solids, TSS	37	35	35–18,000	680	2,700

BDL, below detection limit.

Source: U.S. EPA, Development Document for Effluent Limitations Guidelines and Standards for the Metal Finishing Point Source Category, Report EPA-440/ 1-80/091, U.S. EPA, Washington, DC, 1980.

These solvents are predominately composed of compounds classified by the U.S. EPA as toxic pollutants. Spent solvents should be segregated, hauled for disposal or reclamation, or reclaimed on site. Solvents that are mixed with other wastewaters tend to appear in the common metals or oily wastestreams.

9.3 SOURCE REDUCTION

It is not currently feasible to achieve a zero discharge of chemical pollutants from metal finishing operations. However, substantial reductions in the type and volume of hazardous chemicals wasted from most metal finishing operations are possible.[8] Because end-of-pipe waste detoxification is costly for small- and medium-sized metal finishers, and the cost and liability of residuals disposal have increased for all metal finishers, management and production personnel may be more willing to consider production process modifications to reduce the amount of chemicals lost to waste.

This section provides guidance for reducing water-borne wastes from metal finishing operations in order to avoid or reduce the need for waste detoxification and the subsequent off-site disposal of detoxification residuals. Waste reduction practices may take the form of the following[5]:

1. Chemical substitution
2. Waste segregation
3. Process modifications to reduce dragout loss
4. Capture/concentration techniques

9.3.1 Chemical Substitution

The incentive for substituting process chemicals containing nonpolluting materials has only been present in recent years with the advent of pollution control regulations. Chemical manufacturers are gradually introducing such substitutes. By eliminating polluting process materials such as hexavalent chromium and cyanide-bearing cleaners and deoxidizers, the treatments required to detoxify these wastes are also eliminated. It is particularly desirable to eliminate processes using hexavalent chromium and cyanide, because special equipment is needed to detoxify them.

Substituting nonpolluting cleaners for cyanide cleaners can avoid cyanide treatment entirely. For a 7.6 L/min rinsewater flow, this means a savings of about USD 18,400 in equipment costs and USD 10 per kilogram of cyanide treatment chemical costs. In this case, treatment chemical costs are about four times the cost of the raw sodium cyanide cleaner.

There can be disadvantages in using nonpolluting chemicals. Before making a decision the following questions should be asked of the chemical supplier[5]:

- Are substitutes available and practical?
- Will substitution solve one problem but create another?

- Will tighter chemical controls be required of the bath?
- Will product quality or production rate be affected?
- Will the change involve any cost increases or decreases?

Based on a survey of chemical suppliers and electroplaters who use nonpolluting chemicals, some commonly used chemical substitutes are summarized in Table 9.10.

The chemical supplier can also identify any regulated pollutants in the facility's treatment chemicals and offer available substitutes. The federally regulated pollutants are cyanide, chromium, copper, nickel, zinc, lead, cadmium, and silver. Local or state authorities may regulate other substances, such as tin, ammonia, and phosphate. The current status of cyanide and noncyanide substitute plating processes is shown in Table 9.11.

9.3.2 Waste Segregation

After eliminating as many pollutants as possible, the next step is for polluting streams to be segregated from nonpolluting streams. Nonpolluting streams can go directly to the sewer, although pH adjustment may be necessary. The segregation process will likely require some physical re-layout or re-piping of the shop. These potentially nonpolluting rinse streams represent about one-third of all plating process water. Caution must be exercised to make certain that so-called nonpolluting baths contain no dissolved metal. The cost savings in segregating polluting from nonpolluting

TABLE 9.10
Chemical Substitutes

Polluting	Substitute	Comments
Fire dip (NaCN)	Muriatic acid with additives	Slower acting than + H_2O_2 traditional fire dip
Heavy copper cyanide plating bath	Copper sulfate	Excellent throwing power with a bright, smooth, rapid finish
		A copper cyanide strike may still be necessary for steel, zinc, or tin–lead base metals
		Requires good pre-plate cleaning
		Noncyanide process eliminates carbonate buildup in tanks
Chromic acid pickles, deoxidizers and bright dips	Sulfuric acid and hydrogen perioxide	Nonchrome substitute
		Nonfuming
Chrome-based antitarnish	Benzotriazole (0.1–1.0% solution in methanol) or water-based proprietaries	Nonchrome substitute
		Extremely reactive, requires ventilation
Cyanide cleaner	Trisodium-phosphate or ammonia	Noncyanide cleaner
		Good degreasing when hot and in an ultrasonic bath
		Highly basic
		May complex with soluble metals if used as an intermediate rinse between plating baths where metal ion may be dragged into the cleaner and cause wastewater treatment problems
Tin cyanide	Acid tin chloride	Works faster and better

Source: U.S. EPA, Meeting Hazardous Waste Requirements for Metal Finishers, Report EPA/625/4-87/018, U.S. EPA, Cincinnati, OH, 1987.

TABLE 9.11
Cyanide and Noncyanide Plating Processes

Metal	Cyanide	Noncyanide
Brass	Proven	No
Bronze	Proven	No
Cadmium	Proven	Yes
Copper	Proven	Proven
Gold	Proven	Developing
Indium	Proven	Yes
Silver	Proven	Developing
Zinc	Proven	Proven

Source: U.S. EPA, Managing Cyanide in Metal Finishing, Capsule Report EPA 625/R-99/009, U.S. EPA, Cincinnati, OH, December 2000.

streams is realized through wastewater treatment equipment and operating costs. The remaining polluting sources, which require some form of control, include all dumped spent solutions (e.g., tumble finishing and burnishing washes), cyanide cleaner rinses, plating rinses, rinses after "bright dips," and aggressive cleaning solutions.

9.3.3 Process Modifications to Reduce Drag-Out Loss

Plating solution that is wasted by being carried over into the rinsewater as a workpiece emerges from the plating bath is known as dragout, and is the largest-volume source of chemical pollutant in the electroplating shop. Numerous techniques have been developed to control dragout; the effectiveness of each method varies as a function of the plating process, operator cooperation, racking, barrel design, transfer dwell time, and plated part configuration.

Wetting agents and longer workpiece withdrawal/drainage times are two techniques that significantly control drag-out. These and other techniques are discussed below.

9.3.3.1 Wetting Agents

Wetting agents lower the surface tension of process baths. To remove plating solution dragged out with the plated part, gravity-induced drainage must overcome the adhesive force between the solution and the metal surface. The drainage time required for racked parts is a function of the surface tension of the solution, part configuration, and orientation. Lowering the surface tension reduces the drainage time and also minimizes the edge effect (the bead of liquid adhering to the part edge), leading to less drag-out. Plating baths such as nickel and heavy copper cyanide also use wetting agents to maintain grain quality and provide improved coverage. The chemical supplier should be asked if the baths he supplies contain wetting agents and, if not, whether wetting agents can be added. In some baths the use of wetting agents has the potential to reduce drag-out by 50%.

9.3.3.2 Longer Drain Times

By using slower withdrawal rates or longer drain times, the drag-out of process solutions can be reduced by up to 50%. Where high-temperature plating solutions are used, slow withdrawal of the rack may also be necessary to prevent evaporative "freezing," which can actually increase drag-out. In the extreme case, too rapid a withdrawal rate causes "sheeting," where huge volumes of drag-out are lost to waste. Figure 9.1 shows the drainage rates for plain and bent pieces. Drainage for all shapes is almost complete within 15 sec of withdrawal, indicating that this is an optimum drain time for most pieces.

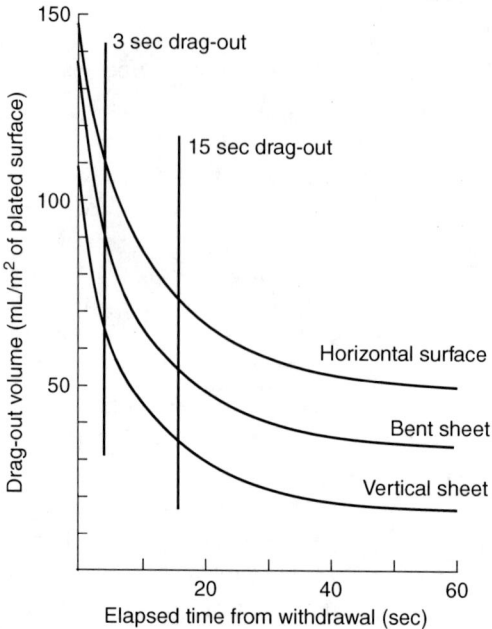

FIGURE 9.1 Typical drag-out drainage rates. (From U.S. EPA, Meeting Hazardous Waste Requirements for Metal Finishers, Report EPA/625/4-87/018, U.S. EPA, Cincinnati, OH, 1987.)

One of the best ways to control drag-out loss from rack plating on hand lines is to provide drain bars over the tank, from which the rack can be hung to drain for a brief period. Hanging and removing the racks from the drain bars ensures an adequate drain time. Slightly jostling the racks helps shake off adhering solution.

In barrel plating, the barrel should be rotated for a time just above the plating tank in order to reduce the volume of dragged-out chemical. Holes in the barrels should be as large as possible to improve solution drainage while still containing the pieces. A fog spray directed at the barrel or its contents can also help drag out drainage. Deionized water is recommended to minimize bath contamination.

The combined application of wetting agents and longer withdrawal/drainage times can significantly reduce the amount of drag-out for many cleaning or plating processes. For example, a typical nickel drag-out can be reduced from 1 L/h to 1/4 L/h by these techniques.

9.3.3.3 Other Drag-Out Reduction Techniques

Rinse elimination
The rinse between a soak cleaner and an electrocleaner may be eliminated if the two baths are compatible.

Low-concentration plating solutions
Low-concentration plating solutions reduce the total mass of chemicals being dragged out. The mass of chemicals removed from a bath is a function of the solution concentration and the volume of solution carried from the bath. Traditionally, bath concentration is maintained at a midpoint within a range of operating conditions. In contrast with the high cost of replacement, treatment, and disposal of dragged-out chemicals, the economics of low-concentration baths are favorable.

As an illustration, a typical nickel plating operation with five nickel tanks has an annual nickel drag-out of about 10,000 L. Assuming the nickel baths are maintained at the midpoint operating

TABLE 9.12
Standard Nickel Solution Concentration Limits

Chemical	Concentration Range (g/L)	Midpoint Operating Condition (g/L)	Modified Operating Condition (g/L)
Nickel sulfate			
$NiSO_4 \cdot 6H_2O$	300–375	338	308
as $NiSO_4$	—	200	182
Nickel chloride			
$NiCl_2 \cdot 6H_2O$	60–90	75	64
as $NiCl_2$	—	41	35
Boric acid, H_3BO_3	45–49	47	46

Source: U.S. EPA, Meeting Hazardous Waste Requirements for Metal Finishers, Report EPA/625/4-87/018, U.S. EPA, Cincinnati, OH, 1987.

concentration, as shown in Table 9.12, the annual cost of chemical replacement, treatment, and disposal are about USD 20,700 (in 2007 dollars). If the bath is converted to the modified operating condition as shown in the table, the annual cost of chemical replacement, treatment, and disposal are approximately USD 18,700, a saving of about USD 2,000 per year. Generally, any percent decrease in bath chemical concentration results in the same percent reduction in the mass of chemicals lost in the drag-out. The disadvantage of low-concentration baths may be lowered plating efficiencies, which may require higher current densities and closer process control in order to compensate. The reduction in plating chemical replacement, treatment, and disposal costs could be partially offset by the added labor and power costs associated with the use of the lower concentration baths.

Clean plating baths
Contaminated plating baths, for example through carbonate buildup in cyanide baths, can increase drag-out by as much as 50% by increasing the viscosity of the bath. Excessive impurities also make the application of recovery technology difficult, if not impossible.

Low-viscosity conducting salts
Bath viscosity indexes are available from chemical suppliers. As bath viscosity increases, drag-out volume also increases.

High-temperature baths
These reduce surface tension and viscosity, thus decreasing drag-out volume. Disadvantages to be considered are more rapid solution decomposition, higher energy consumption, and possible dry-on pattern on the workpiece.

No unnecessary components
Additional bath components (chemicals) tend to increase both viscosity and drag-out.

Fog sprays or air knives
These may be used over the bath to remove drag-out from pieces as they are withdrawn. The spray of deionized water or air removes plating solution from the part and returns as much as 75% of the drag-out back to the plating tank. Fog sprays, located just above the plating bath surface, dilute and drain the adhering drag-out solution, thus reducing the concentration and mass of chemicals lost. Fog sprays are best when tank evaporation rates are sufficient to accommodate the added volume of spray water. Air knives, also located just above the plating bath surface, reduce the volume of drag-out by mechanically scouring the adhering liquid from the workpiece. The drag-out concentration remains constant, but the mass of chemicals lost is reduced. Air knives are best when the surface

evaporation rates of the bath are too low to allow additional spray water. In some cases, use of supplementary atmospheric evaporators may be justified by economic considerations.

Air knives can be installed for about USD 750 to 800 per bath if an oil-free, compressed air source is available. Fog sprays can be installed for the same amount per bath if a deionized water source is available. The spray should be actuated only when work is in the spraying position. Properly designed spray nozzles distribute the water evenly over the work, control the volume of water used, and avoid snagging workpieces as they are withdrawn from the tank.

Proper racking

Every piece has at least one racking position in which drag-out will be at a minimum. In general, to minimize drag-out, the following should be considered:

1. Parts should be racked with major surfaces vertically oriented.
2. Parts should not be racked directly over one another.
3. Parts should be oriented so that the smallest surface area of the piece leaves the bath surface last.

The optimum orientation will provide faster drainage and less drag-out per piece. However, in some cases this may reduce the number of pieces on a rack, or the optimum draining configuration may not be the optimum plating configuration. In addition, the user should maintain rack coatings, replace rack contacts when broken, strip racks before plating buildup becomes excessive, and ensure that all holes on racks are covered or filled.

9.3.3.4 Capture/Concentration Techniques

Capture/concentration with full reuse of drag-out

The pioneer in simple, low-cost methods of reducing waste in the plating shop was Dr. Joseph B. Kushner.[29] In *Water and Waste Control for the Plating Shop* (1994), he describes a "simple waste recovery system" that captures drag-out in a static tank or tanks for return to the plating bath. The drag-out tanks are followed by a rinse tank, which flows to the sewer with only trace amounts of polluting salts and is often in compliance with sewer discharge standards. A simplified diagram of this reuse system is shown in Figure 9.2. It is not difficult to automate the direct drag-out recovery process, and commercial units are available.

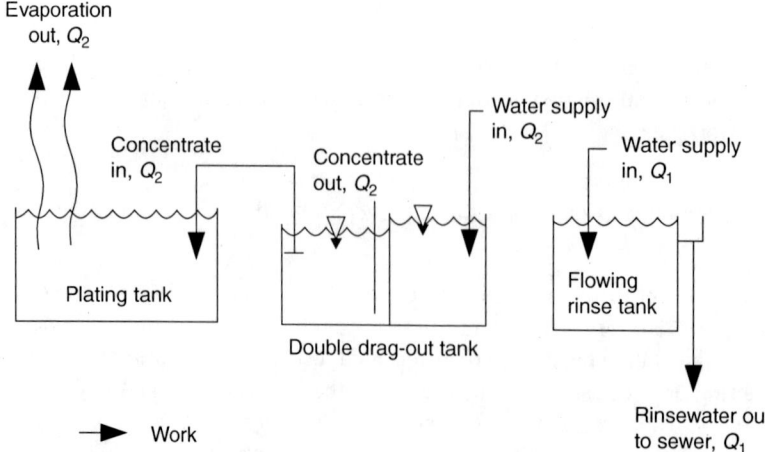

FIGURE 9.2 Kushner method of double drag-out for full reuse. (From U.S. EPA, Meeting Hazardous Waste Requirements for Metal Finishers, Report EPA/625/4-87/018, U.S. EPA, Cincinnati, OH, 1987.)

Treatment of Metal Finishing Industry Wastes

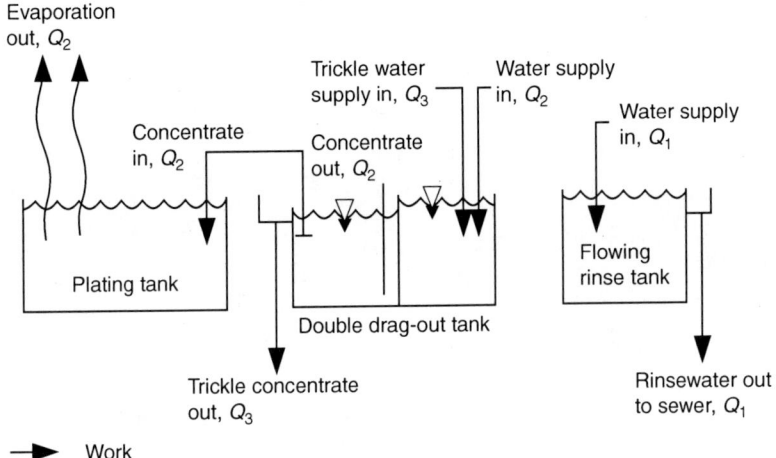

FIGURE 9.3 Modified method of double drag-out for partial reuse. (From U.S. EPA, Meeting Hazardous Waste Requirements for Metal Finishers, Report EPA/625/4-87/018, U.S. EPA, Cincinnati, OH, 1987.)

The Kushner concept is easily applicable to hot plating baths where the bath evaporation rate equals or exceeds the pour-back rate, Q_2. The drag-out concentration depends on the bath drag-out rate, the number of drag-out tanks, the rinsewater flow rate, Q_2, the plating bath evaporation rate, and drag-out return rate. The number of drag-out tanks must be based on the available space. The higher the number of counterflowed drag-out tanks, the smaller will be the return rate necessary to obtain good rinsing. The Kushner multiple drag-outs are not feasible if there is no room for the required drag-out tanks. If there is little or no evaporation from the bath, supplementary evaporation should be considered. Bath contamination must be minimized by using reverse osmosis (RO) purified water for Q_2.

Capture/concentration with partial reuse of drag-out
By adding a trickling water supply and drain, Q_3, to the drag-out tank, the application of Kushner's concept can be extended to other metal finishing processes that may not be amenable to full reuse but can allow partial reuse. Figure 9.3 depicts the partial reuse scheme. The trickle concentrate can also be batch-treated in a small volume on site, recycled at a central facility, or mixed with Q_1 for discharge, if the combined metal content is below sewer discharge standards.

9.3.4 Waste Reduction Costs and Benefits

The benefits of waste reduction in the metal finishing shop include the following:

1. Reduced chemical cost
2. Reduced water cost
3. Reduced volume of "hazardous" residuals
4. Reduced pretreatment cost

The benefits of saving valuable chemicals and water and reducing sludge disposal costs can best be illustrated by an example. An electroplating operation discharges 98,400 L/d of wastewater containing 0.91 kg copper, 1.14 kg nickel, and 0.91 kg cyanide. The shop can reduce its generation of cyanide and copper waste by about 50% by eliminating cyanide cleaners and utilizing pour-back of copper cyanide solution. The generation of nickel waste can be reduced by 90% by pour-back of the nickel solution. Reducing wasted salts also allows a reduced rinsewater flow rate, thus saving water and sewer use fees. The chemical costs of treatment are given in Table 9.13 and the annual replacement

TABLE 9.13
Chemical Costs of Treatment and Disposal

Pollutant	Chemical Cost (2007 USD/kg)[a]	
	Treatment[b]	Disposal[c]
Nickel	2.73	6.70
Copper	2.73	6.70
Cyanide	17.63	NA

[a] Costs were converted from 1979 USD to 2007 USD using the U.S. ACE Yearly Average Cost Index for Utilities.
[b] Cost of NaOH @ USD 1.00/kg and NaOCl @ USD 2.35/kg.
[c] Cost of disposal @ USD 1.84/kg of sludge (USD 400/drum) @ 30% solids content.
Source: U.S. EPA, Meeting Hazardous Waste Requirements for Metal Finishers, Report EPA/625/4-87/018, U.S. EPA, Cincinnati, OH, 1987.

costs of chemicals are given in Figure 9.4. Calculations of the annual dollar savings are shown in Table 9.14. All costs have been converted into 2007 USD using the U.S. ACE Yearly Average Cost Index for Utilities.[9]

9.4 POLLUTANT REMOVABILTY

This section reviews the technologies currently available used to remove or recover pollutants from the wastewater generated in the metal-finishing industry.[5–7,10] Treatment options are presented for each subcategory within the metal finishing industry. Table 9.15 lists the treatment techniques available for treating wastes from each subcategory.

9.4.1 COMMON METALS

The treatment methods used to treat wastes within the "common metals" subcategory fall into two groups:

1. Recovery techniques
2. Solids removal techniques

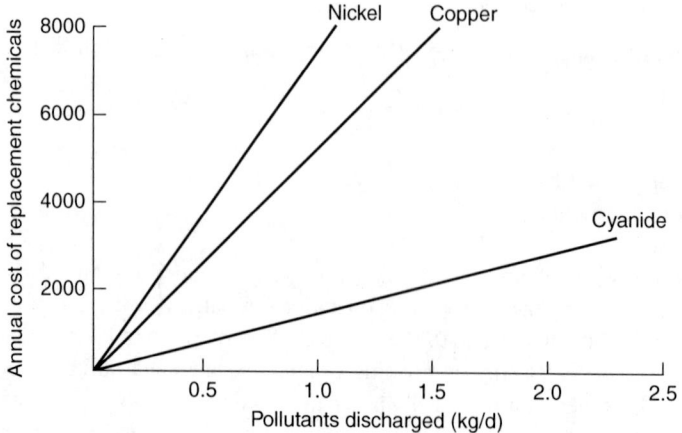

FIGURE 9.4 Annual replacement cost of chemicals in 2007 USD. (From U.S. EPA, Meeting Hazardous Waste Requirements for Metal Finishers, Report EPA/625/4-87/018, U.S. EPA, Cincinnati, OH, 1987.)

TABLE 9.14
Illustration of Annual Cost Savings for Waste Reduction

Item	Cost Saving[a] 2007 USD
Process chemical savings[b]	
Copper	2,425
Cyanide	485
Nickel	7,760
Treatment chemical saving[c]	
Copper	310
Cyanide	2,000
Nickel	700
Reduced treatment sludge disposal[c]	
Copper	760
Cyanide	0
Nickel	1,700
Water and sewer use fee reduction[d]	4,360
Total annual savings	20,500

[a] Costs were converted from 1979 USD to 2007 USD using the U.S. ACE Yearly Average Cost Index for Utilities.
[b] From Figure 9.4.
[c] From Table 9.12 and Figure 9.4.
[d] USD 0.77/m^3.
Source: U.S. EPA, Meeting Hazardous Waste Requirements for Metal Finishers, Report EPA/625/4-87/018, U.S. EPA, Cincinnati, OH, 1987.

TABLE 9.15
Treatment Methods in Current Use or Available for Use in the Metal Finishing Industry

Subcategory/Technology	Number of Plants
Common Metals	
Hydroxide followed by sedimentation	103
Hydroxide followed by sedimentation and filtration	30
Evaporation (metal recovery, bath concentrates, rinse waters)	41
Ion exchange	63
Electrolytic recovery	11
Electrodialysis	3
Reverse osmosis	8
Post-adsorption	0
Insoluble starch xanthate	2
Sulfide precipitation	3
Flotation	29
Membrane flotation	7
Precious Metals	
Evaporation	1
Ion exchange	NR
Electrolytic recovery	NR
Complexed Metals	
High-pH precipitation with sedimentation	NR
High-pH precipitation with sedimentation	NR

Continued

TABLE 9.15 (continued)

Subcategory/Technology	Number of Plants
Hexavalent Chromium	
Chemical chrome reduction	343
Electrochemical chromium reduction	2
Electrochemical chromium regeneration	0
Advanced electrodialysis	NR
Evaporation	1
Ion exchange	1
Cyanide	
Oxidation by chlorine	201
Oxidation by ozone	2
Oxidation by ozone with UV radiation	NR
Oxidation by hydrogen peroxide	3
Electrochemical cyanide oxidation	4
Chemical precipitation	3
Reverse osmosis	NR
Evaporation	NR
Oils (Segregated)	
Emulsion breaking	28
Skimming	94
Emulsion breaking and skimming	NR
Ultrafiltration	20
Reverse osmosis	3
Carbon adsorption	10
Coalescing	3
Flotation	29
Centrifugation	5
Integrated adsorption	0
Resin adsorption	0
Ozonation	0
Chemical oxidation	0
Aerobic decomposition	14
Thermal emulsion breaking	0
Solvent Waste	
Segregation	NR
Contract handling	NR
Sludges	
Gravity thickening	78
Pressure filtration	66
Vacuum filtration	68
Centrifugation	55
Sludge bed drying	77
In-Process Control	
Flow reduction	NR

NR, not reported.

Source: U.S. EPA, Treatability Manual, Vol. II, Industrial Descriptions, Report EPA-600/2-82-001b, U.S. EPA, Washington, DC, September 1981.

Recovery techniques are treatment methods used for the purpose of recovering or regenerating process constituents that would otherwise be discarded. Included in this group are the following[5-7]:

1. Evaporation
2. Ion exchange
3. Electrolytic recovery
4. Electrodialysis
5. Reverse osmosis

Solids removal techniques are used to remove metals and other pollutants from process wastewaters to make these waters suitable for reuse or discharge. These methods include the following[5-7]:

1. Hydroxide and sulfide precipitation
2. Sedimentation
3. Diatomaceous earth filtration
4. Membrane filtration
5. Granular bed filtration
6. Peat adsorption
7. Insoluble starch xanthate treatment
8. Flotation

Three treatment options are used in treating common metals wastes:

1. The *Option 1* system consists of hydroxide precipitation[11] followed by sedimentation.[12] This system accomplishes end-of-pipe metals removal from all common metals-bearing wastewater streams that are present at a facility. The recovery of precious metals, the reduction of hexavalent chromium, the removal of oily wastes, and the destruction of cyanide must be accomplished prior to common metals removal.
2. The *Option 2* system is identical to the Option 1 treatment system but with the addition of filtration devices[13] after the primary solids removal devices. The purpose of these filtration units is to remove suspended solids such as metal hydroxides that do not settle out in the clarifiers. The filters also act as a safeguard against pollutant discharge should an upset occur in the sedimentation device. Filtration techniques applicable to Option 2 systems are diatomaceous earth and granular bed filtration.[14,15]
3. The *Option 3* treatment system for common metals wastes consists of the Option 2 end-of-pipe treatment system plus the addition of in-plant controls for lead and cadmium. In-plant controls would include evaporative recovery, ion exchange, and recovery rinses.[15]

In addition to these three treatments, there are several alternative treatment technologies applicable to the treatment of common metals wastes. These technologies include electrolytic recovery, electrodialysis, reverse osmosis, peat adsorption, insoluble starch xanthate treatment, sulfide precipitation, flotation, and membrane filtration.[14,15]

9.4.2 PRECIOUS METALS

Precious metal wastes can be treated using the same treatment alternatives as those described for the treatment of common metals wastes. However, due to the intrinsic value of precious metals, every effort should be made to recover them. The treatment alternatives recommended for precious metal wastes are the recovery techniques of evaporation, ion exchange, and electrolytic recovery.

9.4.3 COMPLEXED METAL WASTES

Complexed metal wastes within the metal finishing industry are a product of electroless plating, immersion plating, etching, and the manufacture of printed circuit boards. The metals in these wastestreams are tied up or complexed by particular complexing agents whose function is to prevent metals from coming out of solution. This counteracts the technique used by most conventional solids removal methods. Therefore, segregated treatment of these wastes is necessary. The treatment method most suited to treating complexed metal wastes is high-pH precipitation. An alternative method is membrane filtration[16], which is primarily used in place of sedimentation for solids removal.

9.4.4 HEXAVALENT CHROMIUM

Hexavalent chromium-bearing wastewaters are produced in the metal finishing industry in chromium electroplating, in chromate conversion coatings, in etching with chromic acid, and in metal finishing operations carried out on chromium as a basis material.

The selected treatment option involves the reduction of hexavalent chromium to trivalent chromium either chemically or electrochemically. The reduced chromium can then be removed using a conventional precipitation-solids removal system. Alternative hexavalent chromium treatment techniques include chromium regeneration, electrodialysis, evaporation, and ion exchange.[15]

9.4.5 CYANIDE

Cyanides are introduced as metal salts for plating and conversion coating or are active components in plating and cleaning baths. Cyanide is generally destroyed by oxidation. Chlorine, in either elemental or hypochlorate form, is the primary oxidation agent used in industrial waste treatment to destroy cyanide. Alternative treatment techniques for the destruction of cyanide include oxidation by ozone, ozone with ultraviolet radiation (oxyphotolysis), hydrogen peroxide, and electrolytic oxidation.[17] Treatment techniques that remove cyanide but do not destroy it include chemical precipitation, reverse osmosis, and evaporation.[15,17]

9.4.6 OILS

Oily wastes and toxic organics that combine with the oils during manufacturing include process coolants and lubricants, wastes from cleaning operations, wastes from painting processes, and machinery lubricants. Oily wastes are generally of three types: free oils, emulsified or water-soluble oils, and greases. Oil removal techniques commonly employed in the metal finishing industry include skimming, coalescing, emulsion breaking, flotation, centrifugation, ultrafiltration, reverse osmosis, carbon adsorption, and aerobic decomposition.[17–19]

Because emulsified oils and processes that emulsify oils are used extensively in the metal finishing industry, the exclusive occurrence of free oils is nearly nonexistent.

Treatment of oily wastes can be carried out most efficiently if oils are segregated from other wastes and treated separately. Segregated oily wastes originate in the manufacturing areas and are collected in holding tanks and sumps. Systems for treating segregated oily wastes consist of the separation of oily wastes from the water. If oily wastes are emulsified, techniques such as emulsion breaking or dissolved air flotation (DAF)[20] with the addition of chemicals are necessary to remove the oil. Once the oil–water emulsion is broken, the oily waste is physically separated from the water by decantation or skimming. Following oil–water separation, the water is sent to the precipitation/sedimentation unit used for metals removal. There are three options for oily waste removal:

1. The *Option 1* system involves the emulsion breaking process followed by surface skimming (gravity separation is adequate if only free oils are present).
2. The *Option 2* system consists of the Option 1 system followed by ultrafiltration.

3. The *Option 3* treatment system consists of the Option 2 system with the addition of either carbon adsorption or reverse osmosis.

In addition to these three treatment options, several alternative technologies are applicable to the treatment of oily wastewater. These include coalescing, flotation, centrifugation, integrated adsorption, resin adsorption, ozonation, chemical oxidation, aerobic decomposition, and thermal emulsion breaking.[17-19]

9.4.7 Solvents

Spent degreasing solvents should be segregated from other process fluids to maximize the value of the solvents, to preclude contamination of other segregated wastes, and to prevent the discharge of priority pollutants to any wastewaters. This segregation may be accomplished by providing and identifying the necessary storage containers, establishing clear disposal procedures, training personnel in the use of these techniques, and checking periodically to ensure that proper segregation is occurring. Segregated waste solvents are appropriate for on-site solvent recovery or may be contract hauled for disposal or reclamation.

Alkaline cleaning is the most feasible substitute for solvent degreasing. The major advantage of alkaline cleaning over solvent degreasing is the elimination or reduction in the quantity of priority pollutants being discharged. Major disadvantages include high energy consumption and the tendency to dilute oils removed and to discharge these oils as well as the cleaning additive.

9.5 TREATMENT TECHNOLOGIES

9.5.1 Neutralization

One technique that is used in a number of facilities that utilize molten salt for metal surface treatment prior to pickling takes advantage of the alkaline values generated in the molten salt bath in treating other wastes generated in the plant. When the bath is determined to be spent, it is in many instances manifested, hauled off site, and land disposed. One technique is to take the solidified spent molten salt (molten salt is solid at ambient temperatures) and circulate acidic wastes generated in the facility over the material prior to entry to the waste-treatment system. This in effect neutralizes the acid wastes and eliminates the requirements of manifesting and land disposal.

9.5.2 Cyanide-Containing Wastes

There are eight methods applicable to the treatment of cyanide wastes for metal finishing[5,21]:

1. Alkaline chlorination
2. Electrolytic decomposition
3. Ozonation
4. UV/ozonation
5. Hydrogen peroxide
6. Thermal oxidation
7. Acidification and acid hydrolysis
8. Ferrous sulfate precipitation

Alkaline chlorination is the most widely applied in the metal finishing industry. A schematic for cyanide reduction via alkaline chlorination is provided in Figure 9.5. This technology is generally applicable to wastes containing less than 1% cyanide, generally present as free cyanide. It is conducted in two stages. The first stage is operated at a pH greater than 10 and the second stage with a pH in the range of 7.5 to 8. Alkaline chlorination is performed using sodium hypochlorite and chlorine.

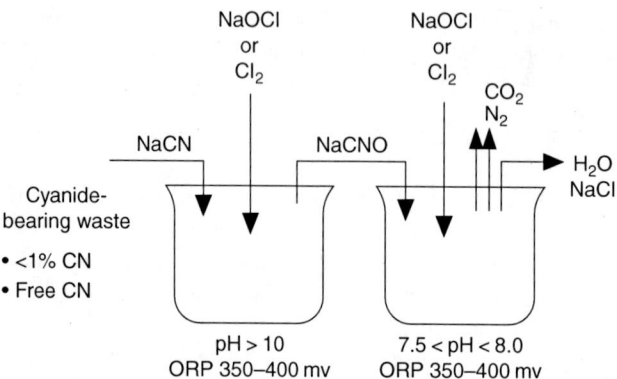

FIGURE 9.5 Cyanide reduction via alkaline chlorination. (From U.S. EPA, Meeting Hazardous Waste Requirements for Metal Finishers, Report EPA/625/4-87/018, U.S. EPA, Cincinnati, OH, 1987.) ORP, oxidation-reduction potential; mv, millivolt.

Electrolytic decomposition technology was applied to cyanide-containing wastes in the early part of this century, but it fell from favor as alkaline chlorination came into use at large-scale facilities. However, as wastes become more concentrated, because this technology is applicable to wastes containing cyanide in excess of 1%, it may find more widespread applications in the future. The basis of this technology is electrolytic decomposition of the cyanide compounds at an elevated temperature (200 °F) to yield nitrogen, CO_2, ammonia, and amines (see Figure 9.6).

Ozonation treatment can be used to oxidize cyanide, thereby reducing the concentration of cyanide in wastewater. Ozone, with an electrode potential of +1.24 V in alkaline solutions, is one of the most powerful oxidizing agents known. Cyanide oxidation with ozone is a two-step reaction similar to alkaline chlorination.[21] Cyanide is oxidized to cyanate, and the ozone is reduced to oxygen according to the following equation:

$$CN^- + O_3 \rightarrow CNO^- + O_2 \tag{9.1}$$

The cyanate is then hydrolyzed in the presence of excess ozone to bicarbonate and nitrogen, and oxidized according to the following reaction:

$$2\,CNO^- + 3\,O_3 + H_2O \rightarrow N_2 + 2\,HCO_3^- + 3\,O_2 \tag{9.2}$$

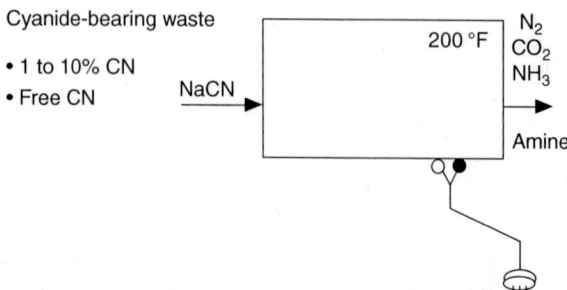

FIGURE 9.6 Cyanide reduction via electrolytic decomposition. (From U.S. EPA, Meeting Hazardous Waste Requirements for Metal Finishers, Report EPA/625/4-87/018, U.S. EPA, Cincinnati, OH, 1987.)

Treatment of Metal Finishing Industry Wastes

The reaction time for complete cyanide oxidation is rapid in a reactor system, with retention times of 10 to 30 min being typical. The second-stage reaction is much slower than the first-stage reaction. The reaction is typically carried out in the pH range 10 to 12, where the reaction rate is relatively constant. Temperature does not influence the reaction rate significantly.

One interesting variation on ozonation technology is augmentation with UV radiation. This is a technology that has been applied on wastes in the coke byproduct manufacturing industry. A significant development has been made that has resulted in a significant reduction in ozone consumption—the use of UV radiation. UV absorption has the following effects:

1. Ozone and cyanide are raised to a higher energy status.
2. Free radicals are formed.
3. There is more rapid reaction.
4. Less ozone is required.

Cyanide reduction with hydrogen peroxide is effective in reducing cyanide. It has been applied on a less frequent basis within this industry, due to the fact that there are high operating costs associated with the generation of hydrogen peroxide. The reduction of cyanide with peroxide occurs in two steps and yields CO_2 and ammonia:

$$NaCN + H_2O_2 \rightarrow NaCNO + H_2O \tag{9.3}$$

$$NaCNO + 2 H_2O \rightarrow CO_2 + NH_3 + NaOH \tag{9.4}$$

Thermal oxidation is another alternative for destroying cyanide. Thermal destruction of cyanide can be accomplished through either high-temperature hydrolysis or combustion. At temperatures between 140°C and 200°C and pH 8, cyanide hydrolyzes quite rapidly to produce formate and ammonia.[22,23] Pressures up to 100 bar are required, but the process can effectively treat wastestreams over a wide concentration range and is applicable to both rinsewater and concentrated solutions.[21] The process involves the following reaction:

$$CN^- + 2 H_2O \rightarrow HCOO^- + NH_3 \tag{9.5}$$

In the presence of nitrites, formate and ammonia can be destroyed in another reactor at 150°C, according to the following equations:

$$NH_4^+ + NO_2^- \rightarrow N_2 + 2 H_2O \tag{9.6}$$

$$3 HCOOH + 2 NO_2^- + 2 H^+ \rightarrow 3 CO_2 + 4 H_2O + N_2 \tag{9.7}$$

Direct acidification of cyanide wastestreams was once a relatively common treatment. Cyanide is acidified in a sealed reactor that is vented to the atmosphere through an air emission control system. Cyanide is converted to gaseous hydrogen cyanide, treated, vented, and dispersed.

Acid hydrolysis of cyanates is still commonly used, following a first-stage cyanide oxidation process. At pH 2 the reaction proceeds rapidly, but at pH 7 cyanate may remain stable for weeks. This treatment process requires specially designed reactors to ensure that HCN is properly vented and controlled. The hydrolysis mechanisms are as follows.

In an acid medium:

$$H_2O + HOCN + H^+ \rightarrow NH_4^+ + CO_2 \text{ (rapid)} \tag{9.8}$$

$$HOCN + H_2O \rightarrow NH_3 + CO_2 \text{ (slow)} \tag{9.9}$$

In strongly alkaline medium:

$$NCO^- + 2\,H_2O \rightarrow NH_3 + HCO_3^- \text{ (very slow)} \qquad (9.10)$$

Each of the technologies described above is effective in treating wastes containing free cyanides; that is, cyanides present as CN in solution. There are instances in metal finishing facilities where complex cyanides are present in wastes. The most common are complexes of iron, nickel, and zinc. A technology that has been applied to remove complex cyanides from aqueous wastes is ferrous sulfate precipitation. The technology involves a two-stage operation in which ferrous sulfate is first added at pH 9 to complex any trace amounts of free cyanide. In the second stage, the complex cyanides are precipitated through the addition of ferrous sulfate or ferric chloride in a pH range of 2 to 4.[5]

9.5.3 Chromium-Containing Wastes

There are three treatment methods applicable to wastes containing hexavalent chromium. Wastes containing trivalent chromium can be treated using chemical precipitation and sedimentation, which is discussed below. The three methods applicable to the treatment of hexavalent chromium use the following:

1. Sulfur dioxide
2. Sodium metabisulfite
3. Ferrous sulfate

Hexavalent chromium reduction through the use of sulfur dioxide and sodium metabisulfite has found the widest application in the metal finishing industry. It is not truly a treatment step, but a conversion process in which the hexavalent chromium is converted to trivalent chromium. The hexavalent chromium is reduced through the addition of the reductant at a pH in the range 2.5 to 3 with a retention time of approximately 30 to 40 min (see Figure 9.7).

Ferrous sulfate has not been as widely applied. However, it is particularly applicable in facilities where ferrous sulfate is produced as part of the process, or is readily available. The basis for this technology is that the hexavalent chromium is reduced to trivalent chromium and the ferrous iron is oxidized to ferric iron.

9.5.4 Arsenic- and Selenium-Containing Wastes

It may be necessary to segregate wastestreams containing elevated concentrations of arsenic and selenium, especially wastestreams with concentrations in excess of 1 mg/L for these pollutants.

$$SO_2 + H_2O \rightarrow H_2SO_3$$
$$2\,H_2CrO_4 + 3\,H_2SO_3 \rightarrow Cr_2(SO_4)_3 + 5\,H_2O$$

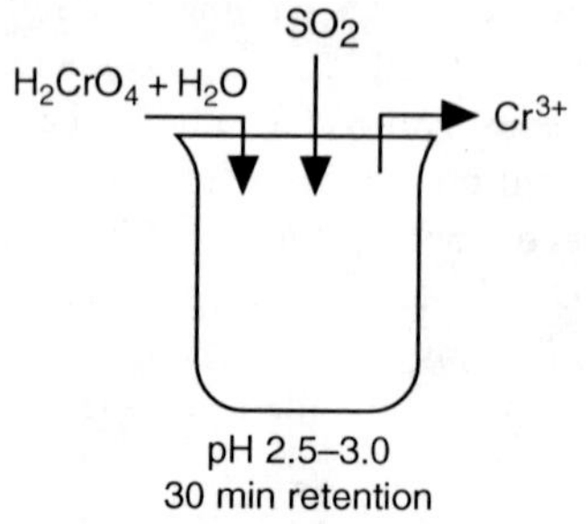

FIGURE 9.7 Hexavalent chromium reduction. (From U.S. EPA, Meeting Hazardous Waste Requirements for Metal Finishers, Report EPA/625/4-87/018, U.S. EPA, Cincinnati, OH, 1987.)

Treatment of Metal Finishing Industry Wastes

Arsenic and selenium form anionic acids in solution (most other metals act as cations) and require special preliminary treatment prior to conventional metals treatment. Lime, a source of calcium ions, is effective in reducing arsenic and selenium concentrations when the initial concentration is below 1 mg/L. However, preliminary treatment with sodium sulfide[21] at a low pH (i.e., 1 to 3) may be required for wastestreams with concentrations in excess of 1 mg/L. The sulfide reacts with the anionic acids to form insoluble sulfides, which are readily separated by means of filtration.

9.5.4.1 Chemical Precipitation and Sedimentation

The most important technology in metals treatment is chemical precipitation and sedimentation. It is accomplished through the addition of a chemical reagent to form metal precipitants, which are then removed as solids in a sedimentation step. The options available to a facility as precipitation reagents are lime [$Ca(OH)_2$], caustic (NaOH), carbonate ($CaCO_3$ and Na_2CO_3), sulfide (NaHS and FeS), and sodium borohydride ($NaBH_4$). The advantages and disadvantages of these reagents are summarized in the following[21]:

1. *Lime*
   a) It is the least expensive precipitation reagent.
   b) It generates the highest sludge volume.
   c) The sludges generally cannot be sold to smelters or refiners.
2. *Caustic*
   a) It is more expensive than lime.
   b) It generates a smaller volume of sludge.
   c) The sludges can be sold to smelter and refiners.
3. *Carbonates*
   a) These may be used for metals where solubility within a pH range is not sufficient to meet treatment standards.

Lime is the least expensive reagent, but it generates the highest volume of residue. It also generates a residue that cannot be resold to smelters and refiners for reclaiming because of the presence of the calcium ion. Caustic is more expensive than lime, but it generates a smaller volume of residue. One key advantage to caustic is that the resulting residues can be readily reclaimed. Carbonates are particularly appropriate for metals where solubility within a pH range is not sufficient to meet a given set of treatment standards. The sulfides offer the benefit of achieving effective treatment at lower concentrations due to the lower solubilities of the metal sulfides. Sodium borohydride has application where small volumes of sludge that are suitable for reclamation are desired.

It is appropriate to look at reagent use in the context of the current regulatory framework under HSWA. Historically, lime has been the reagent of choice. It was relatively inexpensive and simple to handle. The phrase "lime and settle" refers to the application of lime precipitation and sedimentation technology. In the 1970s, new designs made use of caustic as the precipitation reagent because of the reduction in residue volume realized and the possibility of reclamation. In the 1980s, a return to lime and the use of combined reagent techniques came into use.

One obvious question is "Why return to lime as a treatment reagent, given that caustic results in a smaller residue volume and a waste that can undergo reclamation?" The answer lies in the three points that result from the implementation of the HSWA hierarchy. As source reduction and material reuse and recovery techniques are applied, facilities will be generating wastes with the following characteristics:

- Greater concentration
- A varied array of constituents
- A greater degree of complexation

9.5.4.2 Complexation

Complexation is a phenomenon that involves a coordinate bond between a central atom (the metal) and a ligand (the anions). In a coordinate bond, the electron pair is shared between the metal and the ligand. A complex containing one coordinate bond is referred to as a monodentate complex. Multiple coordinate bonds are characteristic of polydentate complexes. Polydentate complexes are also referred to as chelates. An example of a monodentate-forming ligand is ammonia. Examples of chelates are oxylates (bidentates) and EDTA (hexadentates).

The reason for the return to lime is the calcium ion present in lime. The calcium ion that is present in solution on the addition of lime is very effective in competing with the ligand for the metal ion. The sodium ion contributed by caustic is not effective. As such, lime dramatically reduces complexation and is therefore more effective in treating complexed wastes. The term "high lime treatment" is used in cases where excess calcium ions are introduced into solution. This is accomplished through the addition of lime to raise the pH to ca. 11.5 or through the addition of calcium chloride (which has a greater solubility than lime).

The use of combinations of precipitation reagents has been most effective in taking advantage of the attributes of caustic as well as the advantages of lime. As an example, a system may use caustic in a first stage to make a coarse pH adjustment, followed by the addition of lime to make a fine adjustment. This achieves an overall reduction in the sludge volume through the use of the caustic, and more effective metal removal through the use of lime. Sulfide reagents are used in a similar fashion in combination with caustic or lime to provide additional metal removal, taking advantage of the lower solubility of the metal sulfides. Sulfides are also applicable to wastes containing elevated concentrations (i.e., in excess of 2 mg/L) of selenium and arsenic compounds.[21]

9.5.5 OTHER METALS WASTES

There are three techniques applicable to managing solids generated in metal finishing:

1. Dewatering
2. Stabilization
3. Incineration

There are four dewatering techniques (centrifugation, vacuum filtration, belt filtration, and evaporation/drying) that have been applied in metal processing. The most widely used are vacuum and belt filtration.[24,30] They have a higher relative capital cost, but generally have a lower relative operating cost. Plate and frame filter presses have experienced less widespread application. Belt filters generally have a lower relative capital cost and have higher relative operating costs in comparison with other dewatering techniques. The higher operating costs are due to the fact that the units are more labor-intensive to operate. Centrifuges[24] have been applied in specific instances, but are more difficult to operate when a widely varying mix of wastes is treated.

Experience has shown that companies are most successful in applying a dewatering technique that they have successfully designed and operated in similar applications within the company. As an example, many companies operate plate and frame filter presses as a part of metal manufacturing operations. The knowledge gained in metal processing had been successfully transferred to treatment of metal finishing wastes.

There are many stabilization techniques currently available; however, only two of these have found widespread application. These are cementation and stabilization through the addition of lime and fly ash.[24,25,30] Developmental work is currently being undertaken to make use of bitumen, paraffin, and polymeric materials to reduce the degree to which metals can be taken into solution. Encapsulation with inert materials is also under development.

Treatment of Metal Finishing Industry Wastes

9.6 COSTS

The investment, operating and maintenance,[26,27] and energy costs for the application of control technologies to the wastewaters of the metal finishing industry have been analyzed. These costs were developed to reflect the conventional use of technologies in this industry. A detailed presentation of the cost methodology and cost data is available in a U.S. EPA publication.[6] The available industry-specific cost information is characterized in the following.

9.6.1 Typical Treatment Options

Several unit operation/unit process configurations have been analyzed for their cost of application to the wastewater of this industry. The components included in these configurations are as follows:

1. *Option 1* includes emulsion breaking and oil separation by skimming, cyanide oxidation, chromium reduction, chemical precipitation and sedimentation, and sludge drying beds.
2. *Option 2* includes all of Option 1, plus multimedia filtration.
3. *Option 3* includes all of Option 2, plus ultrafiltration and carbon adsorption for oily waste, and achieving zero discharge of any processes using either cadmium or lead by using an evaporative system.

A flow diagram for suggested Option 1 is shown in Figure 9.8. Flow diagram for the other options would be similar.

9.6.2 Costs

Cost estimates for the treatment technologies commonly used in this industry are described briefly in the following. More details on the factors considered in the cost analyses are available from the source document.[6]

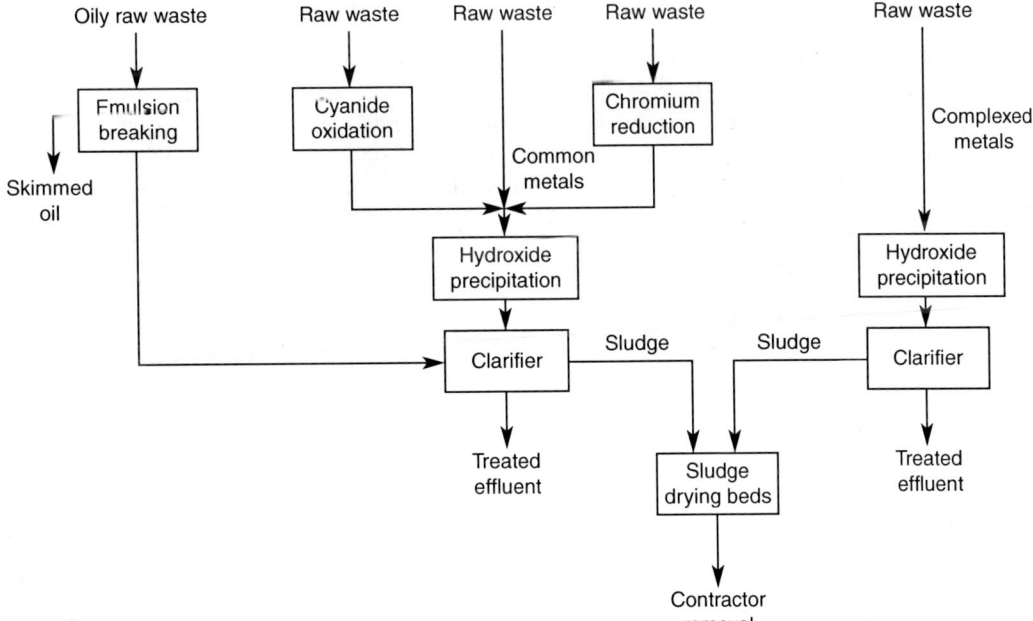

FIGURE 9.8 Metal finishing wastewater treatment flow diagram. (From U.S. EPA, Treatability Manual, Vol. II, Industrial Descriptions, Report EPA-600/2-82-001b, U.S. EPA, Washington, DC, September 1981.)

9.6.2.1 Emulsion Breaking and Oil Separation

1. *Method.* The emulsion is broken by mixing oily waste with alum in a chemical emulsion breaker, followed by gravity oil separation in a tank.
2. *System components.* These comprise a small mixing tank, two chemical feed tanks, a mixer, and a large tank equipped with an oil skimmer and a sludge pump. The mixing tank has a retention time of 15 min and the oil skimming tank a retention time of 2.5 h.

9.6.2.2 Cyanide Oxidation

1. *Method.* Cyanide is destroyed by reaction with sodium hypochlorite under alkaline conditions.
2. *System components.* These comprise reaction tanks, a reagent storage and feed system, mixers, sensors, and controls. Two identical reaction tanks sized as above ground cylindrical tank with a retention time of 4 h. The chemical storage consists of covered concrete tanks able to store a 60 d supply of sodium hypochlorite and 90 d supply of sodium hydroxide.

9.6.2.3 Chromium Reduction

1. *Method.* This involves the chemical reduction of hexavalent chromium by sulfur dioxide under acid conditions for continuous operating systems and by sodium bisulfite under acid conditions for batch operating systems. The reduced trivalent form of chromium is subsequently removed by precipitation as the hydroxide.
2. *System components.* These comprise reaction tanks, a reagent storage and feed system, mixers, sensors, and controls for continuous chromium reduction. A single above-ground concrete tank with retention time of 45 min is provided. For batch operation, dual above-ground concrete tanks with 4 h retention time are provided.

9.6.2.4 Chemical Coagulation, Precipitation and Clarification

1. *Method.* This involves the chemical coagulation/precipitation of dissolved and complexed metals by reaction with lime, alum and polyelectrolyte and subsequent removal of the precipitated solids by gravity settling or dissolved air flotation (DAF) in a clarifier.
2. *System components.* This is a continuous treatment system including reagent storage and feed equipment, a mix tank for reagent feed addition, sensors and controls, and a clarification basin with associated sludge rakes and pumps. Lime is fed as 30% lime slurry prepared by using hydrated lime. The mix tank is sized for a retention time of 45 min, and the setting clarifier is sized for hydraulic loading of 1360 L/m^2 and a retention time of 4 h. Batch setting treatment includes dual reaction-settling tanks sized for 8 h retention time and sludge pumps. The retention time of a DAF clarifier is in the range of 20–60 min.[14,20]

9.6.2.5 Sludge Drying Beds

1. *Method.* Sludge is dewatered by means of gravity drainage and natural evaporation.
2. *System components.* Beds of highly permeable gravel and sand with underlying drain pipes.[28]

9.6.2.6 Multimedia Filter

1. *Method.* This involves a polishing treatment after chemical precipitation and sedimentation by filtration through a bed of particles of several distinct size ranges.
2. *System components.* These comprise filter beds, media, backwash mechanism, pumps, and controls. Filter beds are sized for hydraulic loading of 81 L/min/m^2 (2 gpm/ft^2).

TABLE 9.16
Total Annual Unit Cost (USD/m³ in 2007 Dollars)[a]

Flow (m³/h)	Option 1		Option 2		Option 3	
	Continuous	Batch	Continuous	Batch	Continuous	Batch
2.36	—	14.28	—	23.94	—	28.35
11.81	6.09	5.04	9.66	8.4	11.34	10.29
59.07	2.52	—	4.62	—	5.25	—
118.16	2.10	2.10	3.57	3.78	4.20	4.41

[a] Costs were converted from 1979 USD to 2007 USD using the U.S. ACE Yearly Average Cost Index for Utilities.

Source: From U.S. EPA, Treatability Manual, Vol. II, Industrial Descriptions, Report EPA-600/2-82-001b, U.S. EPA, Washington, DC, September 1981.

9.6.2.7 Ultrafiltration

1. *Method.* This process is used for oily wastestreams after emulsion breaking–gravity oil separation.
2. *System components.* These comprise filter modules sized on the basis of a hydraulic loading of 1 L/min/m².

9.6.2.8 Carbon Adsorption

1. *Method.* This is a packed-bed throwaway system to remove organic pollutants from oily wastestreams.
2. *System component.* These comprise a contactor system, and a pump station designed for a contact time of 30 min and hydraulic loading of 162 L/min/m² (4 gpm/ft²).

Unit costs shown in Table 9.16 are for the complete treatment options described previously. Unit costs are computed for a model plant where flows have contributions from several wastestreams:

1. 30% oily wastestream
2. 4% cyanide wastestream
3. 9% chromium wastestream
4. 52.5% common metals stream
5. 4.5% complex metal stream

REFERENCES

1. Federal Register, Resource Conservation and Recovery Act (RCRA), 42 U.S. Code s/s 6901 et seq. 1976, United States Government, Public Laws, January 2004. Available at www.access.gpo.gov/uscode/title42/chapter82_.html.
2. U.S. EPA, Resource Conservation and Recovery Act (RCRA)—Orientation Manual, U.S. EPA, Report EPA530-R-02-016, Washington, DC, January 2003.
3. U.S. EPA, Federal Hazardous and Solid Wastes Amendments (HSWA), U.S. EPA, Washington, DC, November 1984. Available at http://www.epa.gov/osw/laws-reg.htm.
4. Federal Register, Clean Water Act (CWA), 33 U.S.C. ss/1251 et seq. (1977), U.S. Government, Public Laws, May 2002. Available at www.access.gpo.gov/uscode/title33/chapter26_.html.
5. U.S. EPA, Meeting Hazardous Waste Requirements for Metal Finishers, Report EPA/625/4-87/018, U.S. EPA, Cincinnati, OH, 1987.

6. U.S. EPA, Development Document for Effluent Limitations Guidelines and Standards for the Metal Finishing Point Source Category, Report EPA-440/1-80/091, U.S. EPA, Washington, DC, 1980.
7. U.S. EPA, Treatability Manual, Vol. II, Industrial Descriptions, Report EPA-600/2-82-001b, U.S. EPA, Washington, DC, September 1981.
8. PRC Environmental Management, *Hazardous Waste Reduction in the Metal Finishing Industry*, Noyes Data Corporation, Park Ridge, NJ, 1989.
9. U.S. ACE, Yearly average cost index for utilities, in *Civil Works Construction Cost Index System Manual*, 110-2-1304, U.S. Army Corps of Engineers, Washington, DC, 2007. Available at http://www.nww.usace.army.mil/cost.
10. Patterson, J.W., *Industrial Wastewater Treatment Technology*, 2nd ed., Butterworths, Boston, MA, 1985.
11. Wang, L.K., Vaccari, D.A., Li, Y., and Shammas, N.K., Chemical precipitation, in *Physicochemical Treatment Processes*, Wang, L.K., Hung, Y.T., and Shammas, N.K., Eds., Humana Press, Totowa, NJ, 2005, pp. 141–198.
12. Shammas, N.K., Kumar, I.J., Chang, S.Y., and Shammas, N.K., Sedimentation, in *Physicochemical Treatment Processes*, Wang, L.K., Hung, Y.T., and Shammas, N.K., Eds., Humana Press, Totowa, NJ, 2005, pp. 379–430.
13. Chen, J.P., Chang, S.Y., Huang, J.Y.C., Baumann, E.R., and Hung, Y.T., Gravity filtration, in *Physicochemical Treatment Processes*, Wang, L.K., Hung, Y.T., and Shammas, N.K., Eds., Humana Press, Totowa, NJ, 2005, pp. 501–544.
14. Wang, L.K., Hung, Y.T., and Shammas, N.K., Eds., *Physicochemical Treatment Processes*, Humana Press, Totowa, NJ, 2005, 723 p.
15. Wang, L.K., Hung, Y.T., and Shammas, N.K., Eds., *Advanced Physicochemical Treatment Processes*, Humana Press, Totowa, NJ, 2006, 690 p.
16. Chen, J.P., Mou, H., Wang, L.K., and Matsuura, T., Membrane filtration, in *Advanced Physicochemical Treatment Technologies*, Humana Press, Totowa, NJ, 2007, pp. 203–260.
17. Wang, L.K., Hung, Y.T., and Shammas, N.K., Eds., *Advanced Physicochemical Treatment Technologies*, Humana Press, Totowa, NJ, 2007, 710 p.
18. Wang, L.K., Pereira, N., Hung, Y.T., Eds., and Shammas, N.K., Consulting Ed., *Biological Treatment Processes*, Humana Press, Totowa, NJ, 2008.
19. Wang, L.K., Shammas, N.K., and Hung, Y.T., Eds., *Advanced Biological Treatment Processes*, Humana Press, Totowa, NJ, 2009.
20. Wang, L.K., Fahey, E.M., and Wu, Z., Dissolved air flotation, in *Physicochemical Treatment Processes*, Wang, L.K., Hung, Y.T., and Shammas, N.K., Eds., Humana Press, Totowa, NJ, 2005, pp. 431–500.
21. U.S. EPA, Managing Cyanide in Metal Finishing, Capsule Report EPA 625/R-99/009, U.S. EPA, Cincinnati, OH, December 2000.
22. Hartinger, L., *Handbook of Effluent Treatment and Recycling for the Metal Finishing Industry*, 2nd ed., Finishing Publications, Stevenage, UK, 1994.
23. Eilbeck, W.J. and Mattock, G., *Chemical Processes in Wastewater Treatment*, Ellis Horwood Ltd., New York, 1987.
24. Wang, L.K., Shammas, N.K., and Hung, Y.T., Eds., *Biosolids Treatment Processes*, Humana Press, Totowa, NJ, 2007, 820 p.
25. Singh, I.B., Chaturvedi, K., Singh, D.R., and Yegneswaran, A.H., Thermal stabilization of metal finishing waste with clay, *Environ. Technol.*, 26, 877–884, 2005.
26. Roy, C.H., *Operation and Maintenance of Surface Finishing Wastewater Treatment Systems*, American Electroplaters and Surface Finishers Society, Washington, DC, 1988.
27. Altmayer, F., *Plating and Surface Finishing, Advice & Council*, AESF, Orlando, FL, 1997.
28. Wang, L.K., Li, Y., Shammas, N.K., and Sakellaropoulos, G.P., Drying beds, in *Biosolids Treatment Processes*, Wang, L.K., Shammas, N.K., and Hung, Y.T., Eds., Humana Press, Totowa, NJ, 2007.
29. Kushner, J.B. *Water and Waste Control for the Plating Shop*, 3rd ed., Gardner Publications Inc., Cincinnati, OH, 1994.
30. Wang, L.K., Shammas, N.K., Hung, Y.T., Eds., *Biosolids Engineering and Management*, Humana Press, Totowa, NJ, 2008, 788 p.

10 Removal of Heavy Metals from Soil

Nazih K. Shammas

CONTENTS

10.1	Introduction	382
10.2	Overview of Metals and Their Compounds	383
	10.2.1 Overview of Physical Characteristics and Mineral Origins	383
	10.2.2 Overview of Behavior of As, Cd, Cr, Pb, and Hg	383
10.3	Description of Superfund Soils Contaminated with Metals	385
10.4	Soil Cleanup Goals and Technologies for Remediation	385
10.5	Containment	388
	10.5.1 Process Description	388
	10.5.2 Site Requirements	389
	10.5.3 Applicability	390
	10.5.4 Performance and BDAT Status	390
	10.5.5 SITE Program Demonstration Projects	391
10.6	Solidification–Stabilization Technologies	391
	10.6.1 Process Description	392
	10.6.2 Site Requirements	393
	10.6.3 Applicability	393
	10.6.4 Performance and BDAT Status	395
	10.6.5 SITE Program Demonstration Projects	397
	10.6.6 Cost of S/S	398
10.7	Vitrification	398
	10.7.1 Process Description	399
	10.7.2 Site Requirements	399
	10.7.3 Applicability	400
	10.7.4 Performance and BDAT Status	401
	10.7.5 SITE Program Demonstration Projects	401
10.8	Soil Washing	402
	10.8.1 Process Description	402
	10.8.2 Site Requirements	404
	10.8.3 Applicability	404
	10.8.4 Performance and BDAT Status	405
	10.8.5 SITE Demonstrations and Emerging Technologies Program Projects	405
10.9	Soil Flushing	405
	10.9.1 Process Description	406
	10.9.2 Site Requirements	406
	10.9.3 Applicability	406
	10.9.4 Performance and BDAT Status	407
	10.9.5 SITE Demonstration and Emerging Technologies Program Projects	407

10.10	Pyrometallurgy	408
	10.10.1 Process Description	408
	10.10.2 Site Requirements	408
	10.10.3 Applicability	409
	10.10.4 Performance and BDAT Status	409
	10.10.5 SITE Demonstration and Emerging Technologies Program Projects	410
10.11	Electrokinetics	410
	10.11.1 Process Description	410
	10.11.2 Site Requirements	413
	10.11.3 Applicability and Demonstration Projects	413
	10.11.4 Performance and Cost	416
	10.11.5 Summary of Electrokinetic Remediation	417
10.12	Phytoremediation	418
	10.12.1 Process Description	419
	10.12.2 Applicability	422
	10.12.3 Performance and Cost	422
	10.12.4 Summary of Phytoremediation Technology	424
10.13	Use of Treatment Trains	425
10.14	Cost Ranges of Remedial Technologies	425
References		427

10.1 INTRODUCTION

Metals account for much of the contamination found at hazardous waste sites. They are present in the soil and groundwater at approximately 65% of the Superfund or CERCLA (Comprehensive Environmental Response, Compensation, and Liability Act)[1] sites for which the U.S. Environmental Protection Agency (U.S. EPA) has signed records of decisions (RODs).[2] The metals most frequently identified are lead, arsenic, chromium, cadmium, nickel, and zinc. Other metals often identified as contaminants include copper and mercury. In addition to the Superfund program, metals make up a significant portion of the contamination requiring remediation under the Resource Conservation and Recovery Act (RCRA)[3] and contamination present at federal facilities, notably those that are the responsibility of the Department of Defense (DOD) and the Department of Energy (DOE).

This chapter provides remedial project managers, engineers, on-scene coordinators, contractors, and other state or private remediation managers and their technical support personnel with information to facilitate the selection of appropriate remedial alternatives for soil contaminated with arsenic (As), cadmium (Cd), chromium (Cr), mercury (Hg), and lead (Pb).[4–6]

Common compounds, transport, and fate are discussed for each of these five elements. A general description of metal-contaminated Superfund soils is provided. The technologies covered are containment (immobilization), solidification–stabilization, vitrification, soil washing, soil flushing, pyrometallurgy, electrokinetics and phytoremediation. Use of treatment trains and remediation costs are also addressed.

It is assumed that users of this chapter will, as necessary, familiarize themselves with (1) the applicable or relevant and appropriate regulations pertinent to the site of interest, (2) applicable health and safety regulations and practices relevant to the metals and compounds discussed, and (3) relevant sampling, analysis, and data interpretation methods. Information on Pb battery (Pb, As), wood preserving (As, Cr), pesticide (Pb, As, Hg), and mining sites have been addressed in U.S. EPA Superfund documents.[7–12] The greatest emphasis is on remediation of inorganic forms of the metals of interest. Organometallic compounds, organic–metal mixtures, and multimetal mixtures are briefly addressed.

10.2 OVERVIEW OF METALS AND THEIR COMPOUNDS

This section provides a brief, qualitative overview of the physical characteristics and mineral origins of the five metals, and factors affecting their mobility. More comprehensive and quantitative reviews of the behavior of these five metals in soil can be found in readily available U.S. EPA Superfund documents.[4,13,14]

10.2.1 OVERVIEW OF PHYSICAL CHARACTERISTICS AND MINERAL ORIGINS

Arsenic is a semimetallic element or metalloid that has several allotropic forms. The most stable allotrope is a silver-gray, brittle, crystalline solid that tarnishes in air. Arsenic compounds, mainly As_2O_3, can be recovered as a byproduct of processing complex ores mined mainly for Cu, Pb, Zn, Au, and Ag. Arsenic occurs in a wide variety of mineral forms, including arsenopyrite ($FeAsS_4$), which is the main commercial ore of As worldwide.

Cadmium is a bluish-white, soft, ductile metal. Pure Cd compounds are rarely found in nature, although occurrences of greenockite (CdS) and otavite ($CdCO_3$) are known. The main sources of Cd are sulfide ores of lead, zinc, and copper. Cd is recovered as a byproduct when these ores are processed.

Chromium is a lustrous, silver-gray metal. It is one of the less common elements in the Earth's crust, and occurs only in compounds. The chief commercial source of Cr is the mineral chromite ($FeCr_2O_4$). Cr is mined as a primary product and is not recovered as a byproduct of any other mining operation. There are no chromite ore reserves, nor is there primary production of chromite in the U.S.

Mercury is a silvery, liquid metal. The primary source of Hg is cinnabar (HgS), a sulfide ore. In a few cases, Hg occurs as the principal ore product, but it is more commonly obtained as the byproduct of processing complex ores that contain mixed sulfides, oxides, and chloride minerals (these are usually associated with base and precious metals, particularly gold). Native or metallic Hg is found in very small quantities in some ore sites. The current demand for Hg is met by secondary production (i.e., recycling and recovery).

Lead is a bluish-white, silvery, or gray metal that is highly lustrous when freshly cut, but tarnishes when exposed to air. It is very soft and malleable, has a high density (11.35 g/cm^3) and low melting point (327.4°C), and can be cast, rolled, and extruded. The most important Pb ore is galena (PbS). Recovery of Pb from the ore typically involves grinding, flotation, roasting, and smelting. Less common forms of the mineral are cerussite ($PbCO_3$), anglesite ($PbSO_4$), and crocoite ($PbCrO_4$).

10.2.2 OVERVIEW OF BEHAVIOR OF As, Cd, Cr, Pb, AND Hg

As metals cannot be destroyed, remediation of metal-contaminated soil consists primarily of manipulating (i.e., exploiting, increasing, decreasing, or maintaining) the mobility of metal contaminant(s) to produce a treated soil that has an acceptable total or leachable metal content. Metal mobility depends upon numerous factors. Metal mobility in soil-waste systems is determined by the following factors[13]:

1. The type and quantity of soil surfaces present
2. The concentration of the metal of interest
3. The concentration and type of competing ions and complexing ligands, both organic and inorganic
4. pH
5. Redox status

McLean and Bledsoe[13] state that

> Generalization can only serve as rough guides of the expected behavior of metals in such systems. Use of literature or laboratory data that do not mimic the specific site soil and waste system will not

be adequate to describe or predict the behavior of the metal. Data must be site specific. Long term effects must also be considered. As organic constituents of the waste matrix degrade, or as pH or redox conditions change, either through natural processes of weathering or human manipulation, the potential mobility of the metal will change as soil conditions change.

Cd, Cr(III), and Pb are present in cationic forms under natural environmental conditions.[13] These cationic metals are generally not mobile in the environment and tend to remain relatively close to the point of initial deposition. The capacity of soil to adsorb cationic metals increases with increasing pH, cation exchange capacity, and organic carbon content. Under the neutral to basic conditions typical of most soils, cationic metals are strongly adsorbed on the clay fraction of soils and can be adsorbed by the hydrous oxides of Fe, Al, or Mn present in soil minerals. Cationic metals will precipitate as hydroxides, carbonates, or phosphates. In acidic, sandy soils, the cationic metals are more mobile. Under conditions that are atypical of natural soils (e.g., pH < 5 or pH > 9; elevated concentrations of oxidizers or reducers, high concentrations of soluble organic or inorganic complexing or colloidal substances), but may be encountered as a result of waste disposal or remedial processes, the mobility of these metals may be substantially increased. Also, competitive adsorption between various metals has been observed in experiments involving a number of solids with oxide surfaces (γFeOOH, α-SiO_2, and γ-Al_2O_3). In several experiments, Cd adsorption was decreased by the addition of Pb or Cu for all three of these solids. The addition of Zn resulted in the greatest decrease of Cd adsorption. Competition for surface sites occurred when only a few percent of all surface sites were occupied.[15]

The behavior of As, Cr(VI), and Hg differs considerably from that of Cd, Cr(III), and Pb. Typically, As and Cr(VI) exist in anionic forms under environmental conditions. Hg, although it is a cationic metal, has unusual properties (e.g., liquid at room temperature, easily transforms among several possible valence states).

In most As-contaminated sites, As appears as As_2O_3 or as anionic As species leached from As_2O_3, oxidized to As(V), and then sorbed onto iron-bearing minerals in the soil. It may also be present in organometallic forms, such as methylarsenic acid ($H_2AsO_3CH_3$), and dimethylarsenic acid [$(CH_3)_2AsO_2H$], which are active ingredients in many pesticides, as well as the volatile compounds arsine (AsH_3) and its methyl derivatives [i.e., dimethylarsine $HAs(CH_3)_2$ and trimethylarsine, $As(CH_3)_3$]. These As forms illustrate the various oxidation states that As commonly exhibits (–III, 0, III, and V) and the resulting complexity of its chemistry in the environment.

As(V) is less mobile and less toxic than As(III). As(V) exhibits anionic behavior in the presence of water, and hence its aqueous solubility increases with increasing pH, and it does not complex or precipitate with other anions. As(V) can form low solubility metal arsenates. Calcium arsenate [$Ca_3(AsO_4)_2$] is the most stable metal arsenate in well-oxidized and alkaline environments, but it is unstable in acidic environments. Even under initially oxidizing and alkaline conditions, absorption of CO_2 from the air will result in the formation of $CaCO_3$ and the release of arsenate. In sodic soils, sufficient sodium is available such that the mobile compound Na_3AsO_4 can form. The slightly less stable manganese arsenate [$Mn_2(AsO_4)_2$] forms in both acidic and alkaline environments, and iron arsenate is stable under acidic soil conditions. In aerobic environments, $HAsO_4$ predominates at pH < 2 and is replaced by $H_2AsO_4^-$, $HAsO_4^{2-}$, and AsO_4^{3-} as pH increases to about 2, 7, and 11.5, respectively. Under mildly reducing conditions, H_3AsO_3 is a predominant species at low pH, but is replaced by $H_2AsO_3^-$, $HAsO_3^{2-}$, and AsO_3^{3-} as pH increases. Under still more reducing conditions and in the presence of sulfide, As_2S_3 can form. As_2S_3 is a low-solubility, stable solid. AsS_2 and AsS_2^- are thermodynamically unstable with respect to As_2S_3 (ref. 16). Under extreme reducing conditions, elemental As and volatile arsine (AsH_3) can occur. Just as competition between cationic metals affects mobility in soil, competition between anionic species (chromate, arsenate, phosphate, sulfate, etc.) affects anionic fixation processes and may increase mobility.

The most common valence states of Cr in the Earth's surface and near-surface environment are +3 [trivalent or Cr(III)] and +6 [hexavalent or Cr (VI)]. The trivalent Cr (discussed above) is the most thermodynamically stable form under common environmental conditions. Except in leather tanning, industrial applications of Cr generally use the Cr(VI) form. Owing to kinetic limitations, Cr(VI) does not always readily reduce to Cr(III) and can remain present over an extended period of time.

Cr(VI) is present as the chromate (CrO_4^{2-}) or dichromate ($Cr_2O_7^{2-}$) anion, depending on pH and concentration. Cr(VI) anions are less likely to be adsorbed onto solid surfaces than Cr(III). Most solids in soils carry negative charges that inhibit Cr(VI) adsorption. Although clays have a high capacity to adsorb cationic metals, they interact little with Cr(VI) because of the similar charges carried by the anion and clay in the common pH range of soil and groundwater. The only common soil solid that adsorbs Cr(VI) is iron oxyhydroxide. Generally, a major portion of Cr(VI) and other anions adsorbed in soils can be attributed to the presence of iron oxyhydroxide. The quantity of Cr(VI) adsorbed onto the iron solids increases with decreasing pH.

At metal-contaminated sites, Hg can be present in mercuric form (Hg^{2+}), mercurous form (Hg_2^{2+}), elemental form (Hg), or alkylated form (e.g., methyl and ethyl Hg). Hg_2^{2+} and Hg^{2+} are more stable under oxidizing conditions. Under mildly reducing conditions, both organically bound Hg and inorganic Hg compounds can convert to elemental Hg, which can then be readily converted to methyl or ethyl Hg by biotic and abiotic processes. Methyl and ethyl Hg are mobile and toxic forms.

Hg is moderately mobile, regardless of the soil. Both the mercurous and mercuric cations are adsorbed by clay minerals, oxides, and organic matter. Adsorption of cationic forms of Hg increases with increasing pH. Mercurous and mercuric Hg are also immobilized by forming various precipitates. Mercurous Hg precipitates with chloride, phosphate, carbonate, and hydroxide. At concentrations of Hg commonly found in soil, only the phosphate precipitate is stable. In alkaline soils, mercuric Hg precipitates with carbonate and hydroxide to form a stable (but not exceptionally insoluble) solid phase. At lower pH and high chloride concentration, soluble $HgCl_2$ is formed. Mercuric Hg also forms complexes with soluble organic matter, chlorides, and hydroxides, which may contribute to its mobility.[13] In strong reducing conditions, HgS, a very low solubility compound, is formed.

10.3 DESCRIPTION OF SUPERFUND SOILS CONTAMINATED WITH METALS

Soils can become contaminated with metals from direct contact with industrial plant waste discharges, fugitive emissions, or leachate from waste piles, landfills, or sludge deposits. The specific type of metal contaminant expected at a particular Superfund site would obviously be directly related to the type of operation that had occurred there. Table 10.1 lists the types of operations that are directly associated with each of the five metal contaminants.[5]

Wastes at CERCLA sites are frequently heterogeneous on a macro- and micro-scale. Contaminant concentration and the physical and chemical forms of the contaminant and matrix are usually complex and variable. Waste disposal sites collect a wide variety of waste types; therefore, concentration profiles can vary by orders of magnitude through a pit or pile. Limited volumes of high-concentration "hot spots" may develop due to variations in the historical waste disposal patterns or local transport mechanisms. Similar radical variations frequently occur on the scale of particle size too. The waste often consists of a physical mixture of very different solids, for example, paint chips in spent abrasive.

Industrial processes may result in a variety of solid metal-bearing waste materials, including slags, fumes, mold sand, fly ash, abrasive wastes, spent catalysts, spent activated carbon, and refractory bricks.[17] These process solids may be found above ground as waste piles or below ground in landfills. Solid-phase wastes can be dispersed by well-intended but poorly controlled reuse projects. Waste piles can be exposed to natural disasters or accidents, causing further dispersion.

10.4 SOIL CLEANUP GOALS AND TECHNOLOGIES FOR REMEDIATION

Table 10.2 provides an overview of cleanup goals (actual and potential) for both total and leachable metals. Based on an inspection of the total metals cleanup goals, one can see that they vary considerably both within the same metal and between metals.

TABLE 10.1
Principal Sources of As, Cd, Cr, Hg, and Pb Contaminated Soils

Contaminant	Principal Sources
As	Wood preserving
	As-waste disposal
	Pesticide production and application
	Mining
Cd	Plating
	Ni–Cd battery manufacturing
	Cd-waste disposal
Cr	Plating
	Textile manufacturing
	Leather tanning
	Pigment manufacturing
	Wood preserving
	Cr-waste disposal
Hg	Chloralkali manufacturing
	Weapons production
	Copper and zinc smelting
	Gas line manometer spills
	Paint application
	Hg-waste disposal
Pb	Ferrous/nonferrous smelting
	Pb-acid battery breaking
	Ammunition production
	Leaded paint waste
	Pb-waste disposal
	Secondary metals production
	Waste oil recycling
	Firing ranges
	Ink manufacturing
	Mining
	Pb-acid battery manufacturing
	Leaded glass production
	Tetraethyl Pb production
	Chemical manufacturing

Source: U.S. EPA, Technology Alternatives for the Remediation of Soils Contaminated with As, Cd, Cr, Hg, and Pb, EPA/540/S-97/500, U.S. Environmental Protection Agency, Cincinnati, OH, August 1997.

Similar variation is observed in the actual or potential leachate goals. The observed variation in cleanup goals has at least two implications with regard to technology alternative evaluation and selection. First, the importance of identifying the target metal(s), contaminant state (leachable vs. total metal), the specific type of test and conditions, and the numerical cleanup goals early in the remedy evaluation process is made apparent. Depending on which cleanup goal is selected, the required removal or leachate reduction efficiency of the overall remediation can vary by several orders of magnitude.[5,18] Second, the degree of variation in goals both within and between the metals, plus the many factors that affect the mobility of the metals, suggest that generalizations about effectiveness of a technology for meeting total or leachable treatment goals should be viewed with some caution.

TABLE 10.2
Cleanup Goals (Actual and Potential) for Total and Leachable Metals

Description	As	Cd	Cr (Total)	Hg	Pb
Total metals goals (mg/kg)					
Background (mean)	5	0.06	100	0.03	10
Background (range)	1–50	0.01–0.70	1–1000	0.01–0.30	2–200
Superfund site goals from TRD	5–65	3–20	6.7–375	1–21	200–500
Theoretical minimum total metals to ensure TCLP Leachate < threshold (i.e., TCLP × 20)	100	20	100	4	100
California total threshold limit concentration	500	100	500	20	1000
Leachable metals (µg/L)					
TCLP threshold for RCRA waste	5000	1000	5000	200	5000
Extraction procedure toxicity test	5000	1000	5000	200	5000
Synthetic precipitate leachate	—[b]	—	—	—	—
Multiple extraction procedure	—	—	—	—	—
California soluble threshold leachate concentration	5000	1000	5000	200	5000
Maximum contaminant level[a]	50	5	100	2	15
Superfund site goals from TRD	50	—	50	0.05–2	50

TRD, Technical Report Data.

[a] Maximum contaminant level = the maximum permissible level of contaminant in water delivered to any user of a public system.

[b] — indicates no specified level and no example cases identified.

Source: U.S. EPA, Technology Alternatives for the Remediation of Soils Contaminated with As, Cd, Cr, Hg, and Pb, EPA/540/S-97/500, U.S. Environmental Protection Agency, Cincinnati, OH, August 1997.

Technologies potentially applicable for the remediation of soils contaminated with the five metals or their inorganic compounds are listed in Table 10.3.[2,5]

The best demonstrated available technology (BDAT) status refers to the determination under the RCRA of the BDAT for various industry-generated hazardous wastes that contain the metals of interest. Whether the characteristics of a Superfund metal-contaminated soil (or fractions derived from it) are similar enough to the RCRA waste to justify serious evaluation of the BDAT for a specific Superfund soil must be made on a site-specific basis. Other limitations relevant to BDATs include the following:

1. The regulatory basis for BDAT standards focus BDATs on proven, commercially available technologies at the time of the BDAT determination.
2. A BDAT may be identified, but that does not necessarily preclude the use of other technologies.
3. A technology identified as a BDAT may not necessarily be the current technology of choice in the RCRA hazardous waste treatment industry.

The U.S. EPA's Superfund Innovative Technology Evaluation (SITE) program evaluates many emerging and demonstrated technologies in order to promote the development and use of innovative technologies to clean up Superfund sites across the country. The major focus of SITE is the Demonstration Program, which is designed to provide engineering and cost data for selected technologies.

Cost is not discussed in each technology narrative here; however, a summary table is provided at the end of Section 10.14 that illustrates technology cost ranges and treatment train options.

TABLE 10.3
Technologies Potentially Applicable for the Remediation of Contaminated Soils

Technology Class	Specific Technology
Containment	Caps
	Vertical barriers
	Horizontal barriers
Solidification–stabilization	Cement-based
	Polymer microencapsulation
	Vitrification
Separation–concentration	Soil washing
	Soil flushing
	Pyrometallurgy
	Electrokinetics
	Phytoremediation

Source: U.S. EPA, Technology Alternatives for the Remediation of Soils Contaminated with As, Cd, Cr, Hg, and Pb, EPA/540/S-97/500, U.S. Environmental Protection Agency, Cincinnati, OH, August 1997; U.S. EPA, Recent Developments for In Situ Treatment of Metal Contaminated Soils, Contract no. 68-W5-0055 U.S. Environmental Protection Agency, Washington, DC, 1997.

10.5 CONTAINMENT

Containment technologies for application at Superfund sites include landfill covers (caps), vertical barriers, and horizontal barriers.[4] For metal remediation, containment is considered an established technology except for *in situ* installation of horizontal barriers.

10.5.1 Process Description

Containment ranges from a surface cap (which limits infiltration of uncontaminated surface water) to subsurface vertical or horizontal barriers (which restrict lateral or vertical migration of contaminated groundwater). The material provided here is primarily from U.S. EPA references.[5,9]

10.5.1.1 Caps

Capping systems reduce surface water infiltration, control gas and odor emissions, improve aesthetics, and provide a stable surface over the waste. Caps can range from a simple native soil cover to a full RCRA Subtitle C composite cover.

Cap construction costs depend on the number of components in the final cap system (i.e., costs increase with the addition of barrier and drainage components). Additionally, cost escalates as a function of topographic relief. Side slopes steeper than 3 horizontal to 1 vertical can cause stability and equipment problems that dramatically increase the unit cost.[4,19]

10.5.1.2 Vertical Barriers

Vertical barriers minimize the movement of contaminated groundwater off site or limit the flow of uncontaminated groundwater on site. Common vertical barriers include slurry walls in excavated trenches, grout curtains formed by injecting grout into soil borings, vertically injected, cement–bentonite grout-filled borings or holes formed by withdrawing beams driven into the ground, and sheet-pile walls formed of driven steel.

Certain compounds can affect cement–bentonite barriers. The impermeability of bentonite may significantly decrease when it is exposed to high concentrations of creosote, water-soluble salts (Cu, Cr, As), or fire-retardant salts (borates, phosphates, and ammonia). The specific gravity of salt solutions must be greater than 1.2 to impact bentonite.[20,21] In general, soil–bentonite blends resist chemical attack best if they contain only 1% bentonite and between 30 and 40% natural soil fines. Treatability tests should evaluate the chemical stability of the barrier if adverse conditions are suspected.

Carbon steel used in pile walls quickly corrodes in dilute acids, slowly corrodes in brines or salt water, and remains mostly unaffected by organic chemicals or water. Salts and fire retardants can reduce the service life of a steel sheet pile; corrosion-resistant coatings can extend their anticipated life. Major steel suppliers will provide site-specific recommendations for cathodic protection of piling.

Construction costs for vertical barriers are influenced by the soil profile of the barrier material used and by the method of placing it. The most economical shallow vertical barriers are soil–bentonite trenches excavated with conventional backhoes; the most economical deep vertical barriers consist of a cement–bentonite wall placed by a vibrating beam.

10.5.1.3 Horizontal Barriers

In situ horizontal barriers can underlie a sector of contaminated materials on site without removing the hazardous waste or soil. Established technologies use grouting techniques to reduce the permeability of underlying soil layers. Studies performed by the U.S. Army Corps of Engineers[22] indicate that conventional grout technology cannot produce an impermeable horizontal barrier because it cannot ensure uniform lateral growth of the grout. These same studies found greater success with jet grouting techniques in soils that contain fines sufficient to prevent collapse of the wash hole and that present no large stones or boulders that could deflect the cutting jet.

Few *in situ* horizontal barriers have been constructed, so accurate costs have not yet been established. Work performed by the Corps of Engineers for U.S. EPA has shown that it is very difficult to form effective horizontal barriers. The most efficient barrier installation used a jet wash to create a cavity in sandy soils into which cement–bentonite grouting was injected. The costs relate to the number of borings required and each boring takes at least one day to drill.

10.5.2 Site Requirements

In general, the site must be suitable for a variety of heavy construction equipment including bulldozers, graders, backhoes, multishaft drill rigs, various rollers, vibratory compactors, forklifts, and seaming devices.[23,24] When capping systems are being utilized, on-site storage areas are necessary for the materials to be used in the cover. If site soils are adequate for use in the cover, a borrow area needs to be identified and the soil tested and characterized. If site soils are not suitable, it may be necessary to truck in other low-permeability soils.[23] In addition, an adequate supply of water may also be needed in order to achieve the optimum soil density.

The construction of vertical containment barriers, such as slurry walls, requires knowledge of the site, the local soil and hydrogeologic conditions, and the presence of underground utilities.[25] Preparation of the slurry requires batch mixers, hydration ponds, pumps, hoses, and an adequate supply of water. Therefore, on-site water storage tanks and electricity are necessary. In addition, areas adjacent to the trench need to be available for the storage of trench spoils (which could potentially be contaminated) and the mixing of backfill. If excavated soils are not acceptable for use as backfill, suitable backfill must be trucked onto the site.[25]

10.5.3 APPLICABILITY

Containment is most likely to be applicable to the following[5]:

1. Wastes that are low-hazard (e.g., low toxicity or low concentration) or immobile
2. Wastes that have been treated to produce low-hazard or low-mobility wastes for on-site disposal
3. Wastes whose mobility must be reduced as a temporary measure to mitigate risk until a permanent remedy can be tested and implemented

Situations where containment would not be applicable include the following:

1. Wastes for which there is a more permanent and protective remedy that is cost-effective
2. Where effective placement of horizontal barriers below existing contamination is difficult
3. Where drinking water sources will be adversely affected if containment fails, and if there is inadequate confidence in the ability to predict, detect, or control harmful releases due to containment failure

Containment has the following important advantages[5]:

1. Surface caps and vertical barriers are relatively simple and rapid to implement at low cost, and can be more economical than the excavation and removal of waste.
2. Caps and vertical barriers can be applied to large areas or volumes of waste.
3. Engineering control (containment) is achieved, and may be a final action if metals are well immobilized and potential receptors are distant.
4. A variety of barrier materials are available commercially.
5. In some cases it may be possible to create a land surface that can support vegetation or be applicable for other purposes.

Containment also has the following disadvantages[5]:

1. Design life is uncertain.
2. Contamination remains on site, available to migrate should containment fail.
3. Long-term inspection, maintenance, and monitoring are required.
4. The site must be amenable to effective monitoring.
5. The placement of horizontal barriers below existing waste is difficult to implement successfully.

10.5.4 PERFORMANCE AND BDAT STATUS

Containment is widely accepted as a means of controlling the spread of contamination and preventing the future migration of waste constituents. Table 10.4 presents a list of selected sites where containment has been selected for remediating metal-contaminated solids.

The performance of capping systems, once installed, may be difficult to evaluate.[23] Monitoring well systems or infiltration monitoring systems can provide some information, but it is often not possible to determine whether the water or leachate originated as surface water or groundwater.

With regard to slurry walls and other vertical containment barriers, performance may be affected by a number of variables including geographic region, topography, and material availability. A thorough characterization of the site and a compatibility study are highly recommended.[25]

Containment technologies are not considered "treatment technologies" and hence no BDATs involving containment have been established.

TABLE 10.4
Containment Applications at Selected Superfund Sites with Metal Contamination

Site Name and State	Specific Technology	Key Metal Contaminants	Associated Technology
Ninth Avenue Dump, IN	Containment—slurry wall	Pb	Slurry wall/capping
Industrial Waste Control, AK	Containment—slurry wall	As, Cd, Cr, Pb	Capping/French drain
E.H. Shilling Landfill, OH	Containment—slurry wall	As	Capping/clay berm
Chemtronic, NC	Capping	Cr, Pb	Capping
Ordnance Works Disposal, WV	Capping	As, Pb	Capping
Industriplex, MA	Capping	As, Pb, Cr	Capping

Source: U.S. EPA, Technology Alternatives for the Remediation of Soils Contaminated with As, Cd, Cr, Hg, and Pb, EPA/540/S-97/500, U.S. Environmental Protection Agency, Cincinnati, OH, August 1997.

10.5.5 SITE PROGRAM DEMONSTRATION PROJECTS

Ongoing SITE demonstrations applicable to soils contaminated with the metals of interest include the following:

1. Morrison Knudsen Corporation (high clay grouting technology)
2. RKK, Ltd (frozen soil barriers)

10.6 SOLIDIFICATION–STABILIZATION TECHNOLOGIES

The term "solidification–stabilization" refers to a general category of processes that are used to treat a wide variety of wastes, including solids and liquids. Solidification and stabilization are each distinct technologies, as described below.[26]

Solidification refers to processes that encapsulate a waste to form a solid material and to restrict contaminant migration by decreasing the surface area exposed to leaching or by coating the waste with low-permeability materials. Solidification can be accomplished by a chemical reaction between a waste and binding (solidifying) reagents or by mechanical processes. Solidification of fine waste particles is referred to as microencapsulation, and solidification of a large block or container of waste is referred to as macroencapsulation.

Stabilization refers to processes that involve chemical reactions that reduce the leachability of a waste. Stabilization chemically immobilizes hazardous materials (such as heavy metals) or reduces their solubility through a chemical reaction. The physical nature of the waste may or may not be changed by this process.

Solidification–stabilization (S/S) aims to accomplish one or more of the following objectives[4]:

1. To improve the physical characteristics of the waste by producing a solid from liquid or semiliquid wastes
2. To reduce contaminant solubility by formation of sorbed species or insoluble precipitates (e.g., hydroxides, carbonates, silicates, phosphates, sulfates, or sulfides)
3. To decrease the exposed surface area across which mass transfer loss of contaminants may occur by the formation of a crystalline, glassy, or polymeric framework that surrounds the waste particles
4. To limit the contact between transport fluids and contaminants by reducing the material's permeability

S/S technology is usually applied by mixing contaminated soils or treatment residuals with a physical binding agent to form a crystalline, glassy, or polymeric framework surrounding the waste particles. In addition to microencapsulation, some chemical fixation mechanisms may improve the waste's leach resistance. Other forms of S/S treatment rely on macroencapsulation, where the waste is unaltered but macroscopic particles are encased in a relatively impermeable coating,[27] or on specific chemical fixation, where the contaminant is converted to a solid compound resistant to leaching. S/S treatment can be accomplished primarily through the use of either inorganic binders (e.g., cement, fly ash, or blast furnace slag) or by organic binders such as bitumen.[4] Additives may be used, for example, to convert the metal to a less mobile form or to counteract adverse effects of the contaminated soil on the S/S mixture (e.g., accelerated or retarded setting times, and low physical strength). The form of the final product from S/S treatment can range from a crumbly, soil-like mixture to a monolithic block. S/S is more commonly done as an *ex situ* process, but an *in situ* option is available. The full range of inorganic binders, organic binders, and additives is too broad, so the emphasis in this chapter is on *ex situ*, cement-based S/S, which is widely used, *in situ*, cement-based S/S, which has been applied to metals at full scale, and polymer microencapsulation, which appears applicable to certain wastes that are difficult to treat with cement-based S/S.

Additional information and references on solidification–stabilization of metals can be found in U.S. EPA documents.[4,28–30] Innovative S/S technologies (e.g., sorption and surfactant processes, bituminization, emulsified asphalt, modified sulfur cement, polyethylene extrusion, soluble silicate, slag, lime, and soluble phosphates) are addressed in U.S. EPA reports.[31–36]

10.6.1 Process Description

10.6.1.1 *Ex Situ* Cement-Based S/S

Ex situ cement-based S/S is performed on contaminated soil that has been excavated and classified to reject oversize. Cement-based S/S involves mixing contaminated materials with an appropriate ratio of cement or similar binder/stabilizer, and possibly water and other additives. A system is also necessary for delivering the treated wastes to molds, surface trenches, or subsurface injection. Off-gas treatment (if volatiles or dust are present) may be necessary. The fundamental materials used to perform this technology are Portland-type cements and pozzolanic materials. Portland cements are typically composed of calcium silicates, aluminates, aluminoferrites, and sulfates. Pozzolans are very small spheroidal particles that are formed in the combustion of coal (fly ash) and in lime and cement kilns, for example. Pozzolans of high silica content are found to have cement-like properties when mixed with water. Cement-based S/S treatment may involve using only Portland cement, only pozzolanic materials, or blends of both. The composition of the cement and pozzolan, together with the amount of water, aggregate, and other additives, determines the set time, cure time, pour characteristics, and material properties (e.g., pore size, compressive strength) of the resulting treated waste. The composition of cements and pozzolans, including those commonly used in S/S applications, are classified according to American Society for Testing and Materials (ASTM) standards. S/S treatment usually results in an increase (>50% in some cases) in the treated waste volume. *Ex situ* treatment provides high throughput (100 to 200 m^3/d/mixer).

Cement-based S/S reduces the mobility of inorganic compounds by formation of insoluble hydroxides, carbonates, or silicates, substitution of the metal into a mineral structure, sorption, physical encapsulation, and perhaps other mechanisms. Cement-based S/S involves a complex series of reactions, and there are many potential interferences (e.g., coating of particles by organics, excessive acceleration or retardation of set times by various soluble metal and inorganic compounds; excessive heat of hydration; pH conditions that solubilize anionic species of metal compounds) that can prevent attainment of S/S treatment objectives for physical strength and leachability. Although there are many potential interferences, Portland cement is widely used and studied, and a knowledgeable vendor may be able to identify, and confirm through treatability studies, approaches to counteract adverse effects by use of appropriate additives or other changes in formulation.

Removal of Heavy Metals from Soil

10.6.1.2 *In Situ* Cement-Based S/S

In situ cement-based S/S has only two steps: mixing and off-gas treatment. The processing rate for *in situ* S/S is typically considerably lower than for *ex situ* processing. *In situ* S/S has been demonstrated to depths of 10 m and may be able to extend to 50 m. The most significant challenge in applying S/S *in situ* for contaminated soils is achieving complete and uniform mixing of the binder with the contaminated matrix.[37] Three basic approaches are used for *in situ* mixing of the binder with the matrix[5]:

1. Vertical auger mixing.
2. In-place mixing of binder reagents with waste by conventional earthmoving equipment, such as draglines, backhoes, or clamshell buckets.
3. Injection grouting, which involves forcing a binder containing dissolved or suspended treatment agents into the subsurface, allowing it to permeate the soil. Grout injection can be applied to contaminated formations lying well below the ground surface. The injected grout cures in place to produce an *in situ* treated mass.

10.6.1.3 Polymer Microencapsulation S/S

Polymer microencapsulation S/S can include the application of thermoplastic or thermosetting resins. Thermoplastic materials are the most commonly used organic-based S/S treatment materials. Potential candidate resins for thermoplastic encapsulation include bitumen, polyethylene and other polyolefins, paraffins, waxes, and sulfur cement. Of these candidate thermoplastic resins, bitumen (asphalt) is the least expensive and by far the most commonly used.[38] The process of thermoplastic encapsulation involves heating and mixing the waste material and the resin at elevated temperature, typically 130 to 230°C in an extrusion machine. Any water or volatile organics in the waste boil off during extrusion and are collected for treatment or disposal. Because the final product is a stiff, yet plastic resin, the treated material typically is discharged from the extruder into a drum or other container.

S/S process quality control requires information on the range of contaminant concentrations, potential interferences in waste batches awaiting treatment, and treated product properties such as compressive strength, permeability, leachability, and in some instances, toxicity.[28]

10.6.2 SITE REQUIREMENTS

The site must be prepared for the construction, operation, maintenance, decontamination, and decommissioning of the equipment. The size of the area required for the process equipment depends on several factors, including the type of S/S process involved, the required treatment capacity of the system, and site characteristics, especially soil topography and load-bearing capacity. A small mobile *ex situ* unit occupies space for two, standard flatbed trailers. An *in situ* system requires a larger area to accommodate a drilling rig as well as a larger area for auger decontamination.

10.6.3 APPLICABILITY

This section addresses expected applicability based on the chemistry of the metal and the S/S binders. The soil–contaminant–binder equilibrium and kinetics are complicated, and many factors influence metal mobility, so there may be exceptions to the generalizations presented below.

10.6.3.1 Cement-Based S/S

For cement-based S/S, if a single metal is the predominant contaminant in the soil, then Cd and Pb are the most amenable to cement-based S/S. The predominant mechanism for immobilization of metals in Portland and similar cements is precipitation of hydroxides, carbonates, and silicates.

Both Pb and Cd tend to form insoluble precipitates in the pH ranges found in cured cement. They may resolubilize, however, if the pH is not carefully controlled. For example, Pb in aqueous solutions tends to resolubilize as $Pb(OH)_3^-$ around pH 10 and above. Hg, although it is a cationic metal like Pb and cadmium, does not form low-solubility precipitates in cement, so it is difficult to stabilize reliably by cement-based processes, and this difficulty would be expected to be greater with increasing Hg concentration and with organomercury compounds. Owing to its formation of anionic species, As also does not form insoluble precipitates in the high-pH cement environment, and cement-based solidification is generally not expected to be successful. Cr(VI) is difficult to stabilize in cement due to the formation of anions that are soluble at high pH. However, Cr(VI) can be reduced to Cr(III), which does form insoluble hydroxides. Although Hg, As(III), and As(V) are particularly difficult candidates for cement-based S/S, this should not necessarily eliminate S/S (even cement-based) from consideration for the following reasons:

1. As with Cr(VI), it may be possible to devise a multistep process that will produce an acceptable product for cement-based S/S.
2. A non-cement-based S/S process (e.g., lime and sulfide for Hg; oxidation to As(V) and co-precipitation with Fe) may be applicable.
3. The leachable concentration of the contaminant may be sufficiently low that a highly efficient S/S process may not be required to meet treatment goals.

The discussion of applicability above also applies to *in situ*, cement-based S/S. If *in situ* treatment introduces chemical agents into the ground, this chemical addition may cause a pollution problem in itself, and may be subject to additional requirements under the Land Disposal Restrictions.

10.6.3.2 Polymer Microencapsulation

Polymer microencapsulation has been mainly used to treat low-level radioactive wastes. However, organic binders have been tested or applied to wastes containing chemical contaminants such as As, metals, inorganic salts, polychlorinated biphenyls (PCBs), and dioxins.[38] Polymer microencapsulation is particularly well suited to treating water-soluble salts such as chlorides or sulfates that are generally difficult to immobilize in a cement-based system.[39] Characteristics of the organic binder and extrusion system impose compatibility requirements on the waste material. The elevated operating temperatures place a limit on the quantity of water and volatile organic chemicals (VOCs) in the waste feed. Low-volatility organics will be retained in the bitumen but may act as solvents, causing the treated product to be too fluid. The bitumen is a potential fuel source so the waste should not contain oxidizers such as nitrates, chlorates, or perchlorates. Oxidants present the potential for rapid oxidation, causing immediate safety concerns, as well as slow oxidation, which results in waste form degradation.

Cement-based S/S of multiple metal wastes is particularly difficult if a set of treatment and disposal conditions cannot be found that simultaneously produces low-mobility species for all the metals of concern. For example, the relatively high pH conditions that favor Pb immobilization would tend to increase the mobility of As. On the other hand, the various metal species in a multiple metal waste may interact (e.g., the formation of low-solubility compounds by the combination of Pb and arsenate) to produce a low-mobility compound.

Organic contaminants are often present with inorganic contaminants at metal-contaminated sites. S/S treatment of organic-contaminated waste with cement-based binders is more complex than treatment of inorganics alone. This is particularly true with VOCs, where the mixing process and heat generated by cement hydration reactions can increase vapor losses.[40–43] However, S/S can be applied to wastes that contain lower levels of organics, particularly when inorganics are present or the organics are semivolatile or nonvolatile. Also, recent studies indicate that the addition of silicates or modified clays to the binder system may improve S/S performance with organics.[27]

10.6.4 Performance and BDAT Status

Information in 2000 about the use of S/S at Superfund remedial sites has indicated that S/S has been used at 167 sites since FY 1982.[34] Figure 10.1 shows the number of projects by status for the following stages: predesign/design, design completed/being installed, operational, and completed. Data are shown for *in situ* and *ex situ* S/S projects. In addition, information about all source control technologies is provided. With respect to S/S projects, the majority of *in situ* and *ex situ* projects (62%) are completed, followed by projects in the predesign/design stage (21%). Overall, completed S/S projects represent 30% of all completed Superfund projects in which treatment technologies have been used for source control.

Figure 10.2 shows the types of binder materials used for S/S projects at Superfund remedial sites, including inorganic binders, organic binders, and combination organic and inorganic binders. Many of the binders used include one or more proprietary additives. Examples of inorganic binders include cement, fly ash, lime, soluble silicates, and sulfur-based binders; organic binders on the other hand include asphalt, epoxide, polyesters, and polyethylene. More than 90% of the S/S projects used inorganic binders. In general, inorganic binders are less expensive and easier to use than organic binders. Organic binders are generally used to solidify radioactive wastes or specific hazardous organic compounds.

Figure 10.3 shows the types of contaminant groups and combination of contaminant groups treated by S/S at Superfund remedial sites. S/S was used to treat metals only in 56% of the projects, and used to treat metals alone or in combination with organics or radioactive metals at approximately 90% of the sites. S/S was used to treat organics only at 6% of the sites.[34] Figure 10.4 provides a further breakdown of the metals treated by S/S at Superfund remedial sites. The top five metals treated by S/S are Pb, Cr, As, Cd, and Cu.

S/S with cement-based and pozzolan binders is a commercially available, established technology.[5] Table 10.5 presents a list of sites where S/S has been selected for remediating metal-contaminated

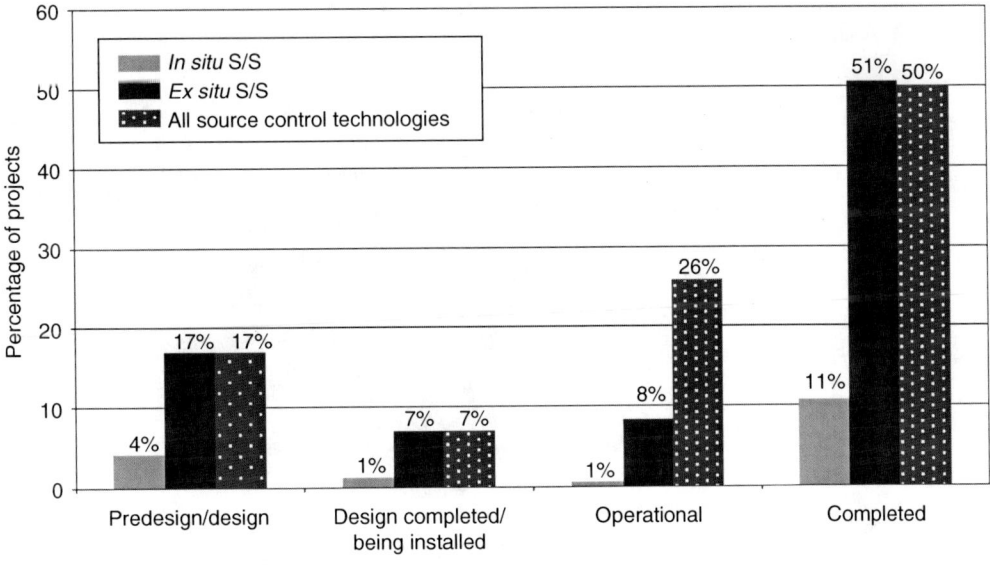

FIGURE 10.1 Percentage of Superfund remedial projects by status. Number of projects: source control, 682; *ex situ* solidification–stabilization (S/S), 139; *in situ* S/S, 28. (From U.S. EPA, Solidification/Stabilization Use at Superfund Sites, EPA-542-R-00-010, U.S. Environmental Protection Agency, Washington, DC, September 2000.)

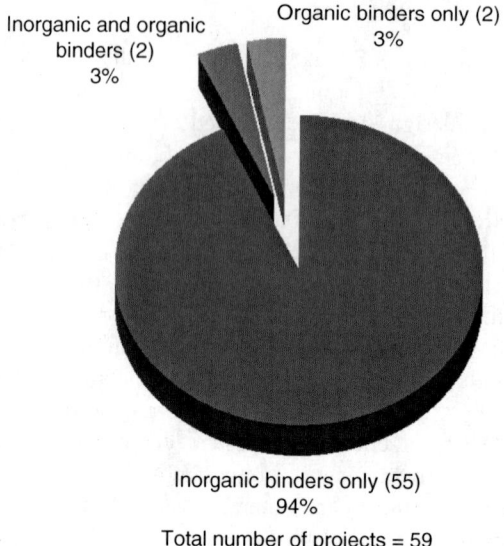

FIGURE 10.2 Binder materials used for solidification–stabilization projects. (From U.S. EPA, Solidification/Stabilization Use at Superfund Sites, EPA-542-R-00-010, U.S. Environmental Protection Agency, Washington, DC, September 2000.)

solids. Note that S/S has been used to treat all five metals (Cr, Pb, As, Hg, and Cd). Although it would not generally be expected that cement-based S/S would be applied to As- and Hg-contaminated soils, it was beyond the scope of the project to examine in detail the characterization data, S/S formulations, and performance data upon which the selections were based, so the selection/implementation data are presented without further comment.

Applications of polymer microencapsulation have been limited to special cases where the specific performance features are required for the waste matrix, and contaminants allow reuse of the treated waste as a construction material.[44]

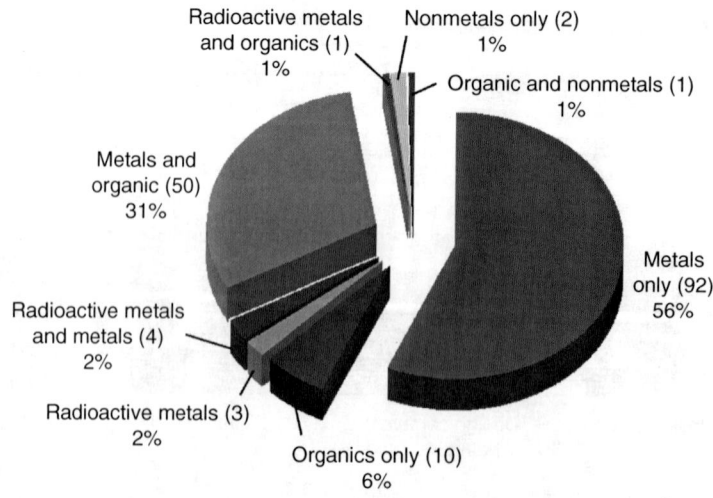

FIGURE 10.3 Contaminant types treated by solidification–stabilization. (From U.S. EPA, Solidification/Stabilization Use at Superfund Sites, EPA-542-R-00-010, U.S. Environmental Protection Agency, Washington, DC, September 2000.)

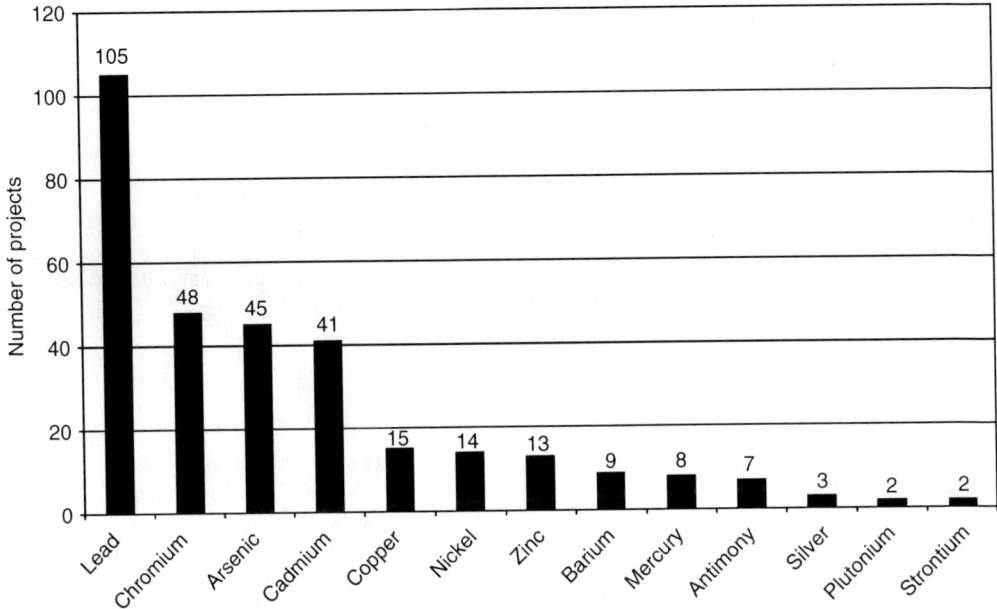

FIGURE 10.4 Number of solidification–stabilization projects treating specific metals. (From U.S. EPA, Solidification/Stabilization Use at Superfund Sites, EPA-542-R-00-010, U.S. Environmental Protection Agency, Washington, DC, September 2000.)

S/S is a BDAT for the following waste types[5]:

1. Cd nonwastewaters other than Cd-containing batteries
2. Cr nonwastewaters following reduction to Cr(III)
3. Pb nonwastewaters
4. Wastes containing low concentrations (<260 mg/kg) of elemental Hg-sulfide precipitation
5. Plating wastes and steel-making wastes

Although vitrification, not S/S, has been selected as BDAT for RCRA As-containing nonwastewaters, U.S. EPA does not preclude the use of S/S for treatment of As (particularly inorganic As) wastes, but recommends that its use be determined on a case-by-case basis. A variety of stabilization techniques including cement, silicate, pozzolan, and ferric co-precipitation were evaluated as candidate BDATs for As. Because of concerns about long-term stability and increase in waste volume, particularly with ferric co-precipitation, stabilization was not accepted as BDAT.

10.6.5 SITE PROGRAM DEMONSTRATION PROJECTS

Completed SITE demonstrations applicable to soils contaminated with the metals of interest include the following[5]:

1. Advanced Remediation Mixing, Inc. (*ex situ* S/S)
2. Funderburk and Associates (*ex situ* S/S)
3. Geo-Con, Inc. (*in situ* S/S)
4. Soliditech, Inc. (*ex situ* S/S)
5. STC Omega, Inc. (*ex situ* S/S)

TABLE 10.5
Solidification–Stabilization (S/S) Applications at Superfund Sites with Metal Contamination

Site Name and State	Specific Technology	Key Metal Contaminants	Associated Technology
DeRewal Chemical, NJ	Solidification	Cr, Cd, Pb	GW pump and treatment
Marathon Battery Co., NY	Chemical fixation	Cd, Ni	Dredging, off-site disposal
Nascolite, Millville, NJ	Stabilization of wetland soils	Pb	On-site disposal of stabilized soils; excavation and off-site disposal of wetland soils
Roebling Steel, NJ	S/S	As, Cr, Pb	Capping
Waldick Aerospace, NJ	S/S	Cd, Cr	Off-site disposal
Aladdin Plating, PA	Stabilization	Cr	Off-site disposal
Palmerton Zinc, PA	Stabilization, fly ash, lime, potash	Cd, Pb	—
Tonolli Corp., PA	S/S	As, Pb	*In situ* chemical limestone barrier
Whitmoyer Laboratories, PA	Oxidation/fixation	As	GW pump and treatment, capping, grading, and revegetation
Bypass 601, NC	S/S	Cr, Pb	Capping, regrading, revegetation, GW pump and treatment
Flowood, MS	S/S	Pb	Capping
Independent Nail, SC	S/S	Cd, Cr	Capping
Pepper's Steel and Alloys, FL	S/S	As, Pb	On-site disposal
Gurley Pit, AR	*In situ* S/S	Pb	
Pesses Chemical, TX	Stabilization	Cd	Concrete capping
E.I. Dupont de Nemours, IA	S/S	Cd, Cr, Pb	Capping, regrading, and revegetation
Shaw Avenue Dump, IA	S/S	As, Cd	Capping, groundwater monitoring
Frontier Hard Chrome, WA	Stabilization	Cr	
Gould Site, OR	S/S	Pb	Capping, regrading, and revegetation

GW, groundwater.

Source: U.S. EPA, Technology Alternatives for the Remediation of Soils Contaminated with As, Cd, Cr, Hg, and Pb, EPA/540/S-97/500, U.S. Environmental Protection Agency, Cincinnati, OH, August 1997.

6. WASTECH Inc. (*ex situ* S/S)
7. Separation and Recovery Systems, Inc. (*ex situ* S/S)
8. Wheelabrator Technologies Inc. (*ex situ* S/S)

10.6.6 Cost of S/S

Information about the cost of using S/S to treat wastes at Superfund remedial sites was reported by U.S. EPA for 29 completed projects in 2000.[34] Total costs[45] in terms of 2007 USD for S/S projects ranged from USD 86,000 to USD 18,000,000, including the cost of excavation, treatment, and disposal (if *ex situ*). The cost ranged from 12 USD/m^3 to approximately 1,800 USD/m^3. The average cost for these projects was 396 USD/m^3, including two projects with relatively high costs (approximately 1,800 USD/m^3). Excluding those two projects, the average cost per cubic meter was USD 291.[34]

10.7 VITRIFICATION

Vitrification applies a high-temperature treatment aimed primarily at reducing the mobility of metals by their incorporation into a chemically durable, leach-resistant, vitreous mass. Vitrification can be carried out on excavated soils as well as *in situ*.

10.7.1 Process Description

During the vitrification process, organic wastes are pyrolyzed (*in situ*) or oxidized (*ex situ*) by the melt front, whereas inorganics, including metals, are incorporated into the vitreous mass. Off-gases released during the melting process, containing volatile components and products of combustion and pyrolysis, must be collected and treated.[4,46,47] Vitrification converts contaminated soils to a stable glass and crystalline monolith.[47] With the addition of low-cost materials such as sand, clay, or native soil, the process can be adjusted to produce products with specific characteristics, such as chemical durability. Waste vitrification may be able to transform the waste into useful, recyclable products such as clean fill, aggregate, or higher valued materials such as erosion-control blocks, paving blocks, and road dividers.

10.7.1.1 *Ex Situ* Vitrification

Ex situ vitrification (ESV) technologies apply heat to a melter through a variety of sources such as combustion of fossil fuels (coal, natural gas, and oil) or input of electric energy by direct joule heat, arcs, plasma torches, and microwaves. Combustion or oxidation of the organic portion of the waste can contribute significant energy to the melting process, thus reducing energy costs. The particle size of the waste may need to be controlled for some of the melting technologies. For wastes containing refractory compounds that melt above the unit's nominal processing temperature, such as quartz or alumina, size reduction may be required to achieve acceptable throughputs and a homogeneous melt. For high-temperature processes using arcing or plasma technologies, size reduction is not a major factor. For intense melters using concurrent gas-phase melting or mechanical agitation, size reduction is needed for feeding the system and for achieving a homogeneous melt.

10.7.1.2 *In Situ* Vitrification

In situ vitrification (ISV) technology is based on electric melter technology, and the principle of operation is joule heating, which occurs when an electrical current is passed through a region that behaves as a resistive heating element. Electrical current is passed through the soil by means of an array of electrodes inserted vertically into the surface of the contaminated soil zone. Because dry soil is not conductive, a starter path of flaked graphite and glass frit is placed in a small trench between the electrodes to act as the initial flow path for electricity. Resistance heating in the starter path transfers heat to the soil, which then begins to melt. Once molten, the soil becomes conductive. The melt grows outward and downward as power is gradually increased to the full constant operating power level. A single melt can treat a region of up to 1000 T. The maximum treatment depth has been demonstrated to be about 6 m. Large contaminated areas are treated in multiple settings, and fuse the blocks together to form one large monolith.[4] Further information on *in situ* vitrification can be found in references 48 to 51.

10.7.2 Site Requirements

The site must be prepared for the mobilization, operation, maintenance, and demobilization of the equipment. Site activities such as clearing vegetation, removing overburden, and acquiring backfill material are often necessary for ESV and ISV. *Ex situ* processes will require areas for storage of excavated, treated, and possibly pretreated materials. The components of one ISV system are contained in three transportable trailers: an off-gas and process control trailer, a support trailer, and an electrical trailer. The trailers are mounted on wheels sufficient for transportation to and over a compacted ground surface.[52]

The field-scale ISV system evaluated in the SITE Program required three-phase electrical power at either 12,500 or 13,800 V, which is usually taken from a utility distribution system.[53] Alternatively, the power may be generated on site by means of a diesel generator. Typical applications require 800 kWh/T to 1000 kWh/T.[48]

10.7.3 APPLICABILITY

Setting cost and implementability aside, vitrification should be most applicable where nonvolatile metal contaminants have glass solubilities exceeding the level of contamination in the soil. Cr-contaminated soil should pose the least difficulties for vitrification, because it has low volatility, and glass solubility between 1% and 3%. Vitrification may or may not be applicable for Pb, As, and Cd, depending on the level of difficulty encountered in retaining the metals in the melt, and controlling and treating any volatile emissions that may occur. Hg clearly poses problems for vitrification due to its high volatility and low glass solubility (<0.1%), but may be allowable at very low concentrations.

Chlorides present in the waste in excess of about 0.5% by weight (wt%) typically will not be incorporated into and discharged with the glass but will fume off and enter the off-gas treatment system. If chlorides are excessively concentrated, salts of alkali, alkaline earths, and heavy metals will accumulate in solid residues collected by off-gas treatment. Separation of the chloride salts from the other residuals may be required before or during the return of residuals to the melter. When excess chlorides are present, there is also a possibility that dioxins and furans may form and enter the off-gas treatment system.

Waste matrix composition affects the durability of the treated waste. Sufficient glass-forming materials, SiO_2 (>30 wt%), and combined alkali (Na + K; >1.4 wt%), are required for vitrification of wastes. If these conditions are not met, frit or flux additives typically are needed. Vitrification is also potentially applicable to soils contaminated with mixed metals and metal–organic wastes.

Specific situations where ESV would not be applicable or would face additional implementation problems include those that involve the following[5]:

1. Wastes containing >25% moisture content, which can cause excessive fuel consumption
2. Wastes where size reduction and classification are difficult or expensive
3. Volatile metals, particularly Cd and Hg, which will vaporize and must be captured and treated separately
4. Arsenic-containing wastes, which may require pretreatment to produce less volatile forms
5. Metal concentrations in soil that exceed their solubility in glass
6. Sites where commercial capacity is not adequate or transportation cost to a fixed facility is unacceptable

Specific situations, in addition to those cited above, where ISV would not be applicable or would face additional implementation problems include the following[5]:

1. Metal-contaminated soil where a less costly and adequately protective remedy exists
2. Projects that cannot be undertaken because of limited commercial availability
3. Contaminated soil <2 m or >6 m below the ground surface
4. The presence of an aquifer with high hydraulic conductivity (e.g., soil permeability >1 × 10^{-5} cm/sec) limits economic feasibility due to excessive energy requirements
5. Contaminated soil mixed with buried metal, which can result in a conductive path causing short circuiting of the electrodes
6. Contaminated soil mixed with loosely packed rubbish or buried coal, which can start underground fires and overwhelm the off-gas collection and treatment system
7. Volatile heavy metals near the surface, which can be entrained in combustion product gases and not retained in the melt
8. Sites where a surface slope >5% may cause melt to flow
9. *In situ* voids >150 m^3, which can interrupt conduction and heat transfer
10. Underground structures and utilities <6 m from the melt zone that must be protected from heat or avoided

Where it can be successfully applied, the advantages of vitrification include the following[5]:

1. The vitrified product is an inert, impermeable solid that should reduce leaching for long periods of time.
2. The volume of the vitrified product will typically be smaller than the initial waste volume.
3. The vitrified product may be usable.
4. A wide range of inorganic and organic wastes can be treated.
5. There is both an *ex situ* and an *in situ* option available.

A particular advantage of *ex situ* treatment is better control of processing parameters. Also, fuel costs may be reduced for *ex situ* vitrification by the use of combustible waste materials. This fuel cost-saving option is not directly applicable for *in situ* vitrification, because combustibles would increase the design and operating requirements for gas capture and treatment.

10.7.4 Performance and BDAT Status

ISV has been implemented at metal-contaminated Superfund sites and has been evaluated under the SITE Program.[54] Some improvements are needed with regard to melt containment and air emission control systems. ISV has been operated at a large scale on many occasions, including two demonstrations on radioactively contaminated sites at the DOE's Hanford Nuclear Reservation.[46,55] Pilot-scale tests have been conducted at Oak Ridge National Laboratory, Idaho National Engineering Laboratory, and Arnold Engineering Development Center. More than 150 tests and demonstrations at various scales have been performed on a broad range of waste types in soils and sludges. The technology has been selected as a preferred remedy at ten private, Superfund, and Department of Defense (DoD) sites.[56] Table 10.6 provides a summary of ISV technology selection/application at metal-contaminated Superfund sites. A number of ESV systems are under development. The technical resource document[27] identified one full-scale *ex situ* melter that was reported to be operating on RCRA organics and inorganics. Vitrification is also a BDAT for As-containing wastes.

10.7.5 SITE Program Demonstration Projects

Completed SITE demonstrations applicable to soils contaminated with the metals of interest include the following[5]:

1. Babcock & Wilcox Co. (cyclone furnace—ESV)
2. Retech, Inc. (Plasma arc—ESV)

TABLE 10.6
In Situ Vitrification Applications at Superfund Sites with Metal Contamination

Site Name and State	Key Metal Contaminants
Parsons Chemical, MI	Hg (low)
Rocky Mountain Arsenal, CO	As, Hg

Source: U.S. EPA, Technology Alternatives for the Remediation of Soils Contaminated with As, Cd, Cr, Hg, and Pb, EPA/540/S-97/500, U.S. Environmental Protection Agency, Cincinnati, OH, August 1997.

3. Geosafe Corporation (ISV)
4. Vortec Corporation (*ex situ* oxidation and vitrification process)

10.8 SOIL WASHING

Soil washing is an *ex situ* remediation technology that uses a combination of physical separation and aqueous-based separation unit operations to reduce contaminant concentrations to site-specific remedial goals.[57] Although soil washing is sometimes used as a stand-alone treatment technology, more often it is combined with other technologies to complete site remediation. Soil-washing technologies have successfully remediated sites contaminated with organic, inorganic, and radioactive contaminants.[57] The technology does not detoxify or significantly alter the contaminant, but transfers the contaminant from the soil into the washing fluid or mechanically concentrates the contaminants into a much smaller soil mass[58] for subsequent treatment (Figure 10.5).

Further information on soil washing can be found in U.S. EPA innovative technology reports and programs.[59,60]

10.8.1 Process Description

Soil-washing systems are quite flexible in terms of the number, type, and order of processes involved. Soil washing is performed on excavated soil and may involve some or all of the following, depending on the contaminant–soil matrix characteristics, cleanup goals, and specific process employed[5,58]:

1. Mechanical screening to remove various oversize materials
2. Crushing to reduce applicable oversize to suitable dimensions for treatment
3. Physical processes (e.g., soaking, spraying, tumbling, and attrition scrubbing) to liberate weakly bound agglomerates (e.g., silts and clays bound to sand and gravel) followed by size classification to generate coarse-grained and fine-grained soil fraction(s) for further treatment
4. Treatment of the coarse-grained soil fraction(s)
5. Treatment of the fine-grained fraction(s)
6. Management of the generated residuals

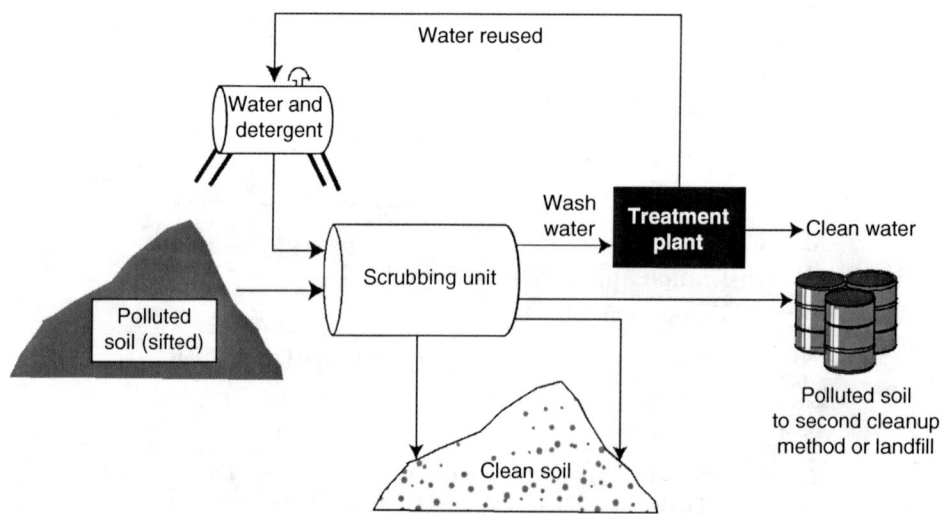

FIGURE 10.5 Soil-washing operation. (From U.S. EPA, A Citizen's Guide to Soil Washing, EPA 542-F-01-008, U.S. Environmental Protection Agency, Washington, DC, May 2001.)

Treatment of the coarse-grained soil fraction typically involves additional application of physical separation techniques and possibly aqueous-based leaching techniques. Physical separation techniques (e.g., sorting, screening, elutriation, hydrocyclones, spiral concentrators, and flotation) exploit physical differences (e.g., size, density, shape, color, and wetability) between contaminated particles and soil particles in order to produce a clean (or nearly clean) coarse fraction and one or more metal-concentrated streams. Many of the physical separation processes listed above involve the use of water as a transport medium, and if the metal contaminant has significant water solubility, then some of the coarse-grained soil cleaning will occur as a result of transfer to the aqueous phase. If the combination of physical separation and unaided transfer to the aqueous phase cannot produce the desired reduction in the soil's metal content, which is frequently the case for metal contaminants, then solubility enhancement is an option for meeting cleanup goals for the coarse fraction. Solubility enhancement can be accomplished in several ways[5,61,62]:

1. Converting the contaminant into a more soluble form (e.g., oxidation/reduction, conversion to soluble metal salts)
2. Using an aqueous-based leaching solution (e.g., acidic, alkaline, oxidizing, reducing) in which the contaminant has enhanced solubility
3. Incorporating a specific leaching process into the system to promote increased solubilization via increased mixing, elevated temperatures, higher solution/soil ratios, efficient solution/soil separation, multiple stage treatment, etc.
4. A combination of the above

After the leaching process is completed on the coarse-grained fraction, it will be necessary to separate the leaching solution and the coarse-grained fraction by settling. A soil rinsing step may be necessary to reduce the residual leachate in the soil to an acceptable level. It may also be necessary to readjust soil parameters such as pH or redox potential before replacement of the soil on the site. The metal-bearing leaching agent must also be treated further to remove the metal contaminant and permit reuse in the process or discharge, and this topic is discussed below under management of residuals.

Treatment of fine-grained soils is similar in concept to the treatment of the coarse-grained soils, but the production rate would be expected to be lower and hence more costly than for the coarse-grained soil fraction. The reduced production rate arises from factors including (1) the tendency of clays to agglomerate, thus requiring time, energy, and high water/clay ratios to produce leachable slurry and (2) slow settling velocities that require additional time or capital equipment to produce acceptable soil/water separation for multibatch or countercurrent treatment, or at the end of treatment. A site-specific determination needs to be made whether the fines should be treated to produce clean fines or whether they should be handled as a residual wastestream.

Management of generated residuals is an important aspect of soil washing. The effectiveness, implementability, and cost of treating each residual stream are important to the overall success of soil washing for the site. Perhaps the most important of the residual streams is the metal-loaded leachant that is generated, particularly if the leaching process recycles the leaching solution. Furthermore, it is often critical to the economic feasibility of the project that the leaching solution be recycled. For these closed- or semiclosed-loop leaching processes, successful treatment of the metal-loaded leachant is imperative to the successful cleaning of the soil. The leachant must (1) have adequate solubility for the metal so that the metal reduction goals can be met without using excessive volumes of leaching solution and (2) be readily, economically, and repeatedly adjustable (e.g., pH adjustment) to a form in which the metal contaminant has very low solubility so that the recycled aqueous phase retains a favorable concentration gradient compared to the contaminated soil. Also, efficient soil–water separation is important prior to recovering metal from the metal-loaded leachant in order to minimize contamination of the metal concentrate. Recycling the leachant reduces logistical requirements and costs associated with makeup water, storage, permitting, compliance analyses, and leaching agents. It also reduces external coordination requirements and eliminates the dependence of the remediation on the ability to meet publicly owned treatment works (POTW) discharge requirements.

Other residual streams that may be generated and require proper handling include the following[5]:

1. Untreatable, uncrushable oversize
2. Recyclable metal-bearing particulates, concentrates, or sludges from physical separation or leachate treatment
3. Nonrecyclable metal-bearing particulates, concentrates, soils, sludges, or organic debris that fail toxicity characteristic leaching procedure (TCLP) thresholds for RCRA hazardous waste
4. Soils or sludges that are not RCRA hazardous wastes but are also not sufficiently clean to permit return to the site
5. Metal-loaded leachant from systems where leachant is not recycled
6. Rinsate from treated soil

10.8.2 SITE REQUIREMENTS

The area required for a unit at a site will depend on the vendor system selected, the amount of soil storage space, or the number of tanks or ponds needed for wash water preparation and wastewater storage and treatment. Typical utilities required are water, electricity, steam, and compressed air; the quantity of each is vendor- and site-specific. It may be desirable to control the moisture content of the contaminated soil for consistent handling and treatment by covering the excavation, storage, and treatment areas. Climatic conditions such as annual or seasonal precipitation cause surface runoff and water infiltration; therefore, runoff control measures may be required. As soil washing is an aqueous-based process, cold weather effects include freezing as well as potential effects on leaching rates.

10.8.3 APPLICABILITY

Soil washing is potentially applicable to soils contaminated with all five metals of interest. Conditions that particularly favor soil washing include the following[5]:

1. There is just a single principal contaminant metal, which occurs in dense, insoluble particles that report to a specific, small mass fraction(s) of the soil.
2. There is just a single contaminant metal and species, which is very water or aqueous leachant soluble and has a low soil/water partition coefficient.
3. The soil contains a high proportion (e.g., >80%) of soil particles >2 mm, which is desirable for efficient contaminant–soil and soil–water separation.

Conditions that clearly do not favor soil washing include the following[5]:

1. Soils with a high (i.e., >40%) silt and clay fraction
2. Soils that vary widely and frequently in significant characteristics such as soil type, contaminant type and concentration, and where blending for homogeneity is not feasible
3. Complex mixtures (e.g., multicomponent, solid mixtures where access of leaching solutions to contaminant is restricted; mixed anionic and cationic metals where pH of solubility maximums are not close)
4. High clay content, cation exchange capacity, or humic acid content, which would tend to interfere with contaminant desorption

5. The presence of substances that interfere with the leaching solution (e.g., carbonaceous soils will neutralize extracting acids; similarly, high humic acid content will interfere with an alkaline extraction)
6. Metal contaminants in a very low-solubility, stable form (e.g., PbS), which may require long contact times and excessive amounts of reagent to solubilize

10.8.4 PERFORMANCE AND BDAT STATUS

Soil washing has been used at waste sites in Europe, in particular in Germany, the Netherlands, and Belgium.[63] Table 10.7 lists selected Superfund sites where soil washing has been selected or implemented. Acid leaching, which is a form of soil washing, is also the BDAT for Hg.

10.8.5 SITE DEMONSTRATIONS AND EMERGING TECHNOLOGIES PROGRAM PROJECTS

SITE demonstrations applicable to soils contaminated with the metals of interest include the following[5]:

1. Bergmann USA (physical separation/leaching) BioGenesis^{SM} (physical separation/leaching)
2. Biotrol, Inc. (physical separation)
3. Brice Environmental Services Corporation (physical separation)
4. COGNIS, Inc. (leaching)
5. Toronto Harbor Commission (physical separation/leaching)

Four SITE Emerging Technologies Program projects have been completed that are applicable to soils contaminated with the metals of interest.

10.9 SOIL FLUSHING

Soil flushing is the *in situ* extraction of contaminants from the soil via an appropriate washing solution. Water or an aqueous solution is injected into or sprayed onto the area of contamination, and the contaminated elutriate is collected and pumped to the surface for removal, recirculation,

TABLE 10.7
Soil-Washing Applications at Selected Superfund Sites with Metal Contamination

Site Name and State	Specific Technology	Key Metal Contaminants	Associated Technology
Ewan Property, NJ	Water washing	As, Cr, Cu, Pb	Pretreatment by solvent extraction to remove organics
GE Wiring Devices, PR	Water with KI solution additive	Hg	Treated residues disposed on site and covered with clean soil
King of Prussia, NJ	Water with washing agent additives	Ag, Cr, Cu	Sludges to be land disposed
Zanesville Well Field, OH	Soil washing	Hg, Pb	SVE to remove organics
Twin Cities Army Ammunition Plant, MN	Soil washing	Cd, Cr, Cu, Hg, Pb	Soil leaching
Sacramento Army Depot Sacramento, CA	Soil washing	Cr, Pb	Off-site disposal of wash liquid

Source: U.S. EPA, Technology Alternatives for the Remediation of Soils Contaminated with As, Cd, Cr, Hg, and Pb, EPA/540/S-97/500, U.S. Environmental Protection Agency, Cincinnati, OH, August 1997.

or on-site treatment and reinjection. The technology is applicable to both organic and inorganic contaminants, and metals in particular[4]. For the purpose of metals remediation, soil flushing has been operated at full scale, but for a small number of sites.

10.9.1 Process Description

Soil flushing uses water, a solution of chemicals in water, or an organic extractant to recover contaminants from the *in situ* material. The contaminants are mobilized by solubilization, formation of emulsions, or a chemical reaction with the flushing solutions. After passing through the contamination zone, the contaminant-bearing fluid is collected by strategically placed wells or trenches and brought to the surface for disposal, recirculation, or on-site treatment and reinjection. During elutriation, the flushing solution mobilizes the sorbed contaminants by dissolution or emulsification.

One key to the efficient operation of a soil-flushing system is the ability to reuse the flushing solution, which is recovered along with groundwater. Various water-treatment techniques can be applied to remove the recovered metals and render the extraction fluid suitable for reuse. Recovered flushing fluids may need treatment to meet appropriate discharge standards prior to release to a POTW or receiving waters. The separation of surfactants from recovered flushing fluid, for reuse in the process, is a major factor in the cost of soil flushing. Treatment of the flushing fluid results in process sludges and residual solids, such as spent carbon and spent ion exchange resin, which must be appropriately treated before disposal. Air emissions of volatile contaminants from recovered flushing fluids should be collected and treated, as appropriate, to meet applicable regulatory standards. Residual flushing additives in the soil may be a concern and should be evaluated on a site-specific basis.[64] Subsurface containment barriers can be used in conjunction with soil-flushing technology to help control the flow of flushing fluids.

Further information on soil flushing can be found in references 59 and 64 to 66.

10.9.2 Site Requirements

Stationary or mobile soil-flushing systems are located on site. The exact area required will depend on the vendor system selected and the number of tanks or ponds needed for wash water preparation and wastewater treatment. Certain permits may be required for operation, depending on the system being utilized. Slurry walls or other containment structures may be needed along with hydraulic controls to ensure capture of contaminants and flushing additives. Impermeable membranes may be necessary to limit infiltration of precipitation, which could cause dilution of the flushing solution and loss of hydraulic control. Cold weather freezing must also be considered for shallow infiltration galleries and aboveground sprayers.[67]

10.9.3 Applicability

Soil flushing may be easy or difficult to apply, depending on the ability to wet the soil with the flushing solution and to install collection wells or subsurface drains to recover all the applied liquids. The achievable level of treatment varies and depends on the contact of the flushing solution with the contaminants and the appropriateness of the solution for contaminants, and the hydraulic conductivity of the soil. Soil flushing is most applicable to contaminants that are relatively soluble in the extracting fluid, and that will not tend to sorb onto soil as the metal-laden flushing fluid proceeds through the soil to the extraction point. Based on the earlier discussion of metal behavior, some potentially promising scenarios for soil flushing would include Cr(VI), As(III), or As(V) in permeable soil with low iron oxide, low clay, and high pH; Cd in permeable soil with low clay, low cation exchange capacity, and moderately acidic pH; and Pb in acid sands. A single target metal would be preferable to multiple metals, due to the added complexity of selecting a flushing fluid that would be reasonably efficient for all contaminants. Also, the flushing fluid must be compatible

Removal of Heavy Metals from Soil

with not only the contaminant, but also the soil. Soils that counteract the acidity or alkalinity of the flushing solution will decrease its effectiveness. If precipitants occur due to interaction between the soil and the flushing fluid, then this could obstruct the soil pore structure and inhibit flow to and through sectors of the contaminated soil. It may take long periods of time for soil flushing to achieve cleanup standards.

A key advantage of soil flushing is that the contaminant is removed from the soil. Recovery and reuse of the metal from the extraction fluid may be possible in some cases, although the value of the recovered metal would not be expected to fully offset the costs of recovery. The equipment used for the technology is relatively easy to construct and operate. It does not involve excavation, treatment, and disposal of the soil, which avoids the expense and hazards associated with these activities.

10.9.4 Performance and BDAT Status

Table 10.8 lists the Superfund sites where soil flushing has been selected or implemented. Soil flushing has a more established history for removal of organics, but has been used for Cr removal (e.g., United Chrome Products Superfund Site, near Corvallis, OR). *In situ* technologies, such as soil flushing, are not considered RCRA BDAT for any of the five metals.[5]

Soil-flushing techniques for mobilizing contaminants can be classified as conventional or unconventional. Conventional applications employ water only as the flushing solution. Unconventional applications that are currently being researched include the enhancement of the flushing water with additives, such as acids, bases, and chelating agents to aid in the desorption/dissolution of the target contaminants from the soil matrix to which they are bound.

Researchers are also investigating the effects of numerous soil factors on heavy metal sorption and migration in the subsurface. Such factors include pH, soil type, soil horizon, particle size, permeability, specific metal type and concentration, and type and concentrations of organic and inorganic compounds in solutions. Generally, as the soil pH decreases, cationic metal solubility and mobility increase. In most cases, metal mobility and sorption are likely to be controlled by the organic fraction in topsoils and the clay content in the subsoils.

10.9.5 SITE Demonstration and Emerging Technologies Program Projects

There are no *in situ* soil-flushing projects reported to be completed either as SITE demonstration or Emerging Technologies Program projects.[67]

TABLE 10.8
Soil-Flushing Applications at Selected Superfund Sites with Metal Contamination

Site Name and State	Specific Technology	Key Metal Contaminants	Associated Technology
Lipari Landfill, NJ	Soil flushing of soil and wastes contained by slurry wall and cap; excavation from impacted wetlands	Cr, Hg, Pb	Slurry wall and cap
United Chrome Products, OR		Cr	Electrokinetic pilot test; considering *in situ* reduction

Source: U.S. EPA, Technology Alternatives for the Remediation of Soils Contaminated with As, Cd, Cr, Hg, and Pb, EPA/540/S-97/500, U.S. Environmental Protection Agency, Cincinnati, OH, August 1997.

10.10 PYROMETALLURGY

Pyrometallurgy is used here as a broad term encompassing elevated temperature techniques for extraction and processing of metals for use or disposal. High-temperature processing increases the rate of reaction and often makes the reaction equilibrium more favorable, lowering the required reactor volume per unit output.[4] Some processes that clearly involve both metal extraction and recovery include roasting, retorting, or smelting. Although these processes typically produce a metal-bearing waste slag, metal is also recovered for reuse. A second class of pyrometallurgical technologies included here is a combination of high-temperature extraction and immobilization. These processes use thermal means to cause volatile metals to separate from the soil and report to the fly ash, but the metal in the fly ash is then immobilized, instead of recovered, and there is no metal recovered for reuse. A third class of technologies includes those that are primarily incinerators for mixed organic–inorganic wastes, but which have the capability of processing wastes containing the metals of interest by either capturing volatile metals in the exhaust gases or immobilizing the nonvolatile metals in the bottom ash or slag. Some of these systems may have applicability to some cases where metals contamination is the primary concern, so a few technologies of this type are noted that are in the SITE Program. Vitrification has already been addressed in a previous section. It is not considered pyrometallurgical treatment as there is typically neither a metal-extraction nor a metal-recovery component in the process.

10.10.1 PROCESS DESCRIPTION

Pyrometallurgical processing is usually preceded by physical treatment[5] to produce a uniform feed material and upgrade the metal content.

Solids treatment in a high-temperature furnace requires efficient heat transfer between the gas and solid phases while minimizing particulate in the off-gas. The particle size range that meets these objectives is limited and is specific to the design of the process. The presence of large clumps or debris slows heat transfer, so pretreatment to either remove or pulverize oversize material is normally required. Fine particles are also undesirable because they become entrained in the gas flow, increasing the volume of dust to be removed from the flue gas. The feed material is sometimes pelletized to give a uniform size. In many cases a reducing agent and flux may be mixed in prior to pelletization to ensure good contact between the treatment agents and the contaminated material and to improve gas flow in the reactor.[4]

Owing to its relatively low boiling point (357°C) and ready conversion at elevated temperature to its metallic form, Hg is commonly recovered through roasting and retorting at much lower temperatures than the other metals. Pyrometallurgical processing to convert compounds of the other four metals to elemental metal requires a reducing agent, fluxing agents to facilitate melting and to slag off impurities, and a heat source. The fluid mass is often called a melt, but the operating temperature, although quite high, is often still below the melting points of the refractory compounds being processed. The fluid forms as a lower-melting-point material due to the presence of a fluxing agent such as calcium. Depending on processing temperatures, volatile metals such as Cd and Pb may fume off and be recovered from the off-gas as oxides. Nonvolatile metals, such as Cr or nickel, are tapped from the furnace as molten metal. Impurities are scavenged by the formation of slag.[4] The effluents and solid products generated by pyrometallurgical technologies typically include solid, liquid, and gaseous residuals. Solid products include debris, oversized rejects, dust, ash, and the treated medium. Dust collected from particulate control devices may be combined with the treated medium or, depending on analyses for carryover contamination, recycled through the treatment unit.

10.10.2 SITE REQUIREMENTS

Few pyrometallurgical systems are available in mobile or transportable configurations. This is typically an off-site technology, so the distance of the site from the processing facility has an important

Removal of Heavy Metals from Soil

influence on transportation costs. Off-site treatment must comply with U.S. EPA's off-site treatment policies and procedures. The off-site facility's environmental compliance status must be acceptable, and the waste must be of a type allowable under their operating permits. In order for pyrometallurgical processing to be technically feasible, it must be possible to generate a concentrate from the contaminated soil that will be acceptable to the processor. The processing rate of the off-site facility must be adequate to treat the contaminated material in a reasonable amount of time. Storage requirements and responsibilities must be determined. The need for air discharge and other permits must be determined on a site-specific basis.

10.10.3 Applicability

With the possible exception of Hg, or a highly contaminated soil, pyrometallurgical processing where metal recovery is the goal would not be applied directly to the contaminated soil, but rather to a concentrate generated via soil washing. Pyrometallurgical processing in conventional rotary kilns, rotary furnaces, or arc furnaces is most likely to be applicable to large volumes of material containing metal concentrations (particularly, Pb, Cd, or Cr) higher than 5 to 20%. Unless a very concentrated feed stream can be generated (e.g., approximately 60% for Pb), there will be a charge, in addition to transportation, for processing the concentrate. Lower metal concentrations can be acceptable if the metal is particularly easy to reduce and vaporize (e.g., Hg) or is particularly valuable (e.g., Au or Pt). Arsenic is the weakest candidate for pyrometallurgical recovery, because there is almost no recycling of As in the U.S. It is also the least valuable of the metals. The price ranges for the five metals[4] are reported here in terms of 2007 USD/T[45]:

As: 300 to 600 (as As trioxide)
Cd: 7320
Cr: 9630
Pb: 860 to 950
Hg: 6500 to 11,000

10.10.4 Performance and BDAT Status

The U.S. EPA technical document of reference 4 contains a list of approximately 35 facilities/addresses/contacts that may accept concentrates of the five metals of interest for pyrometallurgical processing. Sixteen of the 35 facilities are Pb recycling operations, seven facilities recover Hg, and the remainder address a range of RCRA wastes that contain the metals of interest. Owing to the large volume of electric arc furnace emission control waste, extensive processing capability has been developed to recover Cd, Pb, and Zn from solid waste matrices. The available process technologies include the following[5]:

1. Waelz kiln process (Horsehead Resource Development Company, Inc.)
2. Waelz kiln and calcination process (Horsehead Resource Development Company, Inc.)
3. Flame reactor process (Horsehead Resource Development Company, Inc.)
4. Inclined rotary kiln (Zia Technology)

Plasma arc furnaces are successfully treating waste at two steel plants. These are site-dedicated units that do not accept outside material for processing.

Pyrometallurgical recovery is a BDAT for the following waste types[5]:

1. Cd-containing batteries
2. Pb nonwastewaters in the noncalcium sulfate subcategory
3. Hg wastes prior to retorting

4. Pb acid batteries
5. Zn nonwastewaters
6. Hg from wastewater treatment sludge

10.10.5 SITE DEMONSTRATION AND EMERGING TECHNOLOGIES PROGRAM PROJECTS

SITE demonstrations applicable to soils contaminated with the metals of interest include the following[5]:

1. RUST Remedial Services, Inc. (X-Trax Thermal Desorption)
2. Horsehead Resource Development Company, Inc. (Flame Reactor)

10.11 ELECTROKINETICS

Electrokinetic remediation relies on the application of low-intensity direct current between electrodes placed in the soil. Contaminants are mobilized in the form of charged species, particles, or ions.[2] Attempts to leach metals from soils by electro-osmosis date back to the 1930s. In the past, research focused on removing unwanted salts from agricultural soils. Electrokinetics has been used for dewatering of soils and sludges since the first recorded use in the field in 1939.[68] Electrokinetic extraction has been used in the former Soviet Union since the early 1970s to concentrate metals and to explore for minerals in deep soils. By 1979, research had shown that the content of soluble ions increased substantially in electro-osmotic consolidation of polluted dredgings, and metals were not found in the effluent.[69] By the mid-1980s, numerous researchers had realized independently that electrokinetic separation of metals from soils was a potential solution to contamination.[70]

Several organizations are developing technologies for the enhanced removal of metals by transporting contaminants to electrodes, where they are removed and subsequently treated above ground. A variation of the technique involves treatment without removal by transporting contaminants through specially designed treatment zones that are created between electrodes. Electrokinetics can also be used to slow or prevent migration of contaminants by configuring cathodes and anodes in a manner that causes contaminants to flow toward the center of a contaminated area of soil. Performance data illustrate the potential for achieving removals greater than 90% for some metals.[2]

The range of potential metals is broad. Commercial applications in Europe have treated Cu, Pb, Zn, As, Cd, Cr, and Ni. There is also potential applicability for radionuclides and some types of organic compounds. The electrode spacing and duration of remediation is site-specific. The process requires adequate soil moisture in the vadose zone, so the addition of a conducting pore fluid may be required (particularly as there is a tendency for soil drying near the anode). Specially designed pore fluids are also added to enhance the migration of target contaminants. The pore fluids are added at either the anode or cathode, depending on the desired effects.

Table 10.9 presents an overview of two variations of electrokinetic remediation technology. Geokinetics International, Inc., Battelle Memorial Institute, Electrokinetics, Inc., and Isotron Corporation are all developing variations of technologies categorized under Approach 1, "enhanced removal." The consortium of Monsanto, E.I. du Pont de Nemours and Company, General Electric, DOE, and the U.S. EPA Office of Research and Development is developing the Lasagna Process, which is categorized under Approach 2, "treatment without removal."[2]

10.11.1 PROCESS DESCRIPTION

Electrokinetic remediation, also referred to as electrokinetic soil processing, electromigration, electrochemical decontamination, or electroreclamation, can be used to extract radionuclides, metals, and some types of organic wastes from saturated or unsaturated soils, slurries, and sediments.[71]

TABLE 10.9
Overview of Electrokinetic Remediation Technology

General Characteristics
- The depth of soil that is amenable to treatment depends on electrode placement.
- It is best used in homogeneous soils with high moisture content and high permeability.

Approach 1 Enhanced Removal	**Approach 2 Treatment without Removal**
Description	*Description*
Electrokinetic transport of contaminants toward the polarized electrodes to concentrate the contaminants for subsequent removal and *ex situ* treatment.	Electro-osmotic transport of contaminants through treatment zones placed between the electrodes. The polarity of the electrodes is reversed periodically, which reverses the direction of the contaminants back and forth through treatment zones. The frequency with which electrode polarity is reversed is determined by the rate of transport of contaminants through the soil.
Status	*Status*
Demonstration projects using full-scale equipment are reported in Europe. Bench- and pilot-scale laboratory studies are reported in the U.S. and at least two full-scale field studies are ongoing in the U.S.	Demonstrations are ongoing.
Applicability	*Applicability*
Pilot scale: Pb, As, Ni, Hg, Cu, Zn. Laboratory scale: Pb, Cd, Cr, Hg, Zn, Fe, Mg, U, Th, Ra.	Technology developed for organic species and metals.
Comments	*Comments*
Field studies are under evaluation by U.S. EPA, DOE, DOD, and the Electric Power Research Institute (EPRI). The technique primarily would require the addition of water to maintain the electric current and facilitate migration; however, there is ongoing work in application of the technology in partially saturated soils.	This technology is being developed for deep clay formations.

Source: U.S. EPA, Recent Developments for In Situ Treatment of Metal Contaminated Soils, Contract no. 68-W5-0055 U.S. Environmental Protection Agency, Washington, DC, 1997.

This *in situ* soil-processing technology is primarily a separation and removal technique for extracting contaminants from soils.

The principle of electrokinetic remediation relies upon the application of a low-intensity direct current through the soil between two or more electrodes. Most soils contain water in the pores between the soil particles and have an inherent electrical conductivity that results from salts present in the soil.[72] The current mobilizes charged species, particles, and ions in the soil by the following processes[73]:

1. Electromigration (transport of charged chemical species under an electric gradient)
2. Electro-osmosis (transport of pore fluid under an electric gradient)
3. Electrophoresis (movement of charged particles under an electric gradient)
4. Electrolysis (chemical reactions associated with the electric field)

Figure 10.6 presents a schematic diagram of a typical conceptual electrokinetic remediation application.

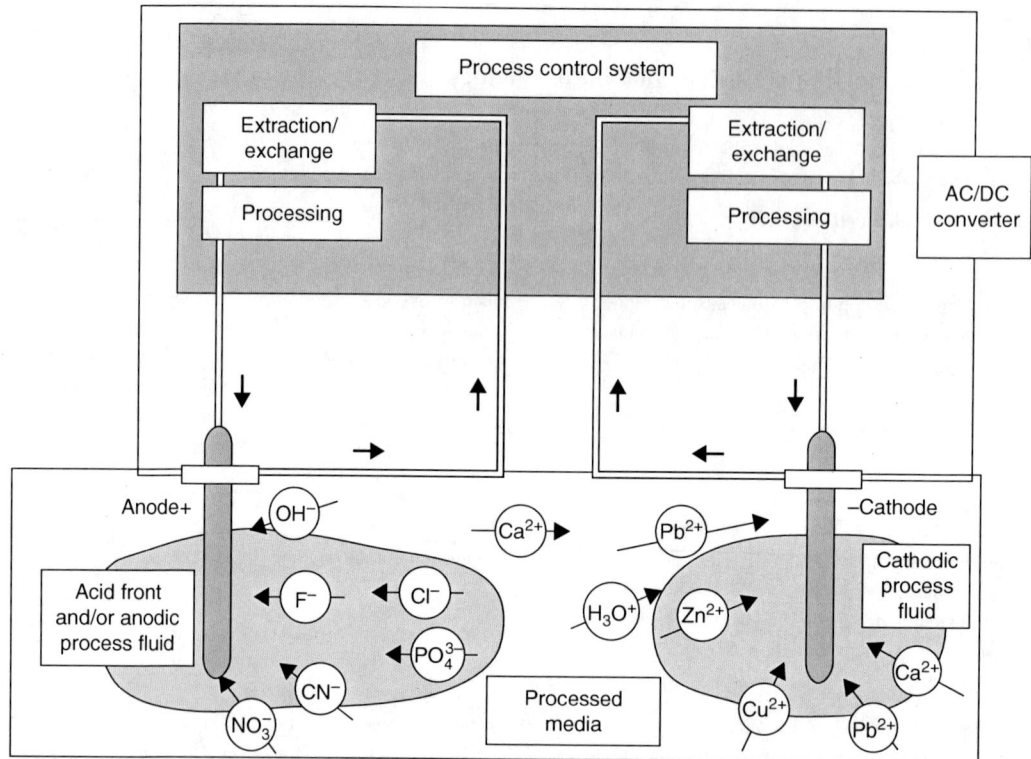

FIGURE 10.6 Diagram of one electrode configuration used in the field implementation of electrokinetics. (From U.S. EPA, Recent Developments for In Situ Treatment of Metal Contaminated Soils, Contract no. 68-W5-0055 U.S. Environmental Protection Agency, Washington, DC, 1997.)

Electrokinetics can be efficient in extracting contaminants from fine-grained, high-permeability soils. A number of factors determine the direction and extent of the migration of the contaminant. Such factors include the type and concentration of the contaminant, the type and structure of the soil, and the interfacial chemistry of the system.[74] Water or some other suitable salt solution may be added to the system to enhance the mobility of the contaminant and increase the effectiveness of the technology (e.g., buffer solutions may change or stabilize pore fluid pH). Contaminants arriving at the electrodes may be removed by any of several methods, including electroplating at the electrode, precipitation or co-precipitation at the electrode, pumping of water near the electrode, or complexing with ion-exchange resins.[74]

Electrochemistry associated with this process involves an acid front that is generated at the anode if water is the primary pore fluid present. The variation of pH at the electrodes results from the electrolysis of the water. The solution becomes acidic at the anode because hydrogen ions are produced and oxygen gas is released, and the solution becomes basic at the cathode, where hydroxyl ions are generated and hydrogen gas is released.[75] At the anode, the pH could drop to below 2, and it could increase at the cathode to above 12, depending on the total current applied. The acid front eventually migrates from the anode to the cathode. Movement of the acid front by migration and advection results in the desorption of contaminants from the soil.[71] The process leads to temporary acidification of the treated soil, and there are no established procedures for determining the length of time needed to reestablish equilibrium. Studies have indicated that metallic electrodes may dissolve as a result of electrolysis and introduce corrosion products into the soil mass. However, if inert electrodes, such as carbon, graphite, or platinum, are used, no residue will be introduced in the treated soil mass as a result of the process.[2]

Removal of Heavy Metals from Soil

10.11.2 Site Requirements

Before electrokinetic remediation is undertaken at a site, a number of different field and laboratory screening tests must be conducted to determine whether the particular site is amenable to the treatment technique:

1. *Field conductivity surveys.* The natural geologic spatial variability should be delineated, because buried metallic or insulating material can induce variability in the electrical conductivity of the soil and, therefore, the voltage gradient. In addition, it is important to assess whether there are deposits that exhibit very high electrical conductivity, at which the technique may be inefficient.
2. *Chemical analysis of water.* The pore water should be analyzed for dissolved major anions and cations, as well as for the predicted concentration of the contaminant(s). In addition, the electrical conductivity and pH of the pore water should be measured.
3. *Chemical analysis of soil.* The buffering capacity and geochemistry of the soil should be determined at each site.
4. *pH effects.* The pH values of the pore water and the soil should be determined because they have a great effect on the valence, solubility, and sorption of contaminant ions.
5. *Bench-scale test.* The dominant mechanism of transport, removal rates, and amounts of contamination left behind can be examined for different removal scenarios by conducting bench-scale tests. Because many of these physical and chemical reactions are interrelated, it may be necessary to conduct bench-scale tests to predict the performance of electrokinetics remediation at the field scale.[70,71]

10.11.3 Applicability and Demonstration Projects

Various methods, developed by combining electrokinetics with other techniques, are being applied for remediation. This section describes different types of electrokinetic remediation methods for use at contaminated sites. The methods discussed were developed by Electrokinetics, Inc., Geokinetics International, Inc., Isotron Corporation, Battelle Memorial Institute, a consortium effort, and P&P Geotechnik GmbH.[2]

10.11.3.1 Electrokinetics, Inc.

Electrokinetics, Inc. operates under a licensing agreement with Louisiana State University. The technology is patented by and assigned to Louisiana State University[76] and a complementing process patent is assigned to Electrokinetics, Inc.[77] As depicted in Figure 10.5, groundwater and/or a processing fluid (supplied externally through the boreholes that contain the electrodes) serve as the conductive medium. The additives in the processing fluid, the products of electrolysis reactions at the electrodes, and the dissolved chemical entities in the contaminated soil are transported across the contaminated soil by conduction under the influence of electric fields. This transport, when coupled with sorption, precipitation/dissolution, and volatilization/complexation, provides the fundamental mechanism that can affect the electrokinetic remediation process. Electrokinetics, Inc. accomplishes extraction and removal by electrodeposition, evaporation/condensation, precipitation, or ion exchange, either at the electrodes or in a treatment unit that is built into the system that pumps the processing fluid to and from the contaminated soil. Pilot-scale testing was carried out with support from the U.S. EPA, which also developed a design and analysis package for the process.[78]

10.11.3.2 Geokinetics International, Inc.

Geokinetics International, Inc. (GII) has obtained a patent for an electroreclamation process. The key claims in the patent are the use of electrode wells for both anodes and cathodes and the management of the pH and electrolyte levels in the electrolyte streams of the anode and the cathode. The patent also includes claims for the use of additives to dissolve different types of contaminants.[79] Fluor Daniel is licensed to operate GII's metal removal process in the U.S.

GII has developed and patented an electrically conductive ceramic material (EBONEX®) that has an extremely high resistance to corrosion. It has a lifetime in soil of at least 45 years and is self-cleaning. GII has also developed a batch electrokinetic remediation (BEK®) process. The process, which incorporates electrokinetic technology, normally requires 24 to 48 h for complete remediation of the substrate. BEK® is a mobile unit that remediates *ex situ* soils on site. GII also has developed a solution treatment technology (EIX®) that allows removal of contamination from the anode and the cathode solutions up to a thousand times faster than can be achieved through conventional means.[2]

10.11.3.3 Isotron Corporation

Isotron Corporation participated in a pilot-scale demonstration of electrokinetic extraction supported by DOE's Office of Technology Development. The demonstration took place at the Oak Ridge K-25 facility in Tennessee. Completed laboratory tests showed that the Isotron process could effect the movement and capture of uranium present in soil from the Oak Ridge site.[80]

Isotron Corporation was also involved with Westinghouse Savannah River Company in a demonstration of electrokinetic remediation. The demonstration, supported by DOE's Office of Technology Development, took place at the old TNX basin at the Savannah River site in South Carolina. Isotron used the Electrosorb® process with a patented cylinder to control buffering conditions *in situ*. An ion-exchange polymer matrix called Isolock® was used to trap metal ions. The process was tested for the removal of Pb and Cr.[80]

10.11.3.4 Battelle Memorial Institute

Another method that uses electrokinetic technology is electroacoustical soil decontamination. This technology combines electrokinetics with sonic vibration. Through the application of mechanical vibratory energy in the form of sonic or ultrasonic energy, the properties of a liquid contaminant in soil can be altered in a way that increases the level of removal of the contaminant. Battelle Memorial Institute of Columbus, OH, developed an *in situ* treatment process that uses both electrical and acoustical forces to remove floating contaminants, and possibly metals, from subsurface zones of contamination. The process was selected for U.S. EPA's SITE Program.[81]

10.11.3.5 Consortium Process

Monsanto Company coined the name, Lasagna™ to identify its products and services that are based on the integrated *in situ* remediation process that has been developed by a consortium. The proposed technology combines electro-osmosis with treatment zones that are installed directly in the contaminated soils to form an integrated *in situ* remedial process, as shown in Figure 10.7. The consortium consists of Monsanto, E.I. du Pont de Nemours and Company (DuPont), and General Electric (GE), with participation by the U.S. EPA Office of Research and Development and DOE.

The *in situ* decontamination process is carried out as follows[2]:

1. Highly permeable zones are created in close proximity, sectioned through the contaminated soil region, and turned into sorption–degradation zones by introducing appropriate materials (sorbents, catalytic agents, microbes, oxidants, buffers, and others).
2. Electro-osmosis is used as a liquid pump to flush contaminants from the soil into the treatment zones of degradation.

Removal of Heavy Metals from Soil

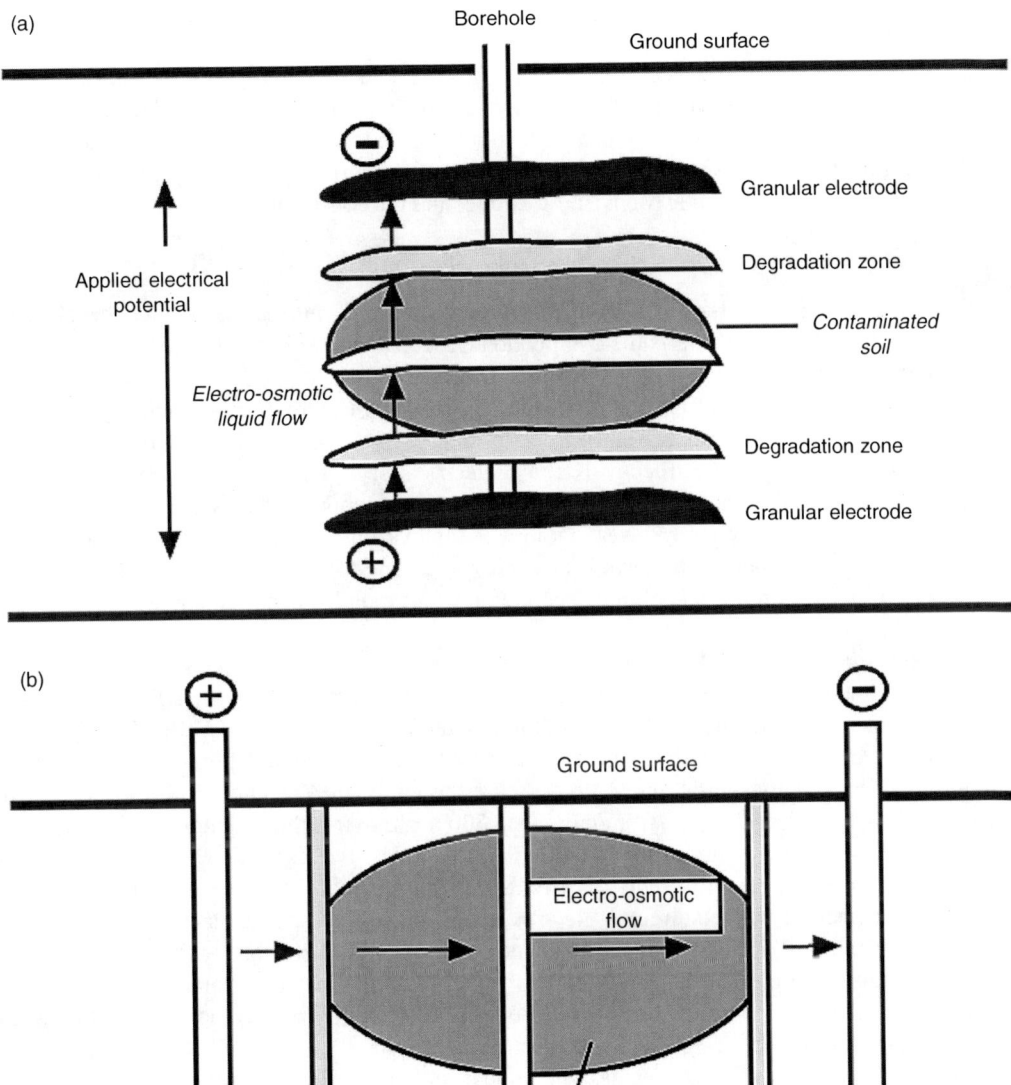

FIGURE 10.7 Schematic diagram of the Lasagna process. (a) Horizontal configuration and (b) vertical configuration. *Note*: Electro-osmotic flow is a reversed upon switching electrical polarity. (From U.S. EPA, Recent Developments for In Situ Treatment of Metal Contaminated Soils, Contract no. 68-W5-0055 U.S. Environmental Protection Agency, Washington, DC, 1997.)

3. Liquid flow is reversed, if desired, by switching the electrical polarity, a mode that increases the efficiency with which contaminants are removed from the soil; this allows repeated passes through the treatment zones for complete sorption.

Initial field tests of the consortium process were conducted at DOE's gaseous diffusion plant in Paducah, KY. The experiment tested the combination of electro-osmosis and *in situ* sorption in treatment zones. Technology development for the degradation processes and their integration into the overall treatment scheme were carried out at bench and pilot scales, followed by field experiments of the full-scale process.[82]

10.11.4 Performance and Cost

Work sponsored by U.S. EPA, DOE, the National Science Foundation, and private industry, when coupled with the efforts of researchers from academic and public institutions, have demonstrated the feasibility of moving electrokinetics remediation to pilot-scale testing and demonstration stages.[71]

This section describes testing and cost summary results reported by Louisiana State University, Electrokinetics, Inc., GII, Battelle Memorial Institute, and the consortium.[2]

10.11.4.1 Louisiana State University–Electrokinetics, Inc.

The Louisiana State University (LSU)–Electrokinetics, Inc. Group has conducted bench-scale testing on radionuclides and on organic compounds. Test results have been reported for Pb, Cd, Cr, Hg, Zn, Fe, and Mg. Radionuclides tested include U, Th, and Ra.

In collaboration with U.S. EPA, the LSU–Electrokinetics, Inc. Group has completed pilot-scale studies of electrokinetic soil processing in the laboratory. Electrokinetics, Inc. carried out a site-specific pilot-scale study of the Electro-Klean™ electrical separation process. Pilot field studies have also been reported in the Netherlands on soils contaminated with Pb, As, Ni, Hg, Cu, and Zn.

A pilot-scale laboratory study investigating the removal of 2000 mg/kg of lead loaded onto kaolinite has been completed. Removal efficiencies of 90 to 95% were obtained. The electrodes were placed 1 in. apart in a 2 T kaolinite specimen for four months, at a total energy cost of about (2007) 22 USD/T.[81]

With the support of DOD, Electrokinetics, Inc. carried out a comprehensive demonstration study of Pb extraction from a creek bed at a U.S. Army firing range in Louisiana. U.S. EPA took part in an independent assessments of the results of that demonstration study under the SITE Program. The soils were contaminated with levels as high as 4500 mg/kg of Pb, and pilot-scale studies have demonstrated that the concentrations of Pb decreased to less than 300 mg/kg in 30 weeks of processing. The TCLP values dropped from more than 300 mg/L to less than 40 mg/L within the same period. At the site of the demonstration study, Electrokinetics, Inc. used the CADEX™ electrode system, which promotes the transport of species into the cathode compartment where they are precipitated or electrodeposited directly. Electrokinetics, Inc. used a special electrode material that is cost-effective and does not corrode. Under the supervision and support of the Electric Power Research Institute and power companies in the southern U.S., a treatability and a pilot-scale field testing study of soils in sites contaminated with As was performed, in a collaborative effort between Southern Company Services Engineers and Electrokinetics, Inc.[2]

The processing cost of a system designed and installed by Electrokinetics, Inc. consists of energy cost, conditioning cost, and fixed costs associated with installation of the system. Power consumption is related directly to the conductivity of the soil across the electrodes. Electrical conductivity of soils can span orders of magnitude, from 30 mhos/cm to more than 3000 μmhos/cm, with higher values in saturated, high-plasticity clays. A mean conductivity value is 500 μmhos/cm. The voltage gradient is held to approximately 1 V/cm in an attempt to prevent adverse effects of temperature increases and for other practical reasons.[71] It may be cost-prohibitive to attempt to remediate high-plasticity soils that have high electrical conductivities. However, for most deposits having conductivities of 500 μmhos/cm, the daily energy consumption will be approximately 12 kWh/m^3/d or about 1.20 USD/m^3/d (0.10 USD/kWh) and 36 USD/m^3/month. The processing time will depend upon several factors, including the spacing of the electrodes and the type of conditioning scheme that will be used. If an electrode spacing of 4 m is selected, it may be necessary to process the site over several months.

Pilot-scale studies using "real-world" soils indicate that the energy expenditures in the extraction of metals from soils may be 500 kWh/m^3 or more at an electrode spacing of between 1.0 m and 1.5 m.[78] The vendor estimates that the direct cost of about 50 USD/m^3 (0.10 USD/kWh) suggested for this energy expenditure, together with the cost of enhancement, could result in direct costs of 100 USD/m^3. If no other efficient *in situ* technology is available to remediate fine-grained and heterogeneous subsurface deposits contaminated with metals, this technique would remain potentially competitive.

10.11.4.2 Geokinetics International, Inc.

GII has successfully demonstrated *in situ* electrochemical remediation of metal-contaminated soils at several sites in Europe. Geokinetics, a sister company of GII, has also been involved in the electrokinetics arena in Europe. Table 10.10 summarizes the physical characteristics of five of the sites, including the size, the contaminant(s) present, and the overall performance of the technology at each site. GII estimates its typical costs for "turn key" remediation projects are in the range of 160–260 USD/m^3 (2007 USD).[2]

10.11.4.3 Battelle Memorial Institute

The technology demonstration through the SITE Program has been completed,[81] and the results indicate that the electroacoustical technology is technically feasible for the removal of inorganic species from clay soils.[83]

10.11.4.4 Consortium Process

The Phase I field test of the Lasagna process has been completed. Scale-up from laboratory units was successfully achieved with respect to electrical parameters and electro-osmotic flow. Soil samples taken throughout the test site before and after the test indicate a 98% removal of trichloroethylene (TCE) from a tight clay soil (i.e., hydraulic conductivity less than 1×10^{-7} cm/sec). TCE soil levels were reduced from the 100 to 500 mg/kg range to an average concentration of 1 mg/kg.[84] Various treatment processes are being investigated in the laboratory to address other types of contaminants, including heavy metals.[84]

10.11.5 Summary of Electrokinetic Remediation

Electrokinetic remediation may be applied to both saturated and partially saturated soils. One problem to overcome when applying electrokinetic remediation to the vadose zone is the drying of

TABLE 10.10
Performance of Electrochemical Soil Remediation Applied at Five Field Sites in Europe

Site Description	Soil Volume (m³)	Soil Type	Contaminant	Initial Concentration (mg/kg)	Final Concentration (mg/kg)
Former paint factory	230	Peat/clay soil	Cu	1220	<200
			Pb	>3780	<280
Operational galvanizing plant	40	Clay soil	Zn	>1400	600
Former timber plant	190	Heavy clay soil	As	>250	<30
Temporary landfill	5440	Argillaceous sand	Cd	>180	<40
Military air base	1900	Clay	Cd	660	47
			Cr	7300	755
			Cu	770	98
			Ni	860	80
			Pb	730	108
			Zn	2600	289

Source: U.S. EPA, Recent Developments for In Situ Treatment of Metal Contaminated Soils, Contract no. 68-W5-0055 U.S. Environmental Protection Agency, Washington, DC, 1997.

soil near the anode. When an electric current is applied to soil, water will flow by electro-osmosis in the soil pores, usually toward the cathode. The movement of the water will deplete soil moisture adjacent to the anode, and moisture will collect near the cathode. However, processing fluids may be circulated at the electrodes. The fluids can serve both as a conducting medium and as a means to extract or exchange the species and introduce other species. Another use of processing fluids is to control, depolarize, or modify either or both electrode reactions. The advance of the process fluid (acid or the conditioning fluid) across the electrodes assists in desorption of species and dissolution of carbonates and hydroxides. Electro-osmotic advection and ionic migration lead to the transport and subsequent removal of the contaminants. The contaminated fluid is then recovered at the cathode.

Spacing of the electrode will depend upon the type and level of contamination and the selected current voltage regime. When higher voltage gradients are generated, the efficiency of the process might decrease because of increases in temperature. A spacing that will generate a potential gradient in the order of 1 V/cm is preferred. The spacing of electrodes generally will be as much as 3 m. The duration of the remediation will be site-specific. The remediation process should be continued until the desired removal is achieved. However, it should be recognized that, in cases in which the duration of treatment is reduced by increasing the electrical potential gradient, the efficiency of the process will decrease.[85,86]

The advantage of the technology is its potential for cost-effective use for both *in situ* and *ex situ* applications. The fact that the technique requires the presence of a conducting pore fluid in a soil mass may have site-specific implications. Also, heterogeneities or anomalies found at sites (such as submerged foundations, rubble, large quantities of iron or iron oxides, large rocks, or gravel) or submerged cover material, such as seashells, are expected to reduce removal efficiencies.[71]

10.12 PHYTOREMEDIATION

This technology is in the stage of commercialization for treatment of soils contaminated with metals, and in the future may provide a low-cost option under specific circumstances. At the current stage of development, this process is best suited for sites with widely dispersed contamination at low concentrations where only treatment of soils at the surface (in other words, within the depth of the root zone) is required.[2]

Phytoremediation is the use of plants to remove, contain, or render harmless environmental contaminants. This definition applies to all biological, chemical, and physical processes that are influenced by plants and that aid in the cleanup of contaminated substances.[87] Plants can be used in site remediation, both to mineralize and immobilize toxic organic compounds at the root zone and to accumulate and concentrate metals and other inorganic compounds from soil into aboveground shoots.[88] Although phytoremediation is a relatively new concept in the waste management community, techniques, skills, and theories developed through the application of well-established agroeconomic technologies are easily transferable. The development of plants for restoring sites contaminated with metals will require the multidisciplinary research efforts of agronomists, toxicologists, biochemists, microbiologists, pest-management specialists, engineers, and other specialists.[87,88] Table 10.11 presents an overview of phytoremediation technology.

Two basic approaches for metals remediation include phytoextraction and phytostabilization. Phytoextraction relies on the uptake of contaminants from the soil and their translocation into aboveground plant tissue, which is harvested and treated. Although hyperaccumulating trees, shrubs, herbs, grasses, and crops have potential, crops seem to be most promising because of their greater biomass production. Ni and Zn appear to be the most easily absorbed, although tests with Cu and Cd are encouraging.[2] Significant uptake of Pb, a commonly occurring contaminant, has not been demonstrated on a large scale. However, some researchers are experimenting with soil amendments that would facilitate uptake of Pb by the plants.

Removal of Heavy Metals from Soil

TABLE 10.11
Overview of Phytoremediation Technology

General characteristics
- It is best used at sites with low to moderate disperse metals content and with soil media that will support plant growth.
- Applications are limited to the depth of the root zone.
- Longer times are required for remediation compared with other technologies.
- Different species have been identified to treat different metals.

Approach 1: Phytoextraction (Harvest)	Approach 2: Phytostabilization (Root-Fixing)
Description Uptake of contaminants from soil into aboveground plant tissue, which is periodically harvested and treated.	*Description* Production of chemical compounds by the plant to immobilize contaminants at the interface of roots and soil. Additional stabilization can occur by raising the pH level in the soil.
Status Field testing for effectiveness on radioactive metals is ongoing in the vicinity of the damaged nuclear reactor in Chernobyl, Ukraine. Field testing is also being conducted in Trenton, NJ, and Butte, MT, and by the Idaho National Engineering Laboratory (INEL) in Fernald, OH.	*Status* Research is ongoing.
Applicability Potentially applicable for many metals. Ni and Zn appear to be most easily absorbed. Preliminary results for absorption of Cu and Cd are encouraging.	*Applicability* Potentially applicable for many metals, especially Pb, Cr, and Hg.
Comments Cost is affected by the volume of biomass produced that may require treatment before disposal. Cost is also affected by the concentration and depth of contamination and the number of harvests required.	*Comments* Long-term maintenance is required.

Source: U.S. EPA, Recent Developments for In Situ Treatment of Metal Contaminated Soils, Contract no. 68-W5-0055 U.S. Environmental Protection Agency, Washington, DC, 1997.

10.12.1 Process Description

Metals considered essential for at least some forms of life include vanadium (V), Cr, Mn, Fe, Co, Ni, Cu, Zn, and Mo.[88] Because many metals are toxic in concentrations above minute levels, an organism must regulate the cellular concentrations of such metals. Consequently, organisms have evolved transport systems to regulate the uptake and distribution of metals. Plants have remarkable metabolic and absorption capabilities, as well as transport systems that can take up ions selectively from the soil. Plants have evolved a great diversity of genetic adaptations to handle potentially toxic levels of metals and other pollutants that occur in the environment. In plants, uptake of metals occurs primarily through the root system, in which the majority of mechanisms to prevent metal toxicity are found.[89] The root system provides an enormous surface area that absorbs and accumulates the water and nutrients essential for growth. In many ways, living plants can be compared to solar-powered pumps that can extract and concentrate certain elements from the environment.[90]

Plant roots cause changes at the soil–root interface as they release inorganic and organic compounds (root exudates) in the area of the soil immediately surrounding the roots (the rhizosphere).[91] Root exudates affect the number and activity of microorganisms, the aggregation and stability of soil particles around the root, and the availability of elements. Root exudates can increase (mobilize) or decrease (immobilize) directly or indirectly the availability of elements in the rhizosphere. Mobilization and immobilization of elements in the rhizosphere can be caused by the following events[92,93]:

1. Changes in soil pH
2. The release of complexing substances, such as metal-chelating molecules
3. Changes in oxidation–reduction potential
4. An increase in microbial activity

Phytoremediation technologies can be developed for different applications in environmental cleanup and are classified into three types:

1. Phytoextraction
2. Phytostabilization
3. Rhizofiltration

10.12.1.1 Phytoextraction

Phytoextraction technologies use hyperaccumulating plants to transport metals from the soil and concentrate them into the roots and aboveground shoots, which can then be harvested.[87,88,91] A plant containing more than 0.1% of Ni, Co, Cu, Cr, or 1% Zn and Mn in its leaves on a dry weight basis is called a hyperaccumulator, regardless of the concentration of metals in the soil.[88,94,95]

Almost all metal-hyperaccumulating species known today were discovered on metal-rich soils, either natural or artificial, often growing in communities with metal excluders.[88,96] In fact, almost all metal-hyperaccumulating plants are endemic to such soils, suggesting that hyperaccumulation is an important ecophysiological adaptation to metal stress and one of the manifestations of resistance to metals. The majority of hyperaccumulating species discovered so far are restricted to a few specific geographical locations.[88,94] For example, Ni hyperaccumulators are found in New Caledonia, the Philippines, Brazil, and Cuba. Ni and Zn hyperaccumulators are found in southern and central Europe and Asia Minor.

Dried or composted plant residues or plant ashes that are highly enriched with metals can be isolated as hazardous waste or recycled as metal ore.[98] The goal of phytoextraction is to recycle as "bio-ores" metals reclaimed from plant ash in the feed stream of smelting processes. Even if the plant ashes do not have enough concentration of metal to be useful in smelting processes, phytoextraction remains beneficial because it reduces by as much as 95% the amount of hazardous waste to be landfilled.[2] Several research efforts in the use of trees, grasses, and crop plants are being pursued to develop phytoremediation as a cleanup technology. The following paragraphs briefly discuss these three phytoextraction techniques.

The use of trees can result in the extraction of significant amounts of metal because of their high biomass production. However, the use of trees in phytoremediation requires long-term treatment and may create additional environmental concerns about falling leaves. When leaves containing metals fall or blow away, recirculation of metals to the contaminated site and migration off site by wind transport or through leaching can occur.[2]

Some grasses accumulate surprisingly high levels of metals in their shoots without exhibiting toxic effects. However, their low biomass production results in relatively low yield of metals. Genetic breeding of hyperaccumulating plants that produce relatively large amounts of biomass could make the extraction process highly effective.[99]

It is known that many crop plants can accumulate metals in their roots and aboveground shoots, potentially threatening the food chain. For example, in May 1980 regulations proposed under RCRA for hazardous waste include limits on the amounts of Cd and other metals that can be applied to crops. Recently, however, the potential use of crop plants for environmental remediation has been under investigation. Using crop plants to extract metals from the soil seems practical because of their high biomass production and relatively fast rate of growth. Other benefits of using crop plants are that they are easy to cultivate and they exhibit genetic stability.[97]

10.12.1.2 Phytostabilization

Phytostabilization uses plants to limit the mobility and bioavailability of metals in soils. Ideally, phytostabilizing plants should be able to tolerate high levels of metals and to immobilize them in the soil by sorption, precipitation, complexation, or the reduction of metal valences. Phytostabilizing plants should also exhibit low levels of accumulation of metals in shoots to eliminate the possibility that residues in harvested shoots might become hazardous wastes.[90] In addition to stabilizing the metals present in the soil, phytostabilizing plants can also stabilize the soil matrix to minimize erosion and migration of sediment. Dr. Gary Pierzynski of Kansas State University is studying phytostabilization in poplar trees, which were selected for the study because they can be deep-planted and may be able to form roots below the zone of maximum contamination.[2]

Because most sites contaminated with metals lack established vegetation, metal-tolerant plants are used to revegetate such sites to prevent erosion and leaching.[100] However, that approach is a containment rather than a remediation technology. Some researchers consider phytostabilization to be an interim measure to be applied until phytoextraction becomes fully developed. However, other researchers are developing phytostabilization as a standard protocol of metal remediation technology, especially for sites at which removal of metals does not seem to be economically feasible. After field applications conducted by a group in Liverpool, England, varieties of three grasses were made commercially available for phytostabilization[90]:

1. *Agrostis tenuis, cv Parys* for Cu wastes
2. *Agrosas tenuis, cv Coginan* for acid Pb and Zn wastes
3. *Festuca rubra, cv Merlin* for calcareous Pb and Zn wastes

10.12.1.3 Rhizofiltration

One type of rhizofiltration uses plant roots to absorb, concentrate, and precipitate metals from wastewater,[90] which may include leachate from soil. Rhizofiltration uses terrestrial plants instead of aquatic plants because the terrestrial plants develop much longer, fibrous root systems covered with root hairs that have extremely large surface areas. This variation of phytoremediation uses plants that remove metals by sorption, which does not involve biological processes. Use of plants to translocate metals to the shoots is a slower process than phytoextraction.[100]

Another type of rhizofiltration, which is more fully developed, involves construction of wetlands or reed beds for the treatment of contaminated wastewater or leachate. The technology is cost-effective for the treatment of large volumes of wastewater that have low concentrations of metals.[100] Because rhizofiltration focuses on the treatment of contaminated water, it is not discussed further in this chapter.

Table 10.12 presents the advantages and disadvantages of each of the types of phytoremediation currently being researched that are categorized as either phytoextraction on phytostabilization.[90]

10.12.1.4 Future Development

Faster uptake of metals and higher yields of metals in harvested plants may become possible through the application of genetic engineering or selective breeding techniques. Recent laboratory-scale testing has revealed that a genetically altered species of mustard weed can uptake mercuric ions from the soil and convert them to metallic mercury, which is transpired through the leaves.[2] Improvements in phytoremediation may be attained through research and a better understanding of

TABLE 10.12
Types of Phytoremediation Technology: Advantages and Disadvantages

Type of Phytoremediation	Advantages	Disadvantages
Phytoextraction by trees	High biomass production	Potential for off-site migration and leaf transportation of metals to surface.
		Metals are concentrated in plant biomass and must be disposed of eventually.
Phytoextraction by grasses	High accumulation	Low biomass production and slow growth rate.
		Metals are concentrated in plant biomass and must be disposed of eventually.
Phytoextraction by crops	High biomass and increased growth rate	Potential threat to the food chain through ingestion by herbivores.
		Metals are concentrated in plant biomass and must be disposed of eventually.
Phytostabilization	No disposal of contaminated biomass required	Remaining liability issues, including maintenance for an indefinite period of time (containment rather than removal).
Rhizofiltration	Readily absorbs metals	Applicable for treatment of water only.
		Metals are concentrated in plant biomass and must be disposed of eventually.

Source: U.S. EPA, Recent Developments for In Situ Treatment of Metal Contaminated Soils, Contract no. 68-W5-0055 U.S. Environmental Protection Agency, Washington, DC, 1997.

the principles governing the processes by which plants affect the geochemistry of their soils. In addition, future testing of plants and microflora may lead to the identification of plants that have metal accumulation qualities that are far superior to those currently known.

10.12.2 APPLICABILITY

Plants have been used to treat wastewater for more than 300 years, and plant-based remediation methods for slurries of dredged material and soils contaminated with metals have been proposed since the mid-1970s.[87,101] Reports of successful remediation of soils contaminated with metals are rare, but the suggestion of such applications is more than two decades old, and progress is being made at a number of pilot test sites.[96] Successful phytoremediation must meet cleanup standards in order to be approved by regulatory agencies.

No full-scale applications of phytoremediation have been reported. One vendor, Phytotech, Inc., is developing phytostabilization for soil remediation applications. Phytotech also has patented strategies for phytoextraction and is conducting several field tests in Trenton, NJ, and in Chernobyl, Ukraine.[97] Also, as was previously mentioned, a group in Liverpool, England, has made three grasses commercially available for the stabilization of Pb, Cu, and Zn wastes.[90]

10.12.3 PERFORMANCE AND COST

A variety of new research approaches and tools are expanding understanding of the molecular and cellular processes that can be employed through phytoremediation.[102]

10.12.3.1 Performance

The potential for phytoremediation (phytoextraction) can be assessed by comparing the concentration of contaminants and volume of soil to be treated with the particular plant's seasonal productivity of

TABLE 10.13
Examples of Metal Hyperaccumulators

Metal	Plant Species	Percentage of Metal in Dry Weight of Leaves (%)	Native Location
Zn	*Thlaspi calaminare*	<3	Germany
	Viola species	1	Europe
Cu	*Aeolanthus biformifolius*	1	Zaire
Ni	*Phyllanthus serpentinus*	3.8	New Caledonia
	Alyssum bertoloni and 50 other species of alyssum	>3	Southern Europe and Turkey
	Sebertia acuminata	25 (in latex)	New Caledonia
	Stackhousia tryonii	4.1	Australia
Pb	*Brassuca juncea*	<3.5	India
Co	*Haumaniastrum robertii*	1	Zaire

Source: U.S. EPA, Recent Developments for In Situ Treatment of Metal Contaminated Soils, Contract no. 68-W5-0055 U.S. Environmental Protection Agency, Washington, DC, 1997.

biomass and ability to accumulate contaminants. Table 10.13 lists selected examples of plants identified as metal hyperaccumulators and their native countries.[2,103] If plants are to be effective remediation systems, one ton of plant biomass, costing from several hundred to a few thousand dollars to produce, must be able to treat large volumes of contaminated soil. For metals that are removed from the soil and accumulated in aboveground biomass, the total amount of biomass per hectare required for soil cleanup is determined by dividing the total weight of metal per hectare to be remediated by the accumulation factor, which is the ratio of the accumulated weight of the metal to the weight of the biomass containing the metal. The total biomass per hectare (T/ha) then can be divided by the productivity of the plant (T/ha/yr) to determine the number of years (yr) required to achieve cleanup standards—a major determinant of the overall cost and feasibility of phytoremediation.[102]

As discussed earlier, the amount of biomass is one of the factors that determine the practicality of phytoremediation. Under the best climatic conditions, with irrigation, fertilization, and other factors, total biomass productivity can approach 100 T/ha/yr. One unresolved issue is the tradeoff between accumulation of toxic elements and productivity.[104] In practice, a maximum harvest biomass yield of 10 to 20 T/ha/yr is likely, particularly for plants that accumulate metals.

These values for productivity of biomass and the metal content of the soil would limit annual capacity for removal of metals to approximately 10 to 400 kg/ha/yr, depending on the pollutant, species of plant, climate, and other factors. For a target soil depth of 30 cm (4000 T/ha), this capacity amounts to an annual reduction of 2.5 to 100 mg/kg of soil contaminants. This rate of removal of contamination often is acceptable, allowing total remediation of a site over a period of a few years to several decades.[102]

10.12.3.2 Cost

The practical objective of phytoremediation is to achieve major reductions in the cost of cleanup of hazardous sites. Salt[90] and others have noted the cost-effectiveness of phytoremediation with an example: Using phytoremediation to clean up one acre of sandy loam soil to a depth of 50 cm will typically cost USD 60,000 to USD 100,000, compared with a cost of at least USD 400,000 for excavation and disposal storage without treatment.[90] One objective of field tests is to use commercially available agricultural equipment and supplies for phytoremediation in order to reduce costs. Therefore, in addition to their remediation qualities, the agronomic characteristics of the plants must be evaluated.

The processing and ultimate disposal of the biomass generated is likely to be a major percentage of overall costs, particularly when highly toxic metals and radionuclides are present at a site. Analysis of the costs of phytoremediation must include the entire cycle of the process, from the growing and harvesting of the plants to the final processing and disposal of the biomass. It is difficult to predict costs of phytoremediation, compared with overall cleanup costs at a site. Phytoremediation may also be used as a follow-up technique after areas having high concentrations of pollutants have been mitigated or in conjunction with other remediation technologies, making cost analysis more difficult.

10.12.3.3 Future Directions

Because metal hyperaccumulators generally produce small quantities of biomass, they are unsuited agronomically for phytoremediation. Nevertheless, such plants are a valuable store of genetic and physiologic material and data.[87] To provide effective cleanup of contaminated soils, it is essential to find, breed, or engineer plants that absorb, translocate, and tolerate levels of metals in the range of 0.1 to 1.0%. It also is necessary to develop a methodology for selecting plants that are native to the area.

Three grasses are commercially available for the stabilization of Pb, Cu, and Zn wastes.[90] An integrated approach that involves basic and applied research, along with consideration of safety, legal, and policy issues, will be necessary to establish phytoremediation as a practicable cleanup technology.[87]

According to a DOE report, three broad areas of research and development can be identified for the *in situ* treatment of soil contaminated with metals[102]:

1. *Mechanisms of uptake, transport, and accumulation.* Research is needed to develop a better understanding of the use of physiological, biochemical, and genetic processes in plants. Research on the uptake and transport mechanisms is providing improved knowledge about the adaptability of those systems and how they might be used in phytoremediation.
2. *Genetic evaluation of hyperaccumulators.* Research is being conducted to collect plants growing in soils that contain high levels of metals and screen them for specific traits useful in phytoremediation. Plants that tolerate and colonize environments polluted with metals are a valuable resource, both as candidates for use in phytoremediation and as sources of genes for classical plant breeding and molecular genetic engineering.
3. *Field evaluation and validation.* Research is being carried out to use early and frequent field testing to accelerate implementation of phytoremediation technologies and to provide data to research programs. Standardization of field-test protocols and subsequent application of test results to real problems are also needed.

Research in these areas is expected to grow as many of the current engineering technologies for cleaning surface soil of metals are costly and physically disruptive. Phytoremediation, when fully developed, could result in significant cost savings and in the restoration of numerous sites by a relatively noninvasive, solar-driven, *in situ* method that, in some forms, can be aesthetically pleasing.[87]

10.12.4 Summary of Phytoremediation Technology

Phytoremediation is in the early stage of development and is being field tested at various sites in the U.S. and overseas for its effectiveness in capturing or stabilizing metals, including radioactive wastes. Limited cost and performance data are currently available. Phytoremediation has the potential to develop into a practicable remediation option at sites at which contaminants are near the surface, are relatively nonleachable, and pose little imminent threat to human health or the environment.[87] The efficiency of phytoremediation depends on the characteristics of the soil and the contaminants; these factors are summarized in the sections that follow.

10.12.4.1 Site Conditions

The effectiveness of phytoremediation is generally restricted to surface soils within the rooting zone. The most important limitation to phytoremediation is rooting depth, which can be 20, 50, or even 100 cm, depending on the plant and soil type. Therefore, one of the favorable site conditions for phytoremediation is contamination with metals that is located at the surface.[102]

The type of soil, as well as the rooting structure of the plant relative to the location of the contaminants, can have a strong influence on the uptake of any metal substance by the plant. Amendment of soils to change soil pH, nutrient compositions, or microbial activities must be selected in treatability studies to govern the efficiency of phytoremediation. Certain generalizations can be made about such cases; however, much work is needed in this area.[87] Because the amount of biomass that can be produced is one of the limiting factors affecting phytoremediation, optimal climatic conditions, with irrigation and fertilization of the site, should be considered to promote increased productivity of the best plants for the site.[102]

10.12.4.2 Waste Characteristics

Sites that have low to moderate contamination with metals might be suitable for growing hyperaccumulating plants, although the most heavily contaminated soils do not allow plant growth without the addition of soil amendments. Unfortunately, one of the most difficult metal cations for plants to translocate is Pb, which is present at numerous sites in need of remediation. Although significant uptake of Pb has not yet been demonstrated, one researcher is experimenting with soil amendments that make lead more available for uptake.[90]

Capabilities to accumulate Pb and other metals are dependent on the chemistry of the soil in which the plants are growing. Most metals, and Pb in particular, occur in numerous forms in the soil, not all of which are equally available for uptake by plants.[87,105] Maximum removal of Pb requires a balance between the nutritional requirements of plants for biomass production and the bioavailability of Pb for uptake by plants. Maximizing the availability of Pb requires low pH and low levels of available phosphate and sulfate. However, limiting the fertility of the soil in such a manner directly affects the health and vigor of plants.[87]

10.13 USE OF TREATMENT TRAINS

Several of the metal remediation technologies discussed are often enhanced through the use of treatment trains. Treatment trains use two or more remedial options applied sequentially to the contaminated soil and often increase the effectiveness while decreasing the cost of remediation. Processes involved in treatment trains include soil pretreatment, physical separation designed to decrease the amount of soil requiring treatment, additional treatment of process residuals or off-gases, and a variety of other physical and chemical techniques, which can greatly improve the performance of the remediation technology. Table 10.14 provides examples of treatment trains used to enhance each of the proved and commercialized metal remediation technologies.[5]

10.14 COST RANGES OF REMEDIAL TECHNOLOGIES

Estimated cost ranges for the basic operation of the technology are presented in Table 10.15. The reader is cautioned that the cost estimates generally do not include pretreatment, site preparation, regulatory compliance costs, costs for additional treatment of process residuals (e.g., stabilization of incinerator ash or disposal of metals concentrated by solvent extraction), or profit.[5,106] Since the actual cost of employing a remedial technology at a specific site may be significantly different than these estimates, data are best used for order-of-magnitude cost evaluations only.

TABLE 10.14
Typical Treatment Trains

	Containment	S/S	Vitrification	Soil Washing	Pyrometallurgical	Soil Flushing
Pretreatment						
Excavation	×	E,P	I,E	×	×	
Debris removal		E,P	E	×	×	
Oversize reduction		E,P	E	×	×	
Adjust pH	×	I,E,P				
Reduction [e.g., Cr(VI) to Cr(III)]	×	I,E				
Oxidation [e.g., As(III) to As (V)]	×	I,E				
Treatment to remove or destroy organics		I,E				
Physical separation of rich and lean fractions		I,E,P	E	×	×	
Dewatering and drying for wet sludge	×	P	E		×	
Conversion of metals to less volatile forms [e.g., As$_2$O$_3$ to Ca$_3$(AsO$_4$)$_2$]			E			
Addition of high-temperature reductants					×	
Pelletizing					×	
Flushing fluid delivery and extraction system						×
Containment barriers	×	I,E,P	I	×		×
Post-treatment/residuals management						
Disposal of treated solid residuals (preferably below the frost line and above the water table)		I,E,P	E		×	
Containment barriers		I,E,P	I,E			×
Off-gas treatment		I,E,P	I,E		×	
Reuse for on-site paving		P				
Metal recovery from extraction fluid by aqueous processing (ion exchange, electrowinning, etc.)				×		
Pyrometallurgical recovery of metal from sludge				×		
Processing and reuse of leaching solution				×	×	
S/S treatment of leached residual				×		
Disposal of solid process residuals (preferably below the frost line and above the water table)				×		
Disposal of liquid process residuals				×		×
S/S treatment of slag or fly ash					×	
Reuse of slag/vitreous product as construction material			E	×		
Reuse of metal or metal compound					×	
Further processing of metal or metal compound					×	
Flushing liquid/groundwater treatment/disposal						×

Technology has been divided into the following categories: I = *in situ* process; E = *ex situ* process; P = polymer microencapsulation (*ex situ*).

Source: U.S. EPA, Technology Alternatives for the Remediation of Soils Contaminated with As, Cd, Cr, Hg, and Pb, EPA/540/S-97/500, U.S. Environmental Protection Agency, Cincinnati, OH, August 1997.

TABLE 10.15
Estimated Cost Ranges of Metals Remediation Technologies

Type of Remediation	Cost Range (2007 USD/T)
Containment[a]	13–120
Solidification–stabilization	80–380
Vitrification	520–1140
Soil washing	80–320
Soil flushing[b]	80–215
Pyrometallurgical	330–730
Electrokinetics[b]	60–160
Phytoremediation[c]	30–50

[a] Includes landfill caps and slurry walls. A slurry wall depth of 6 m is assumed.
[b] Costs reported in USD/m^3, assuming soil specific gravity of 1.6.
[c] Costs reported per acre for a soil depth of 0.50 m.

Source: U.S. EPA, Recent Developments for In Situ Treatment of Metal Contaminated Soils, Contract no. 68-W5-0055 U.S. Environmental Protection Agency, Washington, DC, 1997; U.S. EPA, Technology Alternatives for the Remediation of Soils Contaminated with As, Cd, Cr, Hg, and Pb, EPA/540/S-97/500, U.S. Environmental Protection Agency, Cincinnati, OH, August 1997.

REFERENCES

1. Federal Register, Comprehensive Environmental Response, Compensation, and Liability Act (CERCLA or Superfund) 42 U.S.C. s/s 9601 et seq., United States Government, Public Laws, 1980. Available at www.access.gpo.gov/uscode/title42/chapter103_.html.
2. U.S. EPA, Recent Developments for In Situ Treatment of Metal Contaminated Soils, Contract no. 68-W5-0055 U.S. Environmental Protection Agency, Washington, DC, 1997.
3. Federal Register, Resource Conservation and Recovery Act (RCRA), 42 US Code s/s 6901 et seq., 1976, U.S. Government, Public Laws. Available at www.access.gpo.gov/uscode/title42/chapter82_.html.
4. U.S. EPA, Contaminants and Remedial Options at Selected Metal-Contaminated Sites, EPA/540/R-95/512, U.S. Environmental Protection Agency, Washington, DC, July 1995.
5. U.S. EPA, Technology Alternatives for the Remediation of Soils Contaminated with As, Cd, Cr, Hg, and Pb, EPA/540/S-97/500, U.S. Environmental Protection Agency, Cincinnati, OH, August 1997.
6. U.S. EPA, In Situ Technologies for the Remediation of Soils Contaminated with Metals—Status Report. U.S. Environmental Protection Agency, Cincinnati, OH, July 1996.
7. U.S. EPA, Selection of Control Technologies for Remediation of Lead Battery Recycling Sites, EPA/540/2-91/014, U.S. Environmental Protection Agency, Cincinnati, OH, 1991.
8. U.S. EPA, Engineering Bulletin: Selection of Control Technologies for Remediation of Lead Battery Recycling Site, EPA/540/S-92/011, U.S. Environmental Protection Agency, Cincinnati, OH, 1992.
9. U.S. EPA, Contaminants and Remedial Options at Wood Preserving Sites, EPA 600/R-92/182, U.S. Environmental Protection Agency, Washington, DC, 1992.
10. U.S. EPA, Presumptive Remedies for Soils, Sediments, and Sludges at Wood Treater Sites, EPA/540/R-95/128, U.S. Environmental Protection Agency, Washington, DC, 1995.
11. U.S. EPA, Contaminants and Remedial Options at Pesticide Sites, EPA/600/R-94/202, U.S. Environmental Protection Agency, Washington, DC, 1994.
12. U.S. EPA, Separation/Concentration Technology Alternatives for the Remediation of Pesticide-Contaminated Soil, EPA/540/S-97/503, U.S. Environmental Protection Agency, Washington, DC, 1997.
13. McLean, J.E. and Bledsoe, B.E., Behavior of Metals in Soils, EPA/540/S-92/018, U.S. Environmental Protection Agency, Washington, DC, 1992.
14. Palmer, C.D. and Puls, R.W., Natural Attenuation of Hexavalent Chromium in Ground Water and Soils, EPA/540/S-94/505, U.S. Environmental Protection Agency, Washington, DC, 1994.

15. Benjamin, M.M. and Leckie, J.D., Adsorption of metals at oxide interfaces: Effects of the concentrations of adsorbate and competing metals, in *Contaminant sand Sediments, Volume 2: Analysis, Chemistry, Biology*, Baker R.A., Ed., Ann Arbor Science Publishers, Ann Arbor, MI, 1980, chap. 16.
16. Wagemann, R., Some theoretical aspects of stability and solubility of inorganic As in the freshwater environment, *Water Res.*, 12, 139–145, 1978.
17. Zimmerman, L. and Coles, C., Cement industry solutions to waste management—The utilization of processed waste by-products for cement manufacturing, in Proceedings of the 1st International Conference for Cement Industry Solutions to Waste Management, Calgary, Alberta, Canada, 1992, pp. 533–545.
18. Earth Platform, *Contaminated Soil Remediation*, 2007. Available at http://www.earthplatform.com/contaminated/soil/remediation.
19. Sharma, H.D. and Reddy, K.R., *Geoenvironmental Engineering: Site Remediation, Waste Containment, and Emerging Waste Management Technologies*, John Wiley & Sons, Hoboken, NJ, 2004.
20. Weston, R.F., Installation Restoration General Environmental Technology Development Guidelines for In-Place Closure of Dry Lagoons, U.S. Army Toxic and Hazardous Materials, May 1985.
21. U.S. EPA, Slurry Trench Construction for Pollution Migration Control, EPA/540/2-84/001, U.S. Environmental Protection Agency, Washington, DC, February 1984.
22. U.S. EPA, Grouting Techniques in Bottom Sealing of Hazardous Waste Sites, EPA/600/2-86/020, U.S. Environmental Protection Agency, Washington, DC, 1986.
23. U.S. EPA, Engineering Bulletin: Landfill Covers, EPA/540/S-93/500, U.S. Environmental Protection Agency, Cincinnati, OH, February 1993.
24. FRTR, *Physical Barriers*, Remediation Technologies Screening Matrix and Reference Guide, 2007. Available at http://www.frtr.gov/matrix2/section4/4-53.html.
25. U.S. EPA, Engineering Bulletin: Slurry Walls, EPA/540/S-92/008, U.S. Environmental Protection Agency, Cincinnati, OH, October 1992.
26. U.S. EPA, Solidification/Stabilization Use at Superfund Sites, EPA-542-R-00-010, U.S. Environmental Protection Agency, Washington, DC, September 2000.
27. U.S. EPA, Technical Resource Document: Solidification/Stabilization and Its Application to Waste Materials, EPA/530/R-93/012, U.S. Environmental Protection Agency, Cincinnati, OH, June 1993.
28. U.S. EPA, Engineering Bulletin: Solidification/Stabilization of Organics and Inorganics, EPA/540/S-92/015, U.S. Environmental Protection Agency, Cincinnati, OH, 1992.
29. Conner, J.R., Chemical Fixation and Solidification of Hazardous Wastes, Van Nostrand Reinhold, New York, 1990.
30. U.S. EPA, Solidification/Stabilization and Its Application to Waste Materials, EPA/530/R-93/012, U.S. Environmental Protection Agency, Washington, DC, June 1993.
31. Anderson, W.C., Ed., Innovative site remediation technology: solidification/stabilization, in *Innovative Site Remediation Technology: Phase I (Process Descriptions and Limitations)*, Vol. 4, EPA/542-B-94-001, June 1994.
32. WASTECH, *Solidification/Stabilization*, Waste Technology, American Academy of Environmental Engineers, EPA, printed under license EPA/542-B-94-001, June 1994.
33. U.S. EPA, A Citizen's Guide to Solidification/Stabilization, EPA 542-F-01-024, U.S. Environmental Protection Agency, Washington, DC, December 2001.
34. U.S. EPA, Solidification/Stabilization Use at Superfund Sites, EPA-542-R-00-010, U.S. Environmental Protection Agency, Washington, DC, September 2000.
35. U.S. ACE, Solidification/Stabilization of Contaminated Material, Unified Facility Guide Specification, UFGS-02160a, U.S. Army Corps of Engineers, October 2000.
36. ANL. Fact Sheet—Solidification/Stabilization. Drilling Waste Management Information System, Argonne National Laboratory, 2007. Available at http://web.ead.anl.gov/dwm/techdesc/solid/index.cfm.
37. U.S. EPA, Handbook on In Situ Treatment of Hazardous Waste-Contaminated Soils, EPA/540/2-90/002, U.S. Environmental Protection Agency, Cincinnati, OH, 1990.
38. Arniella, E.F. and Blythe, L.J., Solidifying traps hazardous waste, *Chem. Eng.*, 97, 92–102, 1990.
39. Kalb, P.D., Burns, H.H. and Meyer, M., Thermo-plastic encapsulation treatability study for a mixed waste incinerator off-gas scrubbing solution, in *Third International Symposium on Stabilization/Solidification of Hazardous, Radioactive, and Mixed Wastes*, Gilliam, T.M., Ed., ASTM STP 1240, American Society for Testing and Materials, Philadelphia, PA, 1993.
40. Ponder, T.G. and Schmitt, D., Field assessment of air emission from hazardous waste stabilization operation, in Proceedings of the 17th Annual Hazardous Waste Research Symposium, EPA/600/9-91/002, Cincinnati, OH, 1991.

41. Shukla, S.S., Shukla, A.S. and Lee, K.C., Solidification/stabilization study for the disposal of pentachlorophenol, *J. Hazard. Mater.*, 30, 317–331, 1992.
42. U.S. EPA, Evaluation of Solidification/Stabilization as a Best Demonstrated Available Technology for Contaminated Soils, EPA/600/2-89/013, U.S. Environmental Protection Agency, Cincinnati, OH, 1989.
43. Weitzman, L. and Hamel, L.E., Volatile emissions from stabilized waste, in Proceedings of the 15th Annual Research Symposium, EPA/600/9-90/006, U.S. Environmental Protection Agency, Cincinnati, OH, 1990.
44. Means, J.L., Nehring, K.W. and Heath, J.C., Abrasive blast material utilization in asphalt roadbed material, in *Third International Symposium on Stabilization/Solidification of Hazardous, Radioactive, and Mixed Wastes*, ASTM STP 1240, American Society for Testing and Materials, Philadelphia, PA, 1993.
45. U.S. ACE, Yearly average cost index for utilities, in Civil Works Construction Cost Index System Manual, 110-2-1304, U.S. Army Corps of Engineers, Washington, DC, 2007, p. 44. Available at http://www.nww.usace.army.mil/cost.
46. Buelt, J.L., Timmerman, C.L., Oma, K.H., FitzPatrick, V.F. and Carter, J.G., *In Situ Vitrification of Transuranic Waste: An Updated Systems Evaluation and Applications Assessment*, PNL-4800, Pacific Northwest Laboratory, Richland, WA, 1987.
47. U.S. EPA, Vitrification Technologies for Treatment of Hazardous and Radioactive Waste, EPA/625/R-92/002, U.S. Environmental Protection Agency, Cincinnati, OH, May 1992.
48. U.S. EPA, Engineering Bulletin—In Situ Vitrification Treatment, EPA/540/S-94/504, U.S. Environmental Protection Agency, Cincinnati, OH, October 1994.
49. U.S. EPA, Engineering Bulletin: In Situ Vitrification Treatment, EPA/540/S-94/504, U.S. Environmental Protection Agency, Washington, DC, revised May 2002.
50. FRTR, Solidification/Stabilization—In Situ Soil Remediation Technology, Remediation Technologies Screening Matrix and Reference Guide, 2007. Available at http://www.frtr.gov/matrix2/section4/4-8.html.
51. U.S. EPA, Geosafe Corporation In Situ Vitrification Innovative Technology Evaluation Report, EPA/540/R-94/520, U.S. Environmental Protection Agency, Washington, DC, March 1995.
52. FitzPatrick, V.F., Timmerman, C.L. and Buelt, J.L., In situ vitrification: An innovative thermal treatment technology, in Proceedings of the Second International Conference on New Frontiers for Hazardous Waste Management, 1987, pp. 305–322, EPA/600/9-87/018F, U.S. Environmental Protection Agency, EPA printed under license EPA/542-B-94-001, June 1994.
53. Timmerman, C.L., In Situ Vitrification of PCB Contaminated Soils, EPRI CS-4839, Electric Power Research Institute, Palo Alto, CA, 1986.
54. U.S. EPA, The Superfund Innovative Technology Evaluation Program: Technology Profiles, 4th ed., EPA/540/5-91/008, U.S. Environmental Protection Agency, Washington, DC, 1991.
55. Luey, J., Koegler, S.S., Kuhn, W.L., Lowery, P.S. and Winkelman, R.G., In Situ Vitrification of a Mixed-Waste Contaminated Soil Site, The 116-B-6A Crib at Hanford, PNL-8281. Pacific Northwest Laboratory, Richland, WA, 1992.
56. Hansen, J.E. and FitzPatrick, V.F., *In Situ Vitrification Applications*, Geosafe Corporation, Richland, WA, 1991.
57. U.S. EPA, Engineering Bulletin: Soil Washing Treatment, EPA/540/2-90/017, U.S. Environmental Protection Agency, Cincinnati, OH, 1996.
58. U.S. EPA, A Citizen's Guide to Soil Washing, EPA 542-F-01-008, U.S. Environmental Protection Agency, Washington, DC, May 2001.
59. William, C.A., Ed., *Innovative Site Remediation Technology: Soil Washing/Flushing*, Vol. 3, American Academy of Environmental Engineers (published by EPA under EPA 542-B-93-012), November 1993.
60. U.S. EPA, *Technology Focus—Soil Washing*, Technology Innovation Program, U.S. Environmental Protection Agency, Washington, DC, 2007. Available at http://clu-in.org/techfocus/default.focus/sec/Soil_Washing/cat/Overview.
61. Ehsan, S., Prasher, S.O. and Marshall, W.D., A washing procedure to mobilize mixed contaminants from soil. II. Heavy metals, *J. Environ. Qual.*, 35, 2084–2091, 2006.
62. Fischer, K. and Bipp, H.P., Removal of heavy metals from soil components and soils by natural chelating agents. Part II. Soil extraction by sugar acids, *Water, Air, Soil Pollut.*, 38, 271–288, 2002.
63. U.S. EPA, *Citizens Guide to Soil Washing*, EPA/542/F-92/003, U.S. Environmental Protection Agency, Washington, DC, March 1992.
64. U.S. EPA, Engineering Bulletin: In Situ Soil Flushing, EPA/540/2-91/021, U.S. Environmental Protection Agency, Cincinnati, OH, October 1991.
65. FRTR, Soil Flushing—In Situ Soil Remediation Technology, Remediation Technologies Screening Matrix and Reference Guide, 2007. Available at http://www.frtr.gov/matrix2/section4/4-6.html.

66. CPEO, *Soil Flushing*, Center for Public Environmental Oversight (CPEO), San Francisco, CA, 2007. Available at http://www.cpeo.org/techtree/ttdescript/soilflus.htm.
67. U.S. EPA, Superfund Innovative Technology Evaluation Program: Technology Profiles, 7th ed., EPA/540/R-94/526, U.S. Environmental Protection Agency, Washington, DC, November 1994.
68. Pamukcu, S. and Wittle, J.K., Electrokinetic removal of selected metals from soil, *Environ. Prog.*, II, 241–250, 1992.
69. Acar, Y.B., Electrokinetic cleanups, *Civil Eng.*, October, 58–60, 1992.
70. Mattson, E.D. and Lindgren, E.R., Electrokinetics: an innovative technology for in situ remediation of metals, in Proceedings, National Groundwater Association, Outdoor Acnon Conference, Minneapolis, MN, May 1994.
71. Acar, Y.B. and Gale, R.J. Electrokinetic remediation: basics and technology status, *J. Hazard. Mater.*, 40, 117–137, 1995.
72. Will, F., Removing toxic substances from the soil using electrochemistry, *Chem. Ind.*, May 15, 376–379, 1995.
73. Rodsand, T. and Acar, Y.B., Electrokinetic extraction of lead from spiked Norwegian marine clay, *Geoenvironment 2000*, 2, 1518–1534, 1995.
74. Lindgren, E.R., Kozak, M.W. and Mattson, E.D., Electrokinetic remediation of contaminated soils: an update, *Waste Management 92*, Tuscon, AZ, 1992, p. 1309.
75. Jacobs, R.A. and Sengun, M.Z., Model of experiences on soil remediation by electric fields, *J. Environ. Sci. Health*, 29A, 9, 1994.
76. Acar, Y.B. and Gale, R.J., Electrochemical Decontamination of Soils and Slurries, U.S. Patent 5,137,608, August 15, 1992.
77. Marks, R., Acar, Y.B. and Gale, R.J., *In situ* Bioelectrokinetic Remediation of Contaminated Soils Containing Hazardous Mixed Wastes, U.S. Patent 5,458,747, October 17, 1995.
78. Acar, Y.B. and Alshawabkeh, A.N., Electrokinetic remediation: I. Pilot-scale tests with lead spiked kaolinite, II. Theoretical model, *J. Geotech. Eng.*, 122, 173–196, 1996.
79. Pool, W., Process for the Electroreclamation of Soil Material, U.S. Patent 5,433,829, July 18, 1995.
80. U.S. EPA, *In Situ Remediation Technology Status Report: Electrokinetics*, EPA 542-K-94-007, U.S. Environmental Protection Agency, Washington, DC, 1995.
81. Editorial, Innovative in situ cleanup processes, *The Hazardous Waste Consultant*, September/October, 1992.
82. DOE, Development of an Integrated In-Situ Remediation Technology, Technology Development Data Sheet, DE-AR21-94MC31185, U.S. Department of Energy, 1995.
83. U.S. EPA, Superfund Innovative Technology Evaluation Program Technology Profiles, 7th ed., EPA 540-R-94-526, U.S. Environmental Protection Agency, Washington, DC, 1994.
84. U.S. EPA, Lasagna™ Public–Private Partnership, EPA 542-F-96-010A, U.S. Environmental Protection Agency, Washington, DC, 1996.
85. Szpyrkowicz, L., Radaelli, M., Bertini, S., Daniele, S. and Casarin, F., Simultaneous removal of metals and organic compounds from a heavily polluted soil, *Electrochim. Acta*, 52, 3386–3392, 2007.
86. CPEO, *Electrokinetics*, Center for Public Environmental Oversight (CPEO), San Francisco, CA, 2007. Available at http://www.cpeo.org/techtree/ttdescript/elctro.htm.
87. Cunningham, S.D. and Berti, W.R., Remediation of contaminated soils with green plants: an overview, *In Vitro Cell. Dev. Biol.*, 29, 207–212, 1993.
88. Raskin, I., Bioconcentration of metals by plants, *Environ. Biotechnol.*, 5, 285–290, 1994.
89. Goldsbrough, P., Phytochelatins and metallothioneins: complementary mechanisms for metal tolerance, in Fourteenth Annual Symposium 1995 in Current Topics in Plant Biochemistry, Physiology and Molecular Biology, Columbia, MO, 1995.
90. Salt, D.E., Phytoremediation: A novel strategy for the removal of toxic metals from the environment using plants, *Biotechnology*, 13, 468–474, 1995.
91. Kumar, P.B.A., Phytoextraction: the use of plants to remove metals from soils, *Environ. Sci. Technol.*, 29, 1232–1238, 1995.
92. Durham, S., *Using Plants to Clean Up Soil*, U.S. Department of Agriculture (USDA), 2007. Available at http://www.ars.usda.gov/is/pr/2007/070123.htm.
93. Morel, I.L., Root exudates and metal mobilization, in Fourteenth Annual Symposium 1995 in Current Topics in Plant Biochemistry, Physiology and Molecular Biology, Columbia, MO, 1995.
94. Baker, A.J.M. and Brooks, R.R., Terrestrial higher plants which hyperaccumulate metallic elements—a review of their distribution, ecology, and phytochemistry, *Biorecovery*, 1, 81–126, 1989.

95. Hyperaccumulation in the genus *Alyssum*, in Fourteenth Annual Symposium 1995 in Current Topics in Plant Biochemistry, Physiology and Molecular Biology, Columbia, MO, 1995.
96. Baker, A.J.M., Metal hyperaccumulation by plants: our present knowledge of ecophysiological phenomenon, in Fourteenth Annual Symposium 1995 in Current Topics in Plant Biochemistry, Physiology and Molecular Biology, Columbia, MO, 1995.
97. King Communications Group, Inc., Promise of heavy metal harvest lures venture funds, *The Bioremediation Report*, 4, 1, Washington, DC, 1995.
98. Greger, M. and Landberg, M.T., Improving removal of metals from soil by *Salix*, in Proceedings of the 7th International Conference on the Biogeochemistry of Trace Elements, Uppsala, Sweden, June 15–19, 2003.
99. Chaney, R.L, Malik, M., Li, Y.M., Brown, S.L., Angle, J.S. and Baker, A.J.M., Phytoremediation of soil metals, *Curr. Opin. Biotechnol.*, 8, 279–284, 1997.
100. Ensley, B.D., Will plants have a role in bioremediation?, in Fourteenth Annual Symposium 1995 in Current Topics in Plant Biochemistry, Physiology and Molecular Biology, Columbia, MO, 1995.
101. Cunningham, S.D. and Lee, C.R., Phytoremediation: Plant-based remediation of contaminated soil and sediments, in Proceedings of a Symposium of the Soil Science Society of America, Chicago, IL, November 1994.
102. DOE, Summary Report of a Workshop on Phytoremediation Research Needs, U.S. Department of Energy, Santa Rosa, CA, July 24–26, 1994.
103. Baker, A.J.M., Brooks, R.R. and Reeves, R.D., Growing for gold ... and copper ... and zinc, *New Scientist*, 1603, 44–48, 1989.
104. Parry, I., Plants absorb metals, *Pollut. Eng.*, February, 40–41, 1995.
105. USDA, Acidifying soil helps plant remove cadmium, zinc metals, Agricultural Research Service, *Science Daily*, 2007. Available at http://www.sciencedaily.com/releases/2005/06/050619192657.htm.
106. Hyman, M. and Dupont, R.R., *Groundwater and Soil Remediation: Process Design and Cost Estimating of Proven Technologies*, ASCE Publications, Reston, VA, 2001, p. 534.

11 Cleanup of Metal Finishing Brownfield Sites

Nazih K. Shammas

CONTENTS

11.1	Introduction		434
	11.1.1	Background	434
	11.1.2	Metals and Metalloids	435
	11.1.3	Purpose	436
11.2	Industrial Processes and Contaminants at Metal Finishing Sites		437
	11.2.1	Surface Preparation Operations	437
	11.2.2	Metal Finishing Operations	437
		11.2.2.1 Anodizing Operations	438
		11.2.2.2 Chemical Conversion Coating	439
		11.2.2.3 Electroplating	439
		11.2.2.4 Electroless and Immersion Plating	439
		11.2.2.5 Painting	440
		11.2.2.6 Other Metal Finishing Techniques	440
	11.2.3	Auxiliary Activity Areas and Potential Contaminants	440
		11.2.3.1 Wastewater Treatment	440
		11.2.3.2 Sunken Wastewater Treatment Tank	440
		11.2.3.3 Chemical Storage Area	440
		11.2.3.4 Disposal Area	441
		11.2.3.5 Other Considerations	441
11.3	Site Assessment		441
	11.3.1	The Central Role of the State Agencies	442
		11.3.1.1 State Voluntary Cleanup Programs	442
		11.3.1.2 Levels of Contaminant Screening and Cleanup	442
	11.3.2	Performing a Phase I Site Assessment: Obtaining Facility Background Information from Existing Data	442
		11.3.2.1 Facility Records	444
		11.3.2.2 Other Sources of Recorded Information	444
		11.3.2.3 Identifying Migration Pathways and Potentially Exposed Populations	445
	11.3.3	Gathering Topographic Information	445
	11.3.4	Gathering Soil and Subsurface Information	446
	11.3.5	Gathering Groundwater Information	446
		11.3.5.1 Identifying Potential Environmental and Human Health Concerns	446
		11.3.5.2 Community Involvement	447
		11.3.5.3 Conducting a Site Visit	448
		11.3.5.4 Conducting Interviews	448
		11.3.5.5 Developing a Report	448

11.3.6 The Triad Approach: Streamlining Site Investigations and
Cleanup Decisions 449
11.3.7 Performing a Phase II Site Assessment: Sampling the Site 450
11.3.7.1 Setting Data Quality Objectives 450
11.3.7.2 Screening Levels 453
11.3.7.3 Environmental Sampling and Data Analysis 454
11.3.7.4 Levels of Sampling and Analysis 454
11.3.8 Increasing the Certainty of Sampling Results 454
11.3.9 Site Assessment Technologies 455
11.3.9.1 Field versus Laboratory Analysis 455
11.3.9.2 Sample Collection and Analysis Technologies 455
11.3.10 Additional Considerations for Assessing Metal Finishing Sites 455
11.3.10.1 Where to Sample 456
11.3.10.2 How Many Samples to Collect 460
11.3.10.3 What Types of Analysis to Perform 460
11.3.11 General Sampling Costs 461
11.3.11.1 Soil Collection Costs 461
11.3.11.2 Groundwater Sampling Costs 461
11.3.11.3 Surface Water and Sediment Sampling Costs 461
11.3.11.4 Sample Analysis Costs 461
11.4 Site Cleanup 462
11.4.1 Developing a Cleanup 462
11.4.1.1 Institutional Controls 463
11.4.1.2 Containment Technologies 463
11.4.1.3 Types of Cleanup Technologies 464
11.4.2 Keys to Technology Selection and Acceptance 464
11.4.3 Summary of Technologies for Treating Metals/Metalloids
at Brownfield Sites 466
11.4.4 Cleanup Technologies Options for Metal Finishing Sites 467
11.4.5 Post-Construction Care 473
11.5 Conclusions 473
Acronyms 474
References 475

11.1 INTRODUCTION

11.1.1 BACKGROUND

The Comprehensive Environmental Response, Compensation, and Liability Act (CERCLA) or Superfund[1] defines brownfield sites as "real property, the expansion, redevelopment, or reuse of which may be complicated by the presence or potential presence of a hazardous substance, pollutant, or contaminant." According to the U.S. EPA, brownfield sites are abandoned, idled, or underused industrial and commercial facilities where expansion or redevelopment is complicated by real or perceived environmental contamination.[2] Concerns about liability, cost, and potential health risks associated with brownfields sites often prompt businesses to migrate to "greenfields" outside the city. Left behind are communities burdened with environmental contamination, declining property values, and increased unemployment.

U.S. EPA's Brownfields Economic Redevelopment Initiative was established to enable states, site planners, and other community stakeholders to work together in a timely manner to prevent, assess, safely clean up, and sustainably reuse brownfield sites.[3] With the enactment of the Small

Business Liability Relief and Brownfields Revitalization Act in 2002, U.S. EPA assistance was expanded to provide greater support for brownfields cleanup and reuse. Many states and local jurisdictions also help businesses and communities adapt environmental cleanup programs to the special needs of brownfield sites.

Preparing brownfield sites for productive reuse requires integration of many elements—financial issues, community involvement, liability considerations, environmental assessment and cleanup, regulatory requirements, and more—as well as coordination among many groups of stakeholders.[4] The assessment and cleanup of a site must be carried out in a way that integrates all these factors into the overall redevelopment process. In addition, the cleanup strategy will vary from site to site. At some sites, cleanup will be completed before the properties are transferred to new owners. At other sites, cleanup may take place simultaneously with construction and redevelopment activities.

Regardless of when and how cleanups are accomplished, the challenge to any brownfields program is to clean up sites in accordance with redevelopment goals. Such goals may include cost-effectiveness, timeliness, avoidance of adverse effects to site structures and neighboring communities, and redevelopment of land in a way that benefits communities and local economies. Regulators and site managers are increasingly recognizing the value of implementing a more dynamic approach to streamline assessment and cleanup activities at brownfield sites. This approach, referred to as "the triad," is flexible and recognizes site-specific decisions and data needs.[4]

The triad approach focuses on management of decision uncertainty by incorporating systematic project planning; dynamic work planning strategies; and use of real-time measurement technologies, including innovative technologies, to accelerate and improve the cleanup process. The Triad approach can reduce costs, improve decision certainty, expedite site closeout, and positively affect regulatory and community acceptance. This approach is well aligned with brownfield site priorities, which are affected by the economics of redevelopment, community involvement, and liability considerations.

Numerous technology options are available to assist those involved in brownfields cleanup. U.S. EPA's Office of Superfund Remediation and Technology Innovation (OSRTI) encourages use of smarter solutions for characterizing and cleaning up contaminated sites by advocating more effective, less costly technological approaches. Use of innovative technologies to characterize and clean up brownfield sites provides opportunities for stakeholders to reduce cleanup costs and accelerate cleanup schedules. Often, innovative approaches are also more acceptable to communities.

The cornerstone of U.S. EPA's Brownfields Initiative is the Pilot Program. Under this program, U.S. EPA is funding more than 200 brownfields assessment pilot projects in states, cities, towns, counties, and tribes across the country.[2] The pilots, each funded at up to USD 200,000 over two years, are bringing together community groups, investors, lenders, developers, and other affected parties to address the issues associated with assessing and cleaning up contaminated brownfields sites and returning them to appropriate, productive use. U.S. EPA's regional brownfields coordinators can provide communities with technical assistance such as targeted brownfields assessments. In addition to the hundreds of brownfield sites being addressed by these pilots, over 40 states have established brownfields or voluntary cleanup programs to encourage municipalities and private sector organizations to assess, clean up, and redevelop brownfields sites.

11.1.2 Metals and Metalloids

Metals are one of the three groups of elements distinguished by their ionization and bonding properties, along with metalloids and nonmetals. Metals have certain characteristic physical properties: they are usually shiny, have a high density, are ductile and malleable, usually have a high melting point, are usually hard, and conduct electricity and heat well. Metalloids have properties that are intermediate between those of metals and nonmetals. There is no unique way of distinguishing a metalloid from a true metal, but the most common way is that metalloids are usually semiconductors rather than conductors.[4]

TABLE 11.1
Typical Metals and Metalloids at Brownfield Sites

Metals and Metalloids

Aluminum	Calcium	Mercury	Tin
Antimony	Chromium	Molybdenum	Titanium
Arsenic	Cobalt	Nickel	Vanadium
Barium	Copper	Potassium	Zinc
Beryllium	Iron	Selenium	Zirconium
Bismuth	Lead	Silver	
Boron	Magnesium	Sodium	
Cadmium	Manganese	Thallium	

Source: From U.S. EPA, Road Map to Understanding Innovative Technology Options for Brownfields Investigation and Cleanup, EPA 542-B-05-001, 4th ed., U.S. EPA, Washington, DC, September 2005.

Locations where metals and metalloids may be found include artillery and small arms impact areas, battery disposal areas, burn pits, chemical disposal areas, contaminated marine sediments, disposal wells and leach fields, electroplating and metal finishing shops, firefighting training areas, landfills and burial pits, leaking storage tanks, radioactive and mixed waste disposal areas, oxidation ponds and lagoons, paint stripping and spray booth areas, sand blasting areas, surface impoundments, and vehicle maintenance areas. Typical metals and metalloids encountered at many sites include those listed in Table 11.1.

11.1.3 PURPOSE

U.S. EPA has developed a set of technical guides to assist communities, states, municipalities, and the private sector to more effectively address brownfield sites. Each guide in this series contains information on a different type of brownfield site (classified according to former industrial use). In addition, a supplementary guide contains information on cost-estimating tools and resources for brownfield sites.[4–6]

The overview of the technical process involved in assessing and cleaning up brownfield sites can assist planners in making decisions at various stages of the project. An understanding of land use and industrial processes conducted in the past at a site can help the planner to conceptualize the site and identify likely areas of contamination that may require cleanup. Numerous resources are suggested to facilitate characterization of the site and consideration of cleanup technologies.[2–6]

Specifically, the objective of this chapter is to provide decision-makers with the following:

1. An understanding of common industrial processes at metal finishing facilities and the relationship between such processes and potential releases of contaminants to the environment.
2. Information on the types of contaminants likely to be present at a metal finishing site.
3. A discussion of site assessment (also known as site characterization), screening and cleanup levels, and cleanup technologies that can be used to assess and cleanup the types of contaminants likely to be present at metal finishing sites.
4. A conceptual framework for identifying potential contaminants at the site, pathways by which contaminants may migrate offsite, and environmental and human health concerns.

5. Information on developing an appropriate cleanup plan for metal finishing sites where contamination levels must be reduced to allow a site's reuse.
6. A discussion of pertinent issues and factors should be considered when developing a site assessment and cleanup plan and selecting appropriate technologies for brownfields, given time and budget constraints.

11.2 INDUSTRIAL PROCESSES AND CONTAMINANTS AT METAL FINISHING SITES

Understanding the industrial processes used during a metal finishing facility's active life and the types of contaminants that may be present provides important information to guide planners in the assessment, cleanup, and restoration of the site to an acceptable condition for sale or reuse. This section provides a general overview of the processes, chemicals, and contaminants used or found at metal finishing sites. Specific metal finishing brownfields sites may have had a different combination of these processes, chemicals, and contaminants. Therefore, this information can be used only to develop a framework of likely past activities. Planners should obtain facility-specific information on industrial processes at their site whenever possible. Site-specific information is also important to obtain because the site may have been used for other industrial purposes at other times in the past.

This section describes waste-generating surface preparation operations; metal finishing operations and the types of wastestreams and specific contaminants associated with each process; auxiliary areas at metal finishing sites that may produce contaminants and nonprocess-related contamination problems associated with metal finishing sites. Figure 11.1 presents typical metal finishing processes and land areas, along with the types of wastestreams associated with each area.[7] Table 11.2 lists the specific contaminants associated with each wastestream.[2]

11.2.1 SURFACE PREPARATION OPERATIONS

Metal finishing processes are typically housed within one structure. The surface of metal products generally requires preparation (i.e., cleaning) prior to applying a finish. An initial set of degreasing tanks ([A] in Figure 11.1) are used to remove oils, grease, and other foreign matter from the surface of the metal so that a coating can be applied. Metal finishing facilities may use solvents or emulsion solutions (i.e., solvents dispersed in an aqueous medium with the aid of an emulsifying agent) in the degreasing tanks to clean and prepare the surfaces of metal parts. Wastewaters generated from cleaning operations are primarily rinsewaters, which are usually combined with other metal finishing wastewaters and treated onsite by conventional chemical precipitation. These wastewaters may contain solvents, as listed in Table 11.2. Solid wastes such as wastewater treatment sludges, still bottoms, and cleaning tank residues may also be generated.

11.2.2 METAL FINISHING OPERATIONS

Metal finishing operations are typically performed in a series of tanks (baths) followed by rinsing cycles. Acid or alkaline baths "pickle" the surface of the steel to improve the adherence of the coating. After the pickling baths, the metal products are moved to plating tanks, where the final coat is applied. Wastes generated during finishing operations derive from the solvents and cleansers applied to the surface and the metal-ion-bearing aqueous solutions used in acid/alkaline rinsing and bathing operations. Common metal finishing operations include anodizing, chemical conversion coating, electroplating, electroless plating, and painting. Common wastestreams include metals and acids in the wastewater; metals in sludges and solid waste; and solvents from painting operations, as listed in Table 11.2. If these wastes were managed or disposed of onsite, it is possible that pollutants were released into the environment. Even at facilities where wastes were not stored on site, releases may have occurred during the handling and use of chemicals. Metal finishing operations are described in the following.[2]

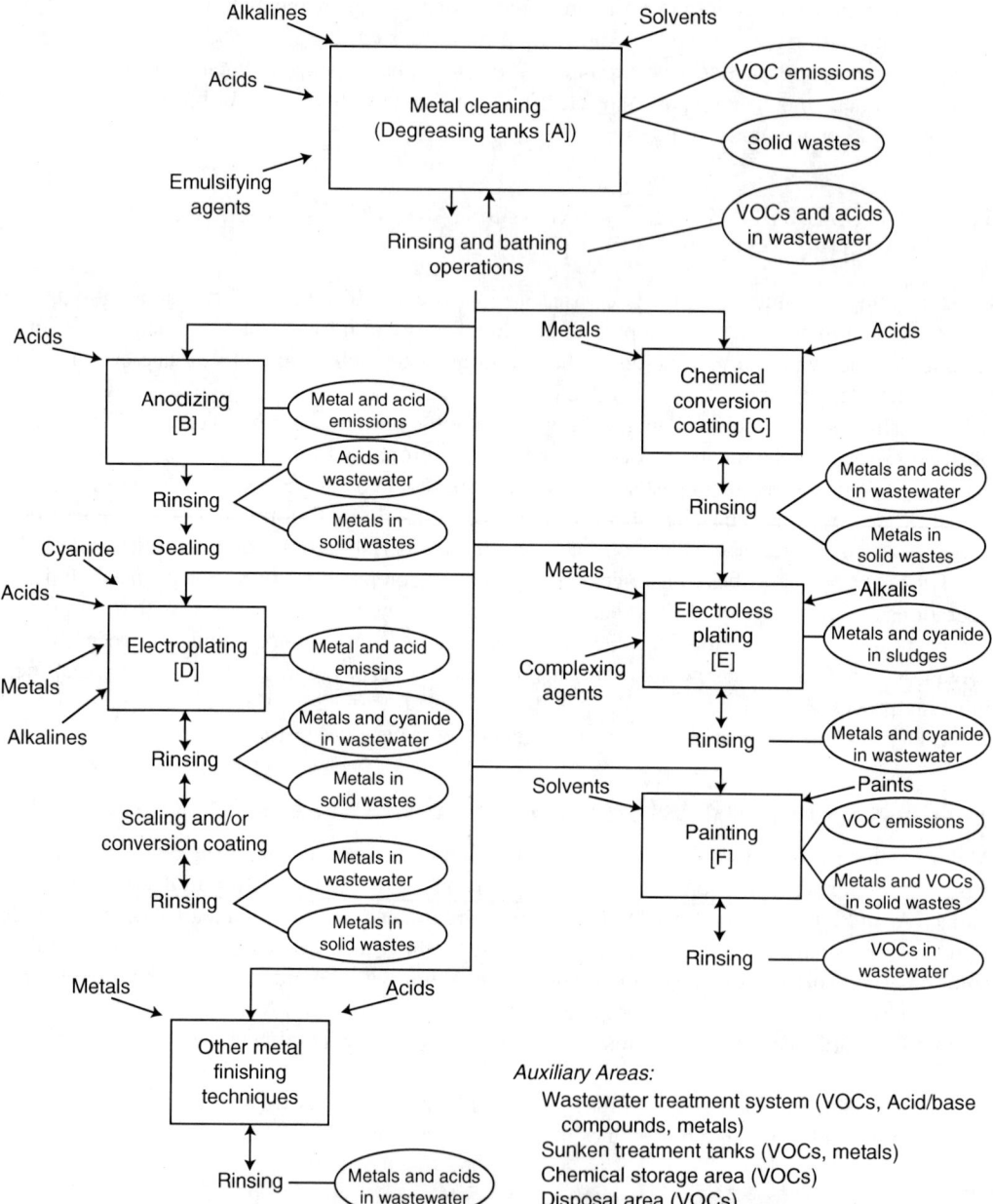

FIGURE 11.1 Typical metal finishing facility. (From U.S. EPA, Brownfields and Land Revitalization—Tools and Technical Information, U.S. EPA, Washington, DC, available at http://www.epa.gov/brownfields/toolsandtech.htm, 2007.)

11.2.2.1 Anodizing Operations

Anodizing is an electrolytic process that uses acids from the combined electrolytic solution/acid bath tank to convert the metal surface into an insoluble oxide coating ([B] in Figure 11.1). After anodizing, metal parts are typically rinsed and then sealed. Anodizing operations produce contaminated wastewaters and solid wastes.

TABLE 11.2
Common Contaminants at Metal Finishing Sites

Contaminant Group	Contaminant Name
Volatile organic compounds (VOCs)	Acetone, benzene, isopropyl alcohol, 2-dichlorobenzene, 4-trimethylbenzene, dichloromethane, ethyl benzene, freon 113, methanol, methyl isobutyl ketone, methyl ethyl ketone, phenol, tetrachloroethylene, toluene, trichloroethylene, xylene (mixed isomers)
Metals/inorganics	Aluminum, antimony, arsenic, asbestos (friable), barium, cadmium, chromium, cobalt, copper, lead, cyanide, manganese, mercury, nickel, silver, zinc
Acids	Hydrochloric acid, nitric acid, phosphoric acid, sulfuric acid

Source: From U.S. EPA, Technical Approaches to Characterizing and Cleaning up Metal Finishing Sites under the Brownfields Initiative, EPA/625/R-98/006, U.S. EPA, Cincinnati, OH, March 1999.

11.2.2.2 Chemical Conversion Coating

Chemical conversion coating ([C] in Figure 11.1) includes the following processes:

1. *Chromating.* Chromate conversion coatings are produced on various metals by chemical or electrochemical treatment. Acid solutions react with the metal surface to form a layer of a complex mixture of the constituent compounds, including chromium and the base metal.
2. *Phosphating.* Phosphate conversion coating involves the immersion of steel, iron, or zinc plated steel into a dilute solution of phosphate salts, phosphoric acid, and other reagents to condition the surfaces for further processing.
3. *Metal coloring.* Metal coloring involves chemically converting the metal surface into an oxide or similar metallic compound to produce a decorative finish.
4. *Passivating.* Passivating is the process of forming a protective film on metals by immersing them in an acid solution (usually nitric acid or nitric acid with sodium dichromate).

Pollutants associated with chemical conversion processes enter the wastestream through rinsing and batch dumping of process baths. Wastewaters containing chromium are usually pretreated; this process generates a sludge that is sent offsite for metals reclamation or disposal.

11.2.2.3 Electroplating

Electroplating is the production of a surface coating of one metal upon another by electrodeposition ([D] in Figure 11.1). In electroplating, metal ions (in either acid, alkaline, or neutral solutions) are reduced on the cathodic surfaces of the workpieces being plated. Electroplating operations produce contaminated wastewaters and solid wastes. Contaminated wastewaters result from workpiece rinsing and process cleanup waters. Rinsewaters from electroplating are usually combined with other metal finishing wastewaters and treated onsite by conventional chemical precipitation, which results in wastewater treatment sludges. Other wastes generated from electroplating include spent process solutions and quench baths, which may be discarded periodically when the concentrations of contaminants inhibit their proper function.

11.2.2.4 Electroless and Immersion Plating

Electroless plating involves chemically depositing a metal coating onto a plastic object by immersing the object in a plating solution ([E] in Figure 11.1). Immersion plating produces a thin metal deposit, commonly zinc or silver, by chemical displacement. Both produce contaminated wastewater and solid wastes. Facilities generally treat spent plating solutions and rinsewaters chemically to precipitate the toxic metals; however, some plating solutions can be difficult to treat because of the presence of chelates. Most waste sludges resulting from electroless and immersion plating contain significant concentrations of toxic metals.

11.2.2.5 Painting

Painting is the application of predominantly organic coatings for protective or decorative purposes ([F] in Figure 11.1). Paint is applied in various forms, including dry powder, solvent diluted formulations, and waterborne formulations, most commonly by spray painting and electrodeposition. Painting operations may result in solvent-containing waste and the direct release of solvents, paint sludge wastes, and paint-bearing wastewaters. Paint cleanup operations also may contribute to the release of chlorinated solvents. Discharge from water curtain booths generates the most wastewater. Onsite wastewater treatment processes generate a sludge that is taken offsite for disposal. Other sources of wastes include emission control devices (e.g., paint booth collection systems, ventilation filters) and discarded paints. Sandblasting may be performed to remove paint and to clean metal surfaces for painting or resurfacing; this practice may be of particular concern if the paint being removed contains lead.

11.2.2.6 Other Metal Finishing Techniques

Polishing, hot dip coating, and etching are other processes used to finish metal. Wastewaters are often generated during these processes. For example, after polishing operations, area cleaning and washdown can produce metal-bearing wastewaters. Hot dip coating techniques, such as galvanizing, use water for rinses following precleaning and for quenching after coating. Hot dip coatings also generate a solid waste, oxide dross that is periodically skimmed off the heated tank. Etching solutions are composed of strong acids or bases, which may result in etching solution wastes that contain metals and acids.

11.2.3 Auxiliary Activity Areas and Potential Contaminants

11.2.3.1 Wastewater Treatment

Many of the operations involved in metal finishing produce wastewaters, which usually are combined and treated onsite, often by conventional chemical precipitation. Even though the facility would have been required to meet state wastewater discharge standards before releasing wastes, spills of process wastewater may have occurred in the area. At abandoned sites, any remaining wastewater left in tanks or floor drains could contain solvents, metals, and acids, such as those listed in Table 11.2. In addition, it is possible that wastewater sludges, which can contain metals, were left at the site in baths or tanks.

11.2.3.2 Sunken Wastewater Treatment Tank

Some metal finishing facilities have wastewater treatment tanks sunk into the concrete slab to rest on the underlying soils. This is done by design to aid facility operators in accessing the tanks. If these tanks develop leaks, the lost material, which may contain VOCs and metals, may be released directly to the soils beneath the building.

11.2.3.3 Chemical Storage Area

At most metal finishing sites an area for storing chemicals used in the various operations was designated. Bulk containers stored in these areas may have leaked or spilled, resulting in discharges to floor drains or cracks in the floor. VOCs such as those listed in Table 11.2 may be found in such areas. Acids and alkaline reagents may also be found in this area.

11.2.3.4 Disposal Area

Materials, both liquid and solid, from process baths may have been disposed of at a designated area at the site. Such areas may be identified by stained soils or a lack of vegetation. These areas may contain VOCs, such as those listed in Table 11.2.

11.2.3.5 Other Considerations

Not all releases are related to the industrial processes described above. Some releases result from the associated services required to maintain the industrial processes. For example, electroplating facilities are large consumers of electricity, which requires a number of transformers. At older facilities, these transformers may have been disposed of in unmarked areas of the facility, which makes it difficult to know where leaks of polychlorinated biphenyl (PCB)-laden oils used as coolants may have occurred. Similarly, large machinery used to move metal pieces requires periodic maintenance. In the past, chemicals used for maintenance operations, such as solvents, oils, and grease, may have been flushed down drain and sumps after use. Stormwater runoff from paved areas such as parking lots may contain petroleum hydrocarbons and oils, which can contaminate areas located downgradient. When conducting initial site evaluations, planners should expand their investigations to include these types of activities.

In addition, metal finishing facilities may have been located in older buildings that contain lead paint and asbestos insulation and tiling. Any structure built before 1970 should be assessed for the presence of these materials. They can cause significant problems during demolition or renovation of the structures for reuse. Special handling and disposal requirements under state and federal laws can significantly increase the cost of construction.

11.3 SITE ASSESSMENT

The site investigation phase focuses on confirming whether any contamination exists at a site, locating any contamination, and characterizing the nature and extent of that contamination.[8] It is essential that an appropriately detailed study of the site be performed to identify the cause, nature, and extent of contamination and the possible threats to the environment or to any people living or working nearby. For brownfield sites, the results of such a study can be used in determining goals for cleanup, quantifying risks, determining acceptable and unacceptable risk, and developing effective cleanup plans that minimize delays or costs in the redevelopment and reuse of property. To ensure that sufficient information is obtained to support future decisions, the proposed cleanup measures and the proposed end use of the site should be considered when identifying data needs during the site investigation.[4]

The elements of a site assessment are designed to help planners build a conceptual framework of the facility, which will aid site characterization efforts.[9] The conceptual framework should identify the following[2]:

1. Potential contaminants that remain in and around the facility
2. Pathways along which contaminants may move
3. Potential risks to the environment and human health that exist along the migration pathways

This section highlights the key role that state environmental agencies usually play in brownfield projects. The types of information that planners should attempt to collect to characterize the site in a Phase I site assessment (i.e., the facility's history) are discussed. Information is presented about where to find and how to use this information to determine whether or not contamination is likely. Additionally, this section provides information to assist planners in conducting a Phase II site assessment, including sampling the site and determining the magnitude of contamination. Other considerations in assessing iron and steel sites are also discussed, and general sampling costs are included. The linking of the decision to be taken to the collected data and technologies is illustrated in Figure 11.2.

11.3.1 THE CENTRAL ROLE OF THE STATE AGENCIES

A brownfields redevelopment project involves partnerships among site planners (whether private or public sector), state and local officials, and the local community. State environmental agencies often are key decision-makers and a primary source of information for brownfields projects. Brownfield sites are generally cleaned up under state programs, particularly state voluntary cleanup or brownfields programs; thus, planners will need to work closely with state program managers to determine their particular state's requirements for brownfields development. Planners may also need to meet additional federal requirements. Key state functions include the following[2]:

1. Overseeing brownfield site assessment and cleanup processes, including the management of voluntary cleanup programs
2. Providing guidance on contaminant screening levels
3. Serving as a source of site information, as well as legal and technical guidance

11.3.1.1 State Voluntary Cleanup Programs

State Voluntary Cleanup Programs (VCPs) are designed to streamline brownfields redevelopment, reduce transaction costs, and provide state liability protection for past contamination. Planners should be aware that state cleanup requirements vary significantly and should contact the state brownfield manager; brownfields managers from state agencies will be able to identify their state requirements for planners and will clarify how their state requirements relate to federal requirements.

11.3.1.2 Levels of Contaminant Screening and Cleanup

Identifying the level of site contamination and determining the risk, if any, associated with that contamination level is a crucial step in determining whether cleanup is needed. Some state environmental agencies, as well as federal and regional U.S. EPA offices, have developed screening levels for certain contaminants, which are incorporated into some brownfields programs. Screening levels represent breakpoints in risk-based concentrations of chemicals in soil, air, or water. If contaminant concentrations are below the screening level, no action is required; above the level, further investigation is needed.

In addition to screening levels, U.S. EPA regional offices and some states have developed cleanup standards; if contaminant concentrations are above cleanup standards, cleanup must be pursued. The section on "Performing a Phase II Site Assessment" in this chapter provides more information on screening levels and the section on "Site Cleanup" provides more information on cleanup standards.

11.3.2 PERFORMING A PHASE I SITE ASSESSMENT: OBTAINING FACILITY BACKGROUND INFORMATION FROM EXISTING DATA

Planners should compile a history of the iron and steel manufacturing facility to identify likely site contaminants and their probable locations. Financial institutions typically require a Phase I site assessment prior to lending money to potential property buyers to protect the institution's role as

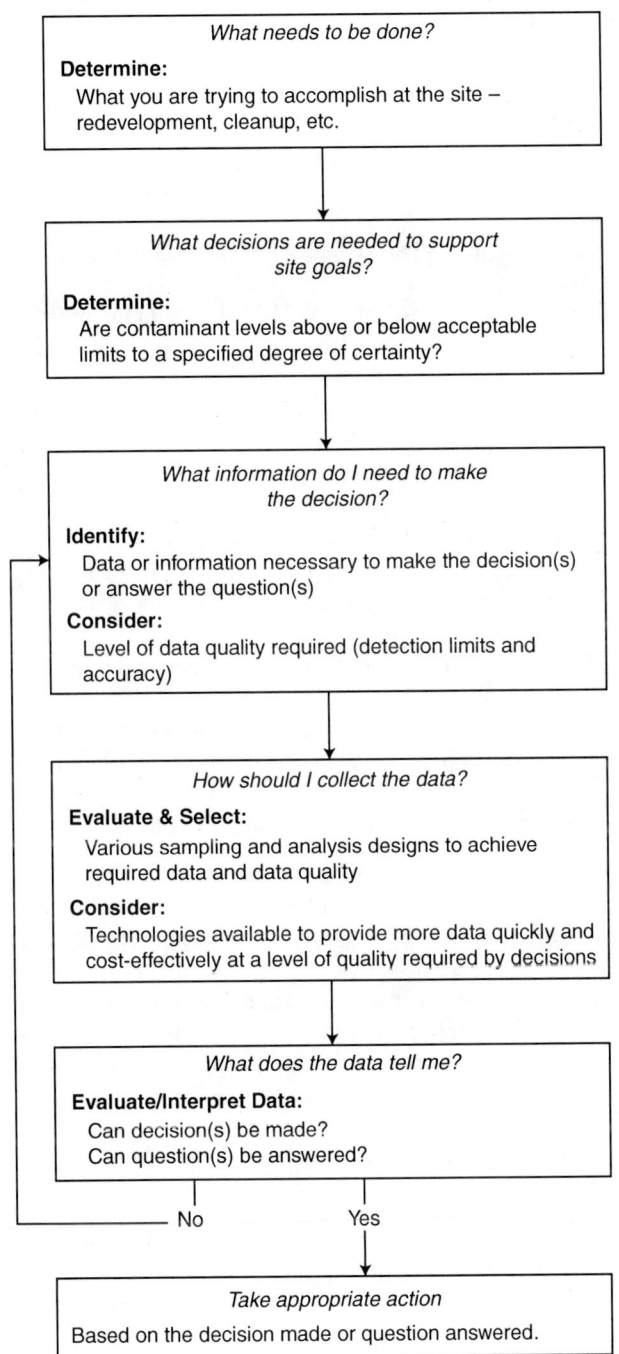

FIGURE 11.2 Linking the decision, data, and technology. (From U.S. EPA, Road Map to Understanding Innovative Technology Options for Brownfields Investigation and Cleanup, EPA 542-B-05-001, 4th ed., U.S. EPA, Washington, DC, September 2005.)

mortgage holder.[10] In addition, parties involved in the transfer, foreclosure, leasing, or marketing of properties recommend some form of site evaluation. The site history should include the following:

1. A review of readily available records (e.g., former site use, building plans, records of any prior contamination events)
2. A site visit to observe the areas used for various industrial processes and the condition of the property
3. Interviews with knowledgeable people (e.g., site owners, operators, and occupants; neighbors; local government officials)
4. A report that includes an assessment of the likelihood that contaminants are present at the site

The Phase I site assessment should be conducted by an environmental professional, and may take three to four weeks to complete. Site evaluations are required in part as a response to concerns over environmental liabilities associated with property ownership. A property owner needs to perform "due diligence," that is, fully enquire into the previous ownership and uses of a property to demonstrate that all reasonable efforts to find site contamination have been made. Because brownfield sites often contain low levels of contamination and pose low risks, due diligence through a Phase I site assessment will help to answer key questions about the levels of contamination. Several federal and state programs exist to minimize owner liability at brownfield sites and facilitate cleanup and redevelopment; planners should contact the state environmental or regional U.S. EPA office for further formation.

Information on how to review records, conduct site visits and interviews, and develop a report during a Phase I site assessment is provided below.

11.3.2.1 Facility Records

Facility records are often the best source of information on former site activities. If past owners are not initially known, a local records office should have deed books that contain ownership history. Generally, records pertaining specifically to the site in question are adequate for review purposes. In some cases, however, records of adjacent properties may also need to be reviewed to assess the possibility of contaminants migrating from or to the site, based on geologic or hydrogeologic conditions. If the brownfields property resides in a low-lying area, in close proximity to other industrial facilities or formerly industrialized sites, or downgradient from current or former industrialized sites, an investigation of adjacent properties is warranted.

11.3.2.2 Other Sources of Recorded Information

Planners may need to use other sources in addition to facility records to develop a complete history. ASTM Standard 1527 identifies standard sources such as historical aerial photographs, fire insurance maps, property tax files, recorded land title records, topographic maps, local street directories, building department records, zoning/land use records, and newspaper archives.[10]

Some metal finishing site managers may have worked with state environmental regulators; these offices may be key sources of information. Federal (e.g., U.S. EPA) records may also be useful. The types of information provided by regulators may include facility maps that identify activities and disposal areas, lists of stored pollutants, and the types and levels of pollutants released. State offices and other sources where planners can search for site-specific information are presented in the following:

1. The state offices responsible for industrial waste management and hazardous waste should have a record of any emergency removal actions at the site (e.g., the removal of leaking drums that posed an "imminent threat" to local residents); any Resource Conservation and

Cleanup of Metal Finishing Brownfield Sites

Recovery Act (RCRA)[11] permits issued at the site; notices of violations issued; and any environmental investigations.

2. The state office responsible for discharges of wastewater to water bodies under the National Pollutant Discharge Elimination System (NPDES)[12] program will have a record of any permits issued for discharges into surface water at or near the site. The local publicly owned treatment works (POTW) will have records for permits issued for indirect discharges into sewers (e.g., floor drain discharges to a sanitary sewer).
3. The state office responsible for underground storage tanks may also have records of tanks located at the site, as well as records of any past releases.
4. The state office responsible for air emissions may be able to provide information on air pollutants associated with particular types of onsite contamination.
5. U.S. EPA's Comprehensive Environmental Response, Compensation, and Liability Information System (CERCLIS)[13] of potentially contaminated sites should have a record of any previously reported contamination at or near the site.
6. U.S. EPA Regional Offices can provide records of sites that have hazardous substances. Information is available from the Federal National Priorities List (NPL) and lists of treatment, storage, and disposal (TSD) facilities subject to corrective action under RCRA. RCRA non-TSD facilities, RCRA generators, and Emergency Response Notification System (ERNS) information on contaminated or potentially contaminated sites can help to determine if neighboring facilities are recorded as having released hazardous substances into the immediate environment.
7. State and local records may indicate any permit violations or significant contaminant releases from or near the site.
8. Residents and former employees may be able to provide useful information on waste management practices, but these reports should be substantiated.
9. Local fire departments may have responded to emergency events at the facility. Fire departments or city halls may have fire insurance maps or other historical maps or data that indicate the location of hazardous waste storage areas at the site.
10. Local waste haulers may have records of the facility's disposal of hazardous or other waste materials.
11. Utility records.
12. Local building permits.

11.3.2.3 Identifying Migration Pathways and Potentially Exposed Populations

Offsite migration of contaminants may pose a risk to human health and the environment; planners should gather as much readily available information on the physical characteristics of the site as possible. Migration pathways, that is, soil, groundwater, and air, will depend on site-specific characteristics such as geology and the physical characteristics of the individual contaminants (e.g., mobility). Information on the physical characteristics of the general area can play an important role in identifying potential migration pathways and focusing environmental sampling activities, if needed. Planners should collect three types of information to obtain a better understanding of migration pathways, including topographic, soil and subsurface, and groundwater data, as described below.[14–17]

11.3.3 GATHERING TOPOGRAPHIC INFORMATION

In this preliminary investigation, topographic information will be helpful in determining whether the site may be subject to contamination by adjoining properties or may be the source of contamination of other properties. Topographic information will help planners identify low-lying areas of the facility where rain and snowmelt (and any contaminants in them) may collect and contribute both water and contaminants to the underlying aquifer or surface runoff to nearby areas. The U.S.

Geological Survey (USGS) of the Department of the Interior has topographic maps for nearly every part of the country.

11.3.4 GATHERING SOIL AND SUBSURFACE INFORMATION

Planners should know about the types of soils at the site from the ground surface extending down to the water table because soil characteristics play a large role in how contaminants move in the environment. For example, clay soils limit downward movement of pollutants into underlying groundwater but facilitate surface runoff. Sandy soils, on the other hand, can promote rapid infiltration into the water table while inhibiting surface runoff. Soil information can be obtained through a number of sources[2]:

1. Local planning agencies should have soil maps to support land use planning activities. These maps provide a general description of the soil types present within a county (or sometimes a smaller administrative unit, such as a township).
2. The Natural Resource Conservation Service and Co-operative Extension Service offices of the U.S. Department of Agriculture (USDA) are also likely to have soil maps.
3. Well-water companies are likely to be familiar with local subsurface conditions, and local water districts and state water divisions may have well-logging information.
4. Local health departments may be familiar with subsurface conditions because of their interest in septic drain fields.
5. Local construction contractors are likely to be familiar with subsurface conditions from their work with foundations.

Soil characteristics can vary widely within a relatively small area, and it is common to find that the top layer of soil in urban areas is composed of fill materials, not native soils. Although local soil maps and other general soil information can be used for screening purposes such as in a Phase I assessment, site-specific information will be needed in the event that cleanup is necessary.

11.3.5 GATHERING GROUNDWATER INFORMATION

Planners should obtain general groundwater information about the site area, including the following:

1. State classifications of underlying aquifers
2. Depth to the groundwater tables
3. Groundwater flow direction and rate

This information can be obtained by contacting state environmental agencies or from several local sources, including water authorities, well drilling companies, health departments, and Agricultural Extension and Natural Resource Conservation Service offices.

11.3.5.1 Identifying Potential Environmental and Human Health Concerns

Identifying possible environmental and human health risks early in the process can influence decisions regarding the viability of a site for cleanup and the choice of cleanup methods used. A visual inspection of the area will usually suffice to identify onsite or nearby wetlands and water bodies that may be particularly sensitive to releases of contaminants during characterization or cleanup activities. Planners should also review available information (e.g., from state and local environmental agencies) to ascertain the proximity of residential dwellings, nearby industrial/commercial activities, and wetlands/water bodies, and to identify people, animals, or plants that might receive migrating contamination; any particularly sensitive populations in the area (e.g., children; endangered species); and whether any major contamination events have occurred previously in the area (e.g., drinking water problems; groundwater contamination).

For environmental information, planners can contact the U.S. Army Corps of Engineers, state environmental agencies, local planning and conservation authorities, the USGS, and the USDA Natural Resource Conservation Service. State and local agencies and organizations can usually provide information on local fauna and the habitats of any sensitive and/or endangered species.

For human health information, planners can contact the following:

1. *State and local health assessment organizations.* Organizations such as health departments should have data on the quality of local well water used as a drinking water source, as well as any human health risk studies that have been conducted. In addition, these groups may have other relevant information, such as how certain types of contaminants (e.g., volatile organics such as benzene and phenols) might pose a health risk (e.g., dermal exposure to volatile organics during site characterization); information on exposures to particular contaminants and potential associated health risks can also be found in health profile documents developed by the Agency for Toxic Substances and Disease Registry (ATSDR). In addition, ATSDR may have conducted a health consultation or health assessment in the area if an environmental contamination event that may have posed a health risk occurred in the past; such an event and assessment should have been identified in the Phase I records review of prior contamination incidents at the site if any occurred.
2. *Local water and health departments.* During the site visit (described below), when visually inspecting the area around the facility, planners should identify any residential dwellings or commercial activities near the facility and evaluate whether people there may come into contact with contamination along one of the migration pathways. Where groundwater contamination may pose a problem, planners should identify any nearby waterways or aquifers that may be affected by groundwater discharge of contaminated water, including any drinking water wells that may be downgradient of the site, such as a municipal well field. Local water departments will have a count of well connections to the public water supply. Planners should also pay particular attention to information on private wells in the area downgradient of the facility, because, depending on their location, they may be vulnerable to contaminants migrating offsite even when the public municipal drinking water supply is not vulnerable. Local health departments often have information on the locations of private wells.

In addition to groundwater sources and migration pathways, surface water sources and pathways should be evaluated because groundwater and surface waters can interface at some (or several) point(s) in the region. Contaminants in groundwater can eventually migrate to surface waters, and contaminants in surface waters can migrate to groundwater.

11.3.5.2 Community Involvement

It is important that brownfields decision-makers encourage acceptance of redevelopment plans and cleanup alternatives by involving members of the community early in the decision-making process through community meetings, newsletters, or other outreach activities. For an individual site, the community should be informed about how the use of a proposed technology might affect redevelopment plans or the adjacent neighborhood.[4] For example, the planting of trees for the use of phytoremediation may create aesthetic or visual improvements; on the other hand, the use of phytoremediation may bring about issues related to site security or long-term maintenance that could affect access to the site.

Community-based organizations represent a wide range of issues, from environmental concerns to housing issues to economic development. These groups can often be helpful in educating planners and others in the community about local brownfield sites, which can contribute to successful brownfield site assessment and cleanup activities. In addition, most state voluntary cleanup programs require that local communities be adequately informed about brownfields cleanup activities. Planners can contact the local Chamber of Commerce, local philanthropic organizations, local service organizations, and neighborhood committees for community input. State and local environmental groups may be able to supply relevant information and identify other appropriate community organizations. Local community involvement in brownfields projects is a key component in the success of such projects.[2]

U.S. EPA can assist members of the brownfields community by directing its members to appropriate resources and providing opportunities to network and participate in the sharing of information. A number of internet sites, databases, newsletters, and reports provide opportunities for brownfields stakeholders to network with other stakeholders to identify information about cleanup and technology options. U.S. EPA's Brownfields and Land Revitalization Technology Support Center is a valuable resource for brownfields decision-makers.

11.3.5.3 Conducting a Site Visit

In addition to collecting and reviewing available records, planners need to conduct a site visit to visually and physically observe the uses and conditions of the property, including both outdoor areas and the interior of any structure or property. Current and past uses involving the use, treatment, storage, disposal, or generation of hazardous substances or petroleum products should be noted. Current or past uses of abutting properties that can be observed readily while conducting the site visit also should be noted. In addition, readily observable geologic, hydrologic, and topographic conditions should be identified, including any possibility of hazardous substances migrating on- or offsite.

Roads, water supplies, and wastewater systems should be identified, as well as any storage tanks, whether above or below ground. If any hazardous substances or petroleum products are found, their type, quantity, and storage conditions should be noted. Any odors, pools of liquids, drums or other containers, and equipment likely to contain PCBs should be noted. Additionally, indoors, heating and cooling systems should be noted, as well as any stains, corrosion, drains, or sumps. Outdoors, any pits, ponds, lagoons, stained soil or pavement, stressed vegetation, solid waste, wastewater, and wells should be noted.[10]

11.3.5.4 Conducting Interviews

In addition to reviewing available records and visiting the site, conducting interviews with the site owner or site manager, site occupants, and local officials is highly recommended to obtain information about the prior and current uses and conditions of the property, and to enquire about any useful documents that exist regarding the property. Such documents include environmental audit reports, environmental permits, registrations for storage tanks, material safety data sheets, community right-to-know plans, safety plans, government agency notices or correspondence, hazardous waste generator reports or notices, geotechnical studies, or any proceedings involving the property.[10] Interviews with at least one staff person from the following local government agencies are recommended: the fire department, health agency, and the agency with authority for hazardous waste disposal or other environmental matters. Interviews can be conducted in person, by telephone, or in writing.

ASTM standard 1528[18] provides a questionnaire that may be appropriate for use in interviews for certain sites. ASTM suggests that this questionnaire be posed to the current property owner, any major occupant of the property (or at least 10% of the occupants of the property if no major occupant exists), or "any occupant likely to be using, treating, generating, storing, or disposing of hazardous substances or petroleum products on or from the property." A user's guide accompanies the ASTM questionnaire to assist the investigator in conducting interviews, as well as researching records and making site visits.

11.3.5.5 Developing a Report

Toward the end of the Phase I assessment, planners should develop a report that includes all of the important information obtained during record reviews, the site visit, and interviews. Documentation,

Cleanup of Metal Finishing Brownfield Sites

such as references and important exhibits, should be included, as well as the credentials of the environmental professional that conducted the Phase I environmental site assessment. The report should include all information regarding the presence or likely presence of hazardous substances or petroleum products on the property and any conditions that indicate an existing, past, or potential release of such substances into property structures or into the ground, groundwater, or surface water of the property.[10] The report should include the environmental professional's opinion of the impact of the presence or likely presence of any contaminants, and a findings and conclusion section that either indicates that the Phase I environmental site assessment revealed no evidence of contaminants in connection with the property, or discusses what evidence of contamination was found.

Additional sections of the report might include a recommendations section (e.g., for a Phase II site assessment, if appropriate), and sections on the presence or absence of asbestos, lead paint, lead in drinking water, radon, and wetlands. Some states or financial institutions may require information on these substances.

If the Phase I site assessment adequately informs state and local officials, planners, community representatives, and other stakeholders that no contamination exists at the site, or that contamination is so minimal that it does not pose a health or environmental risk, then those involved may decide that adequate site assessment has been accomplished and the process of redevelopment may proceed. In some cases where evidence of contamination exists, stakeholders may decide that enough information is available from the Phase I site assessment to characterize the site and determine an appropriate approach for site cleanup of the contamination. In other cases, stakeholders may decide that additional site assessment is warranted, and a Phase II site assessment would be conducted.

11.3.6 THE TRIAD APPROACH: STREAMLINING SITE INVESTIGATIONS AND CLEANUP DECISIONS

The modernization of the collection, analysis, interpretation, and management of data to support decisions about hazardous waste sites rests on U.S. EPA's three-pronged or "triad" approach.[19–21] The introduction of new technologies in a dynamic framework allows project managers to meet clearly defined objectives. Such an approach incorporates the elements described below.[4,19–21]

Systematic planning is a common-sense approach to assuring that the level of detail in project planning matches the intended use of the data being collected. Once cleanup goals have been defined, systematic planning is undertaken to chart a course for the project that is resource effective, as well as technically sound and defensible to reach these project-critical goals. A team of multidisciplinary, experienced technical staff works to translate the project's goals into realistic technical objectives. The Conceptual Site Model (CSM) is the planning tool that organizes the information that is already known about the site; the CSM helps the team identify the additional information that must be obtained. The systematic planning process ties project goals to individual activities necessary to reach these goals by identifying data gaps in the CSM. The team then uses the CSM to direct the gathering of needed information, allowing the CSM to evolve and mature as work progresses at the site.

A dynamic working strategy approach relies on real-time data to reach decision points. The logic for decision-making is identified and responsibilities, authority, and lines of communication are established. Dynamic work strategy implementation relies on and is driven by critical project decisions needed to reach closure. It uses a decision tree and real-time uncertainty management practices to reach critical decision points in as few mobilizations as possible. Success of a dynamic approach depends on the presence of experienced staff in the field empowered to make decisions based on the decision logic and their capability to deal with new data and any unexpected issues, as they arise. Field staff maintains close communication with regulators or others overseeing the project during implementation of dynamic work plans.

The use of onsite analytical tools, rapid sampling platforms, and onsite interpretation and management of data makes dynamic work strategies possible. Such real-time measurement tools are among the key streamlined site investigation tools because they provide the data that are used for onsite decision-making. The tools are a broad category of analytical methods and equipment that can be applied at the sample collection site. They include methods that can be used outdoors with hand-held, portable equipment, as well as more rigorous methods that require the controlled environments of a mobile laboratory (transportable). During the planning process, the team identifies the type, rigor, and quantity of data needed to answer the questions raised by the CSM. Those decisions then guide the design sampling modifications and the selection of analytical tools.

The triad approach enables project managers to minimize uncertainty while expediting site cleanup and reducing project costs. For example, U.S. EPA collaborated with the Town of Greenwich, Connecticut, to implement the triad approach to characterize a former power plant site scheduled for redevelopment as a waterfront park. The triad approach yielded an estimated cost savings of 50 to 60% when compared with a traditional approach involving two mobilizations and comprehensive analytical methods at a fixed laboratory. The City of Trenton, New Jersey, began implementing the triad approach in 2001 as part of its program to redevelop a large number of abandoned industrial sites. Overall, the triad approach eliminated costs associated with follow-on investigation activities while accelerating the redevelopment schedule and reducing decision uncertainty. Additional details about these and other examples are available in the U.S. EPA's Technology News and Trends newsletter.[22]

11.3.7 Performing a Phase II Site Assessment: Sampling the Site

A Phase II site assessment[23] typically involves taking soil, water, and air samples to identify the types, quantity, and extent of contamination in these various environmental media. The types of data used in a Phase II site assessment can vary from existing site data (if adequate), to limited sampling of the site, to more extensive contaminant-specific or site-specific sampling data. Planners should use knowledge of past facility operations whenever possible to focus the site evaluation on those process areas where pollutants were stored, handled, used, or disposed. These will be the areas where potential contamination will be most readily identified. Generally, to minimize costs, a Phase II site assessment will begin with limited sampling (assuming readily available data do not exist that adequately characterize the type and extent of contamination on the site) and will proceed to more comprehensive sampling if needed (e.g., if the initial sampling could not identify the geographical limits of contamination).

This section explains the importance of setting Data Quality Objectives (DQOs) and provides brief guidance for doing so. It also describes screening levels to which sampling results can be compared, and provides an overview of environmental sampling and data analysis, including sampling methods and ways to increase data certainty.

11.3.7.1 Setting Data Quality Objectives

U.S. EPA has developed a guidance document that describes key principals and best practices for brownfield site assessment quality assurance and quality control based on program experience.[24–26]

U.S. EPA has adopted the DQO process[26] as a framework for making decisions. The DQO process is common-sense, systematic planning tool based on the scientific method. Using a systematic planning approach, such as the DQO process, ensures that the data collected to support defensible site decision-making will be of sufficient quality and quantity, as well as be generated through the most cost-effective means possible. DQOs, themselves, are statements that unambiguously communicate the following:

1. The study objective
2. The most appropriate type of data to collect
3. The most appropriate conditions under which to collect the data
4. The amount of uncertainty that will be tolerated when making decisions

It is important to understand the concept of uncertainty and its relationship to site decision-making.[27–29] Regulatory agencies, and the public they represent, want to be as confident as possible about the safety of reusing brownfield sites. Public acceptance of site decisions may depend on the site managers being able to scientifically document the adequacy of site decisions. During negotiations with stakeholders, effective communication about the tradeoffs between project costs and confidence in the site decision can help set the stage for a project's successful completion. When the limits on uncertainty (e.g., only a 5, 10, or 20% chance of a particular decision error is permitted) are clearly defined in the project, subsequent activities can be planned so that data collection efforts will be able to support those confidence goals in a resource-effective manner. On the one hand, a manager would like to reduce the chance of making a decision error as much as possible, but on the other hand, reducing the chance of making that decision error requires collecting more data, which, in itself, is a costly process.

Striking a balance between these two competing goals—more scientific certainty versus less cost—requires careful thought and planning, as well as the application of professional expertise.[27–29]

The following steps are involved in systematic planning:

1. *Agreement on intended land reuse.* All parties should agree early in the process on the intended reuse for the property, because the type of use may strongly influence the choice of assessment and cleanup approaches. For example, if the area is to be a park, removal of all contamination will most likely be needed. If the land will be used for a shopping center, with most of the land covered by buildings and parking lots, it may be appropriate to reduce, rather than totally remove, contaminants to specified levels (e.g., state cleanup levels; see "Site Cleanup" later in this chapter).

2. *Clarification of the objective of the site assessment.* What is the overall decision(s) that must be made for the site? Parties should agree on the purpose of the assessment. Is the objective to confirm that no contamination is present? Or is the goal to identify the type, level, and distribution of contamination above the levels that are specified, based on the intended land use. These are two fundamentally different goals that suggest different strategies. The costs associated with each approach will also vary.

    As noted above, parties should also agree on the total amount of uncertainty allowable in the overall decision(s). Conducting a risk assessment involves identifying the levels of uncertainty associated with characterization and cleanup decisions. A risk assessment involves identifying potential contaminants and analyzing the pathways through which people, other species of concern, or the environment can become exposed to those contaminants. Such an assessment can help identify the risks associated with varying the levels of acceptable uncertainty in the site decision and can provide decision-makers with greater confidence about their choice of land use decisions and the objective of the site assessment. If cleanup is required, a risk assessment can also help determine how clean the site needs to be, based on expected reuse (e.g., residential or industrial), to safeguard people from exposure to contaminants.

3. *Definition of the appropriate type(s) of data that will be needed to make an informed decision at the desired confidence level.* Parties should agree on the type of data to be collected by defining a preliminary list of suspected analytes, media, and analyte-specific action levels (screening levels). Define how the data will be used to make site decisions. For example, data values for a particular analyte may or may not be averaged across the site for the purposes of reaching a decision to proceed with work. Are there maximum values that a contaminant(s) cannot exceed? If found, will concentrations of contaminants above a certain action level (hotspots) be characterized and treated separately? These discussions should also address the types of analyses to be performed at different stages of the project. Planners and regulators can reach an agreement to focus initial characterization efforts in

those areas where the preliminary information indicates potential sources of contamination may be located. It may be appropriate to analyze for a broad class of contaminants by less expensive screening methods in the early stages of the project in order to limit the number of samples needing analysis by higher quality, more expensive methods later. Different types of data may be used at different stages of the project to support interim decisions that efficiently direct the course of the project as it moves forward.

4. *Determination of the most appropriate conditions under which to collect the data.* Parties should agree on the timing of sampling activities, as weather conditions can influence how representative the samples are of actual conditions.
5. *Identification of appropriate contingency plans/actions.* Certain aspects of the project may not develop as planned. Early recognition of this possibility can be a useful part of the DQO process. For example, planners, regulators, and other stakeholders can acknowledge that screening-level sampling may lead to the discovery of other contaminants on the site than were originally anticipated. During the DQO process, stakeholders may specify appropriate contingency actions to be taken in the event that contamination is found. Identifying contingency actions early in the project can help ensure that the project will proceed even in light of new developments. The use of a dynamic workplan combined with the use of rapid-turnaround field analytical methods can enable the project to move forward with a minimum of time delay and wasted effort.
6. *Development of a sampling and analysis plan that can meet the goals and permissible uncertainties described in the preceding steps.* The overall uncertainty in a site decision is a function of several factors: the number of samples across the site (the density of sample coverage), the heterogeneity of analytes from sample to sample (spatial variability of contaminant concentrations), and the accuracy of the analytical method(s). Studies have demonstrated that analytical variability tends to contribute much less to the uncertainty of site decisions than does sample variability due to matrix heterogeneity. Therefore, spending money to increase the sample density across the site will usually (for most contaminants) make a larger contribution to confidence in the site decision, and thus be more cost-effective, than will spending money to achieve the highest data quality possible, but a lower sampling density.

   Examples of important consideration for developing a sampling and analysis plan include the following:
   a) Determining the sampling location placement that can provide an estimate of the matrix heterogeneity and thus address the desired certainty. Is locating hotspots of a certain size important? Can composite sampling be used to increase coverage of the site (and decrease overall uncertainty due to sample heterogeneity) while lowering analytical costs?
   b) Evaluation of the available pool of analytical technologies/methods (both field methods and laboratory methods, which might be implemented in either a fixed or mobile laboratory) for those methods that can address the desired action levels (the analytical methods quantification limit should be well below the action level). Account for possible or expected matrix interferences when considering appropriate methods. Can field analytical methods produce data that will meet all of the desired goals when sampling uncertainty is also taken into account? Evaluate whether a combination of screening and definitive methods may produce a more cost-effective means to generate data. Can economy of scale be used? For example, the expense of a mobile laboratory is seldom cost-effective for a single small site, but might be cost-effective if several sites can be characterized sequentially by a single mobile laboratory.
   c) When the sampling procedures, sample preparation, and analytical methods have been selected, a quality control protocol should be designed for each procedure and method that ensures that the data generated will be of known, defensible quality.

Cleanup of Metal Finishing Brownfield Sites

7. *Through a number of iterations, refining of the sampling and analysis plan to one that can most cost-effectively address the decision-making needs of the site planner.*
8. *Frequent review of agreements.* As more information becomes available, some decisions that were based on earlier, limited information should be reviewed to see if they are still valid. If they are not, the parties can again use the DQO framework to revise and refine site assessment and cleanup goals and activities.

The data needed to support decision-making for brownfield sites are generally not complicated and are less extensive than those required for more heavily contaminated, higher-risk sites (e.g., Superfund sites). But data uncertainty may still be a concern at brownfield sites because knowledge of past activities at a site may be less than comprehensive, resulting in limited site characterization. Establishing DQOs can help address the issue of data uncertainty in such cases. Examples of DQOs include verifying the presence of soil contaminants, and assessing whether contaminant concentrations exceed screening levels.

11.3.7.2 Screening Levels

In the initial stages of a Phase II site assessment an appropriate set of screening levels for contaminants in soil, water, and air should be established. Screening levels are risk-based benchmarks that represent concentrations of chemicals in environmental media that do not pose an unacceptable risk. Sample analyses of soils, water, and air at the facility can be compared with these benchmarks. If onsite contaminant levels exceed the screening levels, further investigation will be needed to determine if and to what extent cleanup is appropriate.

Some states have developed generic screening levels (e.g., for industrial and residential use). These levels may not account for site-specific factors that affect the concentration or migration of contaminants. Alternatively, screening levels can be developed using site-specific factors. Although site-specific screening levels can more effectively incorporate elements unique to the site, developing site-specific standards is a time- and resource-intensive process. Planners should contact their state environmental offices and/or U.S. EPA regional offices for assistance in using screening levels and in developing site-specific screening levels.

Risk-based screening levels are based on calculations/models that determine the likelihood that exposure of a particular organism or plant to a particular level of a contaminant would result in a certain adverse effect. Risk-based screening levels have been developed for tap water, ambient air, fish, and soil. Some states or U.S. EPA regions also use regional background levels (or ranges) of contaminants in soil and maximum contaminant levels (MCLs) in water established under the Safe Drinking Water Act[30] as screening levels for some chemicals. In addition, some states and/or U.S. EPA regional offices have developed equations for converting soil screening levels to comparative levels for the analysis of air and groundwater.

When a contaminant concentration exceeds a screening level, further site assessment (such as sampling the site at strategic locations and/or performing more detailed analysis) is needed to determine that: (1) the concentration of the contaminant is relatively low and/or the extent of contamination is small and does not warrant cleanup for that particular chemical, or (2) the concentration or extent of contamination is high, and that site cleanup is needed.

Using state cleanup standards for an initial brownfields assessment may be beneficial if no industrial screening levels are available or if the site may be used for residential purposes. U.S. EPA's soil screening guidance is a tool developed by U.S. EPA to help standardize and accelerate the evaluation and cleanup of contaminated soils at sites on the NPL where future residential land use is anticipated. This guidance may be useful at corrective action or VCP (Voluntary Cleanup Program) sites where site conditions are similar. However, use of this guidance for sites where residential land use assumptions do not apply could result in overly conservative screening levels.

11.3.7.3 Environmental Sampling and Data Analysis

Environmental sampling and data analysis are integral parts of a Phase II site assessment process. Many different technologies are available to perform these activities, as discussed below.

11.3.7.4 Levels of Sampling and Analysis

There are two levels of sampling and analysis: screening and contaminant-specific. Planners are likely to use both at different stages of the site assessment:

1. *Screening.* Screening sampling and analysis use relatively low-cost technologies to take a limited number of samples at the most likely points of contamination and analyze them for a limited number of parameters. Screening analyses often test only for broad classes of contaminants, such as total petroleum hydrocarbons, rather than for specific contaminants, such as benzene or toluene. Screening is used to narrow the range of areas of potential contamination and reduce the number of samples requiring further, more costly, analysis. Screening is generally performed onsite, with a small percentage of samples (generally 10%) submitted to a state-approved laboratory for a full organic and inorganic screening analysis to validate or clarify the results obtained. Some geophysical methods are used in site assessments because they are noninvasive (i.e., do not disturb environmental media as sampling does). Geophysical methods are commonly used to detect underground objects that might exist at a site, such as underground storage tanks (USTs), dry wells, and drums. The two most common and cost-effective technologies used in geophysical surveys are ground-penetrating radar and electromagnetic.[31]
2. *Contaminant-specific.* For a more in-depth understanding of contamination at a site (e.g., when screening data are not detailed enough), it may be necessary to analyze samples for specific contaminants. With contaminant-specific sampling and analysis, the number of parameters analyzed is much greater than for screening level sampling, and analysis includes more accurate, higher-cost field and laboratory methods. Such analyses may take several weeks.

   Computerization, microfabrication, and biotechnology have permitted the recent development of analytical equipment that can be generated in the field, onsite in a mobile laboratory, and offsite in a laboratory. The same kind of equipment may be used in two or more locations.

11.3.8 INCREASING THE CERTAINTY OF SAMPLING RESULTS

One approach to reducing the level of uncertainty associated with site data is to implement a statistical sampling plan. Statistical sampling plans use statistical principles to determine the number of samples needed to accurately represent the contamination present. With the statistical sampling method, samples are usually analyzed with highly accurate laboratory or field technologies, which increase costs and take additional time. Using this approach, planners can negotiate with regulators and determine in advance specific measures of allowable uncertainty (e.g., an 80% level of confidence with a 25% allowable error).

Another approach to increasing the certainty of sampling results is to use lower-cost technologies with higher detection limits to collect a greater number of samples. This approach would provide a more comprehensive picture of contamination at the site, but with less detail regarding the specific contamination. Such an approach would not be recommended to identify the extent of contamination by a specific contaminant, such as benzene, but may be an excellent approach for defining the extent of contamination by total organic compounds with a strong degree of certainty. Planners will find that there is a tradeoff between scope and detail. Performing a limited number of detailed analyses provides good detail but less certainty about overall contamination, whereas performing a

Cleanup of Metal Finishing Brownfield Sites

larger number of general analyses provides less detail but improves the understanding and certainty of the scope of contamination.

11.3.9 SITE ASSESSMENT TECHNOLOGIES

This section discusses the differences between using field and laboratory technologies and provides an overview of applicable site assessment technologies.[32,33] In recent years, several innovative technologies that have been field-tested and applied to hazardous waste problems have emerged. In many cases, innovative technologies may cost less than conventional techniques and can successfully provide the needed data. Operating conditions may affect the cost and effectiveness of individual technologies.

11.3.9.1 Field versus Laboratory Analysis

The principal advantages of performing field sampling and field analysis are that results are immediately available and more samples can be taken during the same sampling event; also, sampling locations can be adjusted immediately to clarify the first round of sampling results if warranted. This approach may reduce costs associated with conducting additional sampling events after receipt of laboratory analysis. Field assessment methods have improved significantly over recent years; however, although many field technologies may be comparable to laboratory technologies, some field technologies may not detect contamination at levels as low as laboratory methods, and may not be contaminant-specific. To validate the field results or to gain more information on specific contaminants, a small percentage of the samples can be sent for laboratory analysis. The choice of sampling and analytical procedures should be based on DQOs established earlier in the process, which determine the quality (e.g., precision, level of detection) of the data needed to adequately evaluate site conditions and identify appropriate cleanup technologies.

11.3.9.2 Sample Collection and Analysis Technologies

Tables 11.3 and 11.4 list sample collection technologies for oil in subsurface and groundwater that may be appropriate for metal finishing brownfield sites. Technology selection depends on the medium being sampled and the type of analysis required, based on DQOs. Soil samples are generally collected using spoons, scoops, and shovels. The selection of a subsurface sample collection technology depends on the subsurface conditions (e.g., consolidated materials, bedrock), the required sampling depth and level of analysis, and the extent of sampling anticipated. For example, if subsequent sampling efforts are likely, then installing semipermanent well casings with a well drilling rig may be appropriate. If limited sampling is expected, direct push methods, such as cone penetrometers, may be more cost-effective. The types of contaminants will also play a key role in the selection of sampling methods, devices, containers, and preservation techniques.

Table 11.5 lists analytical technologies that may be appropriate for assessing metal finishing sites, the types of contamination they can measure, applicable environmental media, and the relative cost of each. The final two columns of the table contain the applicability (e.g., field or laboratory) of analytical methods and the technology's ability to generate quantitative versus qualitative results. Less expensive technologies that have rapid turnaround times and produce only qualitative results generally should be sufficient for many brownfield sites.

11.3.10 ADDITIONAL CONSIDERATIONS FOR ASSESSING METAL FINISHING SITES

When assessing a metal finishing brownfield site, planners should focus on the most likely areas of contamination. Although the specific locations vary from site to site, this section provides some general guidelines.

TABLE 11.3
Soil and Subsurface Sampling Tools

Technique/ Instrumentation	Media — Soil	Media — Groundwater	Relative Cost per Sample	Sample Quality
Drilling Methods				
Cable	×	×	Mid-range expensive	Soil properties will most likely be altered
Casing advancement	×	×	Most expensive	Soil properties will likely be altered
Direct air rotary with rotary hammer	×	×	Mid-range expensive	Soil properties will most likely be altered
Direct mud rotary	×	×	Mid-range expensive	Soil properties may be altered
Directional drilling	×	×	Most expensive	Soil properties may be altered
Hollow-stem auger	×	×	Mid-range expensive	Soil properties may be altered
Jetting methods	×	×	Least expensive	Soil properties may be altered
Rotary diamond drilling	×	×	Most expensive	Soil properties may be altered
Rotating core	×		Mid-range expensive	Soil properties may be altered
Solid flight and bucket augers	×	×	Mid-range expensive	Soil properties will likely be altered
Sonic drilling	×	×	Most expensive	Soil properties will most likely not be altered
Split and solid barrel	×		Least expensive	Soil properties may be altered
Thin-wall open tube	×		Mid-range expensive	Soil properties will most likely not be altered
Thin-wall piston/ specialized thin wall	×		Mid-range expensive	Soil properties will most likely not be altered
Direct Push Methods				
Cone penetrometer	×	×	Mid-range expensive	Soil properties may be altered
Driven wells		×	Mid-range expensive	Soil properties may be altered
Hand-held Methods				
Augers	×	×	Least expensive	Soil properties may be altered
Rotating core	×		Mid-range expensive	Soil properties may be altered
Scoop, spoons, and shovels	×		Least expensive	Soil properties may be altered
Split and solid barrel	×		Least expensive	Soil properties may be altered
Thin-wall open tube	×		Mid-range expensive	Soil properties will most likely not be altered
Thin-wall piston/ specialized thin wall	×		Mid-range expensive	Soil properties will most likely not be altered
Tubes	×		Least expensive	Soil properties will most likely not be altered

Source: From U.S. EPA, Technical Approaches to Characterizing and Cleaning up Metal Finishing Sites under the Brownfields Initiative, EPA/625/R-98/006, U.S. EPA, Cincinnati, OH, March 1999.

11.3.10.1 Where to Sample

Most metal finishing facilities perform all operations indoors. Consequently, most site assessment activities should focus on contamination inside and underneath the facility. Outdoor assessment activities should evaluate points where drain pipes may have carried contaminated wastewater or spilled materials.

The typical metal finishing facility comprises one or more large, warehouse-type buildings that contain the bath tanks, chemical storage areas, and wastewater treatment system. The floors are likely to be a continuous concrete slab containing several drains leading to a central storm drain or

TABLE 11.4
Groundwater Sampling Tools

Technique/Instrumentation	Contaminants	Relative Cost per Sample	Sample Quality
Portable Grab Samplers			
Bailers	Metals, VOCs	Least expensive	Liquid properties may be altered
Pneumatic depth-specific samplers	Metals, VOCs	Mid-range expensive	Liquid properties will most likely not be altered
Portable In Situ Groundwater Samplers/Sensors			
Cone penetrometer samplers	Metals, VOCs	Least expensive	Liquid properties will most likely not be altered
Direct drive samplers	Metals, VOCs	Least expensive	Liquid properties will most likely not be altered
Hydropunch	Metals, VOCs	Mid-range expensive	Liquid properties will most likely not be altered
Fixed Situ Samplers			
Multilevel capsule samplers	Metals, VOCs	Mid-range expensive	Liquid properties will most likely not be altered
Multiple-port casings	Metals, VOCs	Least expensive	Liquid properties will most likely not be altered
Passive multilayer samplers	VOCs	Least expensive	Liquid properties will most likely not be altered

Source: From U.S. EPA, Technical Approaches to Characterizing and Cleaning up Metal Finishing Sites under the Brownfields Initiative, EPA/625/R-98/006, U.S. EPA, Cincinnati, OH, March 1999.

sewer access. In older facilities, the feed lines from bath to wastewater tanks are under the floor slab. In newer facilities, the bath tanks and/or wastewater tanks will likely be partially submerged in the floor slab and positioned directly on the ground.

A visual inspection of the site should identify the most likely points of potential contaminant releases. These include the areas surrounding the following:

1. Floor drains in chemical storage and process bath areas
2. Sludges left in process bath and wastewater treatment tanks
3. Pipes underneath the floor slab
4. Tanks set through the floor slab
5. Cracks in the floor or stains in low spots in the floor

Solvents can be highly mobile on release, and can seep into and through concrete flooring, which is porous. The inspection of the facility floor should look not only for cracks through which solvents could migrate, but also for stained areas where spilled solvents may have pooled. Wipe samples should be taken along the walls of the facility, as solvent vapors may have penetrated wall materials.

As metal finishing operations are typically conducted inside the facility, outside points of potential release are likely to be limited to the following:

1. Points of discharge from effluent pipes
2. Waterways, canals, and ditches at points of pipe discharge
3. Areas where process bath materials may have been dumped

458 Waste Treatment in the Metal Manufacturing, Forming, Coating, and Finishing Industries

TABLE 11.5
Sample Analysis Technologies

Technique Instrumentation	Analytes	Soil	Media Ground Water	Gas	Relative Cost per Analysis	Application	Produces Quantitative Data
Laser-induced breakdown spectrometry	Metals	X			Least expensive	Usually used in field	Additional effort required
Titrimetry kits	Metals	X	X		Least expensive	Usually used in laboratory	Additional effort required
Particle-induced x-ray emissions	Metals	X	X		Mid-range expensive	Usually used in laboratory	Additional effort required
Atomic adsorption spectrometry	Metals	X*	X	X	Most expensive	Usually used in laboratory	Yes
Inductively coupled plasma-atomic emission spectroscopy	Metals	X	X	X	Most expensive	Usually used in laboratory	Yes
Field bioassessment	Metals	X	X		Most expensive	Usually used in field	No
X-ray fluorescence	Metals	X	X	X	Least expensive	Laboratory and field	Yes (limited)
Chemical calorimetric kits	VOCs	X	X		Least expensive	Can be used in field, usually used in laboratory	Additional effort required
Flame ionization detector (hand-held)	VOCs	X	X	X	Least expensive	Immediate, can be used in field	No
Explosimeter	VOCs	X	X*	X	Least expensive	Immediate, can be used in field	No
Photo ionization detector (hand-held)	VOCs	X	X	X	Least expensive	Immediate, can be used in field	No
Catalytic surface oxidation	VOCs	X*	X	X	Least expensive	Usually used in laboratory	No
Near IR reflectance/ trans spectroscopy	VOCs	X			Mid-range expensive	Usually used in laboratory	Additional effort required

Method	Analyte				Cost	Usage	Portable
Ion mobility spectrometer	VOCs	X*	X	X	Mid-range expensive	Usually used in laboratory	Yes
Raman spectroscopy/SERS	VOCs	X	X	X*	Mid-range expensive	Usually used in laboratory	Additional effort required
Infrared spectroscopy	VOCs	X	X	X	Mid-range expensive	Usually used in laboratory	Additional effort required
Scattering/absorption lidar	VOCs	X*	X	X	Mid-range expensive	Usually used in laboratory	Additional effort required
FTIR spectroscopy	VOCs	X	X	X	Mid-range expensive	Laboratory and field	Additional effort required
Synchronous luminescence/fluorescence	VOCs	X	X		Mid-range expensive	Usually used in laboratory, can be used in field	Additional effort required
Gas chromatography (GC) (can be used with numerous detectors)	VOCs	X*	X	X	Mid-range expensive	Usually used in laboratory, can be used in field	Yes
UV-visible spectrophotometry	VOCs	X	X	X	Mid-range expensive	Usually used in laboratory	Additional effort required
UV fluorescence	VOCs	X	X	X	Mid-range expensive	Usually used in laboratory	Additional effort required
Ion trap	VOCs	X	X*	X	Most expensive	Laboratory and field	Yes
Other chemical reaction-based test papers	VOCs, Metals	X	X		Least expensive	Usually used in field	Yes
Immunoassay and calorimetric kits	VOCs, Metals	X	X		Least expensive	Usually used in laboratory, can be used in filed	Additional effort required

VOCs, volatile organic compounds.

X*, indicates there must be extraction of the sample to gas or liquid phase.

Source: From U.S. EPA, Technical Approaches to Characterizing and Cleaning up Metal Finishing Sites under the Brownfields Initiative, EPA/625/R-98/006, U.S. EPA, Cincinnati, OH, March 1999.

Although discharge points may be visually obvious, areas of dumping may be less apparent. Often these areas are marked by stained soils and a lack of vegetation. Low-lying areas should also be investigated, as they make natural dumping areas and contaminants may drain to these points.

11.3.10.2   How Many Samples to Collect

Samples should be taken in and around the areas of potential release mentioned above.[34] Planners should expect that two to three samples will be required in each area, depending on DQOs. A cost-effective approach is to perform screening analyses using field methods on all samples and then to submit one sample to a laboratory for analysis by an accepted U.S. EPA method. Although the screening analyses can be conducted for broad contaminant groups, such as total organics, a contaminant-specific analysis should be conducted as a full screen for organic and inorganic contaminants and to validate the screening analyses. Contaminant-specific analyses may be conducted either in the field using appropriate technologies and protocols or in a laboratory.

11.3.10.3   What Types of Analysis to Perform

The selection of analytical procedures will be based on the DQOs established. Generally, the following analyses may be appropriate at metal finishing sites:

1. Residuals taken from drain sumps in storage areas should be screened for total organics and acids. Screening analyses for these contaminants can be performed inexpensively using a photo-ionization detector (PID) or flame ionization detector (FID) for total organics.
2. Residuals taken from drains in the process and wastewater treatment areas should be screened for a similar range of organic contaminants, but additional analyses should be performed to screen for the presence of inorganic contaminants, such as the metals used in the metal finishing process. Immunoassays are an inexpensive field technology that can be used to perform the screening analyses for organic contaminants and mercury. X-ray fluorescence (XRF) is another innovative technology that can be used to perform either field or laboratory analyses.
3. Soil gas should be collected at points underneath the floor slab, particularly near any tanks that are set through the floor slab, to detect the presence of solvents and other organic contaminants. These samples can be analyzed with the PID/FID technology described above. Corings of the floor slab may need to be taken and sent to a laboratory to determine if contaminants have penetrated floor slabs.
4. Wipe samples taken from walls should be analyzed for organic compounds. These analyses can be performed using the same technologies that are used to analyze residuals samples.
5. Soils and sediments at points of pipe discharge should be screened for both organic and inorganic contaminants using the PID/FID technology. XRF can be used for field or laboratory analyses.
6. Water samples collected in swales, canals, and ditches should be screened for organics. Inorganic contamination can sometimes be detected in water samples, but conditions do not always allow it.

In addition, as discussed earlier, many older structures contain lead paint and asbestos insulation and tiling. Numerous kits are readily available to test for lead paint. Experienced professionals may be able to visually identify asbestos insulation, but specialized equipment may be needed to confirm the presence of asbestos in other areas. Core or wipe samples can be analyzed for asbestos using polarized light microscopy (PLM). Local and state laws regarding lead and asbestos should be consulted to determine how they may affect the selection of DQOs, sampling, and analysis.

11.3.11 GENERAL SAMPLING COSTS

Site assessment costs vary widely, depending on the nature and extent of the contamination and the size of the sampling area. The sample collection costs discussed below are based on an assumed labor rate of USD 40/hr plus USD 12 per sample for shipping and handling. All costs have been updated to 2007 USD using U.S. ACE Yearly Average Cost Index for Utilities.[36]

11.3.11.1 Soil Collection Costs

Surface soil samples can be collected with tools as simple as a stainless steel spoon, shovel, or hand auger. Samples can be collected using hand tools in soft soil for as low as USD 12 per sample (assuming that a field technician can collect 10 samples/hr). When soils are hard, or deeper samples are required, a hammer-driven split spoon sampler or a direct push rig is needed. Using a drill rig equipped with a split spoon sampler or a direct push rig typically costs more than USD 700/d for rig operation,[35] with the cost per sample exceeding USD 35 (assuming that a field technician can collect 2 samples/hr). Labor costs generally increase when heavy machinery is needed.

11.3.11.2 Groundwater Sampling Costs

Groundwater samples can be extracted through conventional drilling of a permanent monitoring well or using the direct push methods listed in Table 11.3. The conventional, hollow stem auger-drilled monitoring well is more widely accepted but generally takes more time than direct push methods. Typical quality assurance protocols for the conventional monitoring well require the well to be drilled, developed, and allowed to achieve equilibrium for 24 to 48 hr. After the development period, a groundwater sample is extracted. With the direct push sampling method, a probe is either hydraulically pressed or vibrated into the ground, and groundwater percolates into a sampling container attached to the probe. The direct push method costs are contingent upon the hardness of the subsurface, depth to the water table, and permeability of the aquifer. Costs for both conventional and direct push techniques are generally more than USD 47 per sample (assuming that a field technician can collect 1 sample/hr); well installation costs must be added to that number.

11.3.11.3 Surface Water and Sediment Sampling Costs

Surface water and sediment sampling costs depend on the location and depth of the required samples. Obtaining surface water and sediment samples can cost as little as USD 35 per sample (assuming that a field technician can collect 2 samples/hr). Sampling sediment in deep water or sampling a deep level of surface water, however, requires the use of larger equipment, which drives up the cost. Also, if surface water presents a hazard during sampling and protective measures are required, costs will increase greatly.

11.3.11.4 Sample Analysis Costs

Costs for analyzing samples in any medium can range from as little as USD 32 per sample for a relatively simple test (e.g., an immunoassay test for metals) to more than USD 470 per sample for a more extensive analysis (e.g., for semivolatiles) and up to USD 1400 per sample for dioxins.[32] Major factors that affect the cost of sample analysis include the type of analytical technology used, the level of expertise needed to interpret the results, and the number of samples to be analyzed. Planners should make sure that laboratories that have been certified by state programs are used.

For information on costs for brownfields cleanup, the reader is referred to U.S. EPA document,[37] guide,[38] and remediation cost compendium.[39]

11.4 SITE CLEANUP

The purpose of this section is to guide planners in the selection of appropriate cleanup technologies. The principal factors that will influence the selection of a cleanup technology include the following[2]:

1. Types of contamination present
2. Cleanup and reuse goals
3. Length of time required to reach cleanup goals
4. Post-treatment care needed
5. Budget

The selection of appropriate cleanup technologies often involves a tradeoff between time and cost. The U.S. EPA document on cost-estimating tools and resources[37] provides information on cost factors and developing cost estimates. In general, the more intensive the cleanup approach, the more quickly the contamination will be mitigated and the more costly the effort. In the case of brownfields cleanup, this can be a major point of concern, considering the planner's desire to return the facility to the point of reuse as quickly as possible. Thus, the planner may wish to explore a number of options and weigh carefully the costs and benefits of each. One effective method of comparison is the cleanup plan, as discussed below. Planners should involve stakeholders in the community in the development of the cleanup plan.

The intended future use of a brownfield site will drive the level of cleanup needed to make the site safe for redevelopment and reuse. Brownfield sites are by definition not Superfund NPL (National Priorities List) sites; that is, brownfield sites usually have lower levels of contamination present and therefore generally require less extensive cleanup efforts than Superfund NPL sites. Nevertheless, all potential pathways of exposure, based on the intended reuse of the site, must be addressed in the site assessment and cleanup; if no pathways of exposure exist, less cleanup (or possibly none) may be required.

Some regional U.S. EPA and state offices have developed cleanup standards for different chemicals, which may serve as guidelines or legal requirements for cleanups. It is important to understand that screening levels are different from cleanup levels. Screening levels indicate whether further site investigation is warranted for a particular contaminant. Cleanup levels indicate whether cleanup action is needed and how extensive it needs to be. Planners should check with their state environmental office for guidance and requirements for cleanup standards.

This section contains information on developing a cleanup plan; various alternatives for addressing contamination at the site (i.e., institutional controls and containment and cleanup technologies); using different technologies for cleaning up metal finishing sites; and postconstruction issues that planners need to consider when considering alternatives.

11.4.1 Developing a Cleanup

If the results of the site evaluation indicate the presence of contamination above acceptable levels, planners will need to have a cleanup plan developed by a professional environmental engineer that describes the approach that will be used to contain and possibly clean up the contamination present at the site. In developing this plan, planners and their engineers should consider a range of possible options, with the intent of identifying the most cost-effective approaches for cleaning up the site, given time, and cost concerns. The cleanup plan can include the following elements[2,4,40,41]:

1. A clear delineation of environmental concerns at the site. Areas should be discussed separately if the cleanup approach for an area is different than that for other areas of the site. Clear documentation of existing conditions at the site and a summarized assessment of the nature and scope of contamination should be included.

Cleanup of Metal Finishing Brownfield Sites

2. A recommended cleanup approach for each environmental concern that takes into account expected land reuse plans and the adequacy of the technology selected.
3. A cost estimate that reflects both expected capital and operating/maintenance costs.
4. Postconstruction maintenance requirements for the recommended approach.
5. A discussion of the assumptions made to support the recommended cleanup approach, as well as the limitations of the approach.

Planners can use the framework developed during the initial site evaluation and the controls and technologies described below to compare the effectiveness of the least costly approaches for meeting the required cleanup goals established in the DQOs. These goals should be established at levels that are consistent with the expected reuse plans. A final cleanup plan may include a combination of actions, such as institutional controls, containment technologies, and cleanup technologies, as discussed below.

11.4.1.1 Institutional Controls

Institutional controls may play an important role in returning a metal finishing brownfield site to a marketable condition. Institutional controls are mechanisms that control the current and future use of, and access to, a site. They are established, in the case of brownfields, to protect people from possible contamination. Institutional controls can range from a security fence prohibiting access to a certain portion of the site to deed restrictions imposed on the future use of the site. If the overall cleanup approach does not include the complete cleanup of the facility (i.e., the complete removal or destruction of onsite contamination), a deed restriction will likely be required that clearly states that hazardous waste is being left in place within the site boundaries. Many state brownfields programs include institutional controls.

11.4.1.2 Containment Technologies

Containment technologies, in many instances, will be the likely cleanup approach for landfilled waste and wastewater lagoons (after contaminated wastewaters have been removed) at metal finishing facilities. The purpose of containment is to reduce the potential for offsite migration of contaminants and, possible subsequent exposure. Containment technologies include engineered barriers such as caps[42] for contaminated soils, slurry walls,[43] and hydraulic containment. Often, soils contaminated with metals can be solidified[44,45] by mixing them with cement-like materials, and the resulting stabilized material can be stored onsite in a landfill. Like institutional controls, containment technologies do not remove or destroy contamination, but mitigate potential risk by limiting access to it.

If contamination is found underneath the floor slab at metal finishing facilities, leaving the contaminated materials in place and repairing any damage to the floor slab may be justified. The likelihood that such an approach will be acceptable to regulators will depend on whether potential risk can be mitigated and managed effectively over the long term. In determining whether containment is feasible, planners should consider[2,4] the following:

1. *Depth to groundwater.* Planners should be prepared to prove to regulators that groundwater levels will not rise, due to seasonal conditions, and come into contact with contaminated soils.
2. *Soil types.* If contaminants are left in place, the native soils should not be highly porous, as are sandy or gravelly soils, which enables contaminants to migrate easily. Clay and fine silty soils provide a much better barrier.
3. *Surface water control.* Planners should be prepared to prove to regulators that rainwater and snowmelt cannot infiltrate under the floor slab and flush the contaminants downward.

4. *Volatilization of organic contaminants.* Regulators are likely to require that air monitors be placed inside the building to monitor the level of organics that may be escaping upward through the floor and drains.

11.4.1.3 Types of Cleanup Technologies

Cleanup may be required to remove or destroy onsite contamination if regulators are unwilling to accept the level of contamination present or if the types of contamination are not conducive to the use of institutional controls or containment technologies. Cleanup technologies fall broadly into two categories—ex situ and in situ, as described below.

1. *Ex situ.* An ex situ technology treats contaminated materials after they have been removed and transported to another location. After treatment, if the remaining materials, or residuals, meet cleanup goals, they can be returned to the site. If the residuals do not yet meet cleanup goals, they can be subjected to further treatment, contained onsite, or moved to another location for storage or further treatment. A cost-effective approach to cleaning up a metal finishing brownfield site may be the partial treatment of contaminated soils or groundwater, followed by containment, storage, or further treatment offsite.[2] For example, it is common practice for operating metal finishing facilities to treat wastewaters to an intermediate level and then send the treated water to the local POTW.
2. *In situ.* The use of in situ technologies has increased dramatically in recent years. In situ technologies treat contamination in place and are often innovative technologies. Examples of in situ technologies include bioremediation,[46] soil flushing,[47] oxygen releasing compounds,[48] air sparging,[49] and treatment walls.[50] In some cases, in situ technologies are feasible, cost-effective choices for the types of contamination that are likely at metal finishing sites. Planners, however, do need to be aware that cleanup with in situ technologies is likely to take longer than with ex situ technologies.

Maintenance requirements associated with in situ technologies depend on the technology used and vary widely in both effort and cost. For example, containment technologies such as caps and liners will require regular maintenance, such as maintaining the vegetative cover and performing periodic inspections to ensure the long-term integrity of the cover system. Groundwater treatment systems will require varying levels of postcleanup care. If an ex situ system is in use at the site, it will require regular operations support and periodic maintenance to ensure that the system is operating as designed.

11.4.2 Keys to Technology Selection and Acceptance

Innovative technologies and technology approaches offer many advantages in the cleanup of brownfield sites.[51–56] Stakeholders in such sites, however, first must accept the technology. Brownfields decision-makers should consider the following elements to increase the likelihood that the technology will be accepted, thereby facilitating the cleanup of the site:[4]

1. *Focus on the decisions that support site goals.* The triad approach of systematic planning is an important element of all cleanup activities. Clear and specific planning to meet explicit decision objectives is essential in managing the process of cleaning up contaminated sites: site assessment, site investigation, site monitoring, and remedy selection. With good planning, brownfields decision-makers can establish the cleanup goals for the site, identify the decisions necessary to achieve those goals, and develop and implement a strategy for addressing the decision needs. Technology decisions are made in the context of the requirements for such decisions. All cleanup activities are driven by the project goals.

An explicit statement of the decisions to be made and the way in which the planned approach supports the decisions should be included in the work plan.
2. *Build consensus.* Investing time, before the site work begins, in developing decisions that are acceptable to all decision-makers will foster more efficient site activities and make successful cleanup more likely. Conversely, allowing work to begin at a site before a common understanding and acceptance of the decisions have been established increases the likelihood that the cleanup process will be inefficient, resulting in delays and inefficient use of time and money. Further, decision-makers must understand that there is uncertainty in all scientific and technical decisions. Clearly defining and accepting uncertainty thresholds before making decisions about the site remedy will build consensus. Decisions also should be made in the context of applicable regulatory requirements, political considerations, budget available for the project, and time constraints.
3. *Understand the technology.* A thorough knowledge of a technology's capabilities and limitations is necessary to secure its acceptance. All technologies are subject to limitations in performance. Planning for the strengths and weaknesses of a technology maximizes understanding of its benefits and its acceptance. "Technology approvers," typically regulators, community groups, and financial service providers are likely to be more receptive of a new approach if the proposer provides a clear explanation of the rationale for its use and demonstrates confidence in its applicability to specific site conditions and needs. This latter point underscores the importance of carefully selecting an experienced, multidimensional team of professionals who have the expertise necessary to plan, present, and implement the chosen approach.
4. *Allow flexibility.* Streamlining site activities, whether site assessment, site investigation, removal, treatment, or monitoring, requires a flexible approach. Site-specific conditions, including various physical conditions, contamination issues, stakeholder needs, uses of the site, and supporting decisions, require that all decision-makers understand the need for flexibility. Although presumptive remedies, standard methods, applications at other sites, and program guidance can serve as the basis for designing a site-specific cleanup plan and can help decision-makers avoid "starting from scratch" at each site, decision-makers should be wary of depending too heavily on "boilerplate language" and prescriptive methodologies, as well as standard operating procedures and "accepted" methods. Although such tools provide excellent starting points, they lack the flexibility to meet site-specific goals. To ensure an efficient and effective cleanup, the actual technology approach, whether established or innovative, must focus on decisions specific to the site.
5. *Narrow the list of potential technologies that are most appropriate for addressing the contamination identified at the site and that are compatible with the specific conditions of the site and the proposed reuse of the property:*
   a) Network with other brownfields stakeholders and environmental professionals to learn about their experiences and to tap their expertise
   b) Determine whether sufficient data are available to support identification and evaluation of cleanup alternatives
   c) Evaluate the options against a number of factors, including toxicity levels, exposure pathways, associated risks, future land use, and economic considerations
   d) Analyze the applicability of a particular technology to the contamination identified at a site
   e) Determine the effects of various technology alternatives on redevelopment objectives
6. *Continue to work with appropriate regulatory agencies to ensure that regulatory requirements are addressed properly*:
   a) Consult with the appropriate federal, state, local, and tribal regulatory agencies to include them in the decision-making process as early as possible

b) Contact the U.S. EPA regional brownfields coordinator to identify and determine the availability of U.S. EPA support programs
7. *Integrate cleanup alternatives with reuse alternatives to identify potential constraints on reuse and time schedules and to assess cost and risk factors.*
8. *To provide a measure of certainty and stability to the project, investigate environmental insurance policies.* These include protection against cost overruns, undiscovered contamination, and third-party litigation, and their cost should be integrated into the project financial package.
9. *Select an acceptable remedy.* This should not only achieve cleanup goals and address the risk of contamination, but also best meet the objectives for redevelopment and reuse of the property and be compatible with the needs of the community.
10. *Communicate information.* This includes information about the proposed cleanup option to brownfields stakeholders, including the affected community.

11.4.3 Summary of Technologies for Treating Metals/Metalloids at Brownfield Sites

1. *Chemical treatment.* Also known as chemical reduction/oxidation (redox),[48] it typically involves redox reactions that chemically convert hazardous contaminants into compounds that are nonhazardous, less toxic, more stable, less mobile, or inert. Redox reactions involve the transfer of electrons from one compound to another. Specifically, one reactant is oxidized (loses electrons) and one reactant is reduced (gains electrons). The oxidizing agents used for treatment of hazardous contaminants in soil include ozone, hydrogen peroxide, hypochlorites, potassium permanganate, Fenton's reagent (hydrogen peroxide and iron), chlorine, and chlorine dioxide. This method may be applied in situ or ex situ to soils, sludges, sediments, and other solids and may also be applied to groundwater in situ or ex situ chemical treatment using pump and treat technology. Chemical treatment may also include use of ultraviolet (UV) light in a process known as UV oxidation.
2. *Electrokinetics.* This is based on the theory that a low-density current will mobilize contaminants in the form of charged species.[57,58] A current passed between electrodes is intended to cause aqueous media, ions, and particulates to move through soil, waste, and water. Contaminants arriving at the electrodes can be removed by means of electroplating or electrodeposition, precipitation or coprecipitation, adsorption, complexing with ion exchange resins, or pumping of water (or other fluid) near the electrodes.
3. *Flushing.* For flushing, a solution of water, surfactants, or cosolvents is applied to soil or injected into the subsurface to treat contaminated soil or groundwater.[47] When soil is being treated, injection is often designed to raise the water table into the contaminated soil zone. Injected water and treatment agents are recovered together with flushed contaminants.
4. *Permeable reactive barriers.* These are also known as passive treatment walls and are installed across the flow path of a contaminated groundwater plume, allowing the water portion of the plume to flow through the wall.[50] These barriers allow the passage of water while prohibiting movement of contaminants by means of treatment agents within the wall such as zero-valent metals (usually zero-valent iron), chelators, sorbents, compost, and microbes. The contaminants are either degraded or retained in a concentrated form by the barrier material, which may need to be replaced periodically.
5. *Physical separation.* These processes use physical properties to separate contaminated and uncontaminated media or to separate different types of media.[58–60] For example, different-sized sieves and screens can be used to separate contaminated soil from relatively uncontaminated debris. Another application of physical separation is dewatering of sediments or sludge.

6. *Phytoremediation*. This is a process in which plants are used to remove, transfer, stabilize, or destroy contaminants in soil, sediment, or groundwater. The mechanisms of phytoremediation include enhanced rhizosphere biodegradation (which takes place in soil or groundwater immediately around plant roots), phytoextraction (also known as phytoaccumulation, the uptake of contaminants by plant roots and the translocation and accumulation of contaminants into plant shoots and leaves), phytodegradation (metabolism of contaminants within plant tissues), and phytostabilization (production of chemical compounds by plants to immobilize contaminants at the interface of roots and soil). The term phytoremediation applies to all biological, chemical, and physical processes that are influenced by plants (including the rhizosphere) and that aid in the cleanup of contaminated substances.[61–64] Phytoremediation may be applied in situ or ex situ to soils, sludges, sediments, other solids, or groundwater. Environment Canada[64] studied the effectiveness of phytoremediation in Quebec's climate using herbaceous plants (Indian mustard and fescue) and shrubs (willow) to absorb heavy metals (lead, copper, and zinc). They reported that metal concentration levels in the leaves reached 1500 to 2300 mg/kg and resulted in total extraction of between 2 and 13 kg of metal per ha, per growth period.
7. *Pump and treat*. This involves extraction of groundwater from an aquifer and treatment of the water above the ground. The extraction step is usually conducted by pumping groundwater from a well or trench.[65] The treatment step can involve a variety of technologies such as adsorption, air stripping, bioremediation, chemical treatment, filtration, ion exchange, metal precipitation, and membrane filtration.[58–60]
8. *Soil washing*. For soil washing, contaminants sorbed onto fine soil particles are separated from bulk soil in a water-based system based on particle size.[66] The washwater may be augmented with a basic leaching agent, surfactant, or chelating agent or by adjustment of pH to help remove contaminants. Soils and wash water are mixed ex situ in a tank or other treatment unit. The washwater and various soil fractions are usually separated by means of gravity settling.[58]
9. *Solidification/stabilization* (S/S). This reduces the mobility of hazardous substances and contaminants in the environment through both physical and chemical means.[44,45] The S/S process physically binds or encloses contaminants within a stabilized mass. S/S can be performed both ex situ and in situ. Ex situ S/S requires excavation of the material to be treated, and the treated material must be disposed of. In situ S/S involves the use of auger or caisson systems and injector head systems to add binders to contaminated soil or waste without excavation, and the treated material is left in place.[67,68]
10. *Solvent extraction*. This involves the use of an organic solvent as an extractant to separate contaminants from soil. The organic solvent is mixed with contaminated soil in an extraction unit. The extracted solution is then passed through a separator, where the contaminants and extractant are separated from the soil.[69]
11. *Vitrification*. This involves use of an electric current to melt contaminated soil at elevated temperatures (1600 to 2000°C or 2900 to 3650°F). Upon cooling, the vitrification product is a chemically stable, leach-resistant glass and crystalline material similar to obsidian or basalt rock. The high-temperature component of the process destroys or removes organic materials. Radionuclides and heavy metals are retained within the vitrified product. Vitrification may be conducted in situ or ex situ.[70]

11.4.4 Cleanup Technologies Options for Metal Finishing Sites

Table 11.6 presents the technologies that may be appropriate for use at metal finishing sites. In addition to more conventional technologies, a number of innovative technology options are listed. Many possible cleanup approaches use institutional controls and one or a combination of the

TABLE 11.6
Cleanup Technologies for Metal Finishing Brownfield Sites Sample Analysis Technologies

Applicable Technology	Description	Examples of Applicable Land/Process Areas	Contaminants Treated by this Technology	Limitations
Containment Technologies				
Sheet piling	Steel or iron sheets are driven into the ground to form a subsurface barrier Low-cost containment method Used primarily for shallow aquifers	Metal cleaning, rinsing and bathing operations, chemical storage, wastewater treatment	Not contaminant-specific	Not effective in the absence of a continuous aquitard Can leak at the intersection of the sheets and the aquitard or through pile wall joints
Grout curtain	Grout curtains are injected into subsurface soils and bedrock Forms an impermeable barrier in the subsurface	Metal cleaning, rinsing and bathing operations, chemical storage, wastewater treatment	Not contaminant-specific	Difficult to ensure a complete curtain without gaps through which the plume can escape: however, new techniques have improved continuity of curtain
Slurry walls	Consist of a vertically excavated slurry-filled trench The slurry hydraulically shores the trench to prevent collapse and forms a filtercake to reduce groundwater flow Often used where the waste mass is too large for treatment and where soluble and mobile constituents pose an imminent threat to a source of drinking water Often constructed of a soil, bentonite, and water mixture	Metal cleaning, rinsing and bathing operations, chemical storage, wastewater treatment	Not contaminant-specific	Contains contaminants only within a specified area Soil-bentonite backfills are not able to withstand attack by strong acids, bases, salt solutions, and some organic chemicals Potential for the Slurry walls to degrade or deteriorate over time
Capping	Used to cover buried waste materials to prevent migration Made of a relatively impermeable material that will minimize rainwater infiltration Waste materials can be left in place Requires periodic inspections and routine monitoring Contaminant migration must be monitored periodically	Anodizing, solid wastes from anodizing, electroplating, electroplating wastewaters and solid wastes, finishing wastewaters, chemical conversion coating wastewaters and solid wastes, electroless plating, electroless plating wastewaters, solid wastes from painting, wastewater treatment system, sunken treatment tank	Metals	Costs associated with routine sampling and analysis may be high Long-term maintenance may be required to ensure impermeability May have to be replaced after 20 to 30 years of operation May not be effective if groundwater table is high
Ex situ Technologies				
Excavation/offsite disposal	Removes contaminated material to an EPA-approved landfill	Wastes from painting, wastewater treatment system, sunken treatment tanks, chemical storage, disposal	Not contaminant-specific	Generation of fugitive emissions may be a problem during operations The distance from the contaminated site to the nearest disposal facility will affect cost Depth and composition of the media requiring excavation must be considered

	Description	Source	Metals/VOCs/Cyanide	Limitations
Chemical oxidation/ reduction	Reduction/oxidation (Redox) reactions chemically convert hazardous contaminants to nonhazardous or less toxic compounds that are more stable, less mobile, or inert. Redox reactions involve the transfer of electrons from one compound to another. The oxidizing agents commonly used are ozone, hydrogen peroxide, hypochlorite, chlorine, and chlorine dioxide	Wastes from anodizing, electroplating, finishing, chemical conversion coating, electroless plating, painting, rinsing operations, wastewater treatment system, sunken treatment tank	Metals Cyanide	Transportation of the soil through populated areas may affect community acceptability. Disposal options for certain waste (e.g., mixed waste or transuranic waste) may be limited. There is currently only one licensed disposal facility for radioactive and mixed waste in the U.S. Not cost-effective for high contaminant concentrations because of the large amounts of oxidizing agent required. Oil and grease in the media should be minimized to optimize process efficiency
UV oxidation	Destruction process that oxidizes constituents in wastewater by the addition of strong oxidizers and irradiation with UV light. Practically any organic contaminant that is reactive with the hydroxyl radical can potentially be treated. The oxidation reactions are achieved through the synergistic action of UV light in combination with ozone or hydrogen peroxide. Can be configured in batch or continuous flow models, depending on the throughput rate under consideration	Wastes from metal cleaning, painting, rinsing operations, wastewater treatment system, sunken treatment tank, chemical storage area, disposal area	VOCs	The aqueous stream being treated must provide for good transmission of UV light (high turbidity causes interference). Metal ions in the wastewater may limit effectiveness. VOCs may volatilize before oxidation can occur. Off-gas may require treatment. Costs may be higher than competing technologies because of energy needs. Handling and storage of oxidizers require special safety precautions
Precipitation	Involves the conversion of soluble heavy metal salts to insoluble salts that will precipitate. Precipitate can be removed from the treated water by physical methods such as clarification or filtration. Often used as a pretreatment for other treatment technologies where the presence of metals would interfere with the treatment processes. Primary method for treating metal-laden industrial wastewater	Wastes from anodizing, electroplating, finishing, chemical conversion coating, electroless plating, painting, rinsing operations, wastewater treatment system, sunken treatment tank	Metals	Contamination source is not removed. The presence of multiple metal species may lead to removal difficulties. Discharge standard may necessitate further treatment of effluent. Metal hydroxide sludges must pass TCLP criteria prior to land disposal. Treated water will often require pH adjustment

Continued

TABLE 11.6 (continued)

Applicable Technology	Description	Examples of Applicable Land/Process Areas	Contaminants Treated by this Technology	Limitations
Liquid-phase carbon adsorption	Groundwater is pumped through a series of vessels containing activated carbon, to which dissolved contaminants adsorb Effective for polishing water discharges from other remedial technologies to attain regulatory compliance Can be quickly installed High contaminant-removal efficiencies	Wastes from metal cleaning, painting, rinsing operations, wastewater treatment system, sunken treatment tank, chemical storage area, disposal area	VOCs	The presence of multiple contaminants can affect process performance Metals can foul the system Costs are high if used as the primary treatment on wastestreams with high contaminant concentration levels Type and pore size of the carbon and operating temperature will impact process performance Transport and disposal of spent carbon can be expensive Water soluble compounds and small molecules are not adsorbed well
Air stripping	Contaminants are partitioned from groundwater by greatly increasing the surface area of the contaminated water exposed to air Aeration methods include packed towers, diffused aeration, tray aeration, and spray aeration Can be operated continuously or in a batch mode, where the air stripper is intermittently fed from a collection tank The batch mode ensures consistent air stripper performance and greater efficiency than continuously operated units because mixing in the storage tank eliminates any inconsistencies in feedwater composition	Wastes from metal cleaning, painting, rinsing operations, wastewater treatment system, sunken treatment tank, chemical storage area, disposal area	VOCs	Potential for inorganic (iron greater than 5 ppm, hardness greater than 800 ppm) or biological fouling of the equipment, requiring pretreatment of groundwater or periodic column cleaning Consideration should be given to the Henry's law constant of the VOCs in the water stream and the type and amount of packing used in the tower Compounds with low volatility at ambient temperature may require preheating of the groundwater Off-gases may require treatment based on mass emission rate and state and federal air pollution laws

In situ Technologies

Natural attenuation	Natural subsurface processes such as dilution, volatilization, biodegradation, adsorption, and chemical reactions with subsurface media can reduce contaminant concentrations to acceptable levels Consideration of this option requires modeling and evaluation of contaminant degradation rates and pathways	Metal cleaning, metal cleaning wastewaters, painting, painting wastewaters and solid wastes, wastewater treatment system, sunken treatment tank, chemical storage area, disposal area	VOCs	Intermediate degradation products may be more mobile and more toxic than original contaminants Contaminants may migrate before they degrade The site may have to be fenced and may not be available for reuse until hazard levels are reduced Source areas may require removal for natural attenuation to be effective

Cleanup of Metal Finishing Brownfield Sites

Technology	Description	Contaminants	Limitations
Soil vapor extraction	A vacuum is applied to the soil to induce controlled air flow and remove contaminants from the unsaturated (vadose) zone of the soil The gas leaving the soil amy be treated to recover or denstroy the contaminants The continuous air flow promotes in situ biodegradation of low-volatility organic compounds that may be present	VOCs	Sampling and analyses must be conducted throughout the process to confirm that degradation is proceeding at sufficient rates to meet cleanup objectives Modeling contaminant degradation rates, and sampling and analysis to confirm modeled predictions is extremely expensive Tight or extremely moist content (>50%) has a reduced permeability to air, requiring higher vacuums Large screened intervals are required in extraction wells for soil with highly variable permeabilities Air emissions may require treatment to eliminate possible harm to the public or environment Off-gas treatment residual liquids and spent activated carbon may require treatment or disposal Not effective in the saturated zone
Soil flushing	Extraction of contaminants from the soil with water or other aqueous solutions Accomplished by passing the extraction fluid through in-place soils using injection or infiltration processes Extraction fluids must be recovered with extraction wells from the underlying aquifer and recyled when possible	Metals	Low-permeability soils are difficult to treat Surfactants can adhere to soil and reduce effective soil porosity Reactions of flushing fluids with soil can reduce contaminant mobility Potential of washing the contaminant beyond tghe capture zone and the introduction of sufactants to the subsurface
Air sparging	In situ technology in which air is injected under pressure below the water table to increase groundwater oxygen concentrations and enhance the rate of biological degradation of contaminants by naturally occurring microbes Increases the mixing in the saturated zone, which increases the contact between groundwater and soil Air bubbles traverse horizontally and vertically through the soil column, creating an underground stripper that volatilizes contaminants Air bubbles travel to a soil vapor extraction system Air sparging is effective for facilitating extraction of deep contamination, contamination in low-permeability soils, and contamination in the saturated zone	VOCs	Depth of contaminants and specific site geology must be considered Air flow through the saturated zone may not be uniform A permeability differential such as a clay layer above the air injection zone can reduce the effectiveness Vapors may rise through the vadose zone and be released into the atmosphere Increased pressure in the vadose zone can build up vapors in basements, which are generally low-pressure areas

Continued

TABLE 11.6 (continued)

Applicable Technology	Description	Examples of Applicable Land/Process Areas	Contaminants Treated by this Technology	Limitations
Passive treatment walls	A permeable reaction wall is installed inground, across the flow path of a contaminant plume, allowing the water portion of the plume to passively move through the wall Allows the passage of water while prohibiting the movement of contaminants by employing such agents as iron, chelators (ligands selected for their specificity for a given metal), sorbents, microbes, and others Contaminants are typically completely degraded by the treatment wall	Appropriately selected location for wall	VOCs Metals	The system requires control of pH levels. When pH levels within the passive treatment wall rise, it reduces the reaction rate and can inhibit the effectiveness of the wall Depth and width of the plume For large-scale plumes, installation costs may be high Cost of treatment medium (iron) Biological activity may reduce the permeability of the wall Walls may lose their reactive capacity, requiring replacement of the reactive medium
Biodegradation	Indigenous or introduced microorganisms degrade organic contaminants found in soil and groundwater Used successfully to remediate soils, sludges, and groundwater Especially effective for remediating low-level residual contamination in conjunction with source removal	Metal cleaning, metal cleaning wastewaters, painting, painting wastewaters and solid wastes, wastewater treatment system, sunken treatment tank, chemical storage area, disposal area	VOCs	Cleanup goals may not be attained if the soil matrix prevents sufficient mixing Circulation of water-based solutions through the soil may increase contaminant mobility and necessitate treatment of underlying groundwater Injection wells may clog and prevent adequate flow rates Preferential flow paths may result in nonuniform distribution of injected fluids Should not be used for clay, highly layered, or heterogeneous subsurface environments High concentrations of heavy metals, highly chlorinated organics, long-chain hydrocarbons, or inorganic salts are likely to be toxic to microorganisms Low temperatures slow bioremediation Chlorinated solvents may not degrade fully under certain subsurface conditions

Source: From U.S. EPA, Technical Approaches to Characterizing and Cleaning up Metal Finishing Sites under the Brownfields Initiative, EPA/625/R-98/006, U.S. EPA, Cincinnati, OH, March 1999.

Cleanup of Metal Finishing Brownfield Sites

technologies described in Table 11.6. Whatever cleanup approach is ultimately chosen, planners should explore a number of cost-effective options.

Cleanup at metal finishing facilities will most likely entail removing a complex mix of contaminants, primarily organic solvents and metals. The cleanup will usually require more than one technology, or treatment train, because single technologies tend not to address both metal and organic contaminants. Solidification/stabilization can address metal contamination by limiting mobility (solubility) and thereby limit risk. Approaches at metal finishing sites depend on local conditions. At larger metal finishing sites, one approach may be to excavate and stabilize the contaminated material with either onsite or offsite disposal or treatment of material. Access to contaminated soils may be limited at smaller sites requiring excavation and offsite treatment or disposal. The stabilized material can be placed onsite or sent to an U.S. EPA approved landfill.

11.4.5 POST-CONSTRUCTION CARE

Many of the cleanup technologies that leave contamination onsite, either in containment systems or because of the long periods required to reach cleanup goals, will require long-term maintenance and possibly operation. If waste is left onsite, regulators will likely require long-term-monitoring of applicable media (i.e., soil, water, and air) to ensure that the cleanup approach selected is continuing to function as planned (e.g., residual contamination, if any, remains at acceptable levels and is not migrating). If long-term monitoring is required (e.g., by the state), periodic sampling, analysis, and reporting requirements will also be involved. Planners should be aware of these requirements and provide for them in cleanup budgets. Postconstruction sampling, analysis, and reporting costs in their cleanup budgets can be a significant problem as these costs can be substantial.

11.5 CONCLUSIONS

Brownfields redevelopment contributes to the revitalization of communities across the U.S. Reuse of these abandoned, contaminated sites spurs economic growth, builds community pride, protects public health, and helps maintain our nation's "greenfields," often at a relatively low cost. This chapter provides brownfields planners with the technical methods that can be used to achieve successful site assessment and cleanup, which are two key components in the brownfields redevelopment process.

While the general guidance provided in this chapter will be applicable to many brownfields projects, it is important to recognize the heterogeneous nature of brownfields work. That is, no two brownfields sites will be identical, and planners will need to base site assessment and cleanup activities on the conditions at their particular site. Some of the conditions that may vary by site include the type of contaminants present, the geographic location and extent of contamination, the availability of site records, hydrogeological conditions, and state and local regulatory requirements. Based on these factors, as well as financial resources and desired timeframes, planners will find different assessment and cleanup approaches appropriate.

Consultation with state and local environmental officials and community leaders, as well as careful planning early in the project, will assist planners in developing the most appropriate site assessment and cleanup approaches. Planners should also determine early on if they are likely to require the assistance of environmental engineers. A site assessment strategy should be agreeable to all stakeholders and should address the following:

1. The type and extent of contamination, if any, present at the site
2. The types of data needed to adequately assess the site
3. Appropriate sampling and analytical methods for characterizing contamination
4. An acceptable level of data uncertainty

When used appropriately, the site assessment methods described in this chapter will help to ensure that a good strategy is developed and implemented effectively.

Once the site has been assessed and stakeholders agree that cleanup is needed, planners will need to consider cleanup options. Many different types of cleanup technologies are available. The guidance provided in this chapter on selecting appropriate methods directs planners to base cleanup initiatives on site- and project-specific conditions. The type and extent of cleanup will depend in large part on the type and level of contamination present, reuse goals, and the budget available. Certain cleanup technologies are used onsite, while others require offsite treatment. Also, in certain circumstances, containment of contamination onsite and the use of institutional controls may be important components of the cleanup effort. Finally, planners will need to include budgetary provisions and plans for postcleanup and postconstruction care if required at the brownfield site. By developing a technically sound site assessment and cleanup approach that is based on site-specific conditions and addresses the concerns of all project stakeholders, planners can achieve brownfields redevelopment and reuse goals effectively and safely.

ACRONYMS

ASTM	American Society for Testing and Materials
ATSDR	Agency for Toxic Substances and Disease Registry
BTEX	Benzene, toluene, ethylbenzene, and xylene
CERCLIS	Comprehensive Environmental Response, Compensation, and Liability Information System
DQO	Data Quality Objective
ERNS	Emergency Response Notification System
FID	Flame Ionization Detector
FOIA	Freedom of Information Act
NPDES	National Pollutant Discharge Elimination System
NPL	National Priorities List
O&M	Operations and Maintenance
ORD	Office of Research and Development
OSWER	Office of Solid Waste and Emergency Response
PAH	Polyaromatic hydrocarbon
PCB	Polychlorinated biphenyl
PID	Photoionization detector
PCP	Pentachlorophenol
PLM	Polarized light microscopy
POTW	Publicly owned treatment works
RCRA	Resource Conservation and Recovery Act
SVE	Soil vapor extraction
SVOC	Semivolatile organic compound
TCE	Trichloroethylene
TIO	Technology Innovation Office
TPH	Total petroleum hydrocarbon
TSD	Treatment, storage, and disposal
U.S. EPA	U.S. Environmental Protection Agency
USDA	U.S. Department of Agriculture
USGS	U.S. Geological Survey
UST	Underground storage tank
VCP	Voluntary Cleanup Program
VOC	Volatile organic compound
XRF	X-ray fluorescence

REFERENCES

1. Federal Register, Comprehensive Environmental Response, Compensation, and Liability Act (CERCLA or Superfund) 42 U.S.C. s/s 9601 et seq, 1980, U.S. Government, Public Laws. Available at www.access.gpo.gov/uscode/title42/chapter103_.html, January 2004.
2. U.S. EPA, Technical Approaches to Characterizing and Cleaning up Metal Finishing Sites under the Brownfields Initiative, EPA/625/R-98/006, U.S. EPA, Cincinnati, OH, March 1999.
3. U.S. EPA, Brownfields Home Page, U.S. EPA, available at http://www.epa.gov/brownfields, 2007.
4. U.S. EPA, Road Map to Understanding Innovative Technology Options for Brownfields Investigation and Cleanup, EPA 542-B-05-001, 4th ed., U.S. EPA, Washington, DC, September 2005.
5. U.S. EPA, Brownfields Tool Kit, U.S. EPA, Cincinnati, OH, available at http://www.lacity.org/EAD/labf/toolkit.htm, 2007.
6. U.S. EPA, Brownfields and Land Revitalization—Tools and Technical Information, U.S. EPA, Washington, DC, available at http://www.epa.gov/brownfields/toolsandtech.htm, 2007.
7. U.S. EPA, Profile of the Fabricated Metal Products Industry, EPA 3 10-R-95-007, U.S. EPA, Washington, DC, 1995.
8. Brebbia, C.A., Ed., *Brownfields III: Prevention, Assessment, Rehabilitation and Development of Brownfield Sites*, WIT Transactions on Ecology and the Environment Series, Vol. 94, Wessex Institute of Technology (WIT), UK, 2006.
9. CERP, Brownfields Identification, The Community Environmental Resource Program (CERP), St Louis, MO, available at http://stlcin.missouri.org/cerp/brownfields/identification.cfm, 2007.
10. ASTM, Standard Practice for Environmental Site Assessments: Phase I Environmental Site Assessment Process. E 1527-00, American Society for Testing and Materials, West Conshohocken, PA, 2003.
11. Federal Register, Resource Conservation and Recovery Act (RCRA), 42 US Code s/s 6901 et seq, 1976, U.S. Government, Public Laws, available at www.access.gpo.gov/uscode/title42/chapter82_.html, January 2004.
12. U.S. EPA, National Pollutant Discharge Elimination System (NPDES), U.S. EPA, Washington, DC, NPDES, available at http://cfpub.epa.gov/npdes, 2007.
13. U.S. EPA, Comprehensive Environmental Response, Compensation, and Liability Information System (CERCLIS), U.S. EPA, Washington, DC, available at http://www.epa.gov/superfund/sites/cursites, 2007.
14. ASTM, Standard Guide for Process of Sustainable Brownfields Development, E 1984-03, American Society for Testing and Materials, West Conshohocken, PA, 2003.
15. NEWMOA, Improving Decision Quality: Making the Case for Adopting Next-Generation Site Characterization Practices, Northeast Waste Management Officials' Association, *Remediation*, Spring 2003.
16. U.S. EPA, Quality Assurance Guidance for Conducting Brownfield Site Assessments, EPA 540-R-98-038, U.S. EPA, Washington, DC, 1998.
17. Pediaditi, K., Wehrmeyer, W., and Chenoweth, J. Sustainability Indicators for Brownfield Redevelopment Projects, in *Proceedings of Sustainable Urban Environments: EPSRC Conference*, University of Birmingham, U.K., February 2005.
18. ASTM, ASTM Standard Practice for Environmental Site Assessments: Transaction Screen Process, E 1528-00, American Society for Testing and Materials, West Conshohocken, PA, 2000.
19. U.S. EPA, Brownfields Technology Primer: Using the Triad Approach to Streamline Brownfields Site Assessment and Cleanup, EPA 542-B-03-002, U.S. EPA, Washington, DC, 2003.
20. U.S. EPA, Improving Sampling, Analysis, and Data Management for Site Investigation and Cleanup, EPA 542-F-04-001a, U.S. EPA, Washington, DC, 2004.
21. U.S. EPA, The Triad Resource Center Web site available at www.triadcentral.org, 2007
22. U.S. EPA, Technology News and Trends, www.epa.gov/tio/download/newsltrs/tnandt0704.pdf, 2007.
23. ASTM, ASTM Standard Guide for Environmental Site Assessments: Phase II Environmental Site Assessment Process, E1903-97, American Society for Testing and Materials, West Conshohocken, PA, 2002.
24. U.S. EPA, Quality Assurance Guidance for Conducting Brownfield Site Assessments, EPA 540-R-98-038, U.S. EPA, Washington, DC, 1998.
25. U.S. EPA, Clarifying DQO Terminology Usage to Support Modernization of Site Cleanup Practices, EPA 542-R-01-014, U.S. EPA, Washington, DC, 2001.
26. U.S. EPA, Data Quality Objective Process for Hazardous Waste Site Investigations, EPA 600-R-00-007, U.S. EPA, Washington, DC, 2000.
27. U.S. EPA and U.S.ACE, Managing Uncertainty in Environmental Decisions, *Environmental Science and Technology*, American Chemical Society, October 2001.

28. U.S. ACE, Engineering and Design: Requirements for the Preparation of Sampling and Analysis Plans, EM 200-1-3, U.S. Army Corps of Engineers, February 2001.
29. U.S. EPA-OSRTI. In search of representativeness: evolving the environmental data quality model, *Quality Assurance*, 9, 179–190, 2002.
30. Federal Register, Safe Drinking Water Act (SDWA), 42 U.S.C. s/s 300f et seq, 1974, U.S. Government, Public Laws, available at http://frwebgate.access.gpo.gov/cgi-bin/getdoc.cgi?dbname=browse_usc& docid=Cite:+42USC300f, January 2002.
31. U.S. EPA, Subsurface Characterization and Monitoring Techniques: A Desk Reference Guide, EPA/625/R-93-003a, U.S. EPA, Washington, DC, 1993.
32. Robbat, A., Jr., *Dynamic Workplans and Field Analytics: The Keys to Cost Effective Site Characterization and Cleanup*, Tufts University under Cooperative Agreement with the U.S. EPA, October 1997.
33. U.S. EPA, Field Analytical and Site Characterization Technologies: Summary of Applications, EPA 542-R-97-011, U.S. EPA, Washington, DC, 1997.
34. U.S. EPA, Electroplating, U.S. EPA, Mid-Atlantic Brownfields, http://www.epa.gov/reg3hscd/bfs/regional/industry/electroplating.htm, 2007.
35. Geo-Environmental Solutions, Rental Rate Sheet, Geoprobe Systems, Inc., available at http://www.gesolutions.com/assess.htm, September 15, 1998.
36. U.S. ACE, Yearly Average Cost Index for Utilities, in *Civil Works Construction Cost Index System Manual*, 110-2-1304, U.S. Army Corps of Engineers, Washington, DC, available at http://www.nww.usace.army.mil/cost, 2007.
37. U.S. EPA, Cost Estimating Tools and Resources for Addressing the Brownfields Initiatives, EPA 625-R-99-001, U.S. EPA, Washington, DC, 1999.
38. U.S. EPA, Guide to Documenting and Managing Cost and Performance Information for Remediation Projects, EPA 542-B-98-007, U.S. EPA, Washington, DC, 1998.
39. U.S. EPA, Remediation Technology Cost Compendium—Year 2000, EPA 542-R-01-009, U.S. EPA, Washington, DC, September 2001.
40. Al-Tabbaa, A., Impact of and Response to Climate Change in UK Brownfield Remediation, paper presented to the *Chartered Institute of Water and Environmental Management Hong Kong*, The Hong Kong Institution of Engineers, Hong Kong, May 2007.
41. Catney, P., Yount, K., Henneberry, J., and Meyer, P., Can We Really Compare Brownfield Regulation and Redevelopment in the United States and European Union? *Revit & Cabernet 2nd International Conference on Managing Urban Land*, Theaterhaus Stuttgart, Germany, April 2007.
42. U.S. EPA, Capping, EPA 542-F-01-022, U.S. EPA, Washington, DC, 2001.
43. U.S. EPA, Evaluation of Subsurface Engineered Barriers at Waste Sites, EPA 542-R-98-005, U.S. EPA, Washington, DC, 1998.
44. U.S. EPA, Solidification/Stabilization Use at Superfund Sites, EPA 542-R-00-010, U.S. EPA, Washington, DC, 2000.
45. U.S. EPA, Solidification/Stabilization, EPA 542-F-01-024, U.S. EPA, Washington, DC, 2001.
46. U.S. EPA, Bioremediation, EPA 542-F-01-001, U.S. EPA, Washington, DC, 2001.
47. U.S. EPA, In Situ Flushing, EPA 542-F-01-011, U.S. EPA, Washington, DC, 2001.
48. U.S. EPA, Chemical Oxidation, EPA 542-F-01-013, U.S. EPA, Washington, DC, 2001.
49. U.S. EPA, Soil Vapor Extraction (SVE) and Air Sparging. EPA 542-F-01-006, U.S. EPA, Washington, DC, 2001.
50. U.S. EPA, Permeable Reactive Barriers, EPA 542-F-01-00, U.S. EPA, Washington, DC, 2001.
51. U.S. EPA, Site Remediation Technology InfoBase: A Guide to Federal Programs, Information Resources, and Publications on Contaminated Site Cleanup Technologies, 2nd ed., EPA 542-B-00-005, U.S. EPA, Washington, DC, 2000.
52. U.S. EPA, Innovative Remediation Technologies: Field-Scale Demonstration Projects in North America, 2nd ed., EPA 542-B-00-004, U.S. EPA, Washington, DC, 2000.
53. U.S. EPA, Brownfields Technology Primer: Requesting and Evaluating Proposals that Encourage Innovative Technologies for Investigation and Cleanup, EPA 542-R-01-005, U.S. EPA, Washington, DC, 2001.
54. NATO/CCMS, North Atlantic Treaty Organization/Committee on the Challenges of Modern Society (NATO/CCMS) Pilot Study Evaluation of Demonstrated and Emerging Technologies for the Treatment of Contaminated Land and Groundwater (Phase III), 2002 Annual Report, EPA 542-R-02-010, U.S. EPA, Washington, DC, 2002.
55. U.S. EPA, Treatment Technologies for Site Cleanup: Annual Status Report, 11th ed., EPA 542-R-03-009, U.S. EPA, Washington, DC, 2003.

56. U.S. EPA, Innovative Remediation and Site Characterization Technologies Resources, EPA 542-C-04-002, U.S. EPA, Washington, DC, 2004.
57. USAEC, In-Situ Electrokinetic Remediation of Metal Contaminated Soils Technology Status Report. No. SFIM-AEC-ET-CR-99022, U.S. Army Environmental Center, Washington, DC, July 2000.
58. Wang, L.K., Hung, Y.T., and Shammas, N.K., Eds., *Physicochemical Treatment Processes*, Humana Press, Totowa, NJ, 2005.
59. Wang, L.K., Hung, Y.T., and Shammas, N.K., Eds., *Advanced Physicochemical Treatment Processes*, Humana Press, Totowa, NJ, 2006.
60. Wang, L.K., Hung, Y.T., and Shammas, N.K., Eds., *Advanced Physicochemical Treatment Technologies*, Humana Press, Totowa, NJ, 2007.
61. U.S. EPA, Phytoremediation, EPA 542-F-01-002, U.S. EPA, Washington, DC, 2001.
62. U.S. EPA, Brownfields Technology Primer: Selecting and Using Phytoremediation for Site Cleanup, EPA 542-R-01-006, U.S. EPA, Washington, DC, July 2001.
63. U.S. EPA, Use of Field-Scale Phytotechnology for Chlorinated Solvents, Metals, Explosives and Propellants, and Pesticides—Status Report, EPA 542-R-05-002, U.S. EPA, Washington, DC, April 2005.
64. Environment Canada, Phytoremediation of Soil Containing Heavy Metals and Hydrocarbons, Environmental Protection, Quebec Region, available at http://www.qc.ec.gc.ca/dpe/Anglais/dpe_main_en.asp?innov_cemrs_200409b, 2007.
65. U.S. EPA, Pump and Treat, EPA 542-F-01-025, U.S. EPA, Washington, DC, 2001.
66. U.S. EPA, Soil Washing, EPA 542-F-01-008, U.S. EPA, Washington, DC, 2001.
67. Harbottle, M.J., and Al-Tabbaa, A., Combining Stabilization/Solidification with Biodegradation to Enhance Long-Term Remediation Performance, *Proceedings of the 2nd IASTED International Conference on Advanced Technology in the Environmental Field*, Lanzarote, Spain, pp. 222–227, 2006.
68. Harbottle, M.J., Al-Tabbaa, A., and Evans, C.W., The Technical Sustainability of In-Situ Stabilization/Solidification, in *Proceedings of the International Conference on Stabilization/Solidification Treatment and Remediation*, Al-Tabbaa, A. and Stegemann, J., Eds., Cambridge, U.K., pp. 159–170, April 2005.
69. U.S. EPA, Solvent extraction, EPA 542-F-01-009, U.S. EPA, Washington, DC, 2001.
70. U.S. EPA, Vitrification, EPA 542-F-01-017, U.S. EPA, Washington, DC, 2001.

Index

A

Abrasive jet machining, 347
Acid
    hydrolysis, 371
    pickling, 20, 61, 65, 318
    purification, 67
    recycling, 67
    removal, 64
    reuse, recycle, and recovery systems, 67
    sealing rinse, 266
    treatment, 309, 313
    wastestreams, 21
Acid cleaning
    coil coating industry waste treatment, 263
ADT. *See* Air dissolving tube (ADT)
Agricultural Research Service, 192
Air
    atomized paint spraying, 29
    emission, 163
    knives, 363
    spraying, 310
Air-cooled furnace slag, 169, 174, 182
    resistance, 183
    water absorption, 170
Air dissolving tube (ADT), 253
Airless paint spraying, 29
Air pollution control (APC), 52
Alkaline
    chlorination, 371
    cleaning, 20, 263, 309, 371
    phenolic resin, 159, 160
    solution, 311
    wastestreams, 21
Aluminum, 74–80, 291, 326
    anodes, 78
    casting, 162, 200
    chromate conversion coatings, 265
    chromium reduction, 332
    coil coating industry waste treatment, 269, 278
    degreasing, 202
    effluent limitations, 300–302
    geographic distribution, 74
    ground coats, 312
    industrial process description, 76–78
    industry size, 74
    material inputs, 76–78
    mechanical properties, 161
    performance standards, 300–302
    physical properties, 161
    pollution outputs, 76–78, 80
    porcelain enameling, 311, 314, 324–325, 335–336
    pretreatment standards, 300–302
    processing plants, 216
    process materials inputs, 80
    product characterization, 74–75
    smelting, 81
    surface preparation, 312
    toxic pollutants, 120–122, 274–275
    wastewater pollutants, 97–101, 274–275, 277
Aluminum Association, 74
Aluminum basis material
    BAT application, 300
    BAT economically achievable, 340–341
    BPT application, 300
    definition, 305
    effluent limitations, 300–302, 335, 340
    effluent pretreatment standards, 342
    NSPS, 302, 341
    POTW, 341–342
    PSES, 303
    PSNS, 303
Aluminum forming industry waste treatment, 199–232
    case histories plant-specific description, 216–218
    casting, 200–201
    continuous casting, 200
    direct chill casting, 200, 206
    drawing, 202
    emulsion rolling, 207
    emulsions or soaps drawing, 206, 209
    emulsions rolling, 206
    extrusion, 201, 206, 207
    forging, 202, 206, 207–208
    full-scale treatment U.S. EPA data, 218–220
    heat treatment, 203, 209
    industry and process description, 200–203
    neat oils drawing, 206
    neat oils rolling, 205
    pollutant removability, 221
    rolling, 201
    stationary casting, 200
    subcategory description, 203–206
    surface treatment, 203
    terminologies, 224
    treatment technology costs, 222–223
    waste characterization, 206–216
American Concrete Institute, 189
Ammonia
    removal, 39
    stripping, 69
Ammonium paratungstate (APT), 110
Annealing, 203
Anodes
    reduction process, 77

Anodizing
    definition, 346, 438
Antitrust treatment
    steel, 26
AOD. *See* Argon–oxygen decarburization (AOD)
APC. *See* Air pollution control (APC)
APT. *See* Ammonium paratungstate (APT)
Area coated
    definition, 344
Area processed
    definition, 303
Argon behavior, 383–384
Argon–oxygen decarburization (AOD), 57
Artificial aging, 203
Asphalt concrete test procedures, 181, 182
Assembly
    definition, 349
Atmospheric evaporators, 240

B

Baghouse dust
    metal castings, 168
Barrel finishing
    definition, 347
Base metal surface preparation, 313
Basic oxygen furnace (BOF), 38, 50
Basis material. *See also* Aluminum basis material;
        Cast iron, basis material; Galvanized basis
        material; Steel basis material
    area processed, 342
    copper, 343
    definition, 303, 342
BAT. *See* Best applicable technology (BAT)
Battelle Memorial Institute, 410
Bayer process, 76
    alumina refining, 77
BDAT. *See* Best demonstrated available technology
        (BDAT)
BDL. *See* Below detection limit (BDL)
Belgium
    soil washing, 405
Below detection limit (BDL), 206
Bench-scale tests, 413
Beryllium, 96
Best applicable technology (BAT), 223, 226, 294, 333
    aluminum basis material, 300, 340–341
    PE industrial wastewaters, 332
    porcelain enameling industrial wastes, 333
    U.S. annual cost, 295
Best demonstrated available technology (BDAT), 387
    heavy metal soil removal, 395–396, 401, 405, 407
Best management practices (BMP), 3
    metal finishing, 17
    primary metals, 20
Best practicable technology (BPT), 223, 226,
        293–294, 332
    aluminum basis material, 300
    Federal Register, 73
    galvanized basis material, 298
    limitations, 75
    PE industrial wastewaters, 332
    porcelain enameling industrial wastes, 332
    U.S. annual cost, 294
    U.S. Federal Act Section 304, 293
Bidentates, 376
Binder materials
    solidification–stabilization projects, 395
Blast furnace slag. *See* Furnace slag
Blister copper, 83
BMP. *See* Best management practices (BMP)
BOF. *See* Basic oxygen furnace (BOF)
BPT. *See* Best practicable technology (BPT)
Brazing
    definition, 347
Brownfield site metal finishing cleanup, 433–478
    analysis levels, 454
    analysis technologies, 455
    analysis type, 460
    anodizing operations, 438
    auxiliary activity areas and potential contaminants,
        440–441
    background, 434
    chemical conversion coating, 439
    chemical storage area, 440
    cleanup development, 462–464
    community involvement, 447
    containment technologies, 463–464
    contaminants, 437–440
    contaminant screening and cleanup, 442
    data quality objectives, 450–452
    decisions, 449
    disposal area, 441
    electroless and immersion plating, 439
    electroplating, 439
    environmental sampling and data analysis, 454
    exposed populations identification, 445
    facility records, 444
    field vs. laboratory analysis, 455
    groundwater information, 446–448
    groundwater sampling costs, 461
    health concerns identification, 446
    increasing sampling results certainty, 454
    industrial processes, 437–440
    institutional controls, 463
    interviews, 448
    metals and metalloids, 435, 466
    migration pathways, 445
    operations, 437–440
    painting, 440
    phase I site assessment, 442–445
    phase II site assessment site sampling, 450–454
    post-construction care, 473
    potential environmental identification, 446
    purpose, 436
    recorded information sources, 444
    report, 448
    sample collection, 454, 455
    sampling analysis costs, 461
    sampling costs, 461
    sampling location, 455–459
    sampling numbers, 460
    screening levels, 453
    sediment sampling costs, 461
    selection and acceptance, 464–465
    site assessment, 441–461
    site assessment technologies, 455

Index

site investigations, 449, 455–560
site visit, 448
soil and subsurface information, 446
soil collection costs, 461
state agencies central role, 442
state voluntary cleanup programs, 442
sunken wastewater treatment tank, 440
surface preparation operations, 437
surface water, 461
techniques, 440
technologies, 467–472
topographic information, 445
treatment technologies, 466
wastewater treatment, 440
Burnishing
    definition, 347

C

Cadmium
    behavior, 383–384
    toxic pollutants, 118–119
    wastewater pollutants, 117
Calibration
    definition, 349
California Steel Industries, Inc., 21
Canmaking
    definition, 306
    effluent limitations, 302
    NSPS, 304
    performance standards, 302
    pretreatment standards, 302
    PSES, 305
    PSNS, 305
Capping systems, 388
Carbon dioxide–silica process, 161
Carbon monoxide, 47
Carbon steel
    pile walls, 389
Casting, 57–59. *See also* Metal castings
    aluminum, 200–201, 206
    byproducts, 59
    centrifugal, 158
    continuous, 200, 224
    copper, 162
    die, 157
    direct chill, 200, 201, 206, 209, 224
    effluent, 59
    emissions, 58
    energy facts, 59
    environmental facts, 59
    process description, 58
    process waste sources, 58–59
    treatment techniques, 59
Casting flow, 153–157
    casting cleaning and inspection, 156
    core making, 153
    diagram, 58
    melting and pouring, 155
    molding, 154
    molds and core reclamation, 157
Casting metals, 161–162
    aluminum castings, 162
    copper castings, 162
    iron castings, 161
    steel castings, 162
    wastewater flow characterization, 164
Casting processes, 20, 157–158
    centrifugal casting, 158
    die casting, 157
    inputs, 58
    investment casting, 157
    outputs, 58
    permanent mold casting, 158
    sand casting, 157
    shell casting, 157
Cast iron
    basis material, 338–339
    mechanical properties, 161
    physical properties, 161
    porcelain enameling, 311, 314, 326, 334
    subcategory plants, 328
CDF. *See* Controlled density fill (CDF)
CEB. *See* Chemical emulsion breaking (CEB)
Cement–bentonite barriers, 389
Central nervous system syndrome, 145
Central treatment plan (CTP), 65
Centrifugal casting, 158
CERCLA. *See* Comprehensive Environmental Response, Compensation, and Liabilities Act (CERCLA)
CERCLIS. *See* Comprehensive Environmental Response, Compensation, and Liability Information System (CERCLIS)
Chartered Metal Industries Toolroom, 25
Chemical conversion coatings
    definition, 346
Chemical emulsion breaking (CEB), 218
Chemical flotation–filtration system
    conventional, 251
Chemical innovative flotation–filtration wastewater treatment systems
    nickel-chromium plating wastes treatment, 251
Chemically bonded sand systems
    sand casting systems, 159–160
Chemical milling, 17
Chemical precipitation
    aluminum forming wastewater, 230
Chemical treatment, 466
    nickel-chromium plating wastes treatment, 242–247
Chlorine
    wastestreams chemicals management, 144
Chromate
    coil coating industry waste treatment and management, 265
    conversion coatings, 265, 266
Chromating, 235, 439
Chrome tanning process, 32
Chromic acid, 235
Chromium, 235, 247, 266. *See also* Nickel-chromium plating wastes treatment
    behavior, 383–384
    hexavalent, 243, 245, 247, 249, 350, 374
    plating, 235
    POTW, 236
    recycling, 33
    reduction, 231, 285
    removal treatment, 242

Clarification
    aluminum forming wastewater, 230
Clean Air Act Title V, 295
Cleaner production, 14–16, 296, 304
    audit review, 12–14
    barriers, 14–15
    barriers response, 16
    case studies, 21–35
    goals, 16
    public awareness, 14
Cleaner Production Program, 12
Cleaning
    alkaline, 20, 309, 371
    casting flow, 156
    coil coating industry waste treatment and
        management, 263–264
    effluent-free exhaust, 68
    sodium hydroxide, 264
Cleanup technology, 462, 472, 474
Clean Water Act (CWA), 73
    Effluent Guidelines and Standards for
        Coil Coating, 295
CLSM. *See* Controlled low strength material (CLSM)
Coating operations
    definition, 344
Coil coating industry effluent treatment technologies,
        280–293
    activated carbon filtration, 284–286
    aluminum, 292
    biological treatment, 283
    chromium reduction and chemical
        precipitation, 282
    clarification, 283
    cyanide destruction, 283
    electrochemical chromium regeneration, 282
    granular bed filtration, 284–285
    ion exchange, 281
    membrane processes, 286
    oil skimming, 282
    oil–water separation, 283
    powdered activated carbon adsorption, 283
    steel, 289
    water–solids separation technologies, 286–289
Coil coating industry waste treatment and
        management, 261–306
    acid cleaning, 263
    aluminum, 269, 277
    chromate conversion coatings, 265
    cleaning operation, 263–264
    cold rolled steel and galvanized steel, 279
    complex oxide conversion coatings, 266
    conversion coating process, 264–266
    effluent treatment technologies, 280–293
    liquid effluent limitations, 297–304
    mild alkaline cleaning, 263
    multimedia waste management, 294–297
    no-rinse conversion coatings, 266
    painting operation, 267
    performance standards, 297–298
    phosphate conversion coatings, 265
    plant-specific effluent characterization data, 278
    pretreatment standards, 297–298
    special cleaning, 263
    steel, 268, 269

    strong alkaline cleaning, 263
    subcategories, 267–269
    terminologies, 304–305
    wastewater characterization, 269–277
    wastewater treatment levels vs. costs, 293–294
    zinc coated steel (galvanized steel), 268, 269–276
Coke making, 39–44
    effluents, 43
    emissions, 42
    energy facts, 42
    environmental facts, 42
    flow diagram, 41
    hazardous wastes, 43
    inputs, 42
    outputs, 42
    oven emissions reduction, 69
    plants water consumption, 43
    process description, 39–41
    process wastes sources, 42–43
    treatment techniques, 44
    wastewaters, 43
Cold rolled steel
    coil coating industry waste treatment and
        management, 279
Colloidal graphite
    dies, 202
Columbium, 96
    toxic organic pollutants, 123–124
    toxic pollutants, 102
    wastewater characterization, 101
    wastewater pollutants, 102
Complex oxide conversion coating films
    physical properties, 266
Comprehensive Environmental Response,
        Compensation, and Liabilities Act
        (CERCLA), 180, 434
Comprehensive Environmental Response,
        Compensation, and Liability Information
        System (CERCLIS), 445
Concrete
    aggregate test procedures, 185
    mixes, 186
    paving materials, 184, 185
Contact cooling water, 224
Containment technologies, 390
Contamination
    leachable concentration, 394
    metal, 391, 400–401
    personal property, 449
    soil remediation, 388
Continuous casting, 200, 224
Control authority
    definition, 344
Controlled density fill (CDF), 189
Controlled low strength material (CLSM), 189
Conventional chemical flotation–filtration system
    nickel-chromium plating wastes treatment, 251
Conventional reduction–precipitation system, 248
    wastewater discharge, 257
Copper, 24, 81–85
    basis material, 343
    blister, 83
    castings, 162
    concentration process, 86

Index

industrial process description, 82–84
industry size and geographic distribution, 81
material inputs and pollution outputs, 85
ore grade, 82
plants, 329
pollution outputs, 85
porcelain enameling, 312, 314–317, 326, 337
process materials inputs, 85
product characterization, 81
production process, 83
recovery, 31
recovery process, 24
smelting, 82
toxic organic pollutants, 125–127
toxic pollutants, 105–107
wastestreams chemicals management, 144
wastewater characterization, 101–103
wastewater pollutants, 105–106, 107–109
Copper wire mill pollution abatement, 22–25
areas of application, 25
economics, 25
new process, 23–24
significance, 22
Crushed expanded slag, 170
CTP. See Central treatment plan (CTP)
CWA. See Clean Water Act (CWA)
Cyanate, 372
Cyanide, 370–373
applications, 354
destruction, 283
metal finishing, 355
oxidation, 293
plating processes, 361
reduction, 372, 373
surface-finishing industry, 352
wastestreams, 373
wastewater characterization in metal finishing, 352–355

D

Data Quality Objectives, 450
Degassing, 57, 224
Degreasing, 33–34
aluminum, 202
solvent, 349
Degreasing organic solvent minimization, 33–34
Deionized water
nickel-chromium plating wastes treatment, 242
Denmark
cleaner production, 14
Department of Commerce, 74
Department of Defense (DOD), 382, 401
Department of Energy (DOE), 382
DES. See Direct evacuation system (DES)
Descaling process, 60
Die casting, 157
DIOS. See Direct Iron Ore Smelting (DIOS) process
Dip coating, 310
capital cost, 28
Direct chill casting, 200, 201, 206, 209, 224
Direct evacuation system (DES), 56
Direct Iron Ore Smelting (DIOS) process, 68

Direct reduced iron (DRI), 53
Discharges
waste minimization and cleaner production, 13
Dissolved air flotation, 283, 330
DOD. See Department of Defense (DOD)
DOE. See Department of Energy (DOE)
Domestic battery scrap, 89
Drain boards, 240
DRI. See Direct reduced iron (DRI)

E

EAF. See Electric arc furnace (EAF)
EBONEX. See Electrically conductive ceramic material (EBONEX)
EDR. See Electrodialysis reversal (EDR)
Effluent-free pickling process
fluid bed hydrochloric acid regeneration, 68
Electrical discharge machining
definition, 347
Electrically conductive ceramic material (EBONEX), 414
Electric arc furnace (EAF), 38, 53–56
dust, 56, 94
emissions, 55
operations, 55
process description, 53–54
process waste sources, 55
steel making, 54
treatment techniques, 56
water requirements, 55
Electric induction replacing fossil fuel combustion, 148
Electricity Research Council, 31
Electrochemical machining
definition, 348
Electrochemical soil remediation
performance, 417
Electrochemistry, 412
Electrodialysis
nickel-chromium plating wastes treatment, 241
Electrodialysis reversal (EDR), 286
Electrokinetic remediation technology, 410
Electrokinetics, 466
field implementation, 412
remediation technology, 411
Electrokinetics, Inc.
DOD, 416
Electroless plating, 17
definition, 346
Electrolytic decomposition, 371
Electrolytic process, 31, 92
Electrolytic recovery
nickel-chromium plating wastes treatment, 242
Electrolytic refining, 86
Electron beam machining
definition, 348
Electro-osmosis, 414
advection and ionic migration, 418
Electropainting
definition, 349
Electroplating, 17
definition, 346
waste minimization, 17, 28
wastewater discharge, 256

Electrostatic painting
  definition, 349
  wrap around effect, 30
Emerging Technologies Program projects, 407
Emissions
  charging, 53
  coke making, 42, 69
  composition, 163
  EAF, 55
  forming and finishing, 61
  ironmaking, 46–47
  lead, 148
  nickel, 148
  reduction, 3
  refining and casting, 58
  sinter plants, 46
  sources, 46
  systems, 163
  tapping, 53
  waste, 13
  waste minimization and cleaner production, 13
Emulsions, 225
  aluminum forming industry waste treatment, 206
  breaking, 229
  rolling, 206, 210
Enamel
  adhesion, 311
  application, 311
  heating, 312
  porcelain, 309
  Roman period, 305
End-of-pipe emission control systems
  waste minimization and cleaner production, 13
Environmental awareness
  good housekeeping, 6
Environmental Protection Agency (EPA), 74, 444, 447, 460, 462
  aluminum forming industry, 224
  Paints and Coatings Resource Center, 295–296
  regional offices, 453
  SITE, 387
  triad approach, 449
Environmental regulations
  aluminum basis, 335–336
  cast iron basis, 334
  copper basis, 337
  steel basis, 333
EPA. *See* Environmental Protection Agency (EPA)
Ester-cured alkaline phenolic system, 161
Ester silicate process, 159
ESV. *See Ex situ* vitrification (ESV) technologies
Etching, 17, 440
  definition, 346
Etch line rinses
  pollutant data, 217
Europe
  soil washing, 405
Evaporation
  recovery, 68, 240
Ex situ vitrification (ESV) technologies, 399
Extrinsic waste, 7
Extrusion, 225
  aluminum forming industry waste treatment, 201, 206, 207

  hydraulic press, 201
  subcategory verification data, 211

F

Federal National Priorities List, 445
Ferrous sulfate precipitation, 371
Ferrous sulfide, 247
FID. *See* Flame ionization detectors (FID)
Field analysis, 455
Field conductivity surveys, 413
Field sampling, 455
Fill material test procedures
  embankment, 188
Fine-grained soils
  treatment, 403
Finishing, 60–64
  byproducts, 64
  effluents, 61–63
  emissions, 61
  hazardous wastes, 64
  process overview, 60
  process waste sources, 61–64
  treatment techniques, 64
Flame ionization detectors (FID), 460
Flame spraying
  definition, 347
Flotation clarifiers, 251, 255
Flotation–filtration unit
  clarifier, 251
  operation, 255
  view, 254
  wastewater treatment system, 251, 252, 257
Flotation–precipitation wastewater treatment system
  innovative chemicals, 253
Flowable fill
  mixes, 189
  test procedures, 190
Fluid bed hydrochloric acid regeneration effluent-free pickling process
  process modifications, 68
Fluidized beds, 25–26
  roasters, 92
Flushing, 466
Fluxes
  melting process, 168
Fog sprays, 363
  tanks, 240
Forging, 225
  aluminum forming industry waste treatment, 206
  heat treatment quench, 214–216
  rolled ring, 202
  subcategory verification data, 211–212
Forming, 60–64
  byproducts, 64
  effluents, 61–63
  emissions, 61
  hazardous wastes, 64
  process overview, 60
  process waste sources, 61–64
  treatment techniques, 64
Foundry
  cores, 159

Index

molds, 159
sand sample chemical oxide, 166
solid waste, 178
solid waste program, 179
U.S. EPA promulgated wastewater discharge regulations, 164
waste, 193
Full-scale wastewater treatment plant systems, 291
Fume scrubber water recycle, 67
Furnaces, 159
Furnace slag, 90, 169, 172
air-cooled, 169, 170, 174, 182, 183
composition, 172
expanded, 169
foamed, 169
granulated, 169, 186
metal castings, 168–174
physical properties, 170
properties, 183

G

GAC. *See* Granular activated carbon (GAC)
Galvanized basis material
BPT application, 298
definition, 305
NSPS, 300
PSES, 301
PSNS, 301
Galvanized steel, 26–28, 283
advantages, 27
cleaner production, 27
coil coating industry waste treatment, 268, 269–276, 278
economic benefits, 28
effluent limitations, 298–300
toxic pollutants, 272–273
wastewater pollutants, 272–273
Geokinetics International, Inc., 410
Geosafe Corporation, 402
Germanakos tannery, 33
Germany
soil washing, 405
Good housekeeping, 2–6
environmental awareness, 6
function, 5
methods, 6
recycling, 6
Granular activated carbon (GAC), 284
filtration process, 288
media filter, 331
Green sand system
sand casting systems, 158
Greenwich, Connecticut, 450
Grinding
definition, 346
Groundwater
Brownfield site metal finishing cleanup, 446–448
depth, 463
samples, 461
sampling costs, 461
sampling tools, 457

H

Hall–Heroult process, 76
HAP. *See* Hazardous air pollutants (HAPs)
Hazardous air pollutants (HAPs), 177
Hazardous and Solid Waste Amendments of 1984 (HSWA), 3
Hazardous Substances Data Bank (HSDB), 135
Health assessment organizations, 447
Health departments, 447
Heat treating, 225
definition, 347
quench forging, 214–216
types, 203
Heavy metal soil removal, 381–432
applicability, 390
containment, 388–391
metals and compounds, 383–384
performance and BDAT status, 390
process description, 388
SITE program demonstration projects, 391
site requirements, 389
soil cleanup goals and technologies for remediation, 385–387
soil flushing, 405–407
soil washing, 402–405
solidification–stabilization technologies, 391–398
superfund soils with metal contamination, 385
vitrification, 399–401
Hexadentates, 376
Hexavalent chromium, 247
rate of reduction, 245
reduction, 243, 374
treatment methods, 374
wastes, 249
wastestreams, 350
Hides
tanning, 32
High pressure paint spraying, 29
High-temperature baths, 363
High temperature metals recovery (HTMR), 65
Homogenizing, 203
Hot-box process, 160
Hot dip coating, 440
definition, 348
Hot metal desulfurization, 48
Hot water seal, 225
HSDB. *See* Hazardous Substances Data Bank (HSDB)
HSWA. *See* Hazardous and Solid Waste Amendments of 1984 (HSWA)
HTMR. *See* High temperature metals recovery (HTMR)
Hydraulic extrusion press, 201
Hydraulic flocculater, 253
Hydrochloric acid
regeneration, 67
solution releases, 144
wastestreams chemicals management, 144
Hydrogen peroxide, 371
Hydrometallurgical methods, 84, 85
Hydrometallurgical zinc refining
processing stages, 92
Hydroxide precipitation, 245
Hyperaccumulators.
genetic evaluation, 425

I

ICPAE-MS. *See* Inductively coupled plasma atomic emission mass spectrometry (ICPAE-MS)
ICPIC. *See* International Cleaner Production Information Clearinghouse (ICPIC)
IISI. *See* International Iron and Steel Institute (IISI)
Impact deformation
    definition, 347
Inductively coupled plasma atomic emission mass spectrometry (ICPAE-MS), 167
Industrial wastewater
    BPT, 332
    discharge, 255–256
    treatment costs, 332–333
Industry and Environment Program Activity Centre of UNEP, 16
Industry description
    metal casting wastes management, 152–162
    metal finishing industry wastes treatment, 346–350
    nonferrous metals manufacturing wastes treatment, 72–73
    porcelain enameling industry waste treatment, 308
Influent raw water, 252
In-house process control, 4
Injection grouting, 393
Innovative flotation–filtration wastewater treatment system, 257
Inorganic compounds, 419
In-placing mixing, 393
In-plant demonstrations, 16
In situ soil-processing technology, 411
In situ vitrification (ISV) technology, 399
Insoluble hydroxides, 392
In-tank filtration, 67
Integrated Risk Information system (IRIS), 135
Integration, 16
International Cleaner Production Information Clearinghouse (ICPIC), 11
International Iron and Steel Institute (IISI), 38
Investment casting
    casting processes, 157
Ion exchange, 281
    drawbacks, 241
    nickel-chromium plating wastes treatment, 241
IRIS. *See* Integrated Risk Information system (IRIS)
Iron, 38
    mechanical properties, 161
    physical properties, 161
    plants, 39
    process wastes, 65
    slag, 169
Iron and steel manufacturing industry waste treatment, 37–70
    coke making, 39–44
    electric arc furnace, 53–56
    forming and finishing, 60–64
    industrial processes, 38
    pollution prevention measures, 65–67
    process modifications, 68–69
    refining and casting, 57–59
    steel making, 50–56
Ironmaking, 19, 44–49
    blast furnace, 44, 48
    byproducts, 48
    effluents, 48
    emissions, 46–47
    flow diagram, 46
    inputs, 47
    outputs, 47
    process description, 44–45
    process waste sources, 46–48
    treatment techniques, 49
Iron oxide conversion, 46
Isotron Corporation, 410
ISV. *See In situ* vitrification (ISV) technology

J

Japanese Direct Iron Ore Smelting process, 68

K

Kushner method, 364, 365

L

Ladle metallurgical furnace (LMF), 57
Laminating
    definition, 348
Land reuse, 451
Land Revitalization Technology Support Center, 448
Lasagna process
    schematic diagram, 415
Laser beam machining
    definition, 348
Leachable material, 394, 403
    contaminants, 165
    lead, 56
    metals, 383, 385, 386, 387
    treatment goals, 386
    zinc oxide, 92
Lead, 86–90, 383
    behavior, 383–384
    emissions, 148
    industrial process description, 87–88
    industry size and geographic distribution, 86
    pollution outputs, 90
    process materials inputs, 90
    product characterization, 87
    production, 89
    production process, 87
    raw material inputs and pollution outputs, 89–90
    toxic organic pollutant removal, 128–129
    toxic pollutants, 110, 111–113
    wastestreams chemicals management, 144
    wastewater pollutants, 109, 111
Leather industry chrome recovery and recycling, 32–33
    advantages, 33
    cleaner production, 32
    economic benefits, 33
Lime, 19, 21, 22, 42, 183, 375
    blended cement, 172
    copper processing, 82

Index

metal melting furnaces, 156
precipitation, 119, 216, 219, 221, 282
steel making operations, 50, 51, 54, 57, 65
Liquid–liquid extraction
  nickel, 148
LMF. *See* Ladle metallurgical furnace (LMF)
Low-melting-point alloys
  die, 157
Low-viscosity salts, 363

M

MACT. *See* Maximum Achievable Control Technology (MACT)
Maximum Achievable Control Technology (MACT), 65
Maximum contaminant levels (MCL), 453
MCL. *See* Maximum contaminant levels (MCL)
Melting flow chart, 155
Membrane filtration
  components, 331
  processes, 286
Membrane ultrafiltration, 289
  process, 290
Mercury, 383
  behavior, 383–384
Metal(s)
  behavior, 383–384
  gas-phase heat treatment, 25–26
  ionization groups, 435
  physical characteristics and mineral origins, 383
  primary, 19–20
  recovery techniques, 72, 242
  remediation technologies, 425
  scraps, 168
Metal castings
  agricultural applications, 191–192
  baghouse dust, 168
  flow chart, 153
  industry, 152, 157, 161
  process, 153, 154
  wastewater, 163–164
Metal castings civil engineering reuse, 180–190
  asphalt concrete, 181–182
  embankment or fill material, 187–188
  flowable fill, 189
  hydraulic barrier, 190
  landfill liner and cover, 190
  Portland cement, 186
  Portland cement concrete, 183–185
Metal castings furnace slag, 168–174
  chemical compositions, 171–173
  mechanical properties, 174
  origin, 168–169
  physical properties, 170
  thermal properties, 175
Metal castings solid waste reuse barriers, 193–194
  economics, 194
  education, 193
  environmental regulation, 193
  guidelines, 194
  market potential, 194
  procedures, 194
  specifications, 194
Metal castings solid waste reuse technologies, 178–192
  agricultural applications, 191–192
  civil engineering reuse, 180–190
  growing amendments, 192
  topsoil, 191
Metal castings spent foundry sand, 164–167
  chemical compositions, 166
  mechanical properties, 166
  physical properties, 165
  trace element characterization, 166
Metal castings waste characterization, 162–175
  air emission, 163
  baghouse dust, 168
  furnace slag, 168–174
  spent foundry sand, 164–167
Metal castings waste management, 151–198
  agricultural applications, 191–192
  air emission, 163
  baghouse dust, 168
  casting metals, 161–162
  casting processes, 157–158
  chemical substitution or minimization, 175
  civil engineering reuse, 180–190
  crushing, 192
  design and construction, 192
  economics, 194
  education, 193
  emission reduction, 177
  environmental regulation, 193
  furnace slag, 168–174
  general processes, 192
  growing amendments, 192
  guidelines, 194
  industry description, 152–162
  in-plant reclamation, 176
  market potential, 194
  minimization and recycling, 151–198
  procedures, 194
  reuse evaluation framework, 178–180
  sand casting systems, 158–159
  screening, 192
  solid waste reuse barriers, 193–194
  solid waste reuse technologies, 178–192
  source reduction, 175–177
  specifications, 194
  spent foundry sand, 164–167
  storage, 192
  topsoil, 191
  unresolved issues, 193
  waste characterization, 162–175
  waste segregation, 177
  wastewater, 163–164
Metal coloring, 439
Metal contamination, 398, 404–407
  applications, 391, 405, 407
  ISV, 401
  soil, 400
  soil-flushing, 407
  soil-washing, 405
  solidification–stabilization, 398
Metal finishing, 17–18. *See also* Brownfield site metal finishing cleanup
  BMP, 17
  contaminants, 439

Metal finishing (*Continued*)
  cyanide wastes, 371
  electroplating waste minimization, 17
  facilities, 438, 441, 456
  industrial processes, 437
  operations, 437, 457
  processes, 437
Metal finishing industry wastes treatment, 345–380
  arsenic-containing wastes, 374
  chemical substitution, 359
  chromium-containing wastes, 374
  common metals, 366–368
  complexed metal wastes, 370
  costs, 377–379
  cyanide, 370
  cyanide-containing wastes, 371–373
  dragout loss reduction, 361–364
  hexavalent chromium, 370
  industry descriptions, 346–350
  neutralization, 371
  oils, 371
  pollutant removability, 366–371
  precious metals, 369–370
  selenium-containing wastes, 374
  solvents, 371
  source reduction, 359–365
  treatment technologies, 371–376
  waste reduction costs and benefits, 365
  waste segregation, 360
  wastewater characterization, 350–358
Metal finishing wastewater characterization, 350–358
Metal hydroxides
  solubility, 246
Metal hyperaccumulators
  examples, 423
  quantities of biomass, 424
Metal painting
  advantages, 35
  cleaner production, 34
  degreasing organic solvent minimization, 33–34
  economic benefits, 35
Metal preparation
  definition, 344
Metering pumps, 249
MF. *See* Microfiltration (MF)
Michigan Department of Environmental Quality, 18
Microfiltration (MF), 286
Mixed liquor volatile suspended solids (MLVSS), 2
MLVSS. *See* Mixed liquor volatile suspended solids (MLVSS)
Modified reduction–precipitation wastewater treatment system, 250
Molding flow chart, 155
Multimedia waste management for coil coating industry, 294–297
  air pollution control, 294
  cleaner production alternatives, 296
  hazardous wastes management and disposal, 296
  roll and coil coating waste minimization, 296
  solid waste management and disposal, 296
  water pollution control, 295

N

Nanofiltration (NF), 286
National Emission Standards for Hazardous Air Pollutants (NESHAPs), 44
National Environmental Policy Act (NEPA), 180
National Pollutant Discharge Elimination System (NPDES), 295, 445
  United States, 3
National Primary Drinking Water Standards, 237
National Priorities List (NPL), 445
Natural sand, 189
  replacement with spent foundry sand, 192
Neat oils, 225
  drawing, 206
  rolling, 205
NEPA. *See* National Environmental Policy Act (NEPA)
NESHAP. *See* National Emission Standards for Hazardous Air Pollutants (NESHAPs)
Netherlands
  cleaner production, 14
  soil washing, 405
Neutralization
  nickel-chromium plating wastes treatment, 242
New Source Performance Standards (NSPS)
  aluminum basis material, 302, 341
  canmaking, 304
  galvanized basis material, 300
  steel basis material, 298, 336
NF. *See* nanofiltration (NF)
Nickel
  aquatic organisms, 236
  deposition, 309, 313
  emissions, 148
  liquid–liquid extraction, 148
  plating, 234, 236, 362
  removal treatment, 242
  solution concentration limits, 363
Nickel-chromium plating wastes treatment, 233–260
  bath life extension, 238
  chemical flotation–filtration wastewater treatment systems, 251
  chemical treatment, 242–247
  chromium environmental impact, 236
  conventional chemical flotation–filtration system, 251
  conventional reduction–precipitation system, 248
  deionized water, 242
  dragout recovery, 239
  dragout reduction, 239
  electrodialysis, 241
  electrolytic recovery, 242
  evaporative recovery, 240
  flotation–filtration system, 252–254
  flotation–filtration wastewater treatment systems, 251–252
  hazardous waste assessment, 237
  hexavalent chromium reduction, 243–244
  ion exchange, 241
  material recovery and recycling, 239–242
  material substitution, 238
  modified reduction–flotation wastewater treatment system, 249–250
  neutralization, 242

Index

nickel-chromium plating process, 234–235
nickel environmental impact, 236
pH adjustment and hydroxide precipitation, 245–246
pollution sources, 235–236
reactive rinses, 239
reduction and flotation combination, 247
reverse osmosis, 240
waste minimization, 237–239
waste segregation, 237
Niobium, 96
Nitric-acid-free pickling
  process modifications, 68
Nonferrous metals manufacturing wastes treatment, 71–150
  aluminum, 74–80
  beryllium, 96
  columbium, 96
  copper, 81–85
  electric induction replacing fossil fuel combustion, 148
  industry description, 72–73
  lead, 86–90
  nonferrous metal processing industry, 74–96
  pollutant removability and treatment, 119
  pollution prevention case studies, 147–149
  processing nonferrous metal hydroxide sludge wastes, 148
  selenium, 96
  silver, 96
  tantalum, 96
  tungsten, 96
  wastestreams chemicals management, 119–149
  wastewater characterization, 97–119
  zinc, 91–95
Nonferrous slag
  chemical composition, 173
  mechanical properties, 175
  physical properties, 171
Nonpolluting chemicals, 359
NPDES. *See* National Pollutant Discharge Elimination System (NPDES)
NPL. *See* National Priorities List (NPL)
NSPS. *See* New Source Performance Standards (NSPS)

O

Office of Superfund Remediation and Technology Innovation (OSRTI), 435
Oil-bearing mill scales, 49
Oil coating systems, 296
Oil emulsion-soap
  pollutant data, 212–213
Oils, 225
  drawing, 206
  rolling, 205
Oil sand, 160
Oil skimming, 282
Oil–water separator, 220
Oily wastes
  removal, 370
  treatment, 370
Organic compounds, 419

Organic contaminants
  volatilization, 464
OSRTI. *See* Office of Superfund Remediation and Technology Innovation (OSRTI)
Ostrowiec Steelworks, 28
Oxidizing, 60
Oxygen furnaces
  pollutants, 55
Ozonation, 371

P

Painting, 33
  air atomized, 29
  definition, 349
  high pressure, 29
  spraying, 29
  stripping, 349
  wastewater, 278
Parallel plate oil–water separator, 220
Passivating, 439
PCB. *See* Polychlorinated biphenyls (PCBs)
Permanent mold casting
  casting processes, 158
Permeable reactive barriers, 466
pH
  effects, 413
  instruments metering pumps, 249
  wastewater, 245
Phenolic-isocyanates, 159
Phenolic-urethane-amine gassed process, 160
Phosphate
  coating, 265
  waste treatment and management, 265
Phosphating, 439
Photo-ionization detector (PID), 460
Physical separation, 466
Phytoextraction, 418, 420
Phytoremediation, 467
  advantages, 421
  applications, 422
  disadvantages, 421
  effectiveness, 425
  objectives, 423
  potential for, 422
  technologies, 419, 420, 422
  types, 421
Phytostabilization, 418
Pickling process
  fluid bed hydrochloric acid regeneration effluent-free, 68
PID. *See* Photo-ionization detector (PID)
Pig iron
  chemical composition, 45
Plant-specific effluent concentrations
  verification data, 278–279
Plasma arc machining
  definition, 348
Plating solution, 361
PLM. *See* Polarized light microscopy (PLM)
Polarized light microscopy (PLM), 460
Polishing, 440
  definition, 346

Pollution prevention, 65–67
    acid reuse, recycle, and recovery systems, 67
    audit, 29
    case studies, 147–149
    countercurrent cascade rinsing, 66
    high-rate recycle, 66
    process solution life extension, 67
Pollution Prevention Act of 1990, 119
Polychlorinated biphenyls (PCBs), 394
Poor housekeeping, 6
Porcelain enameling, 308
    definition, 342
Porcelain enameling industrial plants descriptions, 318–326
    aluminum subcategory, 324–325
    cast iron subcategory, 326
    copper subcategory, 326
    steel subcategory, 323
Porcelain enameling industry subcategory descriptions, 310–312
    aluminum iron, 311
    cast iron, 311
    continuous strip, 312
    copper iron, 312
    steel, 310
Porcelain enameling industry waste treatment, 307–344
    application and firing, 309
    base material surface preparation, 309
    cleaner production, 331
    costs, 332–333
    description, 308
    historical cultural development, 308
    industrial cultural development, 308
    performance standards, 333–337
    plant descriptions, 318–326
    point source discharge effluent limitations, 333–337
    pollutant removability, 329–331
    pollution prevention, 331
    pretreatment standards, 333–337
    process steps, 309–310
    slip preparation, 309
    subcategory descriptions, 310–312
    technical terminologies, 342–343
    wastewater characterization, 312–314
Porcelain enameling industry wastewater pollutant removability, 329–331
    chemical addition, 331
    coagulation, 331
    dewatering, 331
    equalization and neutralization, 329
    flocculation, 331
    granular activated carbon filtration, 331
    granular bed filtration, 331
    membrane filtration, 331
    precipitation, 331
    sedimentation (settling) or flotation clarification, 330
    sludge concentration, 331
Porcelain enameling point source discharge environmental regulations, 333–337
    aluminum, 335–336
    cast iron, 334
    copper, 337
    steel, 333

Porcelain enameling process steps, 309
    application, 309
    base material surface preparation, 309
    firing, 309
    slip preparation, 309
Porcelain enameling wastewater characterization, 312–314
    aluminum subcategory, 314
    cast iron subcategory, 314
    copper subcategory, 314–317
    steel subcategory, 313
Portland cement concrete
    civil engineering reuse for metal castings, 183–185
POTW. See Publicly owned treatment works (POTW)
Powder painting techniques, 34
Praegitzer Industries Inc., 30
Precious metal
    definition, 344
    recovery, 147
Precipitation system, 284
Pressure deformation
    definition, 347
Pretreatment Standards for Existing Sources (PSES)
    aluminum basis material, 303
    canmaking, 305
    galvanized basis material, 301
    steel basis material, 299
Pretreatment Standards for New Sources (PSNS)
    aluminum basis material, 303
    canmaking, 305
    galvanized basis material, 301
    steel basis material, 299
Primary metals, 19–20
    BMP, 20
    industry profile, 19
Primary refining
    transfers, 136–139
Primary smelting
    transfers, 136–139
Printed circuit board etchant copper recovery, 30–32
    advantages, 32
    cleaner production, 31
    economic benefits, 32
Process equipment modification
    wastestreams chemicals management, 147
Processing nonferrous metal hydroxide sludge wastes, 148
Process material
    conservation, 331
    recovery reuse, 239
Process modifications, 68–69
    coke elimination with cokeless technologies, 68
    coke oven emissions reduction, 69
    effluent-free exhaust cleaning, 68
    fluid bed hydrochloric acid regeneration, 68
    nitric-acid-free pickling, 68
    pickling process, 68
Product
    life cycle-assessments, 8
    reformulation opportunities identified, 11
PSES. See Pretreatment Standards for Existing Sources (PSES)

Index

PSNS. *See* Pretreatment Standards for New Sources (PSNS)
Public awareness
    cleaner production, 14
Publicly owned treatment works (POTW), 65, 404
Pyrometallurgical processing, 72
Pyrometallurgy
    processing, 408
    recovery, 409
    systems, 408

R

Raw materials
    first-in, first-out policy, 238
    substitution or elimination, 147
RCRA. *See* Resource Conservation and Recovery Act (RCRA)
Recycling
    good housekeeping, 6
Reduction–precipitation system
    conventional, 248, 257
Refining, 57–59
    byproducts, 59
    effluent, 59
    emissions, 58
    primary, 136–139
    process description, 58
    process waste sources, 58–59
    treatment techniques, 59
Reheating, 57
Resin bonded sand, 176
Resource Conservation and Recovery Act (RCRA), 3, 43, 180, 194, 295, 409, 444–445
    metal finishing industry, 346
    production operations, 73
Reuse evaluation framework for metal castings, 178–180
    environmental issues, 180
    technical implementability, 178–179
Reverse osmosis
    nickel-chromium plating wastes treatment, 240
Rolling, 225
    coating systems, 296
    heat treatment process, 201, 213–214
    pollutant data, 213–214
    ring forging, 202
Rough lead bullion, 88

S

Salt bath descaling
    definition, 348
Sand
    blasting definition, 347
    physical properties, 158
    reclamation systems, 193
Sand casting systems, 158–159
    chemically bonded, 159–160
    green sand, 158
Sandy soils, 446
Scrap charge process, 176

Screening
    definition, 454
Secondary aluminum recovery
    methods, 79
Secondary lead
    air emissions, 91
    specialty products, 88
Secondary scrap aluminum smelting, 81
Secondary smelting
    transfers, 140–144
Sedimentation clarification
    aluminum forming wastewater, 231
Segregated oily wastewater
    definition, 351
Selected unit process operations, 8
Selenium, 96
Shearing
    definition, 347
Shell casting
    casting processes, 157
Shell process, 160
Short-term lead poisoning, 144
SIC. *See* Standard Industrial Classification (SIC)
Silicate bonded sand, 176
Silver, 96
    organic pollutants, 130–131
    toxic pollutants, 113–115
    wastewater pollutants, 113
Sintering, 19
    definition, 348
    energy facts, 47
    environmental facts, 47
    flow diagram, 45
    plant emissions, 46
SITE. *See* Superfund Innovative Technology Evaluation (SITE)
Skimming systems
    pollutant removal, 223
Smelting
    primary, 136–139
    process wastewater sources, 101
Soap, drawing
    aluminum forming industry waste treatment, 206
Sodium borohydride, 375
Sodium hydroxide
    cleaning solution, 264
    precipitation, 221
Soil
    chemical analysis, 413
    ranges, 187
    sampling tools, 456
    treatment, 418
    types, 463
    washing, 402–405, 467
    waste systems, 383
Soil flushing, 464
    system, 406
    techniques, 407
Soil washing for heavy metal soil removal, 402–405
    applicability, 404
    performance and BDAT status, 405
    process description, 402–403

SITE demonstrations and emerging technologies program projects, 405
   site requirements, 404
Soldering
   definition, 347
Solidification–stabilization technologies for heavy metal soil removal, 391–398
   applicability, 393–394
   cost, 398
   performance and BDAT status, 395–396
   process description, 392
   SITE program demonstration projects, 397
   site requirements, 393
Solid removal techniques, 369
Solid waste
   evaluation framework, 178–180
   federal legislation, 163
   management, 163, 193
   regulations, 193
   reuse technologies, 178–180
Solubility enhancement, 403
Solution heat treatment, 203
Solvent degreasing
   definition, 349
Solvent extraction, 467
Solvent extraction electrowinning process (SXEW), 84
Spent foundry sand, 165
   flowable fill, 190
   grain size distribution, 165
   mechanical properties, 167
   origin, 164
   use, 187
Spent green foundry sand, 191
   physical properties, 165
Spent material, 165
Spent pickle liquor, 64
Sputtering
   definition, 348
Stabilization, 391
Standard Industrial Classification (SIC), 346
State Voluntary Cleanup Programs, 442
Stationary casting, 226
   aluminum forming industry waste treatment, 200
Steel. *See also* Iron and steel manufacturing industry waste treatment
   antitrust treatment, 26
   carbon, 389
   castings, 162
   coating operations, 61
   coil coating industry waste treatment, 268, 269
   cold rolled, 279
   effluent limitations, 297
   galvanized, 26–28, 272–273, 279, 283, 299–300
   mechanical properties, 161
   performance standards, 297
   physical properties, 161
   porcelain enameling, 310, 313, 315–317, 323, 333
   pretreatment standards, 297
   surface rust, 264
   toxic pollutants, 270–271
   types, 162
   wastewater pollutants, 270
   wastewater subcategories, 290
Steel basis material
   definition, 305
   effluent limitations, 297–298, 334, 335
   NSPS, 298, 336
   pollutants, 337–339
   PSES, 299
   PSNS, 299
Steel making, 19, 40, 50–56
   EAF, 54
   energy facts, 52
   environmental facts, 52
   flow diagram, 51
   inputs, 51, 54
   outputs, 51, 54
   oxygen furnace process, 50–52
   process description, 50
   process waste sources, 50–51
   treatment techniques, 52
Steel slag, 172
   chemical composition, 173
   magnesia, 183
   mechanical properties, 174
   physical properties, 171
Steelwork painting waste reduction, 28–30
   advantages, 30
   cleaner production, 29
Subsurface sampling tools, 456
Sulfides
   pollutant removal, 222
   precipitation, 369
   solubility, 246
Sulfur dioxide, 47
Sulfuric acid recovery, 67
Sulfur process, 93
Superfund Innovative Technology Evaluation (SITE)
   US EPA, 387
Superfund remedial projects
   percentage, 395
Suppressed combustion, 53
Surface-finishing industry
   cyanide, 352
Surface preparation process
   waste water pollutants, 314
Surface soil samples, 461
Surface water control, 463
SXEW. *See* Solvent extraction electrowinning process (SXEW)

T

Tanning, 32
Tantalum, 96
   toxic organic pollutants, 123–124
   toxic pollutants, 102–104
   wastewater characterization, 101
   wastewater pollutants, 102
TCCD. *See* Tetrachlorodibenzo-p-dioxin (TCCD)
TCE. *See* Trichloroethylene (TCE)
TCLP. *See* Toxicity Characteristic Leaching Procedure (TCLP)
TEB. *See* Thermal emulsion breaking (TEB)
Tetrachlorodibenzo-p-dioxin (TCCD), 206
Textile yarns
   zinc, 22

Index

Thermal emulsion breaking (TEB), 218
Thermal infusion
  definition, 348
Thermal oxidation, 371, 373
Topsoil and metal castings, 191
Total suspended solids (TSS), 59
Total toxic organics (T.O.), 219, 224, 225–226, 297
  definition, 306
Toxicity Characteristic Leaching Procedure (TCLP), 167, 180, 404
Toxic Relief Inventory (TRI), 119
  database, 135
TRI. *See* Toxic Relief Inventory (TRI)
Trichloroethylene (TCE), 417
Trivalent chromium, 32
TSS. *See* Total suspended solids (TSS)
TTO. *See* Total toxic organics (TTO)
Tundish heating, 148
Tungsten, 96
toxic organic pollutant removability, 131–132
toxic pollutants, 116–117
wastewater pollutants, 115

U

Ultrafiltration
  aluminum forming wastewater, 229
  membrane, 290
Ultrasonic machining
  definition, 348
United Kingdom
  cleaner production, 14
Electricity Research Council, 31
United States
ACE Yearly Average Cost Index Utilities, 461
Agricultural Research Service, 192
aluminum environmental regulations, 335–336
cast iron environmental regulations, 334
central nervous system syndrome, 145
cleaner production, 14
coil coating, 268, 269
copper, 81
copper environmental regulations, 337
Department of Commerce, 74
Electroplating and Metal Finishing provisions, 200
environmental regulations for porcelain enameling, 333–337
NPDES, 3
pollution prevention, 146
porcelain enameling environmental regulations, 333–337
porcelain enameling plants, 308
power storage batter industry, 87
steel environmental regulations, 333
support programs, 466
United States Department of Agriculture (USDA), 446
United States Environmental Protection Agency, 74, 444, 447, 460, 462
aluminum forming industry, 224
Paints and Coatings Resource Center, 295–296
regional offices, 453

SITE, 387
triad approach, 449
United States Geological Survey (USGS), 446
USDA. *See* United States Department of Agriculture (USDA)
USGS. *See* United States Geological Survey (USGS)

V

Vacuum metalizing, 349
Vapor plating
  definition, 348
VCP. *See* Voluntary Cleanup Programs (VCPs)
Vertical auger mixing, 393
Vertical containment barriers, 389
Viscose rayon plants two-stage precipitation zinc recycling, 21–22
  economics, 21
  new process, 21
  significance, 21
Vitrification for heavy metal soil removal, 399–401
  applicability, 400
  performance and BDAT status, 401
  process description, 399
  SITE program demonstration projects, 401
  site requirements, 399
VOC. *See* Volatile organic compounds (VOCs)
Volatile organic compounds (VOCs), 177, 394
Volatilization of organic contaminants, 464
Voluntary Cleanup Programs (VCPs), 442
Vortec Corporation, 402

W

Warm-box process, 160
Waste and Energy Management Program, 18
Waste minimization, 1–36
  audit review, 12–14
  case studies, 21–35
  cleaner production, 14–16
  copper wire mill pollution abatement, 22–25
  degreasing organic solvent minimization, 33–34
  discharges, 13
  electroplating waste reduction, 28
  end-of-pipe emission control systems, 13
  final emissions, 13
  good housekeeping, 2–6
  handling, 14
  integrated source control, 12
  leather industry chrome recovery, 32–33
  metal finishing, 17–18
  metal painting, 33–34
  metals gas-phase heat treatment, 25–26
  primary metals, 19–20
  printed circuit board etchant copper recovery, 30–32
  processes, 12
  raw materials and utilities, 12
  recycling, 32–33
  steel galvanizing, 26–28
  steelwork painting waste reduction, 28–30
  storage, 14

493

494 Index

Waste minimization (*Continued*)
  Viscose rayon plants two-stage precipitation zinc
    recycling, 21–22
  waste reduction planning, 9–11
  waste reduction strategy, 7–8
Waste reduction, 29
  annual cost savings, 367
  good housekeeping, 237
  improved housekeeping, 9
  options, 10
  plan, 10
  practices, 359
  process changes to, 11
  strategy, 7–8
Wastestreams chemicals management, 119–149
  chemical release and transfer profile, 135
  chlorine, 144
  copper, 144
  hydrochloric acid, 144
  lead, 144
  pollution prevention opportunities, 145–147
  precious metals recovery, 147
  process equipment modification, 147
  raw materials substitution or elimination, 147
  toxicity, 135–145
  toxic release inventory, 145
  zinc and zinc compounds, 145
Wastewater, 48, 97–119, 163–164
  aluminum, 97–98, 99
  cadmium, 115–118
  characterization, 97–119
  coke making, 43
  columbium, 101
  common metals, 352
  complexed metals, 352, 355
  copper, 101–103, 104–108
  cyanide, 352–355, 356
  discharges, 205
  hexavalent chromium, 356
  lead, 109
  metal finishing, 350–358
  oils, 356, 357–359
  oil separation, 293
  pH, 245
  pollutants, 276–277
  precious metals, 352, 355
  silver, 110

  tantalum, 101
  toxic pollutants, 164
  treatment, 49, 333
  tungsten, 110–114
  zinc, 115–118
  zinc processing, 94
Water
  chemical analysis, 413
  conservation, 4
  consumption, 43
  departments, 447
  recovery, 4
  reuse, 331
  rinsing work pieces, 350
Water and Waste Control for the Plating Shop, 364
Welding
  definition, 347
Wet scrubbers, 226
Wetting agents
  application, 362

Z

Zinc, 91–96
  industrial process description, 91–93
    industry, 91
    industry size and geographic distribution, 91
  material inputs and pollution outputs, 94–95
    melting, 27
  pollution outputs, 95
  process materials inputs, 95
  product characterization, 91
  refining process, 93
  removal, 22
  textile yarns, 22
  toxic organic pollutant removability, 133–134
  toxic pollutants, 118–119
  wastewater pollutants, 117
Zinc chloride
  fumes, 145
Zinc coated steel. *See also* Galvanized steel
  coil coating industry waste treatment, 268–276
Zinc electrowinning, 92
Zincrometal coating, 268